Advances in Biochemistry in Health and Disease

Volume 20

Series Editor

Naranjan S. Dhalla, Institute of Cardiovascular Sciences, St. Boniface Hospital, Winnipeg, MB, Canada

Advances in Biochemistry in Health and Disease focus on the latest developments in biochemical research with implications for health and disease. This book series consists of original edited volumes and monographs, presented by leading experts in the field and provides an up to date and unique source of information for all those interested in the fundamental, biochemical processes of the latest and emerging topics and techniques.

Covering a wide variety of topics, this book series is a valuable source of information from those at the lab bench through to the Health Care workers.

More information about this series at http://www.springer.com/series/7064

Paramjit S. Tappia · Sukhwinder K. Bhullar ·
Naranjan S. Dhalla
Editors

Biochemistry of Cardiovascular Dysfunction in Obesity

Springer

Editors
Paramjit S. Tappia
Asper Clinical Research Institute
St. Boniface Hospital
Winnipeg, MB, Canada

Sukhwinder K. Bhullar
Institute of Cardiovascular Sciences
St. Boniface Hospital
Winnipeg, MB, Canada

Naranjan S. Dhalla
Institute of Cardiovascular Sciences
St. Boniface Hospital
Winnipeg, MB, Canada

Advances in Biochemistry in Health and Disease
ISBN 978-3-030-47335-8 ISBN 978-3-030-47336-5 (eBook)
https://doi.org/10.1007/978-3-030-47336-5

This Springer imprint is published by the registered company Springer Nature Switzerland AG
The registered company address is: Gewerbestrasse 11, 6330 Cham, Switzerland

Preface

According to the World Health Organization, overweight and obesity are defined as the abnormal or excessive accumulation of fat that presents a risk to health in men, women, and children. There is now unequivocal evidence that obesity is a global epidemic in both children and adults and is associated with numerous comorbidities including cardiac and vascular abnormalities as well as reduced quality of life and lower life expectancy. In fact, obesity is an independent risk factor for cardiovascular disease (CVD) in adults as well as in obese children. The economic burden and human costs associated with obesity and related diseases have risen dramatically and are expected to continue to rise.

Although body fatness as percent ideal body weight has been used for assessment of overweight/obesity, another assessment and classification tool in use is the measurement of the Body Mass Index (BMI). BMI (weight in kilograms/height^2 in meters) is frequently used as a surrogate measure of fatness in children and adults. In adults, overweight is defined as a BMI of 25.0–29.9 kg/m^2; obesity is defined as a BMI ≥ 30.0 kg/m^2. While metabolic disturbances are known to occur, adaptations/alterations in cardiac structure and function also occur with excessive accumulation of adipose tissue. On the other hand, obesity may affect the heart as a consequence of other known risk factors including dyslipidemia, hypertension, glucose intolerance, oxidative stress, and inflammation. Being overweight or obese increases the risk for the occurrence of a number of different cardiac complications such as coronary heart disease, heart failure, and sudden death because of their impact on the cardiovascular system.

This book will provide a description of the impact of obesity on the cardiovascular system and increased predisposition to CVD. It will identify the major biochemical mechanisms that lead to the occurrence of myocardial abnormalities and vascular alterations in obesity. The book will also address the epidemic of obesity in both children and adults as well as some consideration of sex differences in the mechanisms of obesity-induced dysfunction of the cardiovascular system. We will also have some discussion on the biochemistry of the so- called obesity paradox, a hypothesis which holds that obesity may, counterintuitively, be protective and associated with greater survival in certain groups of people. The contributors to this book are international experts on obesity and associated cardiovascular complications. This book is also uniquely positioned as it focuses on the biochemistry of obesity-induced cardiovascular dysfunction. There are 20 chapters in 2 different parts in this book, comprising of Part I: Pathophysiology of Cardiovascular Complications in Obesity and Part II: Modification of Cardiovascular Dysfunction in obesity.

The intent of this volume is to provide current and basic understanding of the biochemical mechanisms of obesity-induced cardiovascular dysfunction that will be of value not only to cardiologists and other allied health professionals, but will also stimulate and motivate biomedical researchers and scientists to find the way to prevent the epidemic of obesity- associated cardiovascular abnormalities. Furthermore, this book will serve as a highly useful resource for medical students, fellows, residents, and graduate students with an interest in the cardiovascular system.

In summary, this monograph covers a broad range of biochemical mechanisms of obesity-induced cardiovascular complications. We hope that the reader will understand that obesity is linked to an increase in the risk and occurrence of fatal CVD. Furthermore, the underlying message presented in the monograph is that the cause of obesity-related disorders is complex and that understanding the biochemistry of cardiovascular dysfunction may contribute to the development of novel interventions for the prevention and treatment of obesity-associated comorbidities.

Winnipeg, Canada

Paramjit S. Tappia
Sukhwinder K. Bhullar
Naranjan S. Dhalla

Contents

Contributors

Devendra K. Agrawal Department of Translational Research, Western University of Health Sciences, Pomona, CA, USA

Zoha Ahmed Shobhaben Pratapbhai Patel School of Pharmacy and Technology Management, SVKM's NMIMS, Mumbai, Maharashtra, India

Md. Jahangir Alam Translational Health Science and Technology Institute, NCR Biotech Science Cluster, Faridabad, Haryana, India

Ashim K. Bagchi St. Boniface Hospital Albrechtsen Research Centre, Institute of Cardiovascular Sciences, Winnipeg, MB, Canada

Rushita A. Bagchi Department of Medicine, Division of Cardiology, University of Colorado Anschutz Medical Campus, Aurora, CO, USA

Siddhi Bagwe-Parab Shobhaben Pratapbhai Patel School of Pharmacy and Technology Management, SVKM's NMIMS, Mumbai, Maharashtra, India

Sanjay K. Banerjee Department of Biotechnology, National Institute of Pharmaceutical Education and Research (NIPER), Guwahati, Assam, India

Krishna M. Boini Department of Pharmacological and Pharmaceutical Sciences, College of Pharmacy, University of Houston, Houston, TX, USA

Guadalupe Bravo Departamento de Farmacobiología, Centro de Investigación y de Estudios Avanzados, Instituto Politécnico Nacional, Mexico City, Mexico

Harpal Singh Buttar Department of Pathology and Laboratory Medicine, Faculty of Medicine, University of Ottawa, Ottawa, ON, Canada

Karla Carvajal Laboratorio de Nutrición Experimental, Instituto Nacional de Pediatría, Mexico City, Mexico

Alper Celik Metabolic Surgery Clinic, Sisli, Istanbul, Turkey

Meena Chintamaneni Shobhaben Pratapbhai Patel School of Pharmacy & Technology Management, SVKM's NMIMS, Mumbai, India

Dragan M. Djuric Institute of Medical Physiology "Richard Burian", Faculty of Medicine, University of Belgrade, Belgrade, Serbia

Mahmoud Gheblawi Faculty of Medicine and Dentistry, Department of Physiology, University of Alberta, Edmonton, AB, Canada;
Mazankowski Alberta Heart Institute, University of Alberta, Edmonton, AB, Canada

Sarvpreet Singh Ghuman Department of Gynaecology and Obstetrics, Guru Angad Dev Veterinary and Animal Sciences University, Ludhiana, India

Norma L. Gómez-Viquez Departamento de Farmacobiología, Centro de Investigación y de Estudios Avanzados, Instituto Politécnico Nacional, Mexico City, Mexico

Meghana Gore Department of Internal Medicine, Memorial Hermann Katy Hospital, Katy, TX, USA

Ramesh K. Goyal Delhi Pharmaceutical Sciences Research University, Delhi, India

Rodrigo Guevara Departamento de Farmacobiología, Centro de Investigación y de Estudios Avanzados, Instituto Politécnico Nacional, Mexico City, Mexico;
Maestría en Ciencias Biológicas, Centro Tlaxcala de Biología de la Conducta, Universidad Autónoma de Tlaxcala, Tlaxcala, Mexico

Fahmida Jahan Department of Biochemistry, Microbiology and Immunology, Faculty of Medicine, University of Ottawa, Ottawa, ON, Canada

Amudhan Jyothidasan Department of Cardiology, Cape Fear Valley Hospital, Fayetteville, NC, USA

Mehtap Kacar Faculty of Medicine, Department of Physiology, Yeditepe University, Istanbul, Turkey

Ginpreet Kaur Shobhaben Pratapbhai Patel School of Pharmacy and Technology Management, SVKM's NMIMS, Mumbai, Maharashtra, India

Tajpreet Kaur Department of Pharmacology, Khalsa College of Pharmacy, Amritsar, India

Saisudha Koka Department of Pharmacological and Pharmaceutical Sciences, College of Pharmacy, University of Houston, Houston, TX, USA

Sampath Krishnan Institute of Cardiovascular Sciences, St. Boniface Hospital Albrechtsen Research Centre, Winnipeg, MB, Canada;
Canadian Centre for Agri-Food Research in Health and Medicine, St. Boniface Hospital Albrechtsen, Research Centre, Winnipeg, MB, Canada

Vivek S. Kumawat Shobhaben Pratapbhai Patel School of Pharmacy & Technology Management, SVKM's NMIMS, Mumbai, India

Pin-Lan Li Department of Pharmacology and Toxicology, Virginia Commonwealth University, Richmond, VA, USA

Xavier Lieben Louis Institute of Cardiovascular Sciences, St. Boniface Hospital Albrechtsen Research Centre, Winnipeg, MB, Canada;
Department of Biochemistry and Medical Genetics, University of Manitoba, Winnipeg, MB, Canada

María Luna-Luna Department of Molecular Biology, Instituto Nacional de Cardiología, Mexico City, Mexico

Patrick Mailloux-Salinas Departamento de Farmacobiología, Centro de Investigación y de Estudios Avanzados, Instituto Politécnico Nacional, Mexico City, Mexico

Tamhida Masi Institute of Pharmacy, Nirma University, Ahmedabad, Gujarat, India

Jawahar L. Mehta Central Arkansas Veterans Healthcare System, University of Arkansas for Medical Sciences, Little Rock, AR, USA

Tzindilú Molina-Muñoz Departamento de Farmacobiología, Centro de Investigación y de Estudios Avanzados, Instituto Politécnico Nacional, Mexico City, Mexico

Marie-Philippe Morin Internal Medicine Department, Institut Universitaire de Cardiologie et de Pneumologie de Québec, Québec, Canada

Slavica S. Mutavdzin Institute of Medical Physiology "Richard Burian", Faculty of Medicine, University of Belgrade, Belgrade, Serbia

Vijaiganesh Nagarajan Interventional Cardiology, The University of Texas Health Sciences Center at Houston, Houston, TX, USA

Mojdeh Nasiri Department of Epidemiology and Statistic, Michigan State University, East Lansing, MI, USA

Thomas Netticadan Department of Physiology and Pathophysiology, University of Manitoba, Winnipeg, MB, Canada;
Canadian Centre for Agri-Food Research in Health and Medicine, St. Boniface Hospital Albrechtsen, Research Centre, Winnipeg, MB, Canada;
Agriculture and Agri-Food Canada, Winnipeg, MB, Canada;
Heart Failure Research Laboratory, Canadian Centre for Agri-Food Research in Health and Medicine, Winnipeg, MB, Canada

Gavin Y. Oudit Division of Cardiology, Faculty of Medicine and Dentistry, Department of Medicine, University of Alberta, Edmonton, AB, Canada;
Faculty of Medicine and Dentistry, Department of Physiology, University of Alberta, Edmonton, AB, Canada;
Mazankowski Alberta Heart Institute, University of Alberta, Edmonton, AB, Canada

Antara Pal Shobhaben Pratapbhai Patel School of Pharmacy and Technology Management, SVKM's NMIMS, Mumbai, Maharashtra, India

Bhoomika M. Patel Institute of Pharmacy, Nirma University, Ahmedabad, Gujarat, India

Devendra Pathak Department of Veterinary Anatomy, Guru Angad Dev Veterinary and Animal Sciences University, Ludhiana, India

Óscar Pérez-Méndez Department of Molecular Biology, Instituto Nacional de Cardiología, Mexico City, Mexico;
School of Engineering and Sciences, Tecnológico de Monterrey, Campus Mexico City, Mexico

Marie-Eve Piché Cardiology Department, Institut Universitaire de Cardiologie et de Pneumologie de Québec, Québec, Canada

Paul Poirier Cardiology Department, Institut Universitaire de Cardiologie et de Pneumologie de Québec, Québec, Canada;
Faculty of Pharmacy, Université Laval, Québec, Canada

Sjaak Pouwels Department of Surgery, Haaglanden Medical Center, The Hague, The Netherlands

Mohamed M. Radwan Department of Translational Research, Western University of Health Sciences, Pomona, CA, USA

Negar Salehi Central Arkansas Veterans Healthcare System, University of Arkansas for Medical Sciences, Little Rock, AR, USA

Elijah E. Sanches Department of Surgery, Haaglanden Medical Center, The Hague, The Netherlands

Amrit Pal Singh Department of Pharmaceutical Sciences, Guru Nanak Dev University, Amritsar, Punjab, India

Manjinder Singh Department of Pharmaceutical Sciences, Guru Nanak Dev University, Amritsar, India

Aisha Siraj Central Arkansas Veterans Healthcare System, University of Arkansas for Medical Sciences, Little Rock, AR, USA

Finosh G. Thankam Department of Translational Research, Western University of Health Sciences, Pomona, CA, USA

Besir Topal Department of Cardiothoracic Surgery, Onze Lieve Vrouwe Gasthuis, Amsterdam, The Netherlands

Hardeep Singh Tuli Department of Biotechnology, Maharishi Markandeshwar (Deemed to be University), Ambala, Haryana, India

Belma Turan Department of Biophysics, Faculty of Medicine, Ankara University, Ankara, Turkey

Jeffrey T. Wigle Institute of Cardiovascular Sciences, St. Boniface Hospital Albrechtsen Research Centre, Winnipeg, MB, Canada;
Department of Biochemistry and Medical Genetics, University of Manitoba, Winnipeg, MB, Canada

Nuray Yazihan Interdisciplinary Food, Metabolism and Clinical Nutrition Department, Institute of Health Sciences, Ankara University, Ankara, Turkey;
Faculty of Medicine, Department of Pathophysiology, Ankara University, Ankara, Turkey

Alejandro Zentella-Dehesa Biochemistry Unit, Instituto Nacional de Ciencias Médicas y Nutrición Salvador Zubirán, Mexico City, Mexico;
Department of Genomic Medicine and Environmental Toxicology, Instituto de Investigaciones Biomédicas, Universidad Nacional Auntónoma de México, Mexico City, Mexico

Hao Zhang Division of Cardiology, Faculty of Medicine and Dentistry, Department of Medicine, University of Alberta, Edmonton, AB, Canada;
Mazankowski Alberta Heart Institute, University of Alberta, Edmonton, AB, Canada

Jiu-Chang Zhong Heart Center and Beijing Key Laboratory of Hypertension, Beijing Chaoyang Hospital, Capital Medical University, Beijing, China

Part I
Pathophysiology of Cardiovascular Complications in Obesity

Chapter 1
Different Faces of Obesity in Cardiovascular Diseases: Culprit or Protector

Negar Salehi, Aisha Siraj, Mojdeh Nasiri, and Jawahar L. Mehta

Abstract The incidence of obesity has tripled since 1975, and now it has become a world-wide problem. Initially, obesity was observed only in developed countries, but now it is often seen in less-developed and under-developed countries. Obesity is associated with cardiovascular disease (CVD) risk factors such as hypertension, metabolic syndrome, diabetes mellitus, and dyslipidemia. These risk factors lead through complex pathways lead to evolution and progression of atherosclerosis with subsequent expression of clinical events such CVD. Obesity is also associated with the development of heart failure. Obesity is associated with depression and physical inactivity, both of which lead to worsening of CVD. Fortunately, loss of body weight result in reduction on vascular disease (VD) risk factors. Here we review in detail the metabolic and cellular abnormalities in obesity and CVD.

Keywords Obesity · Cardiovascular disease · Cardiovascular risk factors

Introduction and Epidemiology

Obesity has become a widespread problem worldwide. Based on the World Health Organization (WHO) data in 2016, more than 1.9 billion people age 18 or older have altered weight, with 39% being overweight, and 13% obese. According to this report the prevalence of obesity in men and women was 11% and 15%, respectively [1]. Over 340 million children or adolescents aged 5–19 years, and 41 million children under age 5 years were obese or overweight. Japan is the only exception that has not faced obesity epidemic until 2017 [2].

The prevalence of obesity has tripled since 1975. In the early years since then, obesity was a problem in rich and developed countries, but since 2000, the incidence

N. Salehi · A. Siraj · J. L. Mehta (✉)
Central Arkansas Veterans Healthcare System, University of Arkansas for Medical Sciences, Little Rock, AR 72205, USA
e-mail: MehtaJL@uams.edu

M. Nasiri
Department of Epidemiology and Statistic, Michigan State University, East Lansing, MI, USA

P. S. Tappia et al. (eds.), *Biochemistry of Cardiovascular Dysfunction in Obesity*, Advances in Biochemistry in Health and Disease 20,
https://doi.org/10.1007/978-3-030-47336-5_1

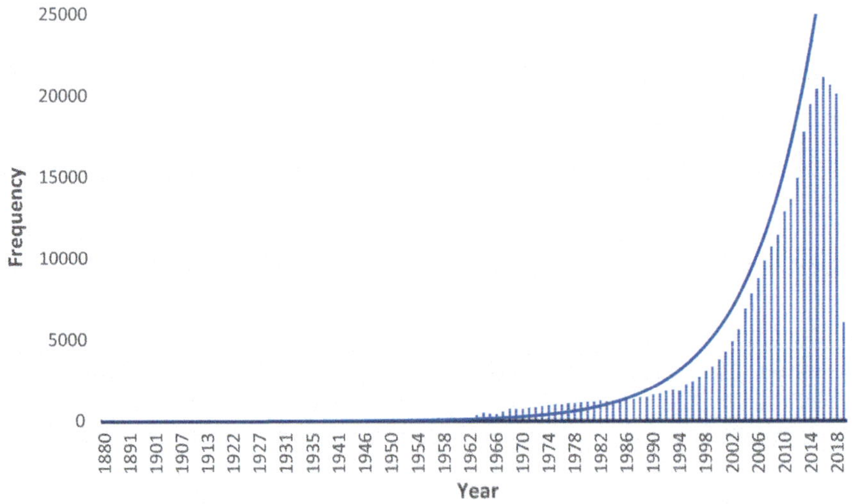

Fig. 1.1 Number of obesity articles in the past four decades. (https://www.ncbi.nlm.nih.gov/pubmed/?term=obesity)

of obesity has increased in developing as well as in under-developed countries. For example, in Africa, the number of obese children has increased by 50% since 2000 [1]. Between 1975 and 2014, body mass index (BMI) has increased by 2.5 kg/m^2 in men and 2.1 kg/m^2 in women worldwide [3] (Fig. 1.2).

As obesity is becoming a leading public health issue, the number of studies that discuss and evaluate different aspects of obesity has increased exponentially. A quick search in PubMed with the keyword "obesity" shows more than 300,000 publications, more than 8,000 of them published in 2019 (until May) alone. As a reference, this topic had less than five articles each year from 1880–1940 (Fig. 1.1).

Definition

Obesity is defined as an excessive accumulation of body fat that compromises health. There are various methods to measure the severity of fatness. Weight is a simple measure. However, it has been replaced by more advanced measures such as BMI (weight [kg]/height^2 [m^2]), waist circumference, waist to hip ratio, percent body fat (BF %), fat mass (FM), and fat-free mass (FFM). Despite its limitations, BMI is still the most commonly used measure of obesity.

According to the WHO, there are different levels of obesity based on nutrition status in adults (Table 1.1) [4]. The WHO defined overweight children age 5–19 years as BMI more than one standard deviation above the WHO growth median reference. Likewise, obesity in this age group is defined when BMI is more than two standard deviations above WHO reference. The definition of obesity and overweight

Table 1.1 WHO classification of overweight and obesity based on BMI in adults

BMI (kg/m^2)	Nutrition status
<18.5	Under weight
18.5–24.9	Normal weight
25–29.9	Pre-obesity
30–34.9	Obesity class I
35–39.9	Obesity class II
≥40	Obesity class III

in children younger than 5 years is based on the weight-height chart. If the weight-height chart is more than two or three standard deviations higher than the growth chart, it is called overweight or obesity, respectively. In adults, the state of being overweight or obesity defined as shown in Table 1.1.

Obesity and Cardiovascular Disease

Cardiovascular diseases, including heart failure (HF), atrial fibrillation (AF), and coronary artery disease (CAD), are often associated with obesity. The duration of obesity has a direct impact on the incidence of cardiovascular disease. With that being said, obesity has specific effects on the cardiac structure and left ventricular systolic and diastolic function [2].

There is an association between adipose cell size and function when body weight exceeds 170% of ideal body weight. With excess caloric intake, adipocyte hyperplasia occurs as a result of an excess in adipogenic progenitors and growth factors like tumor

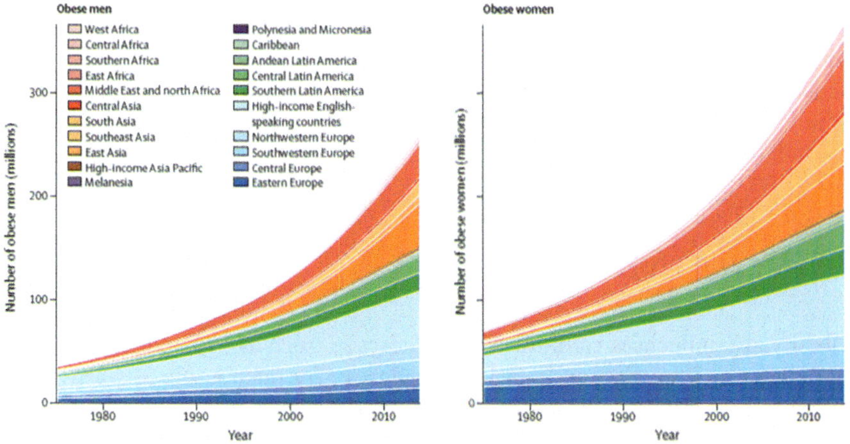

Fig. 1.2 Incidence of obesity worldwide from 1975 to 2014 [3]

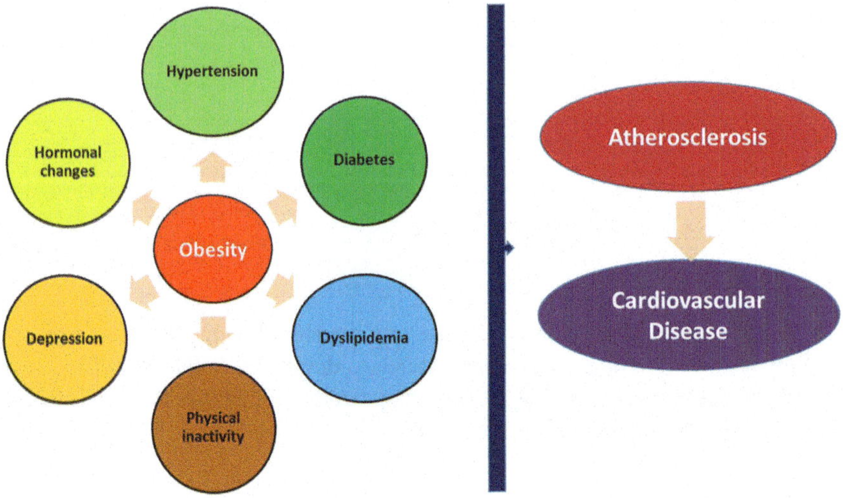

Fig. 1.3 Mechanisms of obesity-induced cardiovascular disease

necrosis factor alpha (TNF-α), angiotensin II, insulin-like growth factor-1, insulin-like growth factor binding proteins, and macrophage colony stimulating factor [4–6]. As obesity progresses, adipocytes undergo apoptosis that induces an inflammatory response and adipose tissue dysfunction. Inflammation, comprising a dysfunction of B-cell, eosinophils, mast cells, macrophages, and neutrophils, plays an essential role in the development of metabolic syndrome in obesity (Fig. 1.3) [2].

The overall effect of obesity on cardiac function includes an increase in stroke volume with reduced peripheral vascular resistance. Duration of obesity has a crucial role in the evolution of these changes; for example, the progression of obesity ultimately causes an increase in ventricular filling pressure, left ventricular hypertrophy, and later dilation. Atrial dilation specifically left atrial dilation, is one of the significant structural changes. Diastolic dysfunction is a result of increased filling pressure and volume overload. All these changes eventually lead to heart failure, either heart failure with reduced ejection fraction [HFrEF] or heart failure with preserved ejection fraction [HEpEF]) [2].

Obesity and CVD Risk Factors

Obesity also influences cardiovascular risk factors, such as hypertension, diabetes, dyslipidemia, depression, and metabolic syndrome. The relationship between obesity and different cardiovascular disease risk factors is shown in Fig. 1.3. Mechanisms by which obesity can lead to the development of atherosclerosis and its manifestations are summarized in Fig. 1.4.

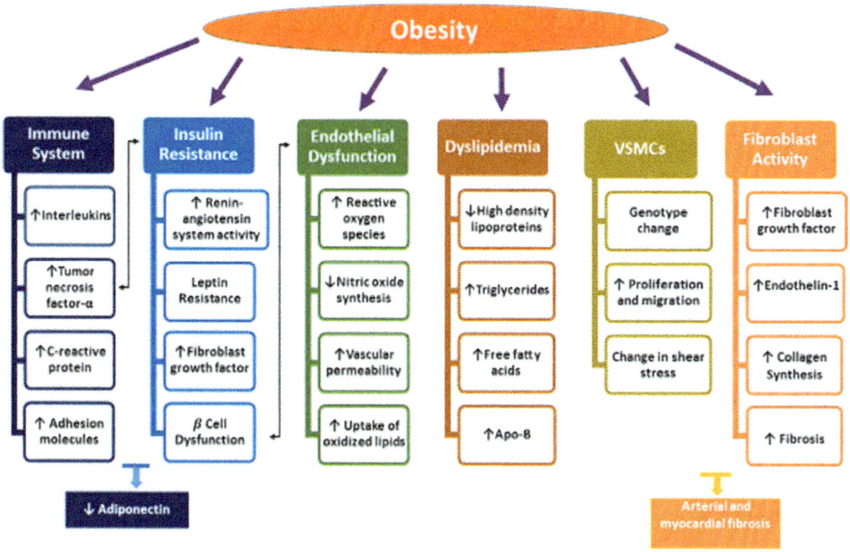

Fig. 1.4 Schematic view of obesity and its effects on pathways leading to cardiovascular disease

Metabolic Syndrome

Several studies have shown that obesity has a direct effect on the incidence of metabolic syndrome. Metabolic syndrome clinically defined by the National Cholesterol Education Program's Adult Treatment Panel III reports, as the presence of at least 2 out of 6 main components. Main components include abdominal obesity, atherogenic dyslipidemia, hypertension, insulin resistance with or without glucose intolerance, proinflammatory, and prothrombotic state [4]. Metabolic syndrome is associated with cellular and inflammatory changes, which result in atherosclerosis. Isomaa et al. [5] showed that women and men with metabolic syndrome have a three times higher risk for CVD and stroke. Some studies indicate that metabolically healthy obese people have a lower risk of CVD mortality, while others have refuted this observation [6–8]. When examined in more detail, cardiovascular fitness has a role in differentiating metabolically healthy obese versus metabolically abnormal obese people in determining the risk for mortality and morbidity [2].

Diabetes Mellitus

Many patients with metabolic syndrome go on to have clinical type 2 diabetes mellitus. Although the pathogenesis of diabetes is not well understood, there is a significant amount of inflammation, excess generation of reactive oxygen species, and abnormalities in glucose metabolism and renin-angiotensin system which are related to the development of diabetes. Diabetes is often associated with an increase in both systolic and diastolic blood pressure, which may contribute to the evolution of atherosclerosis. Inflammation, oxidative stress, and abnormalities in glucose metabolism and renin-angiotensin-aldosterone system are associated with fibroblast growth (via FGF-21) and collagen synthesis, resulting in cardiac, renal, and peripheral arterial fibrosis (Fig. 1.4) [9, 10].

Hypertension

Hypertension is six times more common in obese people [11]. Prevalence of hypertension is about 50% in overweight or obese patients, whereas about 40% in the general population [12]. Gelber et al. showed a strong association between higher BMI and hypertension, (1-unit increase in BMI associated with 8% rise in hypertension prevalence) in a cohort of men followed for 14.5 years [13]. Several mechanisms have been shown to contribute to high blood pressure in obesity, including sodium retention, insulin resistance, oxidative stress, adipokines (such as adiponectin and leptin), renin-angiotensin-aldosterone system, activation of the sympathetic nervous system and hyperinsulinemia [14]. Additionally, there are multiple cardiac structural changes like left ventricular dilation, left ventricular hypertrophy, increased stroke volume, and cardiac output as a result of hypertension. Some investigators suggest that the type of obesity influences left heart remodeling. Central obesity causes mostly concentric left ventricular hypertrophy, while peripheral obesity is associated with eccentric left ventricular dilation with increased left ventricular mass [11, 14].

Studies have demonstrated the effect of weight gain or weight loss on blood pressure. The weight gain in younger population has a higher impact on the blood pressure. A 10 kg increase in body weight will cause 3 and 2.3 mmHg increase in systolic and diastolic blood pressure, respectively [11]. It is of note that long term weight loss can reduce the risks linked with hypertension [15]. Wing et al. found that losing 5 to <10% of body weight can cause 5 mmHg reduction in both systolic and diastolic blood pressure [16].

Dyslipidemia

The unfavorable lipid panel, including high total cholesterol, low-density lipoprotein cholesterol (LDL-C), low high-density lipoprotein (HDL C) and high triglyceride levels, is commonly observed in obese patients [17, 18]. These abnormalities in lipid profile linked with the development of atherosclerosis and resultant CVD (Fig. 1.3).

Release of proinflammatory cytokines in obesity is involved in reducing lipoprotein lipase as a result of increased expression of protein-4 that is lipoprotein lipase inhibitor. Decreased lipoprotein lipase activity delays the clearance of triglyceride-rich lipoprotein, which in turn prompts the rise of triglycerides in obese people [18].

Remarkably, LDL-C levels are generally normal in obese people, but recent studies have shown the role of other lipid abnormalities in developing CVD. Cui et al. showed that non-HDL-C had a robust direct correlation with CVD compared to LDL-C alone. Level of HDL-C also has better prediction than LDL-C. In this study, they clarified that HDL-C and non-HDL-C are better predictors for the CVD mortality [19]. For example, small dense LDL particles are often observed in obesity [19]. Further, the size of the LDL particles has a stronger association with risk of CVD than the level of LDL-C [2]. Some studies showed as much as a 30% decrease in CVD risk over 5 years by treating lipid abnormalities. The key recommendation is lifestyle modification to reduce weight and CVD risk.

Obesity and Atrial Fibrillation

Atrial fibrillation (AF) prevalence has increased significantly over the past few decades and is predicted to rise as much as 2.5 times over the next 30 years [8, 9]. One of the reasons for this increase is the epidemic of obesity. Framingham study showed that each 1-unit increase in BMI would increase the probability of developing AF by 4%; as a result of atrial remodeling secondary to obesity [20]. A large meta-analysis of 16 studies found that obesity causes up to 49% increase in the risk of developing AF among the general population, and the risk is directly related to increase BMI [21].

As mentioned earlier, obesity causes many unfavorable hemodynamic and inflammatory alterations and provides the perfect substrate for the development of AF. The alterations triggered by obesity contribute to atrial enlargement and fibrosis. These changes together can stimulate the development of arrhythmias, especially AF [6, 8, 22, 23].

There is a correlation of left atrial size with a change in BMI. It is noteworthy that left atrial size is related to the occurrence of AF, but not BMI, although left atrial dilation/remodeling has been shown in obesity. Thus, the relationship between obesity and AF appears to be multi-factorial [24].

Obesity and Heart Failure

Obesity and heart failure (HF) association was first reported in 1956 as a case report [25]. The term "Obesity Cardiomyopathy" was first used in 1992, when a study indicated that people with morbid obesity have a higher rate of dilated cardiomyopathy based on the biopsy and right heart catheterization [26]. According to the Framingham study, every 1 kg/m^2 increase in BMI will increase the risk of HF by 7% and 5% in women and men, respectively [27].

Obesity causes a series of changes in the hemodynamic status that result in two different types of HF, HFpEF and HFrEF. Severe systolic dysfunction due to obesity is rare, and its presence should trigger to look for additional etiologies. On the other hand, HFpEF is more common in obesity. As discussed earlier, obesity is associated with hypertension, eccentric or concentric left ventricular hypertrophy, and myocardial fibrosis (Fig. 1.4). All these can lead to diastolic HF [28, 29].

HF can also be a consequence of AF through different pathways. Tachycardia-induced cardiomyopathy can result in HF when AF with a rapid ventricular response is present and persistent. Atrial fibrillation causes atrioventricular dyssynchrony and decreases cardiac output due to loss of atrial contractility knowing as "atrial kick." Left ventricular diastolic dysfunction is another basis for the development of AF [30].

Other features of obesity, including insulin resistance, inflammation, leptin abnormalities, adiponectin deficiency, and volume changes, are related to the development and worsening of HF.

Obesity and Depression and Physical Activity

Although depression has not been recognized as a traditional CVD risk factor, some studies indicated a significant relationship between the evolution of CVD and psychosocial factors. INTERHEART was a case–control study that showed psychological factors, including depression, as one of the substantial risk factors for the CAD, almost same as hypertension, diabetes mellitus, smoking, and hyperlipidemia [31, 32].

This association between obesity and depression may be a response to stress and unhealthy lifestyle or to obesity-related metabolic changes and negative self-image that can result in further worsening of depression. In a meta-analysis, Luppino et al. showed that obese people are at 55% increased risk of depression; on the other hand, people with depression are at a 58% higher risk of developing obesity [33]. This study also confirmed that depression is more common in young females. This high prevalence of depression may be a result of sociocultural status [33]. The association of psychosocial distress, including depression, and CVD, has also been shown in other studies. Pimple et al. showed that for one standard deviation increase in the score of psychological distress, the risk of CVD events increases 1.44 times in women (95% CI, 1.09–1.92), although this association has not been confirmed in men [34].

Obese individuals tend to be physically inactive. Physical inactivity was identified to be a risk factor for the development of myocardial infarction in the INTERHEART study [31]. Other studies have shown poor outcome after myocardial infarction and HF in physically inactive patients (Depressed patients tend to be physically inactive); thus, there appears to be a link between obesity, depression and physical activity and resultant CVD. Liu et al. [35] showed in a metanalysis the salutary effect of cardiac rehabilitation on aerobic endurance, psychosocial well-being, and CVD risk reduction in patients with coronary heart disease [35].

Obesity Paradox

Obesity paradox is defined as the possible protective role of obesity in CVD. Although a vast number of studies have shown obesity as a risk factor for CVD, some investigators have pointed out some protective effects of obesity in CVD. Obesity paradox was initially described in HF patients [36, 37]. Ellis et al. observed obesity paradox around 1996. They showed that overweight and obese patients had lower in-hospital mortality rates compared to normal-weight patients (mortality rates: 2.8% for BMI $\leq$25 kg/m^2, 3.7% for BMI >35 kg/m^2, and 0.9% for BMI 26–34 kg/m^2; $p < 0.001$) [38].

Gurberg and colleagues followed PCI patients for 5 years and found that the rate of major in-hospital complications, including cardiac death was significantly lower in both overweight (0.7%) and obese (0.4%) patients compared to normal weight (1%) people [39]. Further, the rate of in-hospital and one-year mortality in obese and overweight subjects was half of the rate in normal-weight patients (normal BMI cohort 10.6%, overweight cohort 5.7%, and obese cohort 4.9%; $p < 0.0001$) [39].

Femmino et al. [40] have discussed the role of a better clinical status of HF in obese patients than lean heart failure patients. It is possible that obese patients with HF are younger, making age as an important confounder. Another critical factor might well be high levels of "cachectin" in thin heart failure patients. Cachectin, now recognized to be TNF-α, which may have an adverse effect on cardiomyocyte viability and therefore, cardiac function.

However, obesity paradox needs to be studied more extensively.

Conclusions

Obesity is a rapidly growing, global health problem. It is associated with a large number of risk factors that result in CVD. The development of obesity-mediated CVD is multifactorial and include inflammation, oxidative stress, and activation of the renin-angiotensin system. Nonetheless, studies show that a persistent loss in body weight with dietary control and physical activity can result in a decrease in CVD risk and improve outcome of patients CVD.

References

1. Obesity and overweight [Internet] [cited 2019 Apr 30]. https://www.who.int/news-room/fact-sheets/detail/obesity-and-overweight
2. Parto P, Lavie CJ (2017) Obesity and cardiovascular diseases. Curr Probl Cardiol 42(11):376–394
3. Trends in adult body-mass index in 200 countries from 1975 to 2014: a pooled analysis of 1698 population-based measurement studies with 19·2 million participants. Lancet 387(10026):1377–1396 (2016)

4. Grundy SM, Brewer HB, Cleeman JI et al, American Heart Association et al (2004) Definition of metabolic syndrome: report of the National Heart, Lung, and Blood Institute/American Heart Association conference on scientific issues related to definition. Circulation 109(3):433–438
5. Isomaa B, Almgren P, Tuomi T et al (2001) Cardiovascular morbidity and mortality associated with the metabolic syndrome. Diabetes Care 24(4):683–689
6. Lavie CJ, McAuley PA, Church TS et al (2014) Obesity and cardiovascular diseases: implications regarding fitness, fatness, and severity in the obesity paradox. J Am Coll Cardiol 63(14):1345–1354
7. Lavie CJ, Alpert MA, Arena R et al (2013) Impact of obesity and the obesity paradox on prevalence and prognosis in heart failure. JACC Heart Fail 1(2):93–102
8. Lavie CJ, Milani RV, Ventura HO (2009) Obesity and cardiovascular disease: risk factor, paradox, and impact of weight loss. J Am Coll Cardiol 53(21):1925–1932
9. Zick Y (2005) Ser/Thr phosphorylation of IRS proteins: a molecular basis for insulin resistance. Sci STKE Signal Transduct Knowl Environ (268):pe4
10. Saltiel AR, Olefsky JM (2017) Inflammatory mechanisms linking obesity and metabolic disease. J Clin Invest 127(1):1–4
11. Poirier P, Giles TD, Bray GA et al (2006) Obesity and cardiovascular disease: pathophysiology, evaluation, and effect of weight loss: an update of the 1997 American Heart Association Scientific Statement on Obesity and Heart Disease from the Obesity Committee of the Council on Nutrition, Physical Activity, and Metabolism. Circulation 113(6):898–918
12. Elevating hypertension on the public health agenda [Internet]. World Heart Federation, 2018 [cited 2019 Apr 30]. https://www.world-heart-federation.org/news/elevating-hypertension-public-health-agenda/
13. Gelber RP, Gaziano JM, Manson JE et al (2007) A prospective study of body mass index and the risk of developing hypertension in men. Am J Hypertens 20(4):370–377
14. Cohen JB (2017) Hypertension in obesity and the impact of weight loss. Curr Cardiol Rep 19(10):98
15. Moore LL, Visioni AJ, Qureshi MM et al (2005) Weight loss in overweight adults and the long-term risk of hypertension: the Framingham study. Arch Intern Med 165(11):1298–1303
16. Wing RR, Lang W, Wadden TA et al (2011) Benefits of modest weight loss in improving cardiovascular risk factors in overweight and obese individuals with type 2 diabetes. Diabetes Care 34(7):1481–1486
17. Cullen P (2000) Evidence that triglycerides are an independent coronary heart disease risk factor. Am J Cardiol 86(9):943–949
18. Lara-Castro C, Fu Y, Chung BH, Garvey WT (2007) Adiponectin and the metabolic syndrome: mechanisms mediating risk for metabolic and cardiovascular disease. Curr Opin Lipidol 18(3):263–270
19. Cui Y, Blumenthal RS, Flaws JA et al (2001) Non–high-density lipoprotein cholesterol level as a predictor of cardiovascular disease mortality. Arch Intern Med 161(11):1413–1419
20. Wang TJ, Parise H, Levy D et al (2004) Obesity and the risk of new-onset atrial fibrillation. JAMA 292(20):2471–2477
21. Wanahita N, Messerli FH, Bangalore S et al (2008) Atrial fibrillation and obesity—results of a meta-analysis. Am Heart J 155(2):310–315
22. Patel DA, Lavie CJ, Milani RV et al (2012) Association of left ventricular geometry with left atrial enlargement in patients with preserved ejection fraction. Congest Heart Fail Greenwich Conn 18(1):4–8
23. Abed HS, Samuel CS, Lau DH et al (2013) Obesity results in progressive atrial structural and electrical remodeling: implications for atrial fibrillation. Heart Rhythm 10(1):90–100
24. Goudis CA, Korantzopoulos P, Ntalas IV et al (2015) Obesity and atrial fibrillation: a comprehensive review of the pathophysiological mechanisms and links. J Cardiol 66(5):361–369
25. Counihan TB (1956) Heart failure due to extreme obesity. Br Heart J 18(3):425–426
26. Kasper EK, Hruban RH, Baughman KL (1992) Cardiomyopathy of obesity: a clinicopathologic evaluation of 43 obese patients with heart failure. Am J Cardiol 70(9):921–924

27. Kenchaiah S, Evans JC, Levy D et al (2002) Obesity and the risk of heart failure. N Engl J Med 347(5):305–313
28. Kapoor JR, Heidenreich PA (2010) Obesity and survival in patients with heart failure and preserved systolic function: a U-shaped relationship. Am Heart J 159(1):75–80
29. Ather S, Chan W, Bozkurt B et al (2012) Impact of noncardiac comorbidities on morbidity and mortality in a predominantly male population with heart failure and preserved versus reduced ejection fraction. J Am Coll Cardiol 59(11):998–1005
30. Clark DM, Plumb VJ, Epstein AE, Kay GN (1997) Hemodynamic effects of an irregular sequence of ventricular cycle lengths during atrial fibrillation. J Am Coll Cardiol 30(4):1039–1045
31. Yusuf S, Hawken S, Ounpuu S et al (2004) Effect of potentially modifiable risk factors associated with myocardial infarction in 52 countries (the INTERHEART study): a case-control study. Lancet Lond Engl 364(9438):937–952
32. Elderon L, Whooley MA (2013) Depression and cardiovascular disease. Prog Cardiovasc Dis 55(6):511–523
33. Luppino FS, de Wit LM, Bouvy PF et al (2010) Overweight, obesity, and depression: a systematic review and meta-analysis of longitudinal studies. Arch Gen Psychiatry 67(3):220–229
34. Pimple P, Lima BB, Hammadah M et al (2019) Psychological distress and subsequent cardiovascular events in individuals with coronary artery disease. J Am Heart Assoc 8(9):e011866
35. Liu T, Chan AW, Liu YH, Taylor-Piliae RE (2018) Effects of Tai Chi-based cardiac rehabilitation on aerobic endurance, psychosocial well-being, and cardiovascular risk reduction among patients with coronary heart disease: a systematic review and meta-analysis. Eur J Cardiovasc Nurs J Work Group Cardiovasc Nurs Eur Soc Cardiol 17(4):368–383
36. Ortega FB, Lavie CJ, Blair SN (2016) Obesity and cardiovascular disease. Circ Res 118(11):1752–1770
37. Ghoorah K, Campbell P, Kent A et al (2016) Obesity and cardiovascular outcomes: a review. Eur Heart J Acute Cardiovasc Care 5(1):77–85
38. Ellis SG, Elliott J, Horrigan M et al (1996) Low-normal or excessive body mass index: newly identified and powerful risk factors for death and other complications with percutaneous coronary intervention. Am J Cardiol 78(6):642–646
39. Gruberg L, Weissman NJ, Waksman R et al (2002) The impact of obesity on the short-term and long-term outcomes after percutaneous coronary intervention: the obesity paradox? J Am Coll Cardiol 39(4):578–584
40. Femminò S, Pagliaro P, Penna C (2019) Obesity, and cardioprotection. Curr Med Chem

Chapter 2
Obesity-Induced Cardiovascular Complications and Therapeutic Intervention

Md. Jahangir Alam and Sanjay K. Banerjee

Abstract In the last few decades, the prevalence of obesity has increased in a pandemic manner worldwide and remains a growing global health problem. Morbid obesity is a major health risk for different non-communicable diseases such as type 2 diabetes mellitus, fatty liver disease, stroke, dementia, osteoarthritis, and certain malignancies. It is not only associated with these medical conditions but also poses an increased risk of cardiovascular diseases including hypertension, atherosclerosis, and myocardial infarction. Although a plethora of obesity-induced cardiovascular complications has been documented, the underlying mechanisms governing obesity-CVD link are complex and an obesity paradox (protective effect of obesity in patients with CVD) exists. A variety of factors such as sedentary lifestyle, health metrices and different anthropometric indices such as body fat distribution and muscle mass play a critical role in ensuing a cascade of pathophysiological consequences that determines obesity-induced cardiovascular events. Adipose tissues create a microenvironment and act as an endocrine organ by secreting several immune-modulatory molecules. In obese subjects, expanding and dysfunctional adipose tissues undergoes imbalance in the expression of the pro-inflammatory and anti-inflammatory cytokines which ultimately promotes systemic metabolic dysfunction and cardiovascular disease. In this book chapter, we will focus on a comprehensive study of the pathogenic factors and molecular mechanisms associated with obesity, and its link with structural and functional changes of the cardiovascular system. Understanding the mechanistic link between obesity and cardiovascular diseases will provide the basis for therapeutic intervention for obesity-induced cardiovascular complications.

Keywords Obesity · BMI · Cardiovascular disease · Adipose tissue · Adipokines · Inflammation · Heart failure

Md. J. Alam
Translational Health Science and Technology Institute, NCR Biotech Science Cluster, Faridabad 121001, Haryana, India

S. K. Banerjee (✉)
Department of Biotechnology, National Institute of Pharmaceutical Education and Research (NIPER), Guwahati 781101, Assam, India
e-mail: skbanerjee@thsti.res.in; sajayk.banerjee@niperguwahati.ac.in

P. S. Tappia et al. (eds.), *Biochemistry of Cardiovascular Dysfunction in Obesity*, Advances in Biochemistry in Health and Disease 20,
https://doi.org/10.1007/978-3-030-47336-5_2

Abbreviations

11β-HSD2	11β-hydroxysteroid dehydrogenase type 2
ACE	Angiotensin-converting enzyme
AF	Atrial fibrillation
AGT	Angiotensinogen
AMPK	AMP-activated protein kinase
AngII	Angiotensin II
ATBF	Adipose tissue blood flow
BMI	Body mass index
CAC	Coronary artery calcium
CAD	Coronary artery disease
CHD	Congenital heart diseases
CI	Cardiac index
CTRP1	Complement-C1q TNF-related protein 1
CV	Cardiovascular
CVD	Cardiovascular diseases
eCB	Endocannabinoids
ECG	Electrocardiography
HCC	Hepatocellular carcinoma
HF	Heart failure
HFpEF	Heart failure with preserved ejection fraction
hsCRP	High-sensitivity C-reactive protein
IL	Interleukin
IR	Insulin resistance
LDL	Low-density lipoprotein
LV	Left ventricular
LVEF	LV ejection fraction
LVH	Left ventricular hypertrophy
MCP1	Monocyte chemoattractant protein 1
mPTP	Mitochondrial permeability transition pore
mTOR	Mammalian target of rapamycin
NAFLD	Non-alcoholic fatty liver disease
NASH	Non-alcoholic steatohepatitis
NF-κB	Nuclear factor-kappa B
NO	Nitric oxide
NSTEMI	Non-ST segment elevation myocardial infarction
OSA	Obstructive sleep apnea
PCI	Percutaneous coronary intervention
PH	Pulmonary hypertension
RAAS	Renin-angiotensin-aldosterone system
RISK	Reperfusion injury salvage kinases
ROS	Reactive oxygen species
SNS	Sympathetic nervous system

SS	Simple steatosis
STEMI	ST-segment elevation myocardial infarction
SVI	Stroke volume index
SVR	Systemic vascular resistance
T2DM	Type 2 diabetes mellitus
TLR4	Toll-like receptor 4
VEGF	Vascular endothelial growth factor
VSMCs	Vascular smooth muscle cells
WAT	White adipose tissue
WHO	World health organization

Introduction

Non-communicable diseases such as diabetes mellitus, cancer and cardiovascular diseases (CVD) represent the leading cause of morbidity and mortality ($\geq$70%) worldwide [1]. Obesity, being a major risk factor for such diseases [2–4], has been a growing global health concern; affecting both developed and developing countries. The world health organization (WHO) data shows that obesity affects about 13% of the population who are above 18 years of age [5]. Obesity affects both gender, while prevalence is higher in women than men (200 vs. 300 million) which counts sum of 1.5 billion adults worldwide [6].

According to the WHO, obesity is defined as excessive fat accumulation i.e. body mass index (BMI) $\geq$ 30 kg/m^2 that might impair other health conditions [5]. Cardiovascular risk assessment is challenging as the threshold proposed by the WHO is not consistent throughout Asian population due to differences in fat distribution, the existence of obesity paradox [7] and 'metabolically healthy obesity' [8]. Distribution of fat to other organs such as heart and liver plays a key role in the severity of risk posed by obesity. Since BMI does not measure the fat distribution, other parameters like abdominal circumference and the calculation of waist to hip ratio are being used to characterize central or abdominal obesity [9–11].

Accumulating evidence suggests the association between the extent of obesity and other metabolic disorders such as type 2 diabetes mellitus, Non-alcoholic fatty liver disease (NAFLD), a range of cardiovascular diseases (hypertension, myocardial infarction, heart failure, cardiac arrhythmias), neurodegenerative diseases like Alzheimer disease, and some malignancies such as breast cancer, prostate cancer, and hepatocellular carcinoma (HCC) [12]. Now it is clearly evident that obesity is a prominent risk factor for other non-communicable disease where CVDs like dyslipidemia, insulin resistance, hypertension, and atherosclerosis are strongly associated (Fig. 2.1) [13].

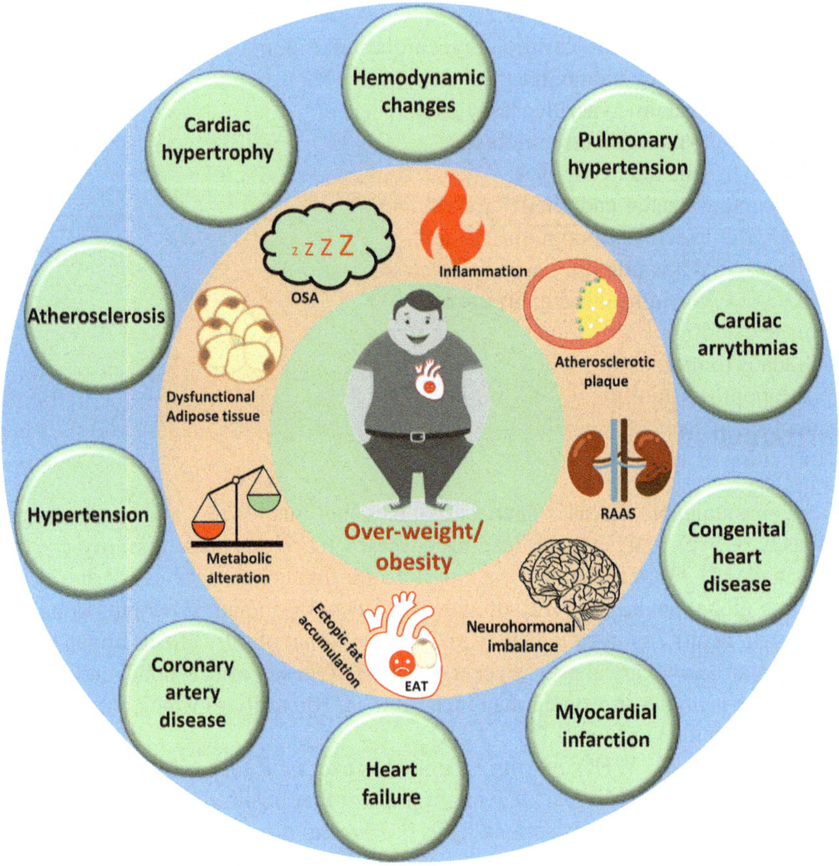

Fig. 2.1 Schematic representation of obesity-induced cardiovascular disease (CVD) and their complex interplay with their risk factors. In obesity or over-weight, various risk factors (middle ring) such as metabolic syndrome, neurohormonal imbalance, inflammation and obstructive sleep apnea (OSA) play a central role in inducing CVDs (outer ring) such as hypertension, coronary artery disease, congenital heart disease and cardiac arrhythmia. (RAAS: Renin-angiotensin-aldosterone system, EAT: Ectopic adipose tissue)

White adipose tissue (WAT), a constituent of adipose tissue, exerts crucial effects on the metabolic and inflammatory pathway by secreting biologically active peptides and proteins such as adiponectin that play an important role in obesity, insulin resistance, inflammation, and cardiovascular diseases [14–16]. When adipose tissues become expanded and dysfunctional in obesity, the adipose tissues cannot secret sufficient level of anti-inflammatory cytokines, thereby aggravate the inflammatory condition, thus obesity is considered as a state of inflammation. This imbalance in the homeostasis of pro- and anti-inflammatory condition eventually starts obesity-induced metabolic alteration not only in the cardiovascular system but also in other tissues such as pancreas, liver, and kidneys [17, 18]. Although it is difficult to decipher the obesity paradox, obesity and its comorbidities have harmful effects on cardiovascular disease through various mechanisms including increase in pro-inflammatory cytokines and decrease in anti-inflammatory factors, ectopic lipid accumulation, and insulin resistance, hyperglycaemia, endoplasmic reticulum stress, generation of reactive oxygen and reactive nitrogen species.

A large number of cross-sectional and meta-analyses studies were performed to understand the association between obesity and various cardiovascular diseases. Although the exact mechanism of link between them is not established well but evidence suggest that there is a close relationship between the increasing BMI and risk of CVD. For instance, in obese individuals, increase in adipose tissue causes insulin resistant and activation of Renin-angiotensin-aldosterone system (RAAS) which leads to remodelling of cardiac structure and function such as cardiac hypertrophy, apoptosis and fibrosis that ultimately leads to heart failure. Multiple factors are involved in the progression of obesity-induced cardiovascular dysfunction such as the formation of atherosclerotic plaques, infiltration of macrophages, neurohormonal and metabolic alteration, hyperglycemia and epigenetic changes. Recent evidence suggests that type 2 diabetes (T2DM) and non-alcoholic and alcoholic fatty liver disease separately or synergistically are higher risk factors for CVD in obese subjects.

To control the risk of obesity-induced cardiovascular diseases, pharmacological intervention alone could not help. Thus losing weight with anti-obesity medications along with lifestyle changes, improvement in diet and comorbid factors are necessary. In this chapter, we are emphasizing the mechanisms of obesity-induced cardiovascular complications and its relation to cardiovascular risks that might help to provide the basis for a rational therapeutic strategy for obesity-induced cardiac complications.

Obesity Paradox

The incidence of obesity is increasing tremendously in the world. As a consequence, there is an increase in the risk of cardiovascular complications, metabolic disorders, and their related risk factors such as fatty liver diseases and type 2 diabetes [19, 20]. While obesity increases the risk for CVD even in the absence of other risk factors, several studies indicate the controversial relationship between obesity and CVD [21]. Several epidemiologic studies have demonstrated the prevalence of potential protective effect of obesity on the cardiovascular complications such as heart failure, pulmonary arterial hypertension and congenital heart defects when they coexist [22–27]; a phenomenon known as "obesity paradox" [28–30]. Recent findings suggest a U-shaped association between BMI categories and mortality in STEMI patients and CVDs. This data showed that the class I obesity (30–34.9) were at lowest risk of mortality rate while normal weight and extremely obese ($\geq$40) patients had higher mortality (Fig. 2.2) [31, 32].

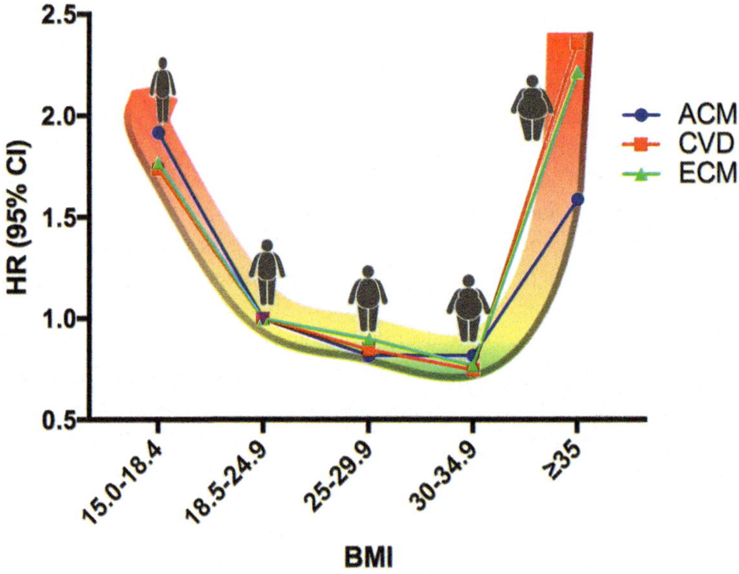

Fig. 2.2 U-shaped associations of BMI with all-cause mortality (ACM), CVD, and expanded CVD mortality (ECM). A U-shaped with flat bottom indicates that not only over-weight or obese individuals are prone to cardiovascular complications but also underweight or low BMI individual has increased risk of CVD. 15.0–18.4: underweight, 18.5–24.9: normal, 25–29.9: overweight, 30–34.9: grade 1 obesity, ≥35: grade 2–3 obesity, HR: Hazard ratio. (Adapted from [32])

Epidemiology of Obesity and Cardiovascular Complication

Over the past few decades, the prevalence of obesity has increased globally in a pandemic manner [5, 33–35]. Recently, a group of investigators from NCD-RisC have demonstrated the trend that how obesity prevalence has increased worldwide in the past 40 years [34]. The proportion of individuals with obesity has been increased from 25.4 to 39.4% between 1980 and 2015, and it is estimated to be 57.8% by 2030 [33, 36]. The prevalence of obesity in 2015 was 10.1% for men and 14.8% for women and estimated to be 18% in men and 21% in women by 2025. Serial National surveys in India have demonstrated an increasing trend in the prevalence of obesity (NFHS 2 and 3) [37, 38]. A population-based study performed by the Indian Council of Medical Research-India Diabetes (ICMR-INDIAB) indicates the prevalence of obesity ranges from 11.8 to 31.3% in India [39]. Prevalence of obesity is higher in women and the urban population as compared to men and the rural population, respectively [40–42].

Obesity-Induced Cardiovascular Complications: Role of Adipose Tissues

In healthy condition, in non-obese subjects, adipose tissues act as an endocrine gland and regulate diverse processes such as maintaining energy homeostasis, insulin sensitivity and inflammation [43]. In obese subjects, a dysfunctional adipose tissue is associated with a plethora of adverse health conditions which ultimately leads to increased risk of mortality at the population level [44, 45]. Obesity and its comorbidities exert its deleterious effects on cardiovascular function via several mechanisms including ectopic fat accumulation, hyperglycemia, and development of the low-grade inflammatory condition, activation of the sympathetic nervous system and neurohormonal imbalance, to name a few. By functioning as an endocrine organ, adipose tissues secrete many peptide (adipokines) and non-peptides (lipokines) hormones, cytokines and growth factors that regulate organ functions and metabolic processes. During uncontrolled and sustained calorie intake adipose tissues expand in size and number, and accompanied by structural and functional remodelling such as adipocyte hypertrophy and fibrosis, local inflammation, infiltration of immune cells, insulin resistance and a altered metabolism. Imbalance in the secretion of adipokine (leptin and adiponectin) leads to the development of a low-grade, chronic inflammatory state that contributes to the development of metabolic and cardiovascular diseases (Fig. 2.3) [46–48].

The adipocytes are enriched with endocrine hormones and immune-modulatory factors which produces many different adipokines at the time of normal physiological state or under stress [49, 50]. In non-obese healthy individuals, adipocytes secrete various anti-inflammatory cytokines such as adiponectin, TGF-β, IL-10, and nitric oxide which show protective effects, normal insulin response (anti-diabetogenic) and anti-atherogenic effects [49, 50]. Adiponectin is the most abundant adipokines secreted by adipocytes. In individuals with high obesity and CVD risk factors, adiponectin concentration drops significantly while the concentration of pro-inflammatory cytokines such as leptin, TNF-α, resistin, and IL-6 increases [51]. Recently Lekva et al. found that Leptin to adiponectin ratio was high during gestational diabetes, a risk factor for obesity-induced CVD [52]. Generally, adiponectin shows anti-inflammatory properties via the inhibition of NF-κB (Nuclear factor-kappa B) expression in macrophages and monocytes. It also inhibits macrophage conversion to foam cells and reduces oxidation of low-density lipoprotein (LDL). Leptin (encoded by ob gene) is the master regulator of food metabolism and energy balance as well as it has various pleiotropic physiological actions such as inflammation, bone metabolism, endocrine function, stimulation of sympathetic nervous system (SNS), cardiovascular regulation and immune responses [53, 54]. Thus it has been shown to increase the level of mediators of vascular inflammation such as IL-2, IL-6, TNFα, MCP-1, reactive oxygen species (ROS), Th1-type cytokines from endothelial cells and peripheral blood mononuclear cells. On the other hand, elevated plasma level of leptin was found to be associated with BMI and a degree of adiposity. However, the satiety effect of leptin is abrogated during high plasma leptin level

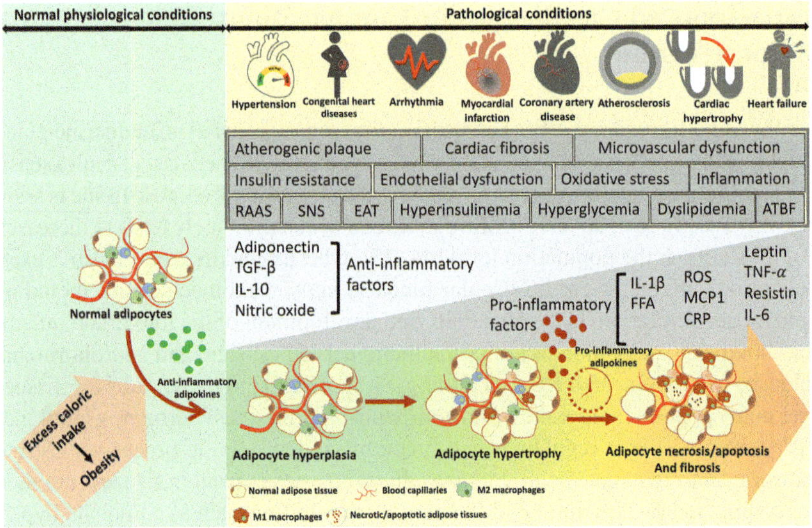

Fig. 2.3 Schematic illustration showing the mechanism of obesity-induced adipose dysfunction and its impact on the cardiovascular system. Excess caloric intake and consequent increase in adiposity lead to adipocyte hyperplasia and hypertrophy. Sustained and chronic adiposity causes necrosis and apoptosis of the cells, which further induce secretion of pro-inflammatory factors such as leptin, TNF-, Resistin, IL-6 and recruitment of M1 macrophages. In contrast to normal physiological condition, obesity is also characterized by the presence of M2 macrophages and a decrease in anti-inflammatory factors such as Adiponectin, TGF-β and IL-10. All the above condition with increased oxidative stress initiates a low-grade systemic pro-inflammatory state and adipose tissue dysfunction, which promotes a number of intermediate risk factors such as insulin resistance, dyslipidaemia, the formation of atherogenic plaque, cardiac fibrosis, microvascular dysfunction, endothelial dysfunction, activation of SNS, alteration in RAAS and reduction in adipose tissue blood flow (ATBF). Ultimately, all these risk factors cause several complications of the cardiovascular system such as hypertension, CHD, arrhythmia, myocardial infarction, CAD and finally heart failure

for the long term, a phenomenon called leptin resistance. Leptin resistance results from several aspects such as (i) interruption of leptin-receptor signalling; (ii) limited permeability to the blood-brain barrier; (iii) down-regulation of leptin receptor in the hypothalamus [55]. Haemostatic imbalance and cardiovascular damage may occur as a result of accumulation of epicardial adipose tissue and concomitant coronary artery disease during hyperleptinemia in obese subjects. Moreover, leptin has been associated with atherosclerosis, hypertension and congestive heart failure. The adipokines action on the cardiovascular system-dependent mainly on two mechanisms either direct to the heart or through interaction with the central nervous system. Leptin acts by stimulating the migration and proliferation of vascular smooth muscle cells (VSMCs) by upregulating the vascular endothelial growth factor (VEGF) expression and the cytoskeleton reorganization [53, 55, 56]. Furthermore, adipose tissue releases plasma endocannabinoids (eCB), such as anandamide and 2-arachidonoylglycerol, which are involved in feeding behaviour and energy metabolism, as well as glucose and lipid metabolism. In obese individuals, insulin resistance and inflammation may increase the level of cannabinoid receptors that causes further increase in visceral fat deposition and concomitant reduction in adiponectin level [57, 58]. Recent findings suggest that endocannabinoids may play an important role in the pathogenesis of obesity-induced cardiovascular complications, such as atherosclerosis [59]. A recent study has also shown that gastric bypass-induced weight loss in obese human improved coronary circulation through the decrease in endocannabinoid levels and a significant increase in adiponectin level [60]. Although there are a large number of biologically active factors secreted from adipose tissue and their complex interactions complicate the understanding of the mechanistic link between obesity and the its impact on the cardiovascular system, the findings from experimental, clinical, and epidemiological data strongly favour the devasting roles of the adipokines on cardiovascular (CV) system.

Adverse Effects of Obesity on the Cardiovascular System

On the other hand, in spite of the existence of obesity paradox or metabolically healthy obesity, large number of studies indicate the potential negative consequences such as hypertension, coronary heart disease, atrial fibrillation and sudden cardiac death [59]. All of the above may occur either independently or with their risk factors i.e. lipid disorder, diabetes mellitus, metabolic syndrome, obesity has adverse effects on various CVD including (Fig. 2.1) [61].

Heart Failure

A number of epidemiological studies have shown that the prevalence of heart failure is increasing. It is one of the major causes of mortality worldwide with a prevalence of approximately 3% in developed countries [62]. Evidence also confirms that there is a close correlation between heart failure and obesity, and obesity is a major risk factor for the development of HF. Framingham Heart Study was done to evaluate the relationship between BMI and incidence of heart failure. According to this study, the rise of one unit BMI (1 kg/m^2) increases the risk of heart failure by 5% for men and 7% in the case of women [63]. Similarly the incidence of heart failure in the obese individuals happens 10 years earlier than in the case of individuals with a normal BMI. Moreover, the duration of morbid obesity increases the prevalence of heart failure by 70% and 90% after 20 and 30 years of obesity, respectively [64]. Severe and long-lasting obesity ultimately causes structural and hemodynamic changes in the heart, which is often referred to as "obesity cardiomyopathy", which leads to congestive heart failure and sudden cardiac death [65]. Evidence on the exact mechanism of obesity cardiomyopathy are sparse but the most important mechanisms include metabolic disturbances, activation of the RAAS and SNS, myocardial remodelling, and endothelial dysfunction [66].

Data suggest that obesity leads to heart failure through several direct and indirect mechanisms including haemodynamic changes, activation of RAAS and SNS, inflammation and other comorbidities. A study showed that an increase in BMI of 5 kg/m^2 involved a 5 mmHg rise in systolic blood pressure. Thus a rise in both cardiac output and blood pressure has been observed [67]. Association between obesity and increased left ventricular (LV) end-diastolic pressure as well as right atrial pressure and pulmonary wedge pressure has been shown in several studies [68, 69]. Increase in blood pressure associated with arterial hypertension and left ventricular afterload in obese individuals [70–72], which may lead to heart failure [73]. In the individual with obesity, an increase in blood volume and high blood-flow is responsible for the increase in ventricular preload and stroke volume [74, 75]. Initially, a persistent increase in ventricular wall tension leads to ventricular dilatation followed by concentric LV hypertrophy and ultimately heart failure [76–78]. Recent data suggest that these cardiac structures and hemodynamic abnormalities can increase the risk of heart failure with preserved LV ejection fraction (LVEF or HFpEF, heart failure with preserved ejection fraction) [79]. Thus, recently, it has been proposed that the development of targeted therapeutics for such patients with increased HFpEF will be helpful [77, 80–82].

Obesity also causes the activation of the renin-angiotensin-aldosterone system and increased activity of the sympathetic nervous system as evident from increase in the aldosterone level and the mineralocorticoid receptor expression. All the changes lead to cardiac fibrosis, platelet aggregation, and endothelial dysfunction [83, 84]. As evident from both human and animal studies, excess visceral fat can induce the synthesis of several pro-inflammatory cytokines and adipokines which can be attributed to the characteristic low-grade systemic inflammation (Fig. 2.3) [85]. Many of the inflammatory mediators and acute-phase proteins such as IL-1β and IL-18, TNF-α enhances myocardial fibrosis, have cardio-depressant properties, and play a central role in the development of heart failure [85–89]. The integrity of skeletal muscle mass is a key factor for maintaining the physical activity. In obesity, myocardium has increased level of triglyceride and muscle atrophy which is mediated by generation of toxic metabolites such as ceramide and diacylglycerol, ultimately causing apoptosis of cardiomyocytes [90–93]. Comorbid health conditions in obese subjects increase the chances of heart failure. In obesity, insulin resistance reduces the contractility of the myocardium [94], atherosclerosis enhances the risk of ischemic cardiomyopathy and coronary artery disease [95] and lipid accumulation in the myocardium enhances fibrosis promotes cardiac arrhythmias, and therefore contribute to the development of heart failure [96, 97].

Atherosclerosis

Atherosclerosis is one of the major causes of morbidity and mortality worldwide and characterized by deposition of cholesterol and formation of the atherosclerotic lesion [98, 99]. In the last 30 years, mechanistic details of the pathophysiology of both obesity and atherosclerosis have been demonstrated as a chronic inflammatory condition, in which the activation of both innate and adaptive immune processes occurs [100, 101]. The association between obesity and atherosclerosis depends on several factors and their underlying mechanisms, including dyslipidaemia, insulin resistance, inflammasome activation, adipocytokines imbalance, and endothelial dysfunction. However, the most important mechanism is the inflammation and NLRP3 inflammasome activation-mediated insulin resistant and endothelial dysfunction, which links obesity to atherosclerosis and other cardiometabolic risks. Several signaling pathways play roles in the pathogenesis of obesity-induced inflammation, such as activation of toll-like receptor 4 (TLR4), activation of protein kinase C or c-JUN N-terminal kinase by fatty acids and their derivatives (diglyceride or ceramide), induction of endoplasmic reticulum stress, increased generation of reactive oxidative species and activation of macrophages by adipocyte death [102].

A number of published results demonstrated that obesity causes a reduction in the level of adiponectin and a rise in inflammatory adipokines such as TNF-α, IL-6, leptin, resistin and CRP which is strongly associated with atherosclerotic risk, myocardial infarction, and diabetes mellitus [103–106]. Recently, Shimobayashi et al. reported that obesity-induced insulin resistance in mice leads to local accumulation of macrophage and causes inflammation in adipose tissue by the production of a chemokine monocyte chemoattractant protein 1 (MCP1). These findings prove that obesity-induced insulin resistance in visceral WAT leads to inflammation rather than vice versa [107]. Similarly, there are other factors contributing to the progression of atherosclerosis in obese individuals viz., autophagy insufficiency, increased oxidative stress and alteration in gut microbiota composition. Various reports showed that autophagy insufficiency may have an impact on metabolic syndrome, may increase generation of pro-oxidants that is linked to atherosclerosis. Similarly, gut microbiota-induced TLR4 activation may be responsible for the low-grade inflammation in the gut [100].

Coronary Artery Disease

According to several epidemiological studies it has been proposed that there is a high prevalence of obesity paradox in patients with coronary artery disease (CAD). Optimal survival was seen in overweight/obese patients undergoing percutaneous coronary intervention, CAD patients with obesity had the lowest risks of CVD mortality. However, there was an increase in total mortality risk in patients with BMI$\geq$ 35 kg/m^2 [7, 108–110]. Although the obesity paradox is controversial till now, overweight and morbid obesity are closely related to risk factors for atherosclerosis and associated CAD.

Obesity increases the formation of atherosclerotic plaques which is characterized by increased infiltration of macrophage and state of low-grade systemic inflammation (Fig. 2.3) [111]. High-sensitivity C-reactive protein (hsCRP) and other pro-inflammatory factors during systemic inflammation has long been attributed to the pathophysiology of atherosclerosis. Recently, an IL-1β targeted anti-inflammatory therapy was proven to be effective in reducing major adverse CVD events in patients with elevated systemic inflammation and established atherosclerotic CVD [112]. However, non-targeted anti-inflammatory therapies may not be efficacious and perhaps could be even detrimental [113]. In addition to the inflammatory hypothesis which may drive obesity to CAD, obesity is also associated with several major risk factors for CAD, like T2DM and dyslipidaemia, which can, in turn, increase the risk for CAD further [21].

Cross-sectional and longitudinal studies demonstrated that the extent and duration of obesity affect the manifestation and risk of coronary heart disease [114, 115]. For instance, atherosclerosis predisposes young patients several decades before the onset of CAD. Patients with increased BMI have more frequent and advanced atherosclerotic lesions compared to that of normal BMI [116]. Increase in body weight also affects the risk of CAD at the rate of 12% per 10kg and the systolic and diastolic blood pressure rises by 3 mmHg and 2.3 mmHg per 10 kg body-weight, respectively [117, 118]. Furthermore, overweight and obesity is also an independent risk factor for ST- and non-ST segment elevation myocardial infarction (STEMI and NSTEMI) patients in young age [108, 119, 120].

Haemodynamic Changes

A range of evidence suggest that obesity results in a variety of hemodynamic changes which may predispose obese individuals to cardiovascular dysfunctions through changes in cardiac structure and ventricular function and finally leads to heart failure, independent of other risk factors. These alterations in obese subjects result from alteration in various neurohormonal and metabolic processes, which may cause LV hypertrophy and impaired LV diastolic function [121–125]. All those above alterations are associated with hypertension [65, 126]. Many of these alterations are reversible with substantial voluntary weight loss.

Excess adipose accumulation during peripheral obesity results into volume stress characterized by an increase in total and central blood volume which leads to structural and functional changes in the heart which in turn predisposes to an increase in cardiac output [65, 75, 127]. As blood volume rises in peripheral obesity, increased stroke volume and cardiac frequency provide the increased cardiac output [65, 127]. Generally, the patients with elevated LV end-diastolic and pulmonary capillary wedge pressures have increased ratio of stroke work index to LV end-diastolic pressure.

Moreover, these obese patients are accompanied by various other altered hemodynamic parameters such as right ventricular pressure, mean right atrial pressure, and pulmonary hemodynamics [65, 128]. On the other hand, a difference has been observed in the central hemodynamics of the patients with central obesity i.e. a lower cardiac output and a higher peripheral vascular resistance [129, 130].

Studies suggest the potential link between adipose tissue blood flow (ATBF; ml min^{-1} per 100 g tissue) and obesity. In both obese human and animals, a 30–40% reduction in ATBF has been found as compared to non-obese subjects [102]. Obese individuals exhibit reduced subcutaneous adipose tissue blood flow (when expressed per 100 g tissue) and increased adipose tissue hyperoxia, as explained by lower adipose tissue oxygen consumption. All of these are accompanied by insulin resistance, reduce capillary density, and increased inflammation [131, 132]. Obese Zucker rats (having inactive leptin receptor) exhibit diverse kind of vascular abnormalities, including high SNS activity, increased vasoconstriction, and impaired vasodilatory mechanisms that cause impaired total blood flow within the skeletal muscle [133].

Recently, Gayda et al. compared cardiovascular hemodynamics and cardiac output during exercise in obese and nonobese individuals. They found a similar cerebral hemodynamic but a higher systolic blood pressure among obese individuals during exercise, which preserves cardiopulmonary and cardiac function during exercise and recovery. Thus, this study suggests that a higher aerobic fitness in obese subjects might have a protective effect through preserving cardiac and pulmonary function during exercise and recovery [134]. Similarly, obesity does not impair myocardial performance and cerebrovascular function as evidenced by a comparable level of Cardiac hemodynamics and cerebral responses irrespective of their low fitness but there was elevated heart rate and VO_2 responses after the two-minute recovery of submaximal effort. However, recovery from a short duration of work was influenced by their fitness level [135]. In another study, cardiac function is significantly altered in morbidly obese pregnant women as reflected by significantly higher heart rate, lower stroke volume index (SVI) and cardiac index (CI) and higher systemic vascular resistance (SVR) [136]. Although cerebral haemodynamics are normal in obese subjects, the cerebral percentage of cardiac output and body oxygen uptake are lower than that of non-obese control subjects. In obese patients, splanchnic blood flow is substantially higher but the renal blood flow is substantially reduced [65].

Hypertension

Currently, hypertension plays a pivotal role in contributing to the global disease burden [137]. It has been estimated that 60–70% of the incidence of hypertension is attributed to overweight and obesity [138]. In the year 2011, around 9.4 million deaths annually are attributable globally for hypertension [62]. According to some studies, the ‘rule of halves’ [139] may be applied to the management of obesity-induced hypertension, which suggest that hypertension may be less well-controlled in obese patients [140–143]. Accumulating evidence suggest that there is a complex interplay between obesity and hypertension. Recent studies added a new paradigm of association between obesity and hypertension which led to a better understanding of the underlying mechanisms such as dysfunctional adipose tissue, imbalance in adipokine synthesis and release, insulin resistance, activation of the RAAS, and increased activity of the SNS, to name a few [144, 145]. Together, these neuroendocrine imbalances contribute to vascular and endothelial dysfunction, impaired pressure natriuresis and sodium excretion, increased cardiac output and changes in systemic vascular resistance and arterial compliance [146]. The accumulation of excess adipose tissues augments a cascade of pathophysiological events that give rise to obesity-induced hypertension, which ultimately increases cardiovascular risks [70, 147, 148]. Retrospective analysis of a clinical study with 202 subjects suggest that with primary hypertension identified that obese patients have increased CI and SVR was the only predictor of the risk of uncontrolled hypertension ($\geq$140 and/or $\geq$90 mmHg) in obese patients compared to non-obese patients [127].

The pro- and anti-inflammatory factors secreted from adipose tissues in pathological or physiological condition plays an important roles in obesity-hypertension link. Insulin also associated with the pathophysiology of obesity-induced hypertension as revealed by an impaired glucose tolerance, hyperinsulinemia, and concomitant insulin resistant [149–152]. Thus increased insulin secretion in obese patients stimulates sympathetic nervous system activity. Obesity induces SNS activity through hypoglycaemia and chronic hyperinsulinemia and linked to heart dysfunction [153]. Furthermore, insulin also favours sodium reabsorption and increase sodium retention by acting directly on the kidney and renal tubules, respectively, which subsequently induces blood pressure during obesity [154–157]. Depending on various pathophysiological states, glucocorticoids and their cognate receptors play a key role in metabolic homeostasis during stress, such as fasting and starvation. During chronic stress, high level of glucocorticoid causes metabolic complications such as insulin resistance, hyperglycaemia, dyslipidaemia, and central obesity. Through their action on adipose tissues, glucocorticoids appear to promote hypertension through increased RAAS activity [70]. In obese individuals, activation of the RAAS plays an central role in elevating blood pressure. This is not only attributable to SNS overactivity and renal compression [70] but also to dysfunctional adipose tissue with increased levels of angiotensin II (AngII) and aldosterone. In obese animal and human, adipose tissues release a significant amount of AngII [158] which ultimately serves as machinery necessary to generate AngII, i.e. angiotensinogen (AGT), renin, and angiotensin-converting enzyme (ACE) at both mRNA and protein level [159–163]. Diet-induced obesity in rats increased both adipose tissue mass and AGT expression in the liver. Moreover, AGT expression also correlates with the level of AGT and AngII in the plasma and blood pressure [164]. Enhanced renal sodium reabsorption and impaired pressure natriuresis are important factors for the increase in blood pressure associated with obesity [70]. Adipocytes of hypertensive patients with high BMI are also capable of aldosterone production [165, 166]. Moreover, AngII, leptin and complement-C1q TNF-related protein 1 (CTRP1) released from the adipose tissue, also stimulate aldosterone release in human adreno-cortical cells [167, 168]. Literature suggests that, both elevated circulating levels of AngII and

Aldosterone can act directly within the brain regions and are capable of stimulating renal sympathetic nerve activity (and thus renin secretion and sodium retention), impaired baroreflex sensitivity, vasopressin release, and elevated blood pressure [71, 148, 169]. Evidence shows that high caloric intake stimulates peripheral α1 and β-adrenoreceptors by increasing peripheral noradrenaline turnover which elevates SNS activity in patients with obesity [170].

Elevated RAAS activity not only induces sodium retention and SNS activation, but may also cause microvascular dysfunction i.e. vascular rarefaction, impaired dilatation, and enhanced constriction) and modulate arterial stiffening, ultimately resulting into obesity-induced hypertension [171, 172]. A characteristic feature of obesity-associated microvascular dysfunction is the microvascular insulin resistance i.e. an impaired ability of insulin to dilate precapillary terminal arterioles and thus causes arterial stiffness and induce capillary recruitment by increasing endothelial nitric oxide (NO) synthesis [154–157, 173–175]. Observation from the Framingham Offspring Study and diet-induced model of obesity also demonstrated that arterial stiffening may precede elevations in systolic blood pressure and incident hypertension [176, 177].

Similar to AngII, aldosterone and insulin, leptin crosses the blood-brain barrier, interacting with the arcuate nucleus, initiates an appetite suppression and increased energy expenditure signal that is mediated through increased SNS activity [178]. Thus elevated circulating level of leptin is also implicated in the obesity-induced hypertension. Recent results from the human study demonstrated that the presence of leptin is not essential for obesity-related hypertension. However, short-term leptin substitution can induce hypertension in these leptin-deficient obese humans, indicating that leptin has a synergistic effect on obesity-associated hypertension by increasing the regional sympathetic tone [179]. Growing evidence suggests that a dysregulated, overstimulated eCB system is implicated both in cardiovascular physiology and the pathogenesis of hypertension, heart disease and atherosclerosis [180, 181].

Pulmonary Arterial Hypertension

Pulmonary hypertension (PH) is a complex and multifactorial disease characterized by pulmonary arterial inflammation, abnormal vasoconstriction and an increase of the mean pulmonary arterial pressure ($\geq$25 mmHg). Recent evidence revealed a strong association between obesity and pulmonary hypertension. Adiponectin has a potential protective role on the pulmonary vasculature by showing various pleiotropic effects on inflammation and cell proliferation. Several in vivo and in vitro studies demonstrated that adiponectin shows its protective roles in preventing endothelial dysfunction and proliferation by an endogenous modulation of nitric oxide production and interfering with AMP-activated protein kinase (AMPK) activation, mammalian target of rapamycin (mTOR) and NF-κβ signalling [182].

The biologically active molecules released from adipose tissue in obese subjects, including adiponectin, leptin, adipsin, visfatin, IL-β, IL-6, and TNF-α play a part in chronic inflammation. It has been demonstrated that leptin and adiponectin are the two most important adipokines in pathogenesis of obesity-induced PH. Normally, leptin has a vasodilator effect and is involved in glucose and lipid metabolism [183]. However, during obesity, its normal physiological level and function get diminished and thus responsible for endothelial dysfunction, and initiates pathogenesis of diseases of the vascular tissues [184]. Evidence revealed that in PH patients, lower leptin levels normalized by BMI were correlated with a higher mortality rate and the leptin/BMI ratio represented a non-linear prognostic value for mortality at two years [185]. Adipokines can also regulate the vascular tone and have direct vasodilator properties. Adiponectin-deficient mice show diminished vasoreactivity and impaired endothelial function, thus these finding supports the association between obesity and PH. The elevated level of insulin in obese subjects may limit the protective effects of adiponectin by decreasing AdipoR1/R2 expression, a phenomenon called adiponectin resistance. Since it has been reported that insulin-resistant can alter hemodynamic parameters, thus insulin resistance can serve as an important linkage between adiponectin and PH [182].

In obese individuals, chronic hyperuricemia has been reported to be an independent risk factor for PH [186] by reducing local flow within the pulmonary vessels as a result of counteracting NO generation [187, 188] and increasing levels of endothelin which ultimately leads to endothelial dysfunction and subsequently increases the pulmonary pressures [189]. Furthermore, ectopic accumulation of triglyceride and free-fatty-acid in the myocardium of obese individuals could lead to the development of eccentric ventricular hypertrophy and diastolic heart failure in severe obesity. As a consequence, a secondary form of PH develops characterized by an increase of left-ventricular filling pressures associated with left-ventricular failure [68]. Moreover, obstructive sleep apnea (OSA)-induced nocturnal hypoxemia observed in obese subjects, could lead to right ventricular hypertrophy and PH [190–192].

Cardiac Hypertrophy

In response to extrinsic or intrinsic stress, heart adaptation involves morphological and structural changes (cardiac remodelling) such as mass and volume (cardiac hypertrophy), diameter and geometry of the cardiomyocytes and cardiac chamber. If this initial compensation mechanism sustains for long time, heart undergoes malfunction processes which are the risks of congestive heart failure and sudden death [193]. Obesity-induced cardiac remodelling is a multi-factorial process that plays a key role in mediating cardiac-related structural and functional changes. The components determining obesity-associated cardiac hypertrophy include hemodynamic alterations, inflammation, neurohormonal imbalance, and metabolic alterations [194–198]. In obese subjects, alterations in these components, mainly hemodynamic alteration, contribute to morphological changes in the cardiovascular system that ultimately predisposes to ventricular dysfunction and heart failure [199, 200].

A recent meta-analysis shows that left ventricular hypertrophy (LVH) is one of the most frequent complications in the obese population [201].

The first observation by Lillington et al. suggests the importance of LVH in evaluating cardiac remodelling in severely obese patients [202]. Later in 1992, Kasper et al. showed that most common histologic abnormality in the obese subjects was characterized predominantly by LVH [69]. Although parameters like neurohormonal and metabolic alteration play important roles in the development of LVH, Strong Heart Study suggests that increased muscle mass is a better predictor of LVH than increased fat mass [203]. More recent reports suggest that concentric LVH occurs more frequently in obese patients than the eccentric LVH. This may be explained by the development of a form of concentric LVH described as "eccentric-concentric LVH" due to hypertension in these obese patients [204–206]. Although some of the studies still report the predominance of concentric LVH in hypertension-adjusted obese patients, the prevalence of development of either kind of LVH i.e. eccentric or concentric is determined by the duration and severity of both hypertension and obesity [206, 207]. Using magnetic resonance imaging and multinuclear spectroscopy on the obese population, Rayner et al. recently found that visceral obesity negatively impacts diastolic function by causing concentric LV remodelling, visceral obesity is also associated with increased triglyceride levels and impaired energetics of the myocardium [208]. Leptin plays a role in cardiovascular disease development such as LVH and fibrosis through various signalling pathways, for instance, activation of a protein Rho-associated protein kinase (ROCK) [209]. Recently, Geng et al. demonstrated that mice fed with high-fat diet induced obesity as well as cardiac hypertrophy, inflammation and oxidative stress. Fibronectin type III domain containing 5 (FNDC5) is a protein that has beneficial roles in metabolic diseases by ameliorating hyperlipemia and increasing lipolysis in adipose tissues. All the changes induced by the high-fat diet were attenuated by FNDC5 by inactivating JAK2/STAT3 [210].

Congenital Heart Disease

Congenital heart diseases (CHDs) is considered as the most prevalent kind of congenital anomalies and the highest prevalence (9.3/1000 live births) was found in Asian sub-continents. It is affecting about one million new-borns annually [211]. Population-based studies reported that CHD occurs in ~1% of live births and 10% of aborted foetuses globally. It is also the leading cause of mortality from birth defects [212]. Between 1970 and 2017, the prevalence of CHD increased by 10% every 5 years. Several analyses including this meta-analysis also showed that there is continued increase in birth prevalence of CHD and estimated that obesity in women will be increased to 21% by 2025 [33, 213]. Although there is inconsistency in results, studies widely reported the link between obesity during pregnancy and CHDs in new-borns. Maternal obesity is also a preventable risk factor for CHDs [214–216].

From recent cross-sectional studies, it is evident that severe obesity in pregnant women has teratogenic effects on the foetus and is associated with several congenital anomalies of CNS, genitourinary system and lymphatic system, atrial and ventricular septum [217, 218]. Although the precise pathophysiological mechanisms behind the obesity-induced CHD is poorly understood, but recent studies using genomic technologies such as next-generation sequencing, single nucleotide polymorphism (SNP) arrays and copy number variation (CNV) analysis enable the identification of genetic causes and suggest a complex cross-talk between genetic and environmental factors, of which the environmental factors is the most prevalent cause [219–222]. Two recent meta-analyses and a study of Persson et al. showed a dose–dependent association between maternal overweight and obesity severity, and overall risk of cardiovascular defects as well as risks of few specific heart defects [223–225]. Some studies reported that clinical outcome of CHD can be decreased if diagnosed prenatally using fetal echocardiography [226, 227], during the neonatal period [228, 229], or later in childhood within the first 3 years of life [230]. Increased fat mass especially in the lower-body compartment in pregnant women, results in the metabolic dysregulation, hyperglycaemia, lipo-toxicity and inflammation and which may influence endothelial function, placental development and pregnancy outcome [231, 232]. Studies have shown that maternal obesity has a tendency to the development of obesity in offspring's later life. A number of evidence showed that maternal obesity may induce epigenetic modifications such as methylation of histone-lysine and DNA, and acetylation of histones during crucial stages of embryonic developmental that result in fetal reprogramming of embryo [233, 234]. The major adipokines, adiponectin and leptin, also play pivotal roles in the development of the functional placenta. As evidenced by recent results, placenta of obese women has decreased level of the two adipokines and has epigenetic changes in their promoters i.e. DNA methylation [235]. Moreover, gestational diabetes is more frequently associated with maternal obesity and is a major risk factor for congenital heart defects in their offspring [236, 237]. Thus, primary prevention aiming at reducing the prevalence of overweight and obesity in women in reproductive age is essential for reducing obesity-related risks of congenital heart defects.

Cardiac Arrhythmias and Sudden Cardiac Death

The conclusion "sudden death is more common in those who are naturally fat than in the lean" derived in the 4th century, is attributed to Hippocrates [238]. A handful of studies have associated obesity as a risk factor for cardiac arrhythmias and sudden cardiac death [125, 239, 240] in both genders [241]. In the Framingham Study, Kannel et al. showed the annual sudden cardiac death rate in the obese subject was approximately 40 times higher than that of the non-obese population [242].

Atrial fibrillation (AF) is the most common and clinically significant form of cardiac arrhythmia. Its incidence and prevalence in the world are still increasing, affecting 1–2% of the adult population and global prevalence of 33.5 million individuals [243, 244]. Numerous studies have reported the association between obesity or increased BMI and AF or its risk factors such as diabetes mellitus, hypertension, myocardial infarction, LVH, left ventricular diastolic dysfunction, and OSA [245]. Epidemiological and meta-analyses studies suggest that the risk of incidence of AF in obese patients increases by 1.52 times and 49%, respectively, as compared to control non-obese patients. A unit rise in BMI causes 3–4.7% increase in the risk of AF. In addition to sudden cardiac death, AF also increases the risk of thromboembolic complications, and heart failure [246–250]. The pathological changes during obesity such as lipid accumulation in the myocardium, infiltration of inflammatory agents and fibrosis are highly associated with decreased conduction velocity and increased AF. A meta-analysis and several cross-sectional studies have demonstrated that accumulation of adipose tissues in the epicardium is associated with prevalence, severity, and recurrence of AF [245]. Together, these factors contribute to [251, 252] development of atrial re-entry and finally AF.

Studies suggest that obesity alters several ECG parameters such as increase in P wave duration, elevated PR duration and QT elongation which independently increases the risk of ventricular arrhythmia [253–255]. An animal study showed that diet-induced obesity may be progressively associated with atrial electro-structural remodelling such as change in atrial conduction and increase in fibrosis markers, which leads to AF [256]. The electrophysiological remodelling in obesity was found to be due to PKD-induced reduction of CREB expression and the resultant decrease in expression of the voltage-dependent potassium channels [253]. Similarly, diet-induced obesity increases SNS activity and affects atrial autonomic control and electrical remodelling of the heart that leads to cardiac arrhythmia [257].

There is an association between atrial fibrillation with low-grade inflammation and oxidative stress, which is mainly observed in relation to obesity [258]. Ectopic adipose tissue-derived adipokines and inflammatory cytokines such as CRP, TNF-α, IL-2, IL-6, IL-8, and MCP1 predispose heart of obese patients to atrial fibrillation [259]. Arrhythmogenic effect of leptin released from adipocytes is mediated by elongation of action potential [126]. Several other factors like rise in the level of atrial natriuretic peptide [260] and activation of the renin-angiotensin system are associated directly or indirectly with atrial fibrillation (Fig. 2.3) [261].

Myocardial Infarction

Obesity also poses a serious risk to another cardiovascular disease i.e. myocardial infarction (MI) [45]. In a study, Smith et al. reported that exogenous leptin, when given at early reperfusion in an isolated mouse heart model, showed cardioprotective effects by reducing infarct size. This action was exerted by leptin is associated with RISK (reperfusion injury salvage kinases) activation and thereby inhibiting mPTP (mitochondrial permeability transition pore) opening in isolated rat cardiomyocytes [262].

It has been demonstrated that overweight and obesity are associated with acute MI (AMI) [45, 263, 264] and may also have an independent relationship between them across age and sex [10, 265]. One study showed that the risk for mortality is lower in obese patients with AMI than patients with normal BMI [266]. Recently Bucholz et al. observed an obesity paradox among patients with acute myocardial infarction. Patients with higher BMIs (median = 28.6) had a 20–68% lower mortality compared with patients with a lower BMI (18.5). This effect was independent of other patient characteristics and was comparable across sex, age, and diabetes subpopulations [267]. Consistent with obesity paradox, Dhoot et al., found that the mortality of patients with morbid obesity was lower than those non-morbid obese patients [268]. In contrast, Fukouoka et al. observed high all-cause mortality in elders with low BMI patients undergoing percutaneous coronary intervention (PCI). Further, young patients with high BMI showed higher all-cause mortality [269].

In contrast, a meta-analysis of previous studies suggested that overweight and obesity are associated with a higher risk of AMI. Although there is an obesity paradox exists in the case of AMI, controlling one's BMI is necessary as overweight and obesity may affect cardiovascular health [270].

NAFLD Is a Risk Factor for Cardiovascular Complications in Obesity

As the incidence of obesity is increasing, it also fuels the prevalence and severity of NAFLD [271]. In NAFLD, hepatic diseases such as cirrhosis and HCC are the factors for mortality but other non-hepatic diseases, including chronic kidney disease, CVD and malignancies play critical risk factors for the mortality [272]. NAFLD is a progressive liver disease and its advancement in pathology is determined by multiple genetic and environmental factors, can be described as "multiple parallel-hit" model [273]. Most of these factors include specific genetic polymorphisms [274], lack of physical activity [275], obesity and insulin resistance (IR) [276], dysregulation of adipokines [277–279], accumulation of toxic lipid metabolites [280], endoplasmic reticulum stress and oxidative stress [281], dysbiosis of the gut microbiota [282] and endocrine disruptors [283]. Excessive accumulation of lipids in the liver and sustained simple steatosis (SS) leads to an intra-hepatic inflammatory process [284] which mimics the low-grade inflammatory state occurring within the adipose tissue of obese individuals [285]. As a result of activation and progressive infiltration of immune cells in the liver [34] and consequent release of cytokines not only intensify the inflammatory process but also contribute to the fibrotic process [286]. Adipokines such as leptin, adiponectin and hormones derived from the adipose tissue, also play crucial roles in contributing SS, Non-alcoholic steatohepatitis (NASH), cirrhosis and carcinogenesis [287]. During the adipose tissue expansion and dysfunction, immune cells produce cytokines such as IL-1, IL-6, TNF-α, which crosstalk with adipokines and shift the liver towards a more steatogenic, inflammatory and fibrogenic profile [276].

Although there are separate effects of obesity and NAFLD on the cardiovascular system, the evidence is sparse regarding the synergistic effects of obesity and

NAFLD on the progression of CVD and other metabolic disorders. Accumulating evidence suggest that the NAFLD is not only a progressive liver disease (ranging from NASH to HCC) [288, 289], but also a risk factor for CVD, T2DM, hypertension and chronic kidney disease [290, 291]. Moreover, long-term prospective studies indicate that majority of patients with NAFLD will die of CVD rather than liver-associated complications (38% versus 8%) [292]. Altered metabolism, including central obesity, dysglycemia, atherogenic dyslipidaemia and hypertension increase the risk of CVD. Thus, as NAFLD progress, most of the death in patients occur due to CVD (~40–45% of the total deaths), followed by non-liver cancers (~20%) and direct liver-related complications (~10% of the total deaths [288, 293, 294].

Mobilization of excess free fatty acid from subcutaneous adipose tissue and their accumulation in the liver leading to SS by storing extra lipids in the form of triglycerides, or in the other organs such as kidney, pancreas, skeletal muscle, and the heart. Thus, evidence suggests that NAFLD and obesity together cause ectopic fat accumulation in other organs such as, in myocardium and epicardium [295]. These epicardial adipose tissues spread into the myocardium and releases altered pattern of various pro-inflammatory adipokines such as TNFα, IL-6, MCP-1 and IL-1β, free fatty acid and other vasoactive mediators, and could cause structural and functional derangements of the myocardium such as mitochondrial dysfunction, cardiac apoptosis, fibrosis, ventricular hypertrophy and contractile dysfunction [296–298]. Several clinical studies have demonstrated associations among epicardial adipose tissue extension, IR and NFLAD [299]. IR may lead to altered metabolic flexibility in the myocardium including fatty acid and glucose metabolism [300], resulting in CVDs [295].

Several studies evaluated the association between NAFLD and CAD. Most results demonstrated a significant increase in coronary artery calcium (CAC) score, a marker of coronary atherosclerosis progression, in the presence of NAFLD [301–304]. On the basis of CAC score, several cross-sectional and population-based studies confirmed association among NAFLD and CAD independent of other non-NAFLD factors [305–310]. Using computerized tomography and by calculating the CAC score, most results suggest a significant increase in coronary atherosclerotic risk in NAFLD patients. Several cross-sectional studies showed an association between the development of aortic valve sclerosis and hepatic steatosis mainly in diabetic patients without other complicating factors [311, 312]. Multiple studies suggest the association between NAFLD with several electrocardiography (ECG) parameters and demonstrated an increased risk of atrial fibrillation and prolonged QTc interval [313].

Although the prevalence of NAFLD and its mortality due to CVD has been increasing worldwide but there is no approved treatment option is available. To prevent progression of NAFLD and NASH-related mortality [314], it has been assumed that the resolution of NASH and fibrosis could be endpoints of successful management [315]. Achieving different level of weight loss i.e. ≥3%, ≥5%, ≥7% and ≥10% results in the resolution of steatosis, inflammation, NASH and fibrosis in patients, respectively [316, 317]. Due to the lack of pharmacological interventions, targeting obesity through lifestyle modification remains the keystone of NAFLD management [318].

Prevention and Treatment Options for the Management of Obesity and CVD

In spite of several paradoxical effects of obesity on CVD (Fig. 2.2), a number of evidence suggest its adverse effects on the cardiovascular system and other comorbidity factors. Thus losing weight should be the key goal and maintenance of the normal weight throughout life must be planned ab initio. Studies demonstrated that taking part in various healthy life style program such as leaving a sedentary lifestyle, regular exercise and increased physical activity, intake of a balanced diet may help to lose an excess body weight.

A number of anti-obesity drugs have been reported to have weight-reducing effects. They improved cardiovascular outcomes as evidenced by double-blind, placebo-controlled, randomized studies (Table 2.1). Insulin-sensitizing effects of leptin may be used for the treatment of obesity-associated cardio-metabolic dysfunctions. Similarly, since adiponectin has several beneficial effects on cardio-metabolic disorders such as insulin resistance and NAFLD; peroxisome proliferator-activated receptors agonists can be used to elevate the level of adiponectin. To ameliorate IR and vascular dysfunction, oral use of adiponectin receptor agonists: Osmotin (a plant defence protein) and "adipoRon" have been shown proven efficacy [43]. Recently, Zhao et al. proposed a new weight-loss strategy in which neutralizing antibodies can be used for the reduction of leptin levels [319]. Further understanding of the pathophysiological and molecular mechanisms of obesity-induced cardiometabolic disorders may help to identify new pharmacological targets. According to recent reports, adiponectin level can also be increased with the help of various drugs such as osartan and simvastatin or fenofibrate which blocks RAAS and cholesterol/triglycerides generation, respectively. Adiponectin can be administrated directly as it has been shown to reduce glucose, lipid, and insulin concentrations and increase insulin receptor expression in obese diabetic mice [59]. Few other anti-obesity drugs and their mechanism of action include, (a) orlistat; a pancreatic and gastric specific inhibitor, (b) liraglutide; an appetite-suppressing human glucagon-like peptide-1 (GLP-1) receptor agonist, (c) lorcaserin; suppresses appetite by inhibiting the serotonin 2C receptor, (d) Sibutramine; increase satiety by inhibiting norepinephrine and serotonin reuptake [13].

Although a number of anti-obesity drugs have been approved and are known to reduce CVD risk factors, but they have encountered several adverse effects. A large number of anti-obesity agents have been withdrawn from the market due to unexpected side effects like valvular abnormalities, increase in cardiovascular events and neuropsychiatric side effects. Thus their toxicity on the CVD outcome must be evaluated for a long duration. Few examples of anti obesity agents withdrawn from the market due to unexpected CVD side effects include fenfluramine/dexfenfluramine, sibutramine, ephedrine, and phenylpropanolamine [320]. To overcome the issue regarding high developmental cost and the adverse effects of anti-obesity drugs, future research should focus on identifying the basic mechanism of action for several adipokines such as chemerin, resistin and apelin [321].

Table 2.1 Mechanisms of action of anti-obesity drugs that show improvement in cardiovascular outcomes

Pharmacotherapy interventions	Mechanism of action	Outcome(s)	Adverse effects	Cardiovascular risks
Orlistat or Tetrahydrolipstatin	Pancreatic and gastric specific lipase inhibitor; prevents absorption of fat in the gastrointestinal tract	Primary: Overweight or obese Others: Reduced blood pressure, LDL cholesterol and fasting glucose in patients with diabetes	Mild gastrointestinal problem	CV safety not established
Liraglutide	Human glucagon like peptide-1 (GLP-1; an incretin hormone that reduces gut motility and promotes satiety) receptor agonist	Primary: Overweight or obese and/or diabetes Others: Lower blood pressure, total cholesterol and triglycerides, HbA1c, fasting glucose, and fasting insulin. Improved fasting lipid levels, CRP, plasminogen activator inhibitor-1 and adiponectin	Nausea, diarrhoea and constipation in transient and mild/moderate intensity	CV safety established in dose indicated for type 2 diabetes
Lorcaserin	A serotonin 2C receptor agonist: Decreases satiety level by activating the anorexigenic POMC pathway	Primary: Overweight or obese with or without CVD Others: Slightly better cardiac risk factors	Hypoglycaemia	Safe for CV problems
Naltrexone/bupropion	Naltrexone: An opioid antagonist, blocks autoinhibitory feedback of β-endorphin Bupropion: Acts on anorexigenic POMC neurons in the hypothalamus to reduce appetite and increase energy expenditure. Inhibits reuptake of norepinephrine and dopamine	Primary: Overweight or obese with controlled hypertension and/or dyslipidemia, obese diabetic Others: Improvements in waist circumference, triglycerides, hsCRP and HDL levels	Nausea, seizures, headache, dizziness, insomnia and vomiting, constipation, upper abdominal pain, migraine and cholecystitis	Elevated blood pressure or myocardial infarction

(continued)

Table 2.1 (continued)

Pharmacotherapy interventions	Mechanism of action	Outcome(s)	Adverse effects	Cardiovascular risks
Phentermine/topiramate (PHEN/TPM)	Phentermine: Sympathomimetic (α-adrenergic receptors antagonist), increases blood leptin level and reduces appetite Topiramate: Targets GABA-mediated pathway to modulate voltage-gated ion channels and inhibits carbonic anhydrase or AMPA/kainate excitatory glutamate receptors	Primary: Overweight or obese with or without CVD Others: An overall significant improvement in markers for cardiovascular risks, such as waist circumference, blood pressure and lipids	Dry mouth, paresthesia, flu, upper respiratory infection, change in taste and insomnia	Potential teratogenic risk, as well as cardiovascular risk with an increase of the heart rate
Sibutramine	An inhibitor of β-phenethylamine or norepinephrine and SRI; increases the levels of endogenous catecholamines and increases satiety	Primary: Overweight or obese	Insomnia, nausea, dry mouth and constipation	Increased blood pressure and pulse rate, tachycardia, hypertension, arrhythmias, nonfatal myocardial infarction and nonfatal stroke

POMC Proopiomelanocortin, *GABA* Gamma-aminobutyric acid

Conclusions

The prevalence of obesity has increased in a pandemic manner worldwide. From the large numbers of epidemiological studies, it is clear that overweight and obesity with increased BMI are associated with an increased risk of cardiovascular diseases. Obesity itself has its adverse effects on the cardiovascular system. Obesity has synergistic effects on the CVD when co-exists with its other risk factors such as diabetes, insulin resistance, NAFLD and hypertension, thereby show more favourable comorbidities. Multiple mechanisms linking obesity and CVD events have been established particularly in terms of (i) release of inflammatory cytokines from the epicardial adipose tissues and (ii) development of dysfunctional subcutaneous adipose tissues that initiates a cascade of pathological events. In obese patients, a low-grade inflammation state, activation of SNS and neurohormonal imbalance play a central role in causing the structural and hemodynamic changes that finally lead to various cardiovascular complications. The pathophysiology of obesity-induced CVD is complex and the exact mechanism of obesity paradox is not clear yet. Considering the above facts, regular screening for cardiovascular complications and early diagnosis of comorbid conditions in obese patients is crucial.

An increasing number of evidence report that anti-obesity agents have several adverse effects on the cardiovascular system. Therefore, a long-term evaluation is needed to investigate the side-effects of such interventions and the development of new effective treatments are required. Although the drugs developed for obesity are quite effective for improving cardiovascular outcomes, accumulating results suggest that drugs or therapy that can decrease the pro-inflammatory and increase the anti-inflammatory response to obesity may represent effective therapeutic strategy. Incorporation of regular exercise and balanced diet along with pharmacological intervention may show more beneficial outcomes in overweight or obese individuals. Treating CVD in obese patients is essential and should be a part of the therapeutic plan of anti-obesity treatment.

Acknowledgements The authors are thankful to the Department of Biotechnology (DBT) for providing funds (Grant number BT/PR22881/BRB/10/1654/2018) and the Translational Health Science and Technology Institute (THSTI), Faridabad, India for providing support. Thanks are due to the Indian Council of Medical Research for supporting Research Associate position to MJA (Grant number 3/1/2(19)/OBS/2019-NCD–II).

References

1. World Health Organization. Noncommunicable diseases progress monitor 2017. WHO
2. Berrington de Gonzalez A, Hartge P, Cerhan JR et al (2010) Body-mass index and mortality among 1.46 million white adults. N Engl J Med 363:2211–2219
3. Fontaine KR, Redden DT, Wang C et al (2003) Years of life lost due to obesity. JAMA 289:187–193
4. Prospective Studies Collaboration, Whitlock G, Lewington S et al (2009) Body-mass index and cause-specific mortality in 900 000 adults: collaborative analyses of 57 prospective studies. Lancet 373:1083–1096
5. World Health Organization. Obesity and overweight. WHO. https://www.who.int/news-room/fact-sheets/detail/obesity-and-overweight. Accessed 27 Nov 2019
6. Finucane MM, Stevens GA, Cowan MJ et al (2011) National, regional, and global trends in body-mass index since 1980: systematic analysis of health examination surveys and epidemiological studies with 960 country-years and 9·1 million participants. Lancet 377:557–567
7. Gruberg L, Weissman NJ, Waksman R et al (2002) The impact of obesity on the short-term and long-term outcomes after percutaneous coronary intervention: the obesity paradox? J Am Coll Cardiol 39:578–584
8. Karelis AD, Faraj M, Bastard J-P et al (2005) The metabolically healthy but obese individual presents a favorable inflammation profile. J Clin Endocrinol Metab 90:4145–4150
9. Ashwell M, Hsieh SD (2005) Six reasons why the waist-to-height ratio is a rapid and effective global indicator for health risks of obesity and how its use could simplify the international public health message on obesity. Int J Food Sci Nutr 56:303–307

10. Yusuf S, Hawken S, Ounpuu S et al (2004) Effect of potentially modifiable risk factors associated with myocardial infarction in 52 countries (the INTERHEART study): case-control study. Lancet 364:937–952
11. Zeller M, Steg PG, Ravisy J et al (2008) Relation between body mass index, waist circumference, and death after acute myocardial infarction. Circulation 118:482–490
12. Blüher M (2019) Obesity: global epidemiology and pathogenesis. Nat Rev Endocrinol 15:288–298
13. Cercato C, Fonseca FA (2019) Cardiovascular risk and obesity. Diabetol Metab Syndr 11:74
14. Achari AE, Jain SK (2017) Adiponectin, a therapeutic target for obesity, diabetes, and endothelial dysfunction. Int J Mol Sci 18
15. Balsan GA, Vieira JL da C, Oliveira AM de et al (2015) Relationship between adiponectin, obesity and insulin resistance. Rev Assoc Med Bras (1992) 61:72–80
16. Unamuno X, Gómez-Ambrosi J, Rodríguez A et al (2018) Adipokine dysregulation and adipose tissue inflammation in human obesity. Eur J Clin Invest 48:e12997
17. Keane KN, Cruzat VF, Carlessi R et al (2015) Molecular events linking oxidative stress and inflammation to insulin resistance and β-cell dysfunction. Oxid Med Cell Longev 181643
18. Mraz M, Haluzik M (2014) The role of adipose tissue immune cells in obesity and low-grade inflammation. J Endocrinol 222:R113–127
19. Iliodromiti S, Celis-Morales CA, Lyall DM et al (2018) The impact of confounding on the associations of different adiposity measures with the incidence of cardiovascular disease: a cohort study of 296 535 adults of white European descent. Eur Heart J 39:1514–1520
20. Khan SS, Ning H, Wilkins JT et al (2018) Association of body mass index with lifetime risk of cardiovascular disease and compression of morbidity. JAMA Cardiol 3:280–287
21. Lavie CJ, Arena R, Alpert MA et al (2018) Management of cardiovascular diseases in patients with obesity. Nat Rev Cardiol 15:45–56
22. Agarwal M, Agrawal S, Garg L et al (2017) Relation between obesity and survival in patients hospitalized for pulmonary arterial hypertension (from a nationwide inpatient sample database 2003 to 2011). Am J Cardiol 120:489–493
23. Brida M, Dimopoulos K, Kempny A et al (2017) Body mass index in adult congenital heart disease. Heart 103:1250–1257
24. Chang VW, Langa KM, Weir D et al (2017) The obesity paradox and incident cardiovascular disease: a population-based study. PLoS ONE 12:e0188636
25. Che L, Xu L, Wang M-Y et al (2018) Obesity paradox among elderly patients with coronary artery disease undergoing non-cardiac surgery. J Geriatr Cardiol 15:598–604
26. Jayedi A, Shab-Bidar S (2018) Nonlinear dose-response association between body mass index and risk of all-cause and cardiovascular mortality in patients with hypertension: a meta-analysis. Obes Res Clin Pract 12:16–28
27. Sandhu RK, Ezekowitz J, Andersson U et al (2016) The "obesity paradox" in atrial fibrillation: observations from the ARISTOTLE (Apixaban for reduction in stroke and other thromboembolic events in atrial fibrillation) trial. Eur Heart J 37:2869–2878
28. Carbone S, Canada JM, Billingsley HE et al (2019) Obesity paradox in cardiovascular disease: where do we stand? Vasc Health Risk Manag 15:89–100
29. Elagizi A, Kachur S, Lavie CJ et al (2018) An overview and update on obesity and the obesity paradox in cardiovascular diseases. Prog Cardiovasc Dis 61:142–150
30. Lavie CJ, De Schutter A, Parto P et al (2016) Obesity and prevalence of cardiovascular diseases and prognosis-the obesity paradox updated. Prog Cardiovasc Dis 58:537–547
31. Neeland IJ, Das SR, Simon DN et al (2017) The obesity paradox, extreme obesity, and long-term outcomes in older adults with ST-segment elevation myocardial infarction: results from the NCDR. Eur Heart J Qual Care Clin Outcomes 3:183–191
32. Wu C-Y, Chou Y-C, Huang N et al (2014) Association of body mass index with all-cause and cardiovascular disease mortality in the elderly. PLoS ONE 9:e102589
33. NCD Risk Factor Collaboration (NCD-RisC) (2016) Trends in adult body-mass index in 200 countries from 1975 to 2014: a pooled analysis of 1698 population-based measurement studies with 19·2 million participants. Lancet 387:1377–1396

34. NCD Risk Factor Collaboration (NCD-RisC) (2017) Worldwide trends in body-mass index, underweight, overweight, and obesity from 1975 to 2016: a pooled analysis of 2416 population-based measurement studies in 128·9 million children, adolescents, and adults. Lancet 390:2627–2642
35. Yanovski JA (2018) Obesity: trends in underweight and obesity—scale of the problem. Nat Rev Endocrinol 14:5–6
36. Kelly T, Yang W, Chen C-S et al (2008) Global burden of obesity in 2005 and projections to 2030. Int J Obes (Lond) 32:1431–1437
37. India. National Family Health Survey, 2005–2006. Key findings
38. India. National Family Health Survey, 2014–2015. Key findings. http://rchiips.org/nfhs/nfhs4.shtml. Accessed 27 Nov 2019
39. Pradeepa R, Anjana RM, Joshi SR et al (2015) Prevalence of generalized & abdominal obesity in urban & rural India–the ICMR-INDIAB study (phase-I) [ICMR- NDIAB-3]. Indian J Med Res 142:139–150
40. Misra A, Khurana L (2008) Obesity and the metabolic syndrome in developing countries. J Clin Endocrinol Metab 93:S9–30
41. Misra A, Shrivastava U (2013) Obesity and dyslipidemia in South Asians. Nutrients 5:2708–2733
42. Shrivastava U, Misra A, Mohan V et al (2017) Obesity, diabetes and cardiovascular diseases in India: public health challenges. Curr Diabetes Rev 13:65–80
43. Scheja L, Heeren J (2019) The endocrine function of adipose tissues in health and cardiometabolic disease. Nat Rev Endocrinol 15:507–524
44. Adams KF, Schatzkin A, Harris TB et al (2006) Overweight, obesity, and mortality in a large prospective cohort of persons 50 to 71 years old. N Engl J Med 355:763–778
45. Yusuf S, Hawken S, Ounpuu S et al (2005) Obesity and the risk of myocardial infarction in 27,000 participants from 52 countries: a case-control study. Lancet 366:1640–1649
46. Emanuela F, Grazia M, Marco DR et al (2012) Inflammation as a link between obesity and metabolic syndrome. J Nutr Metab 476380
47. Esser N, Legrand-Poels S, Piette J et al (2014) Inflammation as a link between obesity, metabolic syndrome and type 2 diabetes. Diabetes Res Clin Pract 105:141–150
48. Saltiel AR, Olefsky JM (2017) Inflammatory mechanisms linking obesity and metabolic disease. J Clin Invest 127:1–4
49. Han SH, Quon MJ, Kim J et al (2007) Adiponectin and cardiovascular disease: response to therapeutic interventions. J Am Coll Cardiol 49:531–538
50. Lim S, Quon MJ, Koh KK (2014) Modulation of adiponectin as a potential therapeutic strategy. Atherosclerosis 233:721–728
51. Fisman EZ, Tenenbaum A (2014) Adiponectin: a manifold therapeutic target for metabolic syndrome, diabetes, and coronary disease? Cardiovasc Diabetol 13:103
52. Lekva T, Michelsen AE, Aukrust P et al (2017) Leptin and adiponectin as predictors of cardiovascular risk after gestational diabetes mellitus. Cardiovasc Diabetol 16:5
53. Berger S, Polotsky VY (2018) Leptin and leptin resistance in the pathogenesis of obstructive sleep apnea: a possible link to oxidative stress and cardiovascular complications. Oxid Med Cell Longev 5137947
54. Francisco V, Pino J, Campos-Cabaleiro V et al (2018) Obesity, fat mass and immune system: role for leptin. Front Physiol 9
55. Cortese L, Terrazzano G, Pelagalli A (2019) Leptin and immunological profile in obesity and its associated diseases in dogs. Int J Mol Sci 20
56. Lau WB, Ohashi K, Wang Y et al (2017) Role of adipokines in cardiovascular disease. Circ J 81:920–928
57. Heyman E, Gamelin F-X, Aucouturier J et al (2012) The role of the endocannabinoid system in skeletal muscle and metabolic adaptations to exercise: potential implications for the treatment of obesity. Obes Rev 13:1110–1124
58. Silvestri C, Di Marzo V (2013) The endocannabinoid system in energy homeostasis and the etiopathology of metabolic disorders. Cell Metab 17:475–490

59. Leggio M, Lombardi M, Caldarone E et al (2017) The relationship between obesity and hypertension: an updated comprehensive overview on vicious twins. Hypertens Res 40:947–963
60. Quercioli A, Montecucco F, Pataky Z et al (2013) Improvement in coronary circulatory function in morbidly obese individuals after gastric bypass-induced weight loss: relation to alterations in endocannabinoids and adipocytokines. Eur Heart J 34:2063–2073
61. Lavie CJ, Milani RV, Ventura HO (2011) Impact of obesity on outcomes in myocardial infarction combating the "obesity paradox". J Am Coll Cardiol 58:2651–2653
62. Mozaffarian D, Benjamin EJ, Go AS et al (2015) Heart disease and stroke statistics–2015 update: a report from the American Heart Association. Circulation 131:e29–322
63. Kenchaiah S, Evans JC, Levy D et al (2002) Obesity and the risk of heart failure. N Engl J Med 347:305–313
64. Alpert MA, Terry BE, Mulekar M et al (1997) Cardiac morphology and left ventricular function in normotensive morbidly obese patients with and without congestive heart failure, and effect of weight loss. Am J Cardiol 80:736–740
65. Alpert MA, Omran J, Bostick BP (2016) Effects of obesity on cardiovascular hemodynamics, cardiac morphology, and ventricular function. Curr Obes Rep 5:424–434
66. Wong C, Marwick TH (2007) Obesity cardiomyopathy: diagnosis and therapeutic implications. Nat Clin Pract Cardiovasc Med 4:480–490
67. Schmieder RE, Messerli FH (1993) Does obesity influence early target organ damage in hypertensive patients? Circulation 87:1482–1488
68. Friedman SE, Andrus BW (2012) Obesity and pulmonary hypertension: a review of pathophysiologic mechanisms. J Obes 505274
69. Kasper EK, Hruban RH, Baughman KL (1992) Cardiomyopathy of obesity: a clinicopathologic evaluation of 43 obese patients with heart failure. Am J Cardiol 70:921–924
70. Hall JE, do Carmo JM, da Silva AA et al (2015) Obesity-induced hypertension: interaction of neurohumoral and renal mechanisms. Circ Res 116:991–1006
71. Kawarazaki W, Fujita T (2016) The role of aldosterone in obesity-related hypertension. Am J Hypertens 29:415–423
72. de Simone G, Palmieri V, Bella JN et al (2002) Association of left ventricular hypertrophy with metabolic risk factors: the HyperGEN study. J Hypertens 20:323–331
73. Benjamin EJ, Virani SS, Callaway CW et al (2018) Heart disease and stroke statistics-2018 update: a report from the American Heart Association. Circulation 137:e67–e492
74. Cepeda-Lopez AC, Zimmermann MB, Wussler S et al (2019) Greater blood volume and Hb mass in obese women quantified by the carbon monoxide-rebreathing method affects interpretation of iron biomarkers and iron requirements. Intl J Obesity 43:999–1008
75. Vricella LK, Louis JM, Chien E et al (2015) Blood volume determination in obese and normal weight gravidas: the Hydroxyethyl Starch Method. Am J Obstet Gynecol 213(408):e1–408.e6
76. Alexander JK, Dennis EW, Smith WG et al (1962) Blood volume, cardiac output, and distribution of systemic blood flow in extreme obesity. Cardiovasc Res Cent Bull 1:39–44
77. Obokata M, Reddy YNV, Pislaru SV et al (2017) Evidence supporting the existence of a distinct obese phenotype of heart failure with preserved ejection fraction. Circulation 136:6–19
78. Packer M (2019) The conundrum of patients with obesity, exercise intolerance, elevated ventricular filling pressures and a measured ejection fraction in the normal range. Eur J Heart Fail 21:156–162
79. Pandey A, LaMonte M, Klein L et al (2017) Relationship between physical activity, body mass index, and risk of heart failure. J Am Coll Cardiol 69:1129–1142
80. Kitzman DW, Shah SJ (2016) The HFpEF obesity phenotype: the elephant in the room. J Am Coll Cardiol 68:200–203
81. Carbone S, Pandey A, Lavie CJ (2018) Editorial commentary: obesity and heart failure with preserved ejection fraction: a single disease or two co-existing conditions? Trends Cardiovasc Med 28:328–329
82. Shah SJ, Kitzman DW, Borlaug BA et al (2016) Phenotype-specific treatment of heart failure with preserved ejection fraction: a multiorgan roadmap. Circulation 134:73–90

83. Ebong IA, Goff DC, Rodriguez CJ et al (2014) Mechanisms of heart failure in obesity. Obes Res Clin Pract 8:e540–548
84. Rabbia F, Silke B, Conterno A et al (2003) Assessment of cardiac autonomic modulation during adolescent obesity. Obes Res 11:541–548
85. Ballak DB, Stienstra R, Tack CJ et al (2015) IL-1 family members in the pathogenesis and treatment of metabolic disease: focus on adipose tissue inflammation and insulin resistance. Cytokine 75:280–290
86. Cavalera M, Wang J, Frangogiannis NG (2014) Obesity, metabolic dysfunction, and cardiac fibrosis: pathophysiological pathways, molecular mechanisms, and therapeutic opportunities. Transl Res 164:323–335
87. Toldo S, Mezzaroma E, O'Brien L et al (2014) Interleukin-18 mediates interleukin-1-induced cardiac dysfunction. Am J Physiol Heart Circ Physiol 306:H1025–1031
88. Toldo S, Mezzaroma E, Mauro AG et al (2015) The inflammasome in myocardial injury and cardiac remodeling. Antioxid Redox Signal 22:1146–1161
89. Van Tassell BW, Toldo S, Mezzaroma E et al (2013) Targeting interleukin-1 in heart disease. Circulation 128:1910–1923
90. Karmazyn M, Purdham DM, Rajapurohitam V et al (2008) Signalling mechanisms underlying the metabolic and other effects of adipokines on the heart. Cardiovasc Res 79:279–286
91. Lipina C, Hundal HS (2017) Lipid modulation of skeletal muscle mass and function. J Cachexia Sarcopenia Muscle 8:190–201
92. Schram K, Sweeney G (2008) Implications of myocardial matrix remodeling by adipokines in obesity-related heart failure. Trends Cardiovasc Med 18:199–205
93. Smith CCT, Yellon DM (2011) Adipocytokines, cardiovascular pathophysiology and myocardial protection. Pharmacol Ther 129:206–219
94. Horwich TB, Fonarow GC (2010) Glucose, obesity, metabolic syndrome, and diabetes relevance to incidence of heart failure. J Am Coll Cardiol 55:283–293
95. Poirier P, Giles TD, Bray GA et al (2006) Obesity and cardiovascular disease: pathophysiology, evaluation, and effect of weight loss. Arterioscler Thromb Vasc Biol 26:968–976
96. Asghar O, Alam U, Hayat SA et al (2012) Obesity, diabetes and atrial fibrillation; epidemiology, mechanisms and interventions. Curr Cardiol Rev 8:253–264
97. Schoonderwoerd BA, Smit MD, Pen L et al (2008) New risk factors for atrial fibrillation: causes of "not-so-lone atrial fibrillation". Europace 10:668–673
98. Libby P, Buring JE, Badimon L et al (2019) Atherosclerosis. Nat Rev Dis Primers 5:56
99. Bäck M, Yurdagul A, Tabas I et al (2019) Inflammation and its resolution in atherosclerosis: mediators and therapeutic opportunities. Nat Rev Cardiol 16:389–406
100. Lovren F, Teoh H, Verma S (2015) Obesity and atherosclerosis: mechanistic insights. Can J Cardiol 31:177–183
101. Rocha VZ, Libby P (2009) Obesity, inflammation, and atherosclerosis. Nat Rev Cardiol 6:399–409
102. Ye J (2009) Emerging role of adipose tissue hypoxia in obesity and insulin resistance. Int J Obes (Lond) 33:54–66
103. Chen L, Chen R, Wang H et al (2015) Mechanisms linking inflammation to insulin resistance. Int J Endocrinol 508409
104. McArdle MA, Finucane OM, Connaughton RM et al (2013) Mechanisms of obesity-induced inflammation and insulin resistance: insights into the emerging role of nutritional strategies. Front Endocrinol (Lausanne) 4:52
105. Shoelson SE, Herrero L, Naaz A (2007) Obesity, inflammation, and insulin resistance. Gastroenterology 132:2169–2180
106. Stafeev IS, Vorotnikov AV, Ratner EI et al (2017) Latent inflammation and insulin resistance in adipose tissue. Int J Endocrinol 5076732
107. Shimobayashi M, Albert V, Woelnerhanssen B et al (2018) Insulin resistance causes inflammation in adipose tissue. J Clin Invest 128:1538–1550

108. Das SR, Alexander KP, Chen AY et al (2011) Impact of body weight and extreme obesity on the presentation, treatment, and in-hospital outcomes of 50,149 patients with ST-Segment elevation myocardial infarction results from the NCDR (National Cardiovascular Data Registry). J Am Coll Cardiol 58:2642–2650
109. Payvar S, Kim S, Rao SV et al (2013) In-hospital outcomes of percutaneous coronary interventions in extremely obese and normal-weight patients: findings from the NCDR (National Cardiovascular Data Registry). J Am Coll Cardiol 62:692–696
110. Romero-Corral A, Montori VM, Somers VK et al (2006) Association of bodyweight with total mortality and with cardiovascular events in coronary artery disease: a systematic review of cohort studies. Lancet 368:666–678
111. De Rosa R, Vasa-Nicotera M, Leistner DM et al (2017) Coronary atherosclerotic plaque characteristics and cardiovascular risk factors—insights from an optical coherence tomography study. Circ J 81:1165–1173
112. Ridker PM, Everett BM, Thuren T et al (2017) Antiinflammatory therapy with canakinumab for atherosclerotic disease. N Engl J Med 377:1119–1131
113. Ridker PM, Everett BM, Pradhan A et al (2019) Low-dose methotrexate for the prevention of atherosclerotic events. N Engl J Med 380:752–762
114. Manson JE, Colditz GA, Stampfer MJ et al (1990) A prospective study of obesity and risk of coronary heart disease in women. N Engl J Med 322:882–889
115. Wilson PWF, D'Agostino RB, Sullivan L et al (2002) Overweight and obesity as determinants of cardiovascular risk: the Framingham experience. Arch Intern Med 162:1867–1872
116. McGill HC, McMahan CA, Herderick EE et al (2002) Obesity accelerates the progression of coronary atherosclerosis in young men. Circulation 105:2712–2718
117. Akil L, Ahmad HA (2011) Relationships between obesity and cardiovascular diseases in four Southern States and Colorado. J Health Care Poor Underserved 22:61–72
118. Din-Dzietham R, Liu Y, Bielo M-V et al (2007) High blood pressure trends in children and adolescents in national surveys, 1963 to 2002. Circulation 116:1488–1496
119. Jamil G, Jamil M, AlKhazraji H et al (2013) Risk factor assessment of young patients with acute myocardial infarction. Am J Cardiovasc Dis 3:170–174
120. Madala MC, Franklin BA, Chen AY et al (2008) Obesity and age of first non-ST-segment elevation myocardial infarction. J Am Coll Cardiol 52:979–985
121. Abel ED, Litwin SE, Sweeney G (2008) Cardiac remodeling in obesity. Physiol Rev 88:389–419
122. Alpert MA, Omran J, Mehra A et al (2014) Impact of obesity and weight loss on cardiac performance and morphology in adults. Prog Cardiovasc Dis 56:391–400
123. Alpert MA, Lavie CJ, Agrawal H et al (2014) Obesity and heart failure: epidemiology, pathophysiology, clinical manifestations, and management. Transl Res 164:345–356
124. Lavie CJ, Alpert MA, Arena R et al (2013) Impact of obesity and the obesity paradox on prevalence and prognosis in heart failure. JACC Heart Fail 1:93–102
125. Poirier P, Giles TD, Bray GA et al (2006) Obesity and cardiovascular disease: pathophysiology, evaluation, and effect of weight loss: an update of the 1997 American Heart Association Scientific Statement on Obesity and Heart Disease from the Obesity Committee of the Council on Nutrition, Physical Activity, and Metabolism. Circulation 113:898–918
126. Csige I, Ujvárosy D, Szabó Z et al (2018) The impact of obesity on the cardiovascular system. J Diabetes Res 3407306
127. Santos C, Marques da Silva P (2018) Hemodynamic patterns in obesity associated hypertension. BMC Obes 5
128. Stelfox HT, Ahmed SB, Ribeiro RA et al (2006) Hemodynamic monitoring in obese patients: the impact of body mass index on cardiac output and stroke volume. Crit Care Med 34:1243–1246
129. Farag NH, Matthews SC, Brzezinski E et al (2004) Relationship between central obesity and cardiovascular hemodynamic indices in postmenopausal women. Fertil Steril 81:465–467
130. Jern S, Bergbrant A, Björntorp P et al (1992) Relation of central hemodynamics to obesity and body fat distribution. Hypertension 19:520–527

131. Goossens GH, Bizzarri A, Venteclef N et al (2011) Increased adipose tissue oxygen tension in obese compared with lean men is accompanied by insulin resistance, impaired adipose tissue capillarization, and inflammation. Circulation 124:67–76
132. Heinonen IHA, Boushel R, Kalliokoski KK (2016) The circulatory and metabolic responses to hypoxia in humans—with special reference to adipose tissue physiology and obesity. Front Endocrinol (Lausanne) 7
133. Limberg JK, Morgan BJ, Schrage WG (2016) Peripheral blood flow regulation in human obesity and metabolic syndrome. Exerc Sport Sci Rev 44:116–122
134. Gayda M, Lapierre G, Dupuy O et al (2017) Cardiovascular and cerebral hemodynamics during exercise and recovery in obese individuals as a function of their fitness status. Physiol Rep 5
135. Cavuoto LA, Maikala RV (2016) Obesity and the role of short duration submaximal work on cardiovascular and cerebral hemodynamics. PLoS One 11
136. Vinayagam D, Gutierrez J, Binder J et al (2017) Impaired maternal hemodynamics in morbidly obese women: a case-control study. Ultrasound Obstet Gynecol 50:761–765
137. Hien HA, Tam NM, Tam V et al (2018) Prevalence, awareness, treatment, and control of hypertension and its risk factors in (Central) Vietnam. Int J Hypertens 2018:6326984
138. Kotchen TA (2010) Obesity-related hypertension: epidemiology, pathophysiology, and clinical management. Am J Hypertens 23:1170–1178
139. Heusch G, Libby P, Gersh B et al (2014) Cardiovascular remodelling in coronary artery disease and heart failure. Lancet 383:1933–1943
140. Banegas JR, López-García E, Dallongeville J et al (2011) Achievement of treatment goals for primary prevention of cardiovascular disease in clinical practice across Europe: the EURIKA study. Eur Heart J 32:2143–2152
141. Bramlage P, Pittrow D, Wittchen H-U et al (2004) Hypertension in overweight and obese primary care patients is highly prevalent and poorly controlled. Am J Hypertens 17:904–910
142. Chopra I, Kamal KM, Candrilli SD (2013) Variations in blood pressure and lipid goal attainment in primary care. Curr Med Res Opin 29:1115–1125
143. Zhou L, Li Y, Guo M et al (2016) Relations of body weight status in early adulthood and weight changes until middle age with hypertension in the Chinese population. Hypertens Res 39:913–918
144. Kang YS (2013) Obesity associated hypertension: new insights into mechanism. Electrolyte Blood Press 11:46–52
145. Landsberg L, Aronne LJ, Beilin LJ et al (2013) Obesity-related hypertension: pathogenesis, cardiovascular risk, and treatment: a position paper of the Obesity Society and the American Society of Hypertension. J Clin Hypertens (Greenwich) 15:14–33
146. Ventura HO, Taler SJ, Strobeck JE (2005) Hypertension as a hemodynamic disease: the role of impedance cardiography in diagnostic, prognostic, and therapeutic decision making. Am J Hypertens 18:26S–43S
147. Aronow WS (2017) Association of obesity with hypertension. Ann Transl Med 5
148. Vaněčková I, Maletínská L, Behuliak M et al (2014) Obesity-related hypertension: possible pathophysiological mechanisms. J Endocrinol 223:R63–78
149. Masuo K, Tuck ML, Lambert GW (2011) Hypertension and diabetes in obesity. Int J Hypertens 695869
150. Sharma AM, Chetty VT (2005) Obesity, hypertension and insulin resistance. Acta Diabetol 42(Suppl 1):S3–8
151. Steinberger J, Daniels SR, American Heart Association Atherosclerosis, Hypertension, and Obesity in the Young Committee (Council on Cardiovascular Disease in the Young) et al (2003) Obesity, insulin resistance, diabetes, and cardiovascular risk in children: an American Heart Association scientific statement from the Atherosclerosis, Hypertension, and Obesity in the Young Committee (Council on Cardiovascular Disease in the Young) and the Diabetes Committee (Council on Nutrition, Physical Activity, and Metabolism). Circulation 107:1448–1453

152. Zhou M-S, Wang A, Yu H (2014) Link between insulin resistance and hypertension: what is the evidence from evolutionary biology? Diabetol Metab Syndr 6:12
153. Thackeray JT, Beanlands RS, DaSilva JN (2012) Altered sympathetic nervous system signaling in the diabetic heart: emerging targets for molecular imaging. Am J Nucl Med Mol Imaging 2:314–334
154. Brands MW (2018) Role of insulin-mediated antinatriuresis in sodium homeostasis and hypertension. Hypertension 72:1255–1262
155. Catena C, Cavarape A, Novello M et al (2003) Insulin receptors and renal sodium handling in hypertensive fructose-fed rats. Kidney Int 64:2163–2171
156. Horita S, Seki G, Yamada H et al (2011) Insulin resistance, obesity, hypertension, and renal sodium transport. Int J Hypertens 391762
157. Tiwari S, Riazi S, Ecelbarger CA (2007) Insulin's impact on renal sodium transport and blood pressure in health, obesity, and diabetes. Am J Physiol Renal Physiol 293:F974–984
158. Harte A, McTernan P, Chetty R et al (2005) Insulin-mediated upregulation of the renin angiotensin system in human subcutaneous adipocytes is reduced by rosiglitazone. Circulation 111:1954–1961
159. Fowler JD, Johnson ND, Haroldson TA et al (2009) Regulated renin release from 3T3-L1 adipocytes. Am J Physiol Endocrinol Metab 296:E1383–E1391
160. Gálvez-Prieto B, Bolbrinker J, Stucchi P et al (2008) Comparative expression analysis of the renin-angiotensin system components between white and brown perivascular adipose tissue. J Endocrinol 197:55–64
161. Giacchetti G, Faloia E, Mariniello B et al (2002) Overexpression of the renin-angiotensin system in human visceral adipose tissue in normal and overweight subjects. Am J Hypertens 15:381–388
162. Lu C, Su L-Y, Lee RMKW et al (2010) Mechanisms for perivascular adipose tissue-mediated potentiation of vascular contraction to perivascular neuronal stimulation: the role of adipocyte-derived angiotensin II. Eur J Pharmacol 634:107–112
163. Yasue S, Masuzaki H, Okada S et al (2010) Adipose tissue-specific regulation of angiotensinogen in obese humans and mice: impact of nutritional status and adipocyte hypertrophy. Am J Hypertens 23:425–431
164. Boustany CM, Bharadwaj K, Daugherty A et al (2004) Activation of the systemic and adipose renin-angiotensin system in rats with diet-induced obesity and hypertension. Am J Physiol Regul Integr Comp Physiol 287:R943–949
165. Briones AM, Nguyen Dinh Cat A, Callera GE et al (2012) Adipocytes produce aldosterone through calcineurin-dependent signaling pathways: implications in diabetes mellitus-associated obesity and vascular dysfunction. Hypertension 59:1069–1078
166. Rossi GP, Belfiore A, Bernini G et al (2008) Body mass index predicts plasma aldosterone concentrations in overweight-obese primary hypertensive patients. J Clin Endocrinol Metab 93:2566–2571
167. Dinh Cat AN, Friederich-Persson M, White A et al (2016) Adipocytes, aldosterone and obesity-related hypertension. J Mol Endocrinol 57:F7–F21
168. Huby A-C, Antonova G, Groenendyk J et al (2015) Adipocyte-derived hormone leptin is a direct regulator of aldosterone secretion, which promotes endothelial dysfunction and cardiac fibrosis. Circulation 132:2134–2145
169. Lohmeier TE, Iliescu R, Liu B et al (2012) Systemic and renal-specific sympathoinhibition in obesity hypertension. Hypertension 59:331–338
170. Lambert EA, Straznicky NE, Dixon JB et al (2015) Should the sympathetic nervous system be a target to improve cardiometabolic risk in obesity? Am J Physiol Heart Circ Physiol 309:H244–258
171. de Jongh RT, Serné EH, IJzerman RG et al (2004) Impaired microvascular function in obesity: implications for obesity-associated microangiopathy, hypertension, and insulin resistance. Circulation 109:2529–2535
172. Sorop O, Olver TD, van de Wouw J et al (2017) The microcirculation: a key player in obesity-associated cardiovascular disease. Cardiovasc Res 113:1035–1045

173. DeMarco VG, Aroor AR, Sowers JR (2014) The pathophysiology of hypertension in patients with obesity. Nat Rev Endocrinol 10:364–376
174. Hornstra JM, Serné EH, Eringa EC et al (2013) Insulin's microvascular vasodilatory effects are inversely related to peripheral vascular resistance in overweight, but insulin-sensitive subjects. Obesity (Silver Spring) 21:2557–2561
175. Li R, Zhang H, Wang W et al (2010) Vascular insulin resistance in prehypertensive rats: role of PI3-kinase/Akt/eNOS signaling. Eur J Pharmacol 628:140–147
176. Kaess BM, Rong J, Larson MG et al (2012) Aortic stiffness, blood pressure progression, and incident hypertension. JAMA 308:875–881
177. Weisbrod RM, Shiang T, Sayah LA et al (2013) Arterial stiffening precedes systolic hypertension in diet-induced obesity. Hypertension 62:1105–1110
178. Haynes WG (2000) Interaction between leptin and sympathetic nervous system in hypertension. Curr Hypertens Rep 2:311–318
179. von Schnurbein J, Manzoor J, Brandt S et al (2019) Leptin is not essential for obesity-associated hypertension. Obes Facts 12:460–475
180. Lobato NS, Filgueira FP, Akamine EH et al (2012) Mechanisms of endothelial dysfunction in obesity-associated hypertension. Braz J Med Biol Res 45:392–400
181. Sierra S, Luquin N, Navarro-Otano J (2018) The endocannabinoid system in cardiovascular function: novel insights and clinical implications. Clin Auton Res 28:35–52
182. Perrotta F, Nigro E, Mollica M et al (2019) Pulmonary hypertension and obesity: focus on adiponectin. Int J Mol Sci 20
183. Huertas A, Tu L, Gambaryan N et al (2012) Leptin and regulatory T-lymphocytes in idiopathic pulmonary arterial hypertension. Eur Respir J 40:895–904
184. Huertas A, Tu L, Thuillet R et al (2015) Leptin signalling system as a target for pulmonary arterial hypertension therapy. Eur Respir J 45:1066–1080
185. Tonelli AR, Aytekin M, Feldstein AE et al (2012) Leptin levels predict survival in pulmonary arterial hypertension. Pulm Circ 2:214–219
186. Dimitroulas T, Giannakoulas G, Dimitroula H et al (2011) Significance of serum uric acid in pulmonary hypertension due to systemic sclerosis: a pilot study. Rheumatol Int 31:263–267
187. Kang D-H, Park S-K, Lee I-K et al (2005) Uric acid-induced C-reactive protein expression: implication on cell proliferation and nitric oxide production of human vascular cells. J Am Soc Nephrol 16:3553–3562
188. Zharikov S, Krotova K, Hu H et al (2008) Uric acid decreases NO production and increases arginase activity in cultured pulmonary artery endothelial cells. Am J Physiol Cell Physiol 295:C1183–1190
189. Simpson CE, Damico RL, Hummers L et al (2019) Serum uric acid as a marker of disease risk, severity, and survival in systemic sclerosis-related pulmonary arterial hypertension. Pulm Circ 9
190. Ahmad M, Makati D, Akbar S (2017) Review of and updates on hypertension in obstructive sleep apnea. Int J Hypertens 1848375
191. Bonsignore MR, Baiamonte P, Mazzuca E et al (2019) Obstructive sleep apnea and comorbidities: a dangerous liaison. Multidiscip Respir Med 14:8
192. Wolk R, Shamsuzzaman ASM, Somers VK (2003) Obesity, sleep apnea, and hypertension. Hypertension 42:1067–1074
193. Azevedo PS, Polegato BF, Minicucci MF et al (2016) Cardiac remodeling: concepts, clinical impact, pathophysiological mechanisms and pharmacologic treatment. Arq Bras Cardiol 106:62–69
194. Berg G, Mesch V, Siseles N (2012) Abdominal obesity and metabolic alterations in the menopausal transition. Curr Obstet Gynecol Rep 1:63–70
195. Canale MP, Manca di Villahermosa S, Martino G et al (2013) Obesity-related metabolic syndrome: mechanisms of sympathetic overactivity. Int J Endocrinol
196. Lambert EA, Esler MD, Schlaich MP et al (2019) Obesity-associated organ damage and sympathetic nervous activity. Hypertension 73:1150–1159

197. Schirone L, Forte M, Palmerio S et al (2017) A review of the molecular mechanisms underlying the development and progression of cardiac remodeling. Oxid Med Cell Longev
198. Wang H-T, Liu C-F, Tsai T-H et al (2012) Effect of obesity reduction on preservation of heart function and attenuation of left ventricular remodeling, oxidative stress and inflammation in obese mice. J Transl Med 10:145
199. Alpert MA, Karthikeyan K, Abdullah O et al (2018) Obesity and cardiac remodeling in adults: mechanisms and clinical implications. Prog Cardiovasc Dis 61:114–123
200. Wang Z, Li L, Zhao H et al (2015) Chronic high fat diet induces cardiac hypertrophy and fibrosis in mice. Metab, Clin Exp 64:917–925
201. Cuspidi C, Rescaldani M, Sala C et al (2014) Left-ventricular hypertrophy and obesity: a systematic review and meta-analysis of echocardiographic studies. J Hypertens 32:16–25
202. Lillington GA, Anderson MW, Brandenburg RO (1957) The cardiorespiratory syndrome of obesity. Dis Chest 32:1–20
203. Bella JN, Devereux RB, Roman MJ et al (1998) Relations of left ventricular mass to fat-free and adipose body mass: the strong heart study. The strong heart study investigators. Circulation 98:2538–2544
204. Messerli FH, Sundgaard-Riise K, Reisin ED et al (1983) Dimorphic cardiac adaptation to obesity and arterial hypertension. Ann Intern Med 99:757–761
205. Woodiwiss AJ, Norton GR (2015) Obesity and left ventricular hypertrophy: the hypertension connection. Curr Hypertens Rep 17:539
206. Woodiwiss AJ, Libhaber CD, Majane OHI et al (2008) Obesity promotes left ventricular concentric rather than eccentric geometric remodeling and hypertrophy independent of blood pressure. Am J Hypertens 21:1144–1151
207. Turkbey EB, McClelland RL, Kronmal RA et al (2010) The impact of obesity on the left ventricle: the Multi-Ethnic Study of Atherosclerosis (MESA). JACC Cardiovasc Imaging 3:266–274
208. Rayner JJ, Banerjee R, Holloway CJ et al (2018) The relative contribution of metabolic and structural abnormalities to diastolic dysfunction in obesity. Int J Obes (Lond) 42:441–447
209. Anaruma CP, Pereira RM, Cristina da Cruz Rodrigues K et al (2019) Rock protein as cardiac hypertrophy modulator in obesity and physical exercise. Life Sci 116955
210. Geng Z, Fan W-Y, Zhou B et al (2019) FNDC5 attenuates obesity-induced cardiac hypertrophy by inactivating JAK2/STAT3-associated inflammation and oxidative stress. J Transl Med 17:107
211. van der Linde D, Konings EEM, Slager MA et al (2011) Birth prevalence of congenital heart disease worldwide: a systematic review and meta-analysis. J Am Coll Cardiol 58:2241–2247
212. Triedman JK, Newburger JW (2016) Trends in congenital heart disease: the next decade. Circulation 133:2716–2733
213. Liu Y, Chen S, Zühlke L et al (2019) Global birth prevalence of congenital heart defects 1970–2017: updated systematic review and meta-analysis of 260 studies. Int J Epidemiol 48:455–463
214. Aviram A, Hod M, Yogev Y (2011) Maternal obesity: implications for pregnancy outcome and long-term risks-a link to maternal nutrition. Int J Gynaecol Obstet 115(Suppl 1):S6–10
215. Flegal KM, Kruszon-Moran D, Carroll MD et al (2016) Trends in obesity among adults in the United States, 2005 to 2014. JAMA 315:2284–2291
216. Liu X, Ding G, Yang W et al (2019) Maternal body mass index and risk of congenital heart defects in infants: a dose-response meta-analysis. Biomed Res Int 1315796
217. Brite J, Laughon SK, Troendle J et al (2014) Maternal overweight and obesity and risk of congenital heart defects in offspring. Int J Obes (Lond) 38:878–882
218. de Moraes CL, Mendonça CR, Melo NCE et al (2019) Prevalence and association of congenital anomalies according to the maternal body mass index: cross-sectional study. Rev Bras Ginecol Obstet 41:280–290
219. Fahed AC, Gelb BD, Seidman JG et al (2013) Genetics of congenital heart disease: the glass half empty. Circ Res 112

220. Hinton RB (2013) Genetic and environmental factors contributing to cardiovascular malformation: a unified approach to risk. J Am Heart Assoc 2:e000292
221. Kalisch-Smith JI, Ved N, Sparrow DB, Environmental risk factors for congenital heart disease. Cold Spring Harb Perspect Biol. https://doi.org/10.1101/cshperspect.a037234
222. Saliba A, Figueiredo ACV, Baroneza JE et al, Genetic and genomics in congenital heart disease: a clinical review. J Pediatr (Rio J). https://doi.org/10.1016/j.jped.2019.07.004
223. Persson M, Cnattingius S, Villamor E et al (2017) Risk of major congenital malformations in relation to maternal overweight and obesity severity: cohort study of 1.2 million singletons. BMJ 357:j2563
224. Zheng Z, Yang T, Chen L et al (2018) Increased maternal body mass index is associated with congenital heart defects: an updated meta-analysis of observational studies. Int J Cardiol 273:112–120
225. Zhu Y, Chen Y, Feng Y et al (2018) Association between maternal body mass index and congenital heart defects in infants: a meta-analysis. Congenit Heart Dis 13:271–281
226. Bakker MK, Bergman JEH, Krikov S et al (2019) Prenatal diagnosis and prevalence of critical congenital heart defects: an international retrospective cohort study. BMJ Open 9:e028139
227. van Velzen CL, Clur SA, Rijlaarsdam MEB et al (2016) Prenatal diagnosis of congenital heart defects: accuracy and discrepancies in a multicenter cohort. Ultrasound Obstet Gynecol 47:616–622
228. Fung A, Manlhiot C, Naik S et al (2013) Impact of prenatal risk factors on congenital heart disease in the current era. J Am Heart Assoc 2:e000064
229. Ghaderian M, Emami-Moghadam A-R, Khalilian M-R et al (2014) Prepregnancy maternal weight and body mass index of children with and without congenital heart disease. Iran J Pediatr 24:313–318
230. Madsen NL, Schwartz SM, Lewin MB et al (2013) Prepregnancy body mass index and congenital heart defects among offspring: a population-based study. Congenit Heart Dis 8:131–141
231. Jarvie E, Hauguel-de-Mouzon S, Nelson SM et al (2010) Lipotoxicity in obese pregnancy and its potential role in adverse pregnancy outcome and obesity in the offspring. Clin Sci 119:123–129
232. Karelis AD, St-Pierre DH, Conus F et al (2004) Metabolic and body composition factors in subgroups of obesity: what do we know? J Clin Endocrinol Metab 89:2569–2575
233. Catalano PM, Shankar K (2017) Obesity and pregnancy: mechanisms of short term and long term adverse consequences for mother and child. BMJ 356:j1
234. Li Y (2018) Epigenetic mechanisms link maternal diets and gut microbiome to obesity in the offspring. Front Genet 9
235. Nogues P, Dos Santos E, Jammes H et al (2019) Maternal obesity influences expression and DNA methylation of the adiponectin and leptin systems in human third-trimester placenta. Clin Epigenetics 11:20
236. Øyen N, Diaz LJ, Leirgul E et al (2016) Prepregnancy diabetes and offspring risk of congenital heart disease: a nationwide cohort study. Circulation 133:2243–2253
237. Persson M, Norman M, Hanson U (2009) Obstetric and perinatal outcomes in type 1 diabetic pregnancies: a large, population-based study. Diabetes Care 32:2005–2009
238. Poirier P, Eckel R (2000) The heart and obesity. In: Fuster V, Alexander R, King S et al (eds) Hurst's the heart. McGraw-Hill Companies, New York, pp 2289–2303
239. Messerli FH, Nunez BD, Ventura HO et al (1987) Overweight and sudden death. Increased ventricular ectopy in cardiopathy of obesity. Arch Intern Med 147:1725–1728
240. Pathak RK, Mahajan R, Lau DH et al (2015) The implications of obesity for cardiac arrhythmia mechanisms and management. Can J Cardiol 31:203–210
241. Rabkin SW, Mathewson FA, Hsu PH (1977) Relation of body weight to development of ischemic heart disease in a cohort of young North American men after a 26 year observation period: the Manitoba Study. Am J Cardiol 39:452–458
242. Kannel WB, Plehn JF, Cupples LA (1988) Cardiac failure and sudden death in the Framingham study. Am Heart J 115:869–875

243. Bhatia S, Sugrue A, Asirvatham S (2018) Atrial fibrillation: beyond rate control. Mayo Clin Proc 93:373–380
244. Cheniti G, Vlachos K, Pambrun T et al (2018) Atrial fibrillation mechanisms and implications for catheter ablation. Front Physiol 9:1458
245. Staerk L, Sherer JA, Ko D et al (2017) Atrial fibrillation: epidemiology, pathophysiology, and clinical outcomes. Circ Res 120:1501–1517
246. Chatterjee NA, Giulianini F, Geelhoed B et al (2017) Genetic obesity and the risk of atrial fibrillation: causal estimates from mendelian randomization. Circulation 135:741–754
247. Gami AS, Hodge DO, Herges RM et al (2007) Obstructive sleep apnea, obesity, and the risk of incident atrial fibrillation. J Am Coll Cardiol 49:565–571
248. Huxley RR, Misialek JR, Agarwal SK et al (2014) Physical activity, obesity, weight change, and risk of atrial fibrillation: the atherosclerosis risk in communities study. Circ Arrhythm Electrophysiol 7:620–625
249. Lavie CJ, Pandey A, Lau DH et al (2017) Obesity and atrial fibrillation prevalence, pathogenesis, and prognosis: effects of weight loss and exercise. J Am Coll Cardiol 70:2022–2035
250. Vyas V, Lambiase P (2019) Obesity and atrial fibrillation: epidemiology, pathophysiology and novel therapeutic opportunities. Arrhythm Electrophysiol Rev 8:28–36
251. Maesen B, Zeemering S, Afonso C et al (2013) Rearrangement of atrial bundle architecture and consequent changes in anisotropy of conduction constitute the 3-dimensional substrate for atrial fibrillation. Circ Arrhythm Electrophysiol 6:967–975
252. Zlochiver S, Muñoz V, Vikstrom KL et al (2008) Electrotonic myofibroblast-to-myocyte coupling increases propensity to reentrant arrhythmias in two-dimensional cardiac monolayers. Biophys J 95:4469–4480
253. Huang H, Amin V, Gurin M et al (2013) Diet-induced obesity causes long QT and reduces transcription of voltage-gated potassium channels. J Mol Cell Cardiol 59:151–158
254. Lalani AP, Kanna B, John J et al (2000) Abnormal signal-averaged electrocardiogram (SAECG) in obesity. Obes Res 8:20–28
255. Magnani JW, Lopez F, Soliman EZ et al (2012) P wave indices, obesity, and the metabolic syndrome: the atherosclerosis risk in communities study. Obesity (Silver Spring) 20:666–672
256. Abed HS, Samuel CS, Lau DH et al (2013) Obesity results in progressive atrial structural and electrical remodeling: implications for atrial fibrillation. Heart Rhythm 10:90–100
257. McCully BH, Hasan W, Streiff CT et al (2013) Sympathetic cardiac hyperinnervation and atrial autonomic imbalance in diet-induced obesity promote cardiac arrhythmias. Am J Physiol Heart Circ Physiol 305:H1530–H1537
258. Karam BS, Chavez-Moreno A, Koh W et al (2017) Oxidative stress and inflammation as central mediators of atrial fibrillation in obesity and diabetes. Cardiovasc Diabetol 16:120
259. Lubbers ER, Price MV, Mohler PJ (2018) Arrhythmogenic substrates for atrial fibrillation in obesity. Front Physiol 9
260. Rossi A, Enriquez-Sarano M, Burnett JC et al (2000) Natriuretic peptide levels in atrial fibrillation: a prospective hormonal and Doppler-echocardiographic study. J Am Coll Cardiol 35:1256–1262
261. Novo G, Guttilla D, Fazio G et al (2008) The role of the renin–angiotensin system in atrial fibrillation and the therapeutic effects of ACE-Is and ARBS. Br J Clin Pharmacol 66:345–351
262. Smith CCT, Mocanu MM, Davidson SM et al (2006) Leptin, the obesity-associated hormone, exhibits direct cardioprotective effects. Br J Pharmacol 149:5–13
263. Oliveira A, Ramos E, Lopes C et al (2009) Self-reporting weight and height: misclassification effect on the risk estimates for acute myocardial infarction. Eur J Public Health 19:548–553
264. Wolk R, Berger P, Lennon RJ et al (2003) Body mass index: a risk factor for unstable angina and myocardial infarction in patients with angiographically confirmed coronary artery disease. Circulation 108:2206–2211
265. Schargrodsky H, Rozlosnik J, Ciruzzi M et al (1994) Body weight and nonfatal myocardial infarction in a case-control study from Argentina. Soz Praventivmed 39:126–133

266. Mehta L, Devlin W, McCullough PA et al (2007) Impact of body mass index on outcomes after percutaneous coronary intervention in patients with acute myocardial infarction. Am J Cardiol 99:906–910
267. Bucholz EM, Rathore SS, Reid KJ et al (2012) Body mass index and mortality in acute myocardial infarction patients. Am J Med 125:796–803
268. Dhoot J, Tariq S, Erande A et al (2013) Effect of morbid obesity on in-hospital mortality and coronary revascularization outcomes after acute myocardial infarction in the United States. Am J Cardiol 111:1104–1110
269. Fukuoka S, Kurita T, Dohi K et al (2019) Untangling the obesity paradox in patients with acute myocardial infarction after primary percutaneous coronary intervention (detail analysis by age). Int J Cardiol 289:12–18
270. Zhu J, Su X, Li G et al (2014) The incidence of acute myocardial infarction in relation to overweight and obesity: a meta-analysis. Arch Med Sci 10:855–862
271. Li L, Liu D-W, Yan H-Y et al (2016) Obesity is an independent risk factor for non-alcoholic fatty liver disease: evidence from a meta-analysis of 21 cohort studies. Obes Rev 17:510–519
272. Rinella ME (2015) Nonalcoholic fatty liver disease: a systematic review. JAMA 313:2263–2273
273. Buzzetti E, Pinzani M, Tsochatzis EA (2016) The multiple-hit pathogenesis of non-alcoholic fatty liver disease (NAFLD). Metab, Clin Exp 65:1038–1048
274. Eslam M, Valenti L, Romeo S (2018) Genetics and epigenetics of NAFLD and NASH: clinical impact. J Hepatol 68:268–279
275. Mahady SE, George J (2016) Exercise and diet in the management of nonalcoholic fatty liver disease. Metab, Clin Exp 65:1172–1182
276. Polyzos SA, Kountouras J, Mantzoros CS (2019) Obesity and nonalcoholic fatty liver disease: from pathophysiology to therapeutics. Metab, Clin Exp 92:82–97
277. Adolph TE, Grander C, Grabherr F et al (2017) Adipokines and non-alcoholic fatty liver disease: multiple interactions. Int J Mol Sci 18
278. Qiu Y, Wang S-F, Yu C et al (2019) Association of circulating adipsin, visfatin, and adiponectin with nonalcoholic fatty liver disease in adults: a case-control study. Ann Nutr Metab 74:44–52
279. Wolfs MGM, Gruben N, Rensen SS et al (2015) Determining the association between adipokine expression in multiple tissues and phenotypic features of non-alcoholic fatty liver disease in obesity. Nutr Diabetes 5:e146
280. Mota M, Banini BA, Cazanave SC et al (2016) Molecular mechanisms of lipotoxicity and glucotoxicity in nonalcoholic fatty liver disease. Metab, Clin Exp 65:1049–1061
281. Maiers JL, Malhi H (2019) Endoplasmic reticulum stress in metabolic liver diseases and hepatic fibrosis. Semin Liver Dis 39:235–248
282. Jasirwan COM, Lesmana CRA, Hasan I et al (2019) The role of gut microbiota in non-alcoholic fatty liver disease: pathways of mechanisms. Biosci Microbiota Food Health 38:81–88
283. Foulds CE, Treviño LS, York B et al (2017) Endocrine-disrupting chemicals and fatty liver disease. Nat Rev Endocrinol 13:445–457
284. Polyzos SA, Kountouras J, Mantzoros CS (2017) Adipose tissue, obesity and non-alcoholic fatty liver disease. Minerva Endocrinol 42:92–108
285. Gómez-Hernández A, Beneit N, Díaz-Castroverde S et al (2016) Differential role of adipose tissues in obesity and related metabolic and vascular complications. Int J Endocrinol 1216783
286. Polyzos SA, Kountouras J, Mantzoros CS (2016) Adipokines in nonalcoholic fatty liver disease. Metab, Clin Exp 65:1062–1079
287. Boutari C, Perakakis N, Mantzoros CS (2018) Association of adipokines with development and progression of nonalcoholic fatty liver disease. Endocrinol Metab (Seoul) 33:33–43
288. Adams LA, Anstee QM, Tilg H et al (2017) Non-alcoholic fatty liver disease and its relationship with cardiovascular disease and other extrahepatic diseases. Gut 66:1138–1153
289. Lonardo A, Nascimbeni F, Mantovani A et al (2018) Hypertension, diabetes, atherosclerosis and NASH: cause or consequence? J Hepatol 68:335–352
290. Ekstedt M, Hagström H, Nasr P et al (2015) Fibrosis stage is the strongest predictor for disease-specific mortality in NAFLD after up to 33 years of follow-up. Hepatology 61:1547–1554

291. Targher G, Byrne CD, Lonardo A et al (2016) Non-alcoholic fatty liver disease and risk of incident cardiovascular disease: a meta-analysis. J Hepatol 65:589–600
292. Angulo P, Kleiner DE, Dam-Larsen S et al (2015) Liver fibrosis, but no other histologic features, is associated with long-term outcomes of patients with nonalcoholic fatty liver disease. Gastroenterology 149:389–397.e10
293. Targher G, Day CP, Bonora E (2010) Risk of cardiovascular disease in patients with nonalcoholic fatty liver disease. N Engl J Med 363:1341–1350
294. Tilg H, Moschen AR, Roden M (2017) NAFLD and diabetes mellitus. Nat Rev Gastroenterol Hepatol 14:32–42
295. Ormazabal V, Nair S, Elfeky O et al (2018) Association between insulin resistance and the development of cardiovascular disease. Cardiovasc Diabetol 17:122
296. Lazzerini PE, Capecchi PL, Laghi-Pasini F (2017) Systemic inflammation and arrhythmic risk: lessons from rheumatoid arthritis. Eur Heart J 38:1717–1727
297. Nyman K, Granér M, Pentikäinen MO et al (2013) Cardiac steatosis and left ventricular function in men with metabolic syndrome. J Cardiovasc Magn Reson 15:103
298. Schulze PC, Drosatos K, Goldberg IJ (2016) Lipid use and misuse by the heart. Circ Res 118:1736–1751
299. Bornachea O, Vea A, Llorente-Cortes V (2018) Interplay between epicardial adipose tissue, metabolic and cardiovascular diseases. Clin Investig Arterioscler 30:230–239
300. Karwi QG, Uddin GM, Ho KL et al (2018) Loss of metabolic flexibility in the failing heart. Front Cardiovasc Med 5:68
301. Kang YM, Jung CH, Cho YK et al (2017) Fatty liver disease determines the progression of coronary artery calcification in a metabolically healthy obese population. PLoS ONE 12:e0175762
302. Wu R, Hou F, Wang X et al (2017) Nonalcoholic fatty liver disease and coronary artery calcification in a Northern Chinese population: a cross sectional study. Sci Rep 7:9933
303. Stefan N (2016) Nonalcoholic fatty liver disease and coronary artery calcification. Clin Gastroenterol Hepatol 14:1345–1346
304. Cho YK, Kang YM, Yoo JH et al (2018) The impact of non-alcoholic fatty liver disease and metabolic syndrome on the progression of coronary artery calcification. Sci Rep 8:12004
305. Chang Y, Ryu S, Sung K-C et al (2019) Alcoholic and non-alcoholic fatty liver disease and associations with coronary artery calcification: evidence from the Kangbuk Samsung Health Study. Gut 68:1667–1675
306. Jung D-H, Lee Y-J, Ahn H-Y et al (2010) Relationship of hepatic steatosis and alanine aminotransferase with coronary calcification. Clin Chem Lab Med 48:1829–1834
307. Lee M-K, Park H-J, Jeon WS et al (2015) Higher association of coronary artery calcification with non-alcoholic fatty liver disease than with abdominal obesity in middle-aged Korean men: the Kangbuk Samsung Health Study. Cardiovasc Diabetol 14:88
308. Kim BJ, Cheong ES, Kang JG et al (2016) Relationship of epicardial fat thickness and nonalcoholic fatty liver disease to coronary artery calcification: from the CAESAR study. J Clin Lipidol 10:619–626.e1
309. Wolff L, Bos D, Murad SD et al (2016) Liver fat is related to cardiovascular risk factors and subclinical vascular disease: the Rotterdam Study. Eur Heart J Cardiovasc Imaging 17:1361–1367
310. Kim MK, Ahn CW, Nam JS et al (2015) Association between nonalcoholic fatty liver disease and coronary artery calcification in postmenopausal women. Menopause 22:1323–1327
311. Markus MRP, Baumeister SE, Stritzke J et al (2013) Hepatic steatosis is associated with aortic valve sclerosis in the general population: the Study of Health in Pomerania (SHIP). Arterioscler Thromb Vasc Biol 33:1690–1695
312. Bonapace S, Valbusa F, Bertolini L et al (2014) Nonalcoholic fatty liver disease is associated with aortic valve sclerosis in patients with type 2 diabetes mellitus. PLoS ONE 9:e88371
313. Ismaiel A, Dumitraşcu DL (2019) Cardiovascular risk in fatty liver disease: the liver-heart axis-literature review. Front Med (Lausanne) 6:202

314. Blond E, Disse E, Cuerq C et al (2017) EASL-EASD-EASO clinical practice guidelines for the management of non-alcoholic fatty liver disease in severely obese people: do they lead to over-referral? Diabetologia 60:1218–1222
315. Polyzos SA, Kountouras J, Zavos C et al (2012) Nonalcoholic fatty liver disease: multimodal treatment options for a pathogenetically multiple-hit disease. J Clin Gastroenterol 46:272–284
316. Hannah WN, Harrison SA (2016) Effect of weight loss, diet, exercise, and bariatric surgery on nonalcoholic fatty liver disease. Clin Liver Dis 20:339–350
317. Vilar-Gomez E, Martinez-Perez Y, Calzadilla-Bertot L et al (2015) Weight loss through lifestyle modification significantly reduces features of nonalcoholic steatohepatitis. Gastroenterology 149:367–378.e5; quiz e14–15
318. Leoni S, Tovoli F, Napoli L et al (2018) Current guidelines for the management of non-alcoholic fatty liver disease: a systematic review with comparative analysis. World J Gastroenterol 24:3361–3373
319. Zhao S, Zhu Y, Schultz RD et al (2019) Partial leptin reduction as an insulin sensitization and weight loss strategy. Cell Metab 30:706–719.e6
320. Koliaki C, Liatis S, Kokkinos A (2019) Obesity and cardiovascular disease: revisiting an old relationship. Metab, Clin Exp 92:98–107
321. El Husseny MWA, Mamdouh M, Shaban S et al (2017) Adipokines: potential therapeutic targets for vascular dysfunction in type II diabetes mellitus and obesity. J Diabetes Res 8095926

Chapter 3
Maternal Obesity: Impacts on the Cardiovascular Health of Mother and Offspring

Fahmida Jahan, Ashim K. Bagchi, and Rushita A. Bagchi

Abstract Maternal obesity serves as a potential risk factor for pregnancy outcomes. Neuroendocrine regulation of energy balance during maternal obesity is associated with childhood health problem. Increased gestational body-fat mass is controlled by maternal neurohormonal factors such as adiponectin, lectin, ghrelin, obestatin and insulin and their imbalance may have association with cardiovascular complications after birth. The long-term effect of maternal obesity-induced neurohormonal factors on their offspring is not well understood. This review focuses on the crosstalk between neurohormonal factors during maternal obesity and their potential outcomes on the mother and offspring's cardiovascular health.

Keywords Obesity · BMI · Preeclampsia · Maternal · Offspring · Hormones · Inflammation · Oxidative stress

Introduction

Obesity poses a serious public health concern and has been associated with reduced life expectancy [1]. It is linked with cardiovascular disease and type-2 diabetes (T2D) which in later life contributes to morbidity and mortality [2–4]. According to the Canadian Health Measures Survey (2007–2017), 34% of the respondents are overweight, and 27% are obese, with a concurrent increase in obesity with aging. Maternal obesity serves as a potential risk factor for complicated pregnancy outcomes. Previous

F. Jahan
Department of Biochemistry, Microbiology and Immunology, Faculty of Medicine, University of Ottawa, Ottawa, ON K1H 8M5, Canada

A. K. Bagchi
St. Boniface Hospital Albrechtsen Research Centre, Institute of Cardiovascular Sciences, Winnipeg, MB R2H 2A6, Canada

R. A. Bagchi (✉)
Department of Medicine, Division of Cardiology, University of Colorado Anschutz Medical Campus, Aurora, CO 80045, USA
e-mail: rushita.bagchi@cuanschutz.edu

P. S. Tappia et al. (eds.), *Biochemistry of Cardiovascular Dysfunction in Obesity*, Advances in Biochemistry in Health and Disease 20,
https://doi.org/10.1007/978-3-030-47336-5_3

surveys indicate that from 1997–2009, the obesity rate among Canadian women of reproductive age has increased from 16% to an alarming 23.9% [5]. According to the US Institute of Medicine (IOM), abnormal weight gain during pregnancy is more prevalent in western countries (approximately 40%) [6]. Studies suggest that maternal pre-pregnancy obesity and excessive gestational weight gain are critical health risk factors for both the mother and infant. Obesity increases the likelihood of developing various pregnancy complications such as hypertension, gestational diabetes, preeclampsia, metabolic disorders, and others [7]. In this chapter, we discuss the negative impacts of obesity on cardiovascular health during pregnancy, and the role of obesity-related neurohormonal regulators and their mechanisms for long-term effects on the mother and the offspring.

Cardiovascular Risks Associated with Maternal Obesity

Pre-existing obesity or excessive gestational weight gain can permanently affect maternal cardiovascular health and fetal development. The following section describes the cardiovascular risks associated with maternal obesity.

Maternal Hypertension and Cardiac Dysfunction

Several epidemiologic studies suggest that pre-pregnancy obesity or excessive gestational weight gain is a crucial risk factor for maternal and fetal complications during pregnancy. These include gestational diabetes, hypertensive disorders of pregnancy, C-section delivery, and stillbirth [2, 8, 9]. Data from a recent study conducted on French and Canadian cohorts revealed that obesity rates were 9.1% and 16% ($p < 0.001$) among 26,973 and 22,046 deliveries, respectively and had a significant correlation with maternal hypertension, C-section delivery and increased fetal weight [5].

Obesity is a nodal risk factor for gestational hypertension and preeclampsia, a hypertensive disorder that arises at or after twenty weeks of gestation [10]. However, the mechanism(s) by which obesity increases the risks for these diseases are not well understood. If a mother's body mass index (BMI) increases from 5 to 7 kg/m^2, the risk for developing preeclampsia increases dramatically by almost two-fold [10]. Obese individuals have been shown to have increased plasma levels of inflammatory molecules [11, 12]. Women with preeclampsia also have elevated levels of these inflammatory markers which may lead to vascular dysfunction and clinical presentation of the disease. Both obesity and preeclampsia are linked to oxidative stress suggesting that perhaps, the pre-existing inflammation and vascular damage due to oxidative stress, makes obesity a predisposing risk factor for developing hypertension during pregnancy [10, 13].

Congenital Heart Defects

Fetal development can be permanently affected due to maternal over-nutrition or under-nutrition as the fetus adapts to its environment. A rodent model study showed that maternal over-nutrition caused interventricular septum thickening, cardiomyocyte hypertrophy and increased left ventricular area in the offspring at the age of 8 weeks [14]. A meta-analysis performed in Sweden among 1,857,822 live single births during 1992–2010 showed that obese and morbidly obese mothers have higher risk of having infants with mild to severe heart defects. Moreover, early gestation obesity was found to be associated with increased infant death mainly due to congenital defects, and neonatal morbidities such as birth asphyxia (reduced oxygen supply), sudden infant death syndrome or infections [15]. In line with these, Stothard and colleagues performed a meta-analysis of 18 different observational studies showing that higher maternal BMI is linked to structural defects in neural tube, spine, heart, orofacial clefts, brain and limb [16]. A large population-based study performed in New York State showed that obese women have a greater chance to give birth to children with fetal congenital heart disease [17].

Several human studies demonstrated that maternal obesity increases placental weight and induces placental inflammation and endothelial dysfunction [18–24]. Increased adiposity reduces placental nutrient transporter activity, such as lowering taurine transporter protein (TauT) activity [19]. In addition, maternal obesity increases placental reactive oxygen species (ROS) levels and causes a decline in placental ATP levels. Cultured primary trophoblasts from obese mothers were shown to have decreased mitochondrial respiration [20, 23]. Together, these studies suggest that placental abnormalities may affect fetal growth and development.

Long-Term Maternal Risk

Excessive gestational weight gain can have a lifelong effect in the mother [25]. A study conducted on 11,006 women with 37 years follow-up suggests that increased maternal BMI during pregnancy has a correlation with increased incidences of death due to coronary heart diseases [26]. Another study performed in the United Kingdom investigated whether maternal obesity in first pregnancy is linked to premature death or later cardiovascular events [25].

As previously mentioned, obese women are more prone to develop gestational hypertension and preeclampsia. These conditions further act as risk factors for developing future cardiovascular diseases due to existing vascular damage. Two studies showed that over 60% of women who experienced gestational hypertension went on to develop hypertension within ten years post-pregnancy [27]. Likewise, several studies showed that preeclamptic women are at higher risk for developing chronic hypertension, ischemic heart disease and stroke later in life [27]. Thus, pre-pregnancy obesity or increased gestational weight gain can lead to serious pregnancy-related complications that may have a lifelong impact on mothers.

Long-Term Fetal Risk

Maternal obesity is not only limited to causing adverse effects on the mother or the developing fetus, but also has a great potential to affect the offspring later in life through developmental programming. Data from animal models and humans suggest that exposure to harmful environmental stimuli during the developmental stages of life can permanently alter metabolism, structure and architecture at the organ level, which can lead to obesity, diabetes and heart disease in adulthood [28–30].

Samuelsson and colleagues studied the impact of maternal obesity on the offspring's cardiovascular health [30]. Female mice were given a palatable obesogenic diet (16% fat, 33% sugar) before mating, and maintained as such during pregnancy and lactation. Although obese dams were fed normal chow diet, they were found to be more hyperphagic and showed less physical movement than control offspring at 4–6 weeks of age. They had decreased skeletal muscle mass, reduced locomotor activity, increased abdominal fat pad mass, elevated fasting insulin levels and increased systolic pressure due to resistant artery endothelial dysfunction at three months of age. Similarly, another animal study showed that obesity during pregnancy led to cardiomyocyte hypertrophy, thickening of the left ventricular wall, and increased heart to body weight ratio in the offspring. Markers of cardiac hypertrophy and pathologic fetal gene reprogramming such as β-myosin heavy chain, *Myh7* and brain natriuretic peptide, Nppb (BNP) were elevated in the dams. Plasma insulin levels were also found to be elevated which led to the activation of several signaling pathways such as cardiac protein kinase B (Akt), mechanistic target of rapamycin (mTOR), extracellular signal-regulated kinase 1/2 (ERK1/2) and mitogen-activated protein kinase (MAPK) pathways in these offspring [31]. A recent study showed that diet-induced obesity during pregnancy causes cardiac dysfunction in offspring. These adult offspring showed ventricular fetal gene reprogramming by expressing higher levels of *Myh7*. Moreover, post-weaning obesogenic diet coupled with maternal obesity resulted in increased skeletal muscle alpha actin (*Acta1*) [14]. Thus, these studies suggest that excess availability of nutrients and hormones may lead to abnormal metabolic and transcriptional profile which ultimately results in cardiac dysfunction.

Maternal pre-existing obesity or excessive weight gain during pregnancy is also associated with macrosomia, infants born large for their gestational age [32, 33]. Many observational studies also showed an association between maternal obesity and risk for neonatal low Apgar score (based on newborns overall health at the time of birth), low blood sugar and a requirement for intensive care upon birth [34]. Results from four meta-analysis showed that maternal obesity increases the risk of their offspring developing obesity or being overweight during their childhood when compared to children born to normal weight mothers (Odds ratio 1.95; 95% CI) [35]. Moreover, another study showed that excessive gestational weight gain increased the risk for childhood obesity by an alarming 33% [36]. A Dutch study among 5908 mother–child pairs revealed that early-gestational weight gain, independent of pre-existing obesity or weight gain later in gestation, negatively affects cardiometabolic profile in offspring during their childhood [37].

Higher maternal pre-pregnancy BMI and excessive gestational weight gain has been shown to have an association with higher adiposity and total body mass in the offspring in adulthood [38–42]. In an Australian study among 2432 mother–child pairs, excessive gestational weight gain independent of pre-existing obesity, was associated with higher BMI and systolic blood pressure in 21 year old offspring [42]. Another study performed in Jerusalem found a correlation between maternal pre-existing overweight and cardiac dysfunction in 32 year old offspring [38]. Another study showed association of premature all-cause mortality and offspring hospitalization due to cardiovascular events with higher maternal BMI [43]. Thus, exposure to an adipogenic diet during development and early stages of life can make individuals more susceptible to metabolic and cardiac diseases in long-term.

Mechanisms of Maternal Obesity and Related Cardiovascular Complications in Their Offspring

The offspring of obese women showed higher risk of developing obesity, insulin resistance and cardiovascular complications [44]. Studies have suggested that left atrial and ventricular dimension are greater in offspring of obese women [45]. Pathogenesis of obesity-induced cardiovascular defects in neonates is mainly controlled by positive energy balance and the rate of adiposity during pregnancy. During pregnancy, increased adiposity with metabolic demand increased blood volume and preload to the fetal heart [46]. Increased blood volume may adversely impact the development of fetal heart leading to cardiovascular complications after birth. Right ventricular outflow tract (RVOT) defect is very common in offspring of obese women. Obesity-related neurohormonal factors such as adiponectin, leptin, ghrelin, obestatin and insulin have been studied broadly (Fig. 3.1). However, molecular interaction and mechanisms of action of these factors are not very clear during maternal obesity and their impact on offspring's health.

Hormone-Mediated Responses

Role of Leptin

Maternal obesity comprises a complex pathophysiological phenomenon involving innumerable molecular, genetic and environmental factors. In general, adipokines and hormones such as leptin, secreted by adipocytes, regulate body-fat mass and maintain energy homeostasis. Physiological leptin replacement ameliorated both the hyperphagia and reversed body-fat mass in leptin-deficient individuals suggesting that leptin regulates normal body weight [47]. In the early days of pregnancy, placenta

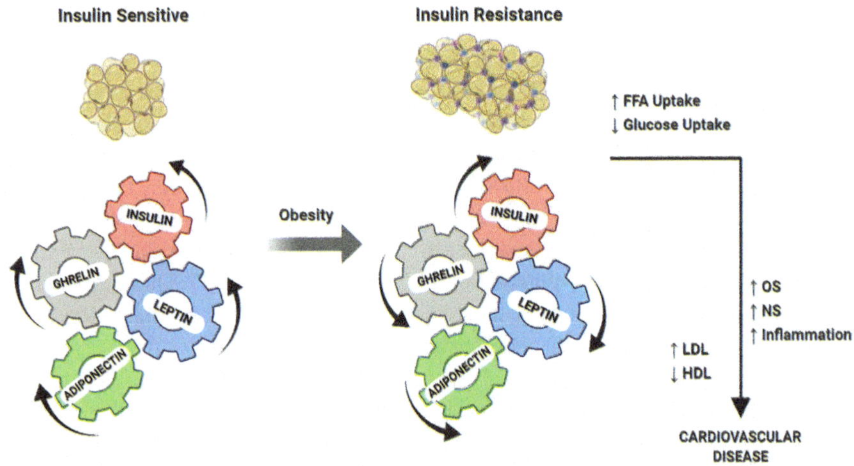

Fig. 3.1 Neurohormonal regulators of energy homeostasis in obesity-associated cardiovascular disease. In this schematic, clockwise and anti-clockwise arrows are symbolic representation of upregulation and downregulation of these regulators, respectively. These regulators have synergistic or antagonistic effect on each other in modulation of obesity. Under normal conditions, ghrelin, secreted by P/D1 cells in fundus or epsilon cells in pancreas is associated with lowering body energy expenditure, while leptin secreted from adipocytes, has the opposite effect. Adiponectin released by adipocytes antagonizes the effect of leptin to control body-fat mass by increasing energy expenditure. It is also suggested that increased ghrelin and adiponectin is associated with insulin sensitivity. Impaired adiponectin, ghrelin and leptin secretion influence insulin sensitivity which is later involved in the pathogenesis of obesity. Increased insulin adversely impacts ghrelin and adiponectin. Nevertheless, decreased adiponectin increases adiposity and food uptake by excessive leptin production as a result of which energy expenditure decreases. In the process of obesity pathogenesis, increased free fatty acid (FFA) oxidation further leads to increase in low density lipoprotein (LDL). Increased adiposity aggravates oxidative stress-mediated inflammatory response. This is responsible for atherosclerosis and other cardiovascular complications

produces higher amounts of leptin causing hyperemesis gravidarum (morning sickness) [48, 49]. Obese pregnancy is associated with hyperleptinemia in the neonate during fetal brain development, where leptin is thought to play a neurotrophic role in establishing the neural circuitry of the hypothalamus, involved in both appetite and blood volume control [50, 51]. Increased blood pressure in prenatal offspring of female obese mice may promote dysregulation of the normal neurotrophic action of leptin leading to leptin insensitivity. In these offspring of obese dams, higher leptin increases the levels of the hypothalamic appetite marker, orexigenic peptide, neuropeptide Y (NPY), agouti-related protein (AgRP), and on the other hand, reduces the levels of proopiomelanocortin (POMC), a precursor for the major anorectic neuropeptide, α-MSH [52] (Fig. 3.2). These adaptive responses play a critical role in the development of hypothalamic leptin resistance via the leptin-dependent signal transducer and activator of transcription 3 (Stat3) pathway [53]. Studies showed an impairment of leptin-signaling which alters neuronal development in hyperphagic ob/ob mice offspring [54]. During maturation of these neonates, body-fat mass

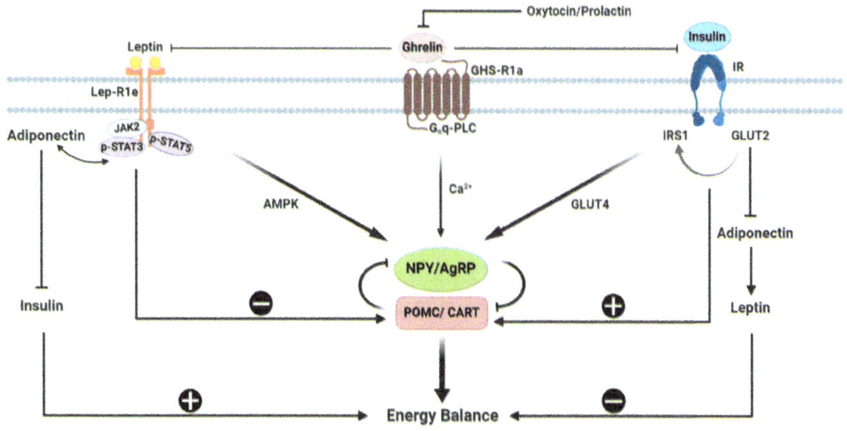

Fig. 3.2 Mechanism of energy balance during obesity, lactating dams and in their offspring. Ghrelin released from neurons in the hypothalamic arcuate nucleus (ARC) potently stimulates the release of growth hormone. Ghrelin receptor R1a (GHS-R1a) triggers transcriptional activation of hypothalamic orexigenic/appetite peptide, Neuropeptide-Y (NPY) agouti-related protein (AgRP), and stimulates food intake. Ghrelin-responsive NPY/AgRP inhibits the levels of proopiomelanocortin (POMC)/ cocaine- and amphetamine-regulated transcript (CART), controlling appetite and energy homeostasis. Gαq-mediated protein lipase C (PLC) activation promotes protein kinase C (PKC) and/or PKA which are required for ghrelin to activate intracellular calcium signaling in the NPY neurons. Under normal condition, ghrelin maintains energy balance by inhibiting excessive leptin and insulin production. However, leptin has opposite effect on ghrelin secretion, thus is able to maintain a negative (−) energy balance. Leptin receptors (Lep-R1e) activation via Janus kinase 2 (JAK2)/signal transducer and activator of transcription 5 (STAT5) regulates neuropeptide POMC activation which cuts-off the neuronal appetite NPY/CART circuit, and regulating hyperphagia and adiposity. Furthermore, leptin attenuates ghrelin-induced $[Ca^{2+}]_i$ increases in ghrelin-responsive NPY neurons. However, during early stage obese pregnancy, over-activation of leptin receptor in placenta as well as in adipose tissues causes STAT3 and 5-adenosine monophosphate-activated protein kinase (AMPK) activation, and thus attains positive (+) energy balance state. Increased insulin levels cause the uptake of glucose into the cells via binding to GLUT4 membrane receptors. On the other hand, insulin activates leptin signaling and regulates glucose uptake via GLUT2 receptor activation. GLUT 2 receptor activation increases phosphorylation of insulin-stimulated insulin receptor substrate–1 (IRS-1) and Akt. Besides, insulin-mediated leptin activation helps in restoration of energy intake via (i) suppression of adiponectin through p38-MAPK pathway and, (ii) overexpressing ghrelin and development of leptin resistance through the leptin-dependent STAT3 pathway. In lactating obese dams, increased oxytocin or prolactin provokes suppression of ghrelin

continues to grow despite a reduction in leptin expression suggesting that plasma leptin is independent of body-fat mass [55].

The mechanisms of leptin action in the development of cardiovascular disease are not well established. Understanding the binding efficacy of leptin to its receptor (LepR) may pave the way for elucidation of the downstream pathways contributing to pathogenesis of obesity. All six isoforms of leptin receptors (LepRa, LepRb, LepRc, LepRd, LepRe, and LepRf), are known to have a common leptin-binding domain [56]. LepRe is a soluble isoform that binds to circulating leptin and inhibits leptin transfer. Once leptin binds to its receptor LepRe, a long isoform Lep-Rb activates Janus Kinase 2 (JAK)/STAT signaling pathway in the hypothalamus. Recruitment of JAK2 and later its phosphorylation at three different sites Y985, Y1077, and Y1138 promotes many downstream pathways [56]. Phosphorylation at Y985 residue promotes mitogen-activated protein kinase (MAPK) pathway leading to a negative regulation of leptin. Phosphorylation at Y1077 residue activates STAT5 and regulates hyperphagia suggesting that JAK2/STAT5 activation prevents adiposity, whereas phosphorylation at Y1138 site controls energy homeostasis via STAT3 recruitment in obese individuals [57, 58]. STAT3 via suppressor of cytokine signaling molecule 3 (SOCS 3) activation inhibits MAPK and promotes neuropeptides such as POMC, AgRP and NPY in the hypothalamus [58, 59]. A study further confirmed that POMC-specific Stat3 female mutant mice exhibited slight increase in total body weight [59]. Obese women usually present higher circulating leptin and develop resistance to leptin by downregulating LepRs in the hypothalamus and other tissues, but not in cardiomyocytes [60, 61]. Expression of cardiac leptin enhances LepR and Stat3 phosphorylation levels in obese wildtype mice, but not in LepR mutant mice suggesting that leptin-LepR mediates cardioprotection via Stat3 in LepRS1138 mice whereas LepR-Tyr1138 promotes cardiac hypertrophy [61].

Role of Insulin

Insulin is more adipogenic than leptin, and shares many similar signaling pathways as leptin, to modulate energy homeostasis [62, 63]. Studies have revealed that leptin infusion enhances insulin-stimulated glucose metabolism in adipose tissues [64, 65]. Insulin signaling is activated by leptin and regulates glycemia via GLUT2 receptor activation [65]. Studies also suggested that GLUT2 downregulates phosphoenolpyruvate carboxykinase (PEPCK) levels and promotes phosphorylation of insulin-stimulated insulin receptor substrate–1 (IRS-1) and Akt at both Thr308 and Ser473 sites (Fig. 3.2). Akt phosphorylation inhibited relative p-phosphatase and tensin (PTEN) expression in leptin-infused rats. Akt is mediator of insulin signaling [66], and its phosphorylation promotes insulin sensitivity [67]. Similarly, insulin intensely increases the production of leptin by adipose tissue [68]. Body-fat mass and positive energy balance in infants is influenced by insulin secretion which is directly associated with adipose tissue leptin mRNA expression. Studies also suggest that offspring from obese dams exhibit increased insulin levels for several days to weeks after birth prior to leptin surge with glucose in the milk [54, 69]. However, no studies were conducted to show if glucose had any feedback effects on insulin or leptin secretion in these offspring. A family-based study on 52 offspring of 22 obese mothers revealed that neonatal obesity and insulin resistance is greatly associated with the maternal obesity [70]. Enzyme required for gluconeogenesis and PEPCK did not increase before birth suggesting a peripheral insulin resistance in the fetus of obese women [71].

Role of Ghrelin

Ghrelin is an orexigenic hormone that controls appetite. In obese patients, serum ghrelin levels were lower compared to lean individuals suggesting that lower ghrelin level is associated with obesity [72, 73]. In contrast, abnormal ghrelin levels in obese individuals may develop due to insensitivity to ghrelin, which inversely affects the insulin levels [74, 75]. Ghrelin is a negative regulator of leptin, and leptin-induced insulin sensitivity is regulated by ghrelin. When ghrelin binds to growth hormone secretagogue receptor (GHS-R), it stimulates the hypothalamus to increase NPY and AgRP, and increases food uptake [76]. The G-protein coupled receptor, GHS-R1a triggers transcriptional activation of these neuropeptides through intracellular calcium signaling via calmodulin-dependent protein kinase-2 (CamK2), and 5-adenosine monophosphate-activated protein kinase (AMPK)-mediated MAPKs pathway through Sirtuin 1 and p53 [77–79]. Effect of ghrelin on neuropeptides suggested that AgRP reduces spontaneous physical activity (SPA) with increased food uptake, however, NPY moderately induces SPA in mice [77]. Peptide YY (PYY) regulates obestatin and inhibits ghrelin action on food intake [76, 80]. Studies suggest that increased insulin levels regulate body weight by down-regulating ghrelin levels in obese women [81, 82]. Maternal high fat diet (HFD) during pregnancy and lactation influences ghrelin in both dams and their offspring which has been shown to increase with age of neonates [80, 83, 84]. It has been demonstrated that ghrelin improves left ventricular (LV) dysfunction and attenuates cardiac cachexia in rats with chronic heart failure (CHF) [85]. Fall in plasma ghrelin levels in lactating obese dams indicated that prolactin or oxytocin release may have suppressed ghrelin during lactation, resulting in cardiac dysfunction in offspring. In contrast, pregnant spontaneous hypertensive rats (SHR) showed an increased plasma levels of ghrelin. However, downregulation of leptin receptors in pregnant SHR presented lower mRNA ghrelin in placenta but not in the stomach [86]. A study conducted on 28 infants with congenital heart defects showed that increased ghrelin level is associated with heart failure in infants [87]. Lower ghrelin levels in obesity may induce oxidative stress and enhance atherosclerosis. Plasma levels of ghrelin has a positive correlation with the development of carotid artery atherosclerosis and other cardiovascular problems.

Like ghrelin, obestatin also plays an important role in the regulation of metabolic functions. Under physiological conditions, obestatin binds to G-protein-coupled receptor GPR39 and regulates expression of ghrelin [88, 89]. Increased maternal obestatin negatively controls ghrelin in pregnancy [90]. Obestatin has been reported to inhibit glucose transport in adipose tissue by downregulating GLUT-4 together with sirtuin 1 to regulate insulin signaling [91]. Moreover, maternal high-fat diet during pregnancy lowers plasma ghrelin/obestatin ratio in offspring [83, 90] which may be linked to the development of diabetic cardiovascular complications [92].

Role of Adiponectin

In addition to leptin and insulin, adipose tissue also produces adiponectin that regulates glucose and fatty acid oxidation. Caloric restriction contributes to elevated plasma levels of adiponectin and increased energy expenditure [93]. Downstream signaling is based on adiponectin's binding affinity to its receptors, AdipoR1 and AdipoR2, which bind with globular adiponectin (gADN) and full length adiponectin (fADN) respectively [94]. AdipoR1 promotes downstream AMPK pathway to control appetite by increasing glucose utilization [95] (Fig. 3.2) and fatty acid oxidation, which is controlled by peroxisome proliferator-activated receptor α (PPARα). Activation of p38-MAPK by AdipoR2 inhibits the insulin-mediated IGF-1 signaling pathway. Adiponectin receptor activation also triggers ERK1/2 signaling in offspring of obese dams infused with adiponectin [96].

Obese individuals showed a decrease in levels of serum adiponectin and had developed coronary artery disease [97, 98]. Low plasma levels of adiponectin in obese pregnant females is associated with increased fetal body weight [96, 99]. Adiponectin secreted from placenta controls placental nutrient transport during fetal development. Low levels of adiponectin in obese pregnant women is also correlated with poor development of placenta. Furthermore, decreased adiponectin is associated with increased insulin resistance via p38-MAPK. Adiponectin supplementation during late gestation period reverses increased insulin resistance and fetal body weight which, in turn, reduces cardiovascular risk suggesting that maternal adiponectin is required to prevent offspring from long-term cardiac dysfunction [100]. Surprisingly, some male offspring of these dams developed cardiac hypertrophy and some females developed cardiac fibrosis. Studies also suggested that leptin-induced adiponectin reverses insulin resistance, T2D and cardiac dysfunction [101].

Inflammation-Mediated Responses

Metabolic events of obesity trigger cellular inflammation. Proinflammatory cells such as macrophages (MQ) and T-cells in response to dietary free fatty acid (FFA) hamper insulin receptor signaling and energy balance [102, 103]. These FFAs are known to be recognized by molecular pattern receptors such as toll-like receptors (TLRs) to activate downstream innate inflammatory signals that induces proinflammatory gene expression in adipocytes. Studies in TLR4 knockout (KO) female mice have shown inflammation to correlate with increased obesity. On the other hand, these mice showed protection against HFD-induced insulin resistance via degradation of IkBα following NF-kB activation. These mice also exhibited an inhibition of tyrosine phosphorylated IRS-1 and docking of p85 to its receptor thus impairing insulin signaling by reducing glucose turnover [103]. TLR4-deficient resident macrophages in adipose tissue lacks TNF-α and IL-6 mRNA suggesting that IL-6 and TNF-α contributes in obesity-induced low-grade inflammation which, in turn, is mediated by SOCS3-mediated TLR4/MyD88 signaling [102].

Leptin and insulin regulate hypothalamic inflammation via activation of suppressor of cytokine signaling (SOCS)-3 and is considered as a negative feedback regulator of inflammation during obesity [104, 105]. HFD supplementation in obese animals increases SOCS3 that inhibits leptin and insulin signaling by direct binding to their cognate receptors to promote IRS-mediated JAK/STAT or IKKβ/NF-κB pathways [105]. Another important adipocyte secretory hormone, resistin is also linked to obesity, insulin resistance and cardiovascular disorders. Overexpression of resistin reduces LC3II-mediated autophagy and regulates innate inflammatory responses. Obesity-mediated inflammation is regulated by direct binding of resistin to innate pattern receptor, TLR4/MyD88 that initiates downstream proinflammatory signaling pathways [106]. Adverse effects by resistin/TLR4-mediated inflammatory signal affecting energy homeostasis in obesity is linked with the progression of metabolic disorders.

Increased adiposity in offspring of obese mothers have correlation with increased insulin resistance and inflammatory markers. Offspring of mothers with diabetes are at risk of obesity in their later life. Studies showed that, at birth, neonates of type 1 diabetic pregnancy have increased placental pro-inflammatory genes [107] suggesting that these offspring have greater risk for the development of type 2 diabetes, obesity and cardiovascular complications. Maternal peripheral blood as well as blood collected from umbilical veins from obese mothers showed increased IL-6 and C-reactive protein (CRP) together with increased leptin levels [71], suggesting that leptin can be a major factor that controls CRP and IL-6 levels. Nonetheless, there was no significant difference in TNF-α and adiponectin levels in these neonates of obese mothers [71, 108]. CRP and IL-6 are positively controlled by adiponectin and insulin, and thus, contribute in the development of neonatal insulin resistance. Differential roles of adiponectin showed that gADN increased secretion of IL-6 and TNF-α in placenta, whereas fADN modulate these IL-6 production with an increased TNF-α [94]. A study reflected that increased CD24 presents on placenta and triggers proinflammatory cytokines and chemokines such as IL-1β and IL8 [109]. These inflammatory responses in neonates of obese women is associated with endothelial cell dysfunction. Upregulation of E-selectin, intercellular adhesion molecule (ICAM-1), and vascular cell adhesion molecule-1 (VCAM-1) promote proinflammatory cytokines and endothelial cell differentiation [110], which is associated with hypertension, atherosclerosis and, later can contribute to development of cardiovascular complications in obese women and a long-term effect in the offspring after birth.

Oxidative Stress-Mediated Responses

Oxidative stress (OS) is one of the most common pathological states that occurs in conditions including cardiovascular disease, diabetes and obesity. During cellular metabolism, reactive oxygen species (ROS) are generated and activate many signaling pathways such as MAPK, Akt/PI3K, PTEN and others [111]. In obesity, there was an increased level of free 8-epi-prostaglandin F(2α) (8-epi-PGF2α) suggesting that obesity is linked to OS. On the other hand, low level of ghrelin in obesity may have direct effect on OS and related heart diseases [112]. Studies have shown that HFD increases ROS together with prostaglandin E2 (PGE2) in hypothalamus [113]. During pregnancy, there are many pathophysiological changes that occur in obese women due to multiple events of OS. Male offspring of these obese rats fed on HFD have shown to affect male fertility due to increased OS [114].

Obesity-induced oxidative stress promotes inflammatory cytokines [115] and cytokine-inducible NOS (iNOS) production [116]. Increased ROS-generation in obese individuals causes higher nitric oxide synthases (NOS) and NADPH oxidase activity [113, 117]. iNOS promotes peroxynitrite (NOO^-) production and causes cardiac dysfunction. Increased cytokine levels together with nitrites in these neonates may cause serious cardiovascular complications. On the other hand, antioxidant enzyme, superoxide dismutase (SOD) activity was also noted higher in obese pregnant mothers and in their offspring [118]. However, lower SOD levels in neonates under intrauterine hypoxia or peroxidation conditions were also reported [119]. Maternal obesity-induced OS may negatively impact fetal development including major heart defects.

Clinical Management of Maternal Obesity

As the obesity epidemic is increasing in women of reproductive age, it is critical to establish an intervention to prevent or reduce maternal cardiovascular risks as well as risks for future generations. Reducing pre-pregnancy BMI will likely benefit both the mother and the offspring by improving pregnancy outcome. As such, the Institute of Medicine (IOM) established a guideline for gestational weight gain for optimizing pregnancy outcome according to pre-pregnancy weight [6].

In addition to optimizing weight gain, increasing physical activity may also reduce maternal and fetal risks. Results from multiple randomized control trial shows that dietary interventions and physical activity has the potential to decrease gestational weight gain and reduces adverse pregnancy outcome [120]. Randomised controlled trials showed that the use of metformin, a drug that lowers blood glucose and increases insulin sensitivity minimally reduced gestational weight gain but failed to decrease macrosomia and did not show any fetal improvement [121]. A recent study showed that there is a decline in plasma adiponectin levels in pregnant women with obesity. In an animal model, low adiponectin levels in obese pregnant mice was linked to placental dysfunction, fetal over-growth, altered metabolism and cardiac dysfunction in the offspring; and boosting adiponectin levels prevented these anomalies [100]. Thus, restoring adiponectin levels in obese mothers has a potential to improve pregnancy outcome.

For management of gestational weight gain, the National Institute for Health and Care Excellence (NICE), UK suggests a balanced diet and moderate physical exercise 30 min daily [122]. Ultimately, raising greater awareness of the risk of maternal obesity, life-style change, having a healthy diet, keeping weight gain under check, seeking doctor's advice; and regular check-ups for diabetes, blood pressure etc. and screening for structural defects in fetus will improve pregnancy outcome.

It is evident from data obtained from studies conducted in both animals and humans that pre-pregnancy as well as gestational obesity can pose deleterious cardiovascular outcomes not just for the newborn, but can continue to generate debilitating health conditions for the mother and growing offspring. While a multitude of factors have been shown to be responsible for these conditions, there still exists a lacunae in our understanding of the mechanisms underlying maternal obesity and its consequences. Further research in this area will warrant the development of highly effective therapies to help maintain a healthier weight during pre- and post-pregnancy stages in the mother and offspring.

Acknowledgements We thank Dr. Jennifer Major (University of Colorado Anschutz Medical Campus) for valuable insight on this manuscript. FJ was supported by a pre-doctoral award from the Canadian Institutes of Health Research (FRN-GSD167027). RAB is supported by a fellowship from the Canadian Institutes of Health Research (FRN-395620). Figures were created using Biorender.com.

References

1. Peeters A, Barendregt JJ, Willekens F, Mackenbach JP, Al Mamun A, Bonneux L, Nedcom tNE, Demography Compression of Morbidity Research G (2003) Obesity in adulthood and its consequences for life expectancy: a life-table analysis. Ann Intern Med 138 (1):24–32. https://doi.org/10.7326/0003-4819-138-1-200301070-00008
2. Grundy SM, Benjamin IJ, Burke GL, Chait A, Eckel RH, Howard BV, Mitch W, Smith SC Jr, Sowers JR (1999) Diabetes and cardiovascular disease: a statement for healthcare professionals from the American Heart Association. Circulation 100(10):1134–1146. https://doi.org/10.1161/01.cir.100.10.1134
3. Mendy VL, Vargas R, Payton M (2017) Trends in mortality rates by subtypes of heart disease in Mississippi, 1980–2013. BMC Cardiovasc Disord 17(1):158. https://doi.org/10.1186/s12872-017-0593-3
4. Cheng YJ, Imperatore G, Geiss LS, Saydah SH, Albright AL, Ali MK, Gregg EW (2018) Trends and disparities in cardiovascular mortality among U.S. adults with and without self-reported diabetes, 1988–2015. Diabetes Care 41(11):2306–2315. https://doi.org/10.2337/dc18-0831
5. Fuchs F, Senat MV, Rey E, Balayla J, Chaillet N, Bouyer J, Audibert F (2017) Impact of maternal obesity on the incidence of pregnancy complications in France and Canada. Sci Rep 7(1):10859. https://doi.org/10.1038/s41598-017-11432-5
6. Rasmussen KM Yaktine AL (eds) (2009) Weight gain during pregnancy: reexamining the guidelines. The national academies collection: reports funded by National Institutes of Health. Washington, DC. https://doi.org/10.17226/12584
7. Leddy MA, Power ML, Schulkin J (2008) The impact of maternal obesity on maternal and fetal health. Rev Obstet Gynecol 1(4):170–178
8. O'Doherty MG, Jorgensen T, Borglykke A, Brenner H, Schottker B, Wilsgaard T, Siganos G, Kavousi M, Hughes M, Muezzinler A, Holleczek B, Franco OH, Hofman A, Boffetta P,

Trichopoulou A, Kee F (2014) Repeated measures of body mass index and C-reactive protein in relation to all-cause mortality and cardiovascular disease: results from the consortium on health and ageing network of cohorts in Europe and the United States (CHANCES). Eur J Epidemiol 29(12):887–897. https://doi.org/10.1007/s10654-014-9954-8
9. Gunnell AS, Knuiman MW, Divitini ML, Cormie P (2014) Leisure time physical activity and long-term cardiovascular and cancer outcomes: the Busselton Health Study. Eur J Epidemiol 29(11):851–857. https://doi.org/10.1007/s10654-014-9963-7
10. O'Brien TE, Ray JG, Chan WS (2003) Maternal body mass index and the risk of preeclampsia: a systematic overview. Epidemiology 14(3):368–374. https://doi.org/10.1097/00001648-200305000-00020
11. Qiu C, Luthy DA, Zhang C, Walsh SW, Leisenring WM, Williams MA (2004) A prospective study of maternal serum C-reactive protein concentrations and risk of preeclampsia. Am J Hypertens 17(2):154–160. https://doi.org/10.1016/j.amjhyper.2003.09.011
12. Kauma S, Takacs P, Scordalakes C, Walsh S, Green K, Peng T (2002) Increased endothelial monocyte chemoattractant protein-1 and interleukin-8 in preeclampsia. Obstet Gynecol 100(4):706–714. https://doi.org/10.1016/s0029-7844(02)02169-5
13. Walsh SW (2007) Obesity: a risk factor for preeclampsia. Trends Endocrinol Metab 18(10):365–370. https://doi.org/10.1016/j.tem.2007.09.003
14. Loche E, Blackmore HL, Carpenter AA, Beeson JH, Pinnock A, Ashmore TJ, Aiken CE, de Almeida-Faria J, Schoonejans JM, Giussani DA, Fernandez-Twinn DS, Ozanne SE (2018) Maternal diet-induced obesity programmes cardiac dysfunction in male mice independently of post-weaning diet. Cardiovasc Res 114(10):1372–1384. https://doi.org/10.1093/cvr/cvy082
15. Johansson S, Villamor E, Altman M, Bonamy AK, Granath F, Cnattingius S (2014) Maternal overweight and obesity in early pregnancy and risk of infant mortality: a population based cohort study in Sweden. BMJ 349:g6572. https://doi.org/10.1136/bmj.g6572
16. Stothard KJ, Tennant PW, Bell R, Rankin J (2009) Maternal overweight and obesity and the risk of congenital anomalies: a systematic review and meta-analysis. JAMA 301(6):636–650. https://doi.org/10.1001/jama.2009.113
17. Mills JL, Troendle J, Conley MR, Carter T, Druschel CM (2010) Maternal obesity and congenital heart defects: a population-based study. Am J Clin Nutr 91(6):1543–1549. https://doi.org/10.3945/ajcn.2009.28865
18. Ouyang F, Parker M, Cerda S, Pearson C, Fu L, Gillman MW, Zuckerman B, Wang X (2013) Placental weight mediates the effects of prenatal factors on fetal growth: the extent differs by preterm status. Obesity (Silver Spring) 21(3):609–620. https://doi.org/10.1002/oby.20254
19. Ditchfield AM, Desforges M, Mills TA, Glazier JD, Wareing M, Mynett K, Sibley CP, Greenwood SL (2015) Maternal obesity is associated with a reduction in placental taurine transporter activity. Int J Obes (Lond) 39(4):557–564. https://doi.org/10.1038/ijo.2014.212
20. Mele J, Muralimanoharan S, Maloyan A, Myatt L (2014) Impaired mitochondrial function in human placenta with increased maternal adiposity. Am J Physiol Endocrinol Metab 307(5):E419-425. https://doi.org/10.1152/ajpendo.00025.2014
21. Huang L, Liu J, Feng L, Chen Y, Zhang J, Wang W (2014) Maternal prepregnancy obesity is associated with higher risk of placental pathological lesions. Placenta 35(8):563–569. https://doi.org/10.1016/j.placenta.2014.05.006
22. Aye IL, Lager S, Ramirez VI, Gaccioli F, Dudley DJ, Jansson T, Powell TL (2014) Increasing maternal body mass index is associated with systemic inflammation in the mother and the activation of distinct placental inflammatory pathways. Biol Reprod 90(6):129. https://doi.org/10.1095/biolreprod.113.116186
23. Malti N, Merzouk H, Merzouk SA, Loukidi B, Karaouzene N, Malti A, Narce M (2014) Oxidative stress and maternal obesity: feto-placental unit interaction. Placenta 35(6):411–416. https://doi.org/10.1016/j.placenta.2014.03.010
24. Zera CA, Seely EW, Wilkins-Haug LE, Lim KH, Parry SI, McElrath TF (2014) The association of body mass index with serum angiogenic markers in normal and abnormal pregnancies. Am J Obstet Gynecol 211 (3):247 e241–247. https://doi.org/10.1016/j.ajog.2014.03.020

25. Lee KK, Raja EA, Lee AJ, Bhattacharya S, Bhattacharya S, Norman JE, Reynolds RM (2015) Maternal obesity during pregnancy associates with premature mortality and major cardiovascular events in later life. Hypertension 66(5):938–944. https://doi.org/10.1161/HYPERTENSIONAHA.115.05920
26. Mongraw-Chaffin ML, Anderson CA, Clark JM, Bennett WL (2014) Prepregnancy body mass index and cardiovascular disease mortality: the Child Health and Development Studies. Obesity (Silver Spring) 22(4):1149–1156. https://doi.org/10.1002/oby.20633
27. Bellamy L, Casas JP, Hingorani AD, Williams DJ (2007) Pre-eclampsia and risk of cardiovascular disease and cancer in later life: systematic review and meta-analysis. BMJ 335(7627):974. https://doi.org/10.1136/bmj.39335.385301.BE
28. Martin-Gronert MS, Ozanne SE (2012) Mechanisms underlying the developmental origins of disease. Rev Endocr Metab Disord 13(2):85–92. https://doi.org/10.1007/s11154-012-9210-z
29. Giussani DA, Davidge ST (2013) Developmental programming of cardiovascular disease by prenatal hypoxia. J Dev Orig Health Dis 4(5):328–337. https://doi.org/10.1017/S204017441300010X
30. Samuelsson AM, Matthews PA, Argenton M, Christie MR, McConnell JM, Jansen EH, Piersma AH, Ozanne SE, Twinn DF, Remacle C, Rowlerson A, Poston L, Taylor PD (2008) Diet-induced obesity in female mice leads to offspring hyperphagia, adiposity, hypertension, and insulin resistance: a novel murine model of developmental programming. Hypertension 51(2):383–392. https://doi.org/10.1161/HYPERTENSIONAHA.107.101477
31. Fernandez-Twinn DS, Blackmore HL, Siggens L, Giussani DA, Cross CM, Foo R, Ozanne SE (2012) The programming of cardiac hypertrophy in the offspring by maternal obesity is associated with hyperinsulinemia, AKT, ERK, and mTOR activation. Endocrinology 153(12):5961–5971. https://doi.org/10.1210/en.2012-1508
32. Gaudet L, Ferraro ZM, Wen SW, Walker M (2014) Maternal obesity and occurrence of fetal macrosomia: a systematic review and meta-analysis. Biomed Res Int 2014:640291. https://doi.org/10.1155/2014/640291
33. Ludwig DS, Currie J (2010) The association between pregnancy weight gain and birthweight: a within-family comparison. Lancet 376(9745):984–990. https://doi.org/10.1016/S0140-6736(10)60751-9
34. Poston L, Harthoorn LF, Van Der Beek EM, Contributors to the IEW (2011) Obesity in pregnancy: implications for the mother and lifelong health of the child. A consensus statement. Pediatr Res 69 (2):175–180. https://doi.org/10.1203/PDR.0b013e3182055ede
35. Yu Z, Han S, Zhu J, Sun X, Ji C, Guo X (2013) Pre-pregnancy body mass index in relation to infant birth weight and offspring overweight/obesity: a systematic review and meta-analysis. PLoS ONE 8(4):e61627. https://doi.org/10.1371/journal.pone.0061627
36. Tie HT, Xia YY, Zeng YS, Zhang Y, Dai CL, Guo JJ, Zhao Y (2014) Risk of childhood overweight or obesity associated with excessive weight gain during pregnancy: a meta-analysis. Arch Gynecol Obstet 289(2):247–257. https://doi.org/10.1007/s00404-013-3053-z
37. Gaillard R, Steegers EA, Franco OH, Hofman A, Jaddoe VW (2015) Maternal weight gain in different periods of pregnancy and childhood cardio-metabolic outcomes. The Generation R Study. Int J Obes (Lond) 39 (4):677–685. https://doi.org/10.1038/ijo.2014.175
38. Hochner H, Friedlander Y, Calderon-Margalit R, Meiner V, Sagy Y, Avgil-Tsadok M, Burger A, Savitsky B, Siscovick DS, Manor O (2012) Associations of maternal prepregnancy body mass index and gestational weight gain with adult offspring cardiometabolic risk factors: the Jerusalem Perinatal Family Follow-up Study. Circulation 125(11):1381–1389. https://doi.org/10.1161/CIRCULATIONAHA.111.070060
39. Schack-Nielsen L, Michaelsen KF, Gamborg M, Mortensen EL, Sorensen TI (2010) Gestational weight gain in relation to offspring body mass index and obesity from infancy through adulthood. Int J Obes (Lond) 34(1):67–74. https://doi.org/10.1038/ijo.2009.206
40. Hrolfsdottir L, Rytter D, Olsen SF, Bech BH, Maslova E, Henriksen TB, Halldorsson TI (2015) Gestational weight gain in normal weight women and offspring cardio-metabolic risk factors at 20 years of age. Int J Obes (Lond) 39(4):671–676. https://doi.org/10.1038/ijo.2014.179

41. Rooney BL, Mathiason MA, Schauberger CW (2011) Predictors of obesity in childhood, adolescence, and adulthood in a birth cohort. Matern Child Health J 15(8):1166–1175. https://doi.org/10.1007/s10995-010-0689-1
42. Mamun AA, O'Callaghan M, Callaway L, Williams G, Najman J, Lawlor DA (2009) Associations of gestational weight gain with offspring body mass index and blood pressure at 21 years of age: evidence from a birth cohort study. Circulation 119(13):1720–1727. https://doi.org/10.1161/CIRCULATIONAHA.108.813436
43. Reynolds RM, Allan KM, Raja EA, Bhattacharya S, McNeill G, Hannaford PC, Sarwar N, Lee AJ, Bhattacharya S, Norman JE (2013) Maternal obesity during pregnancy and premature mortality from cardiovascular event in adult offspring: follow-up of 1 323 275 person years. BMJ 347:f4539. https://doi.org/10.1136/bmj.f4539
44. Tanvig M (2014) Offspring body size and metabolic profile–effects of lifestyle intervention in obese pregnant women. Dan Med J 61(7):B4893
45. Gilboa SM, Correa A, Botto LD, Rasmussen SA, Waller DK, Hobbs CA, Cleves MA, Riehle-Colarusso TJ, National Birth Defects Prevention S (2010) Association between prepregnancy body mass index and congenital heart defects. Am J Obstet Gynecol 202(1):51 e51–51 e10. https://doi.org/10.1016/j.ajog.2009.08.005
46. Alpert MA (2001) Obesity cardiomyopathy: pathophysiology and evolution of the clinical syndrome. Am J Med Sci 321(4):225–236. https://doi.org/10.1097/00000441-200104000-00003
47. Farooqi IS, Jebb SA, Langmack G, Lawrence E, Cheetham CH, Prentice AM, Hughes IA, McCamish MA, O'Rahilly S (1999) Effects of recombinant leptin therapy in a child with congenital leptin deficiency. N Engl J Med 341(12):879–884. https://doi.org/10.1056/NEJM199909163411204
48. Zhao J, Townsend KL, Schulz LC, Kunz TH, Li C, Widmaier EP (2004) Leptin receptor expression increases in placenta, but not hypothalamus, during gestation in Mus musculus and Myotis lucifugus. Placenta 25(8–9):712–722. https://doi.org/10.1016/j.placenta.2004.01.017
49. Aka N, Atalay S, Sayharman S, Kilic D, Kose G, Kucukozkan T (2006) Leptin and leptin receptor levels in pregnant women with hyperemesis gravidarum. Aust N Z J Obstet Gynaecol 46(4):274–277. https://doi.org/10.1111/j.1479-828X.2006.00590.x
50. Bouret SG, Draper SJ, Simerly RB (2004) Formation of projection pathways from the arcuate nucleus of the hypothalamus to hypothalamic regions implicated in the neural control of feeding behavior in mice. J Neurosci 24(11):2797–2805. https://doi.org/10.1523/JNEUROSCI.5369-03.2004
51. Bouret SG, Draper SJ, Simerly RB (2004) Trophic action of leptin on hypothalamic neurons that regulate feeding. Science 304(5667):108–110. https://doi.org/10.1126/science.1095004
52. Chen H, Simar D, Lambert K, Mercier J, Morris MJ (2008) Maternal and postnatal overnutrition differentially impact appetite regulators and fuel metabolism. Endocrinology 149(11):5348–5356. https://doi.org/10.1210/en.2008-0582
53. Ferezou-Viala J, Roy AF, Serougne C, Gripois D, Parquet M, Bailleux V, Gertler A, Delplanque B, Djiane J, Riottot M, Taouis M (2007) Long-term consequences of maternal high-fat feeding on hypothalamic leptin sensitivity and diet-induced obesity in the offspring. Am J Physiol Regul Integr Comp Physiol 293(3):R1056-1062. https://doi.org/10.1152/ajpregu.00117.2007
54. Kirk SL, Samuelsson AM, Argenton M, Dhonye H, Kalamatianos T, Poston L, Taylor PD, Coen CW (2009) Maternal obesity induced by diet in rats permanently influences central processes regulating food intake in offspring. PLoS ONE 4(6):e5870. https://doi.org/10.1371/journal.pone.0005870
55. Ahima RS, Prabakaran D, Flier JS (1998) Postnatal leptin surge and regulation of circadian rhythm of leptin by feeding. Implications for energy homeostasis and neuroendocrine function. J Clin Invest 101 (5):1020–1027. https://doi.org/10.1172/JCI1176
56. Wada N, Hirako S, Takenoya F, Kageyama H, Okabe M, Shioda S (2014) Leptin and its receptors. J Chem Neuroanat 61–62:191–199. https://doi.org/10.1016/j.jchemneu.2014.09.002

57. Lee JY, Muenzberg H, Gavrilova O, Reed JA, Berryman D, Villanueva EC, Louis GW, Leininger GM, Bertuzzi S, Seeley RJ, Robinson GW, Myers MG, Hennighausen L (2008) Loss of cytokine-STAT5 signaling in the CNS and pituitary gland alters energy balance and leads to obesity. PLoS ONE 3(2):e1639. https://doi.org/10.1371/journal.pone.0001639
58. Myers MG, Cowley MA, Munzberg H (2008) Mechanisms of leptin action and leptin resistance. Annu Rev Physiol 70:537–556. https://doi.org/10.1146/annurev.physiol.70.113006.100707
59. Xu AW, Ste-Marie L, Kaelin CB, Barsh GS (2007) Inactivation of signal transducer and activator of transcription 3 in proopiomelanocortin (Pomc) neurons causes decreased pomc expression, mild obesity, and defects in compensatory refeeding. Endocrinology 148(1):72–80. https://doi.org/10.1210/en.2006-1119
60. Holliday R (1991) A re-examination of the effects of ionizing radiation on lifespan and transformation of human diploid fibroblasts. Mutat Res 256(2–6):295–302. https://doi.org/10.1016/0921-8734(91)90020-c
61. Leifheit-Nestler M, Wagner NM, Gogiraju R, Didie M, Konstantinides S, Hasenfuss G, Schafer K (2013) Importance of leptin signaling and signal transducer and activator of transcription-3 activation in mediating the cardiac hypertrophy associated with obesity. J Transl Med 11:170. https://doi.org/10.1186/1479-5876-11-170
62. Pedersen BA, Wang W, Taylor JF, Khattab OS, Chen YH, Edwards RA, Yazdi PG, Wang PH (2015) Hepatic proteomic analysis revealed altered metabolic pathways in insulin resistant Akt1(+/-)/Akt2(-/-) mice. Metabolism 64(12):1694–1703. https://doi.org/10.1016/j.metabol.2015.09.008
63. Niswender KD, Schwartz MW (2003) Insulin and leptin revisited: adiposity signals with overlapping physiological and intracellular signaling capabilities. Front Neuroendocrinol 24(1):1–10. https://doi.org/10.1016/s0091-3022(02)00105-x
64. Cusin I, Zakrzewska KE, Boss O, Muzzin P, Giacobino JP, Ricquier D, Jeanrenaud B, Rohner-Jeanrenaud F (1998) Chronic central leptin infusion enhances insulin-stimulated glucose metabolism and favors the expression of uncoupling proteins. Diabetes 47(7):1014–1019. https://doi.org/10.2337/diabetes.47.7.1014
65. Burgos-Ramos E, Canelles S, Frago LM, Chowen JA, Arilla-Ferreiro E, Argente J, Barrios V (2016) Improvement in glycemia after glucose or insulin overload in leptin-infused rats is associated with insulin-related activation of hepatic glucose metabolism. Nutr Metab (Lond) 13:19. https://doi.org/10.1186/s12986-016-0079-9
66. Lawlor MA, Alessi DR (2001) PKB/Akt: a key mediator of cell proliferation, survival and insulin responses? J Cell Sci 114(Pt 16):2903–2910
67. Li J, An R, Lai S, Li L, Liu S, Xu H (2019) Dysregulation of PP2A-Akt interaction contributes to Sucrose non-fermenting related kinase (SNRK) deficiency induced insulin resistance in adipose tissue. Mol Metab 28:26–35. https://doi.org/10.1016/j.molmet.2019.07.009
68. Lee MJ, Wang Y, Ricci MR, Sullivan S, Russell CD, Fried SK (2007) Acute and chronic regulation of leptin synthesis, storage, and secretion by insulin and dexamethasone in human adipose tissue. Am J Physiol Endocrinol Metab 292(3):E858-864. https://doi.org/10.1152/ajpendo.00439.2006
69. Samuelsson AM, Morris A, Igosheva N, Kirk SL, Pombo JM, Coen CW, Poston L, Taylor PD (2010) Evidence for sympathetic origins of hypertension in juvenile offspring of obese rats. Hypertension 55(1):76–82. https://doi.org/10.1161/HYPERTENSIONAHA.109.139402
70. Mingrone G, Manco M, Mora ME, Guidone C, Iaconelli A, Gniuli D, Leccesi L, Chiellini C, Ghirlanda G (2008) Influence of maternal obesity on insulin sensitivity and secretion in offspring. Diabetes Care 31(9):1872–1876. https://doi.org/10.2337/dc08-0432
71. Catalano PM, Presley L, Minium J, Hauguel-de Mouzon S (2009) Fetuses of obese mothers develop insulin resistance in utero. Diabetes Care 32(6):1076–1080. https://doi.org/10.2337/dc08-2077
72. Abdemur A, Slone J, Berho M, Gianos M, Szomstein S, Rosenthal RJ (2014) Morphology, localization, and patterns of ghrelin-producing cells in stomachs of a morbidly obese population. Surg Laparosc Endosc Percutan Tech 24(2):122–126. https://doi.org/10.1097/SLE.0b013e318290167a

73. Vincent RP, le Roux CW (2008) Changes in gut hormones after bariatric surgery. Clin Endocrinol (Oxf) 69(2):173–179. https://doi.org/10.1111/j.1365-2265.2007.03164.x
74. Mequinion M, Langlet F, Zgheib S, Dickson S, Dehouck B, Chauveau C, Viltart O (2013) Ghrelin: central and peripheral implications in anorexia nervosa. Front Endocrinol (Lausanne) 4:15. https://doi.org/10.3389/fendo.2013.00015
75. Korek E, Krauss H, Gibas-Dorna M, Kupsz J, Piatek M, Piatek J (2013) Fasting and postprandial levels of ghrelin, leptin and insulin in lean, obese and anorexic subjects. Prz Gastroenterol 8(6):383–389. https://doi.org/10.5114/pg.2013.39922
76. Pfluger PT, Castaneda TR, Heppner KM, Strassburg S, Kruthaupt T, Chaudhary N, Halem H, Culler MD, Datta R, Burget L, Tschop MH, Nogueiras R, Perez-Tilve D (2011) Ghrelin, peptide YY and their hypothalamic targets differentially regulate spontaneous physical activity. Physiol Behav 105(1):52–61. https://doi.org/10.1016/j.physbeh.2011.04.025
77. Kohno D, Gao HZ, Muroya S, Kikuyama S, Yada T (2003) Ghrelin directly interacts with neuropeptide-Y-containing neurons in the rat arcuate nucleus: Ca2+ signaling via protein kinase A and N-type channel-dependent mechanisms and cross-talk with leptin and orexin. Diabetes 52(4):948–956. https://doi.org/10.2337/diabetes.52.4.948
78. Minokoshi Y, Kim YB, Peroni OD, Fryer LG, Muller C, Carling D, Kahn BB (2002) Leptin stimulates fatty-acid oxidation by activating AMP-activated protein kinase. Nature 415(6869):339–343. https://doi.org/10.1038/415339a
79. Stevanovic D, Trajkovic V, Muller-Luhlhoff S, Brandt E, Abplanalp W, Bumke-Vogt C, Liehl B, Wiedmer P, Janjetovic K, Starcevic V, Pfeiffer AF, Al-Hasani H, Tschop MH, Castaneda TR (2013) Ghrelin-induced food intake and adiposity depend on central mTORC1/S6K1 signaling. Mol Cell Endocrinol 381(1–2):280–290. https://doi.org/10.1016/j.mce.2013.08.009
80. Taylor VJ, Patterson M, Ghatei MA, Bloom SR, Wilson CA (2009) Ghrelin and peptide YY (PYY) profiles in gastrointestinal tissues and the circulation of the rat during pregnancy and lactation. Peptides 30(12):2213–2220. https://doi.org/10.1016/j.peptides.2009.09.022
81. Soni AC, Conroy MB, Mackey RH, Kuller LH (2011) Ghrelin, leptin, adiponectin, and insulin levels and concurrent and future weight change in overweight, postmenopausal women. Menopause 18(3):296–301
82. Erdmann J, Lippl F, Wagenpfeil S, Schusdziarra V (2005) Differential association of basal and postprandial plasma ghrelin with leptin, insulin, and type 2 diabetes. Diabetes 54(5):1371–1378. https://doi.org/10.2337/diabetes.54.5.1371
83. Slupecka M, Romanowicz K, Wolinski J (2016) Maternal high-fat diet during pregnancy and lactation influences obestatin and ghrelin concentrations in milk and plasma of Wistar rat dams and their offspring. Int J Endocrinol 5739763. https://doi.org/10.1155/2016/5739763
84. Shibata K, Hosoda H, Kojima M, Kangawa K, Makino Y, Makino I, Kawarabayashi T, Futagami K, Gomita Y (2004) Regulation of ghrelin secretion during pregnancy and lactation in the rat: possible involvement of hypothalamus. Peptides 25(2):279–287. https://doi.org/10.1016/j.peptides.2004.01.011
85. Nagaya N, Uematsu M, Kojima M, Ikeda Y, Yoshihara F, Shimizu W, Hosoda H, Hirota Y, Ishida H, Mori H, Kangawa K (2001) Chronic administration of ghrelin improves left ventricular dysfunction and attenuates development of cardiac cachexia in rats with heart failure. Circulation 104(12):1430–1435. https://doi.org/10.1161/hc3601.095575
86. Raso GM, Bianco G, Iacono A, Esposito E, Autore G, Ferrante MC, Calignano A, Meli R (2007) Maternal adaptations to pregnancy in spontaneously hypertensive rats: leptin and ghrelin evaluation. J Endocrinol 194(3):611–619. https://doi.org/10.1677/JOE-07-0159
87. Kitamura T, Shimizu T, Ohkawa N, Oda H, Ohtsuki M, Kishiro M (2010) Effects of ghrelin on growth and cardiac function in infants with congenital heart disease. Pediatr Int 52(3):358–361. https://doi.org/10.1111/j.1442-200X.2009.02966.x
88. Iglesias MJ, Salgado A, Pineiro R, Rodino BK, Otero MF, Grigorian L, Gallego R, Dieguez C, Gualillo O, Gonzalez-Juanatey JR, Lago F (2007) Lack of effect of the ghrelin gene-derived peptide obestatin on cardiomyocyte viability and metabolism. J Endocrinol Invest 30(6):470–476. https://doi.org/10.1007/BF03346330

89. Reinehr T, de Sousa G, Roth CL (2008) Obestatin and ghrelin levels in obese children and adolescents before and after reduction of overweight. Clin Endocrinol (Oxf) 68(2):304–310. https://doi.org/10.1111/j.1365-2265.2007.03042.x
90. Vicennati V, Genghini S, De Iasio R, Pasqui F, Pagotto U, Pasquali R (2007) Circulating obestatin levels and the ghrelin/obestatin ratio in obese women. Eur J Endocrinol 157(3):295–301. https://doi.org/10.1530/EJE-07-0059
91. Pruszynska-Oszmalek E, Szczepankiewicz D, Hertig I, Skrzypski M, Sassek M, Kaczmarek P, Kolodziejski PA, Mackowiak P, Nowak KW, Strowski MZ, Wojciechowicz T (2013) Obestatin inhibits lipogenesis and glucose uptake in isolated primary rat adipocytes. J Biol Regul Homeost Agents 27(1):23–33
92. Cowan E, Burch KJ, Green BD, Grieve DJ (2016) Obestatin as a key regulator of metabolism and cardiovascular function with emerging therapeutic potential for diabetes. Br J Pharmacol 173(14):2165–2181. https://doi.org/10.1111/bph.13502
93. Cawthorn WP, Scheller EL, Learman BS, Parlee SD, Simon BR, Mori H, Ning X, Bree AJ, Schell B, Broome DT, Soliman SS, DelProposto JL, Lumeng CN, Mitra A, Pandit SV, Gallagher KA, Miller JD, Krishnan V, Hui SK, Bredella MA, Fazeli PK, Klibanski A, Horowitz MC, Rosen CJ, MacDougald OA (2014) Bone marrow adipose tissue is an endocrine organ that contributes to increased circulating adiponectin during caloric restriction. Cell Metab 20(2):368–375. https://doi.org/10.1016/j.cmet.2014.06.003
94. Jones HN, Jansson T, Powell TL (2010) Full-length adiponectin attenuates insulin signaling and inhibits insulin-stimulated amino Acid transport in human primary trophoblast cells. Diabetes 59(5):1161–1170. https://doi.org/10.2337/db09-0824
95. Yamauchi T, Kamon J, Minokoshi Y, Ito Y, Waki H, Uchida S, Yamashita S, Noda M, Kita S, Ueki K, Eto K, Akanuma Y, Froguel P, Foufelle F, Ferre P, Carling D, Kimura S, Nagai R, Kahn BB, Kadowaki T (2002) Adiponectin stimulates glucose utilization and fatty-acid oxidation by activating AMP-activated protein kinase. Nat Med 8(11):1288–1295. https://doi.org/10.1038/nm788
96. Aye IL, Powell TL, Jansson T (2013) Review: adiponectin–the missing link between maternal adiposity, placental transport and fetal growth? Placenta 34(Suppl):S40-45. https://doi.org/10.1016/j.placenta.2012.11.024
97. Diez JJ, Iglesias P (2003) The role of the novel adipocyte-derived hormone adiponectin in human disease. Eur J Endocrinol 148(3):293–300. https://doi.org/10.1530/eje.0.1480293
98. Ukkola O, Santaniemi M (2002) Adiponectin: a link between excess adiposity and associated comorbidities? J Mol Med (Berl) 80(11):696–702. https://doi.org/10.1007/s00109-002-0378-7
99. Paulsen ME, Rosario FJ, Wesolowski SR, Powell TL, Jansson T (2019) Normalizing adiponectin levels in obese pregnant mice prevents adverse metabolic outcomes in offspring. FASEB J 33(2):2899–2909. https://doi.org/10.1096/fj.201801015R
100. Vaughan OR, Rosario FJ, Powell TL, Jansson T (2019) Normalisation of circulating adiponectin levels in obese pregnant mice prevents cardiac dysfunction in adult offspring. Int J Obes (Lond). https://doi.org/10.1038/s41366-019-0374-4
101. Yamauchi T, Kamon J, Waki H, Terauchi Y, Kubota N, Hara K, Mori Y, Ide T, Murakami K, Tsuboyama-Kasaoka N, Ezaki O, Akanuma Y, Gavrilova O, Vinson C, Reitman ML, Kagechika H, Shudo K, Yoda M, Nakano Y, Tobe K, Nagai R, Kimura S, Tomita M, Froguel P, Kadowaki T (2001) The fat-derived hormone adiponectin reverses insulin resistance associated with both lipoatrophy and obesity. Nat Med 7(8):941–946. https://doi.org/10.1038/90984
102. Fessler MB, Rudel LL, Brown JM (2009) Toll-like receptor signaling links dietary fatty acids to the metabolic syndrome. Curr Opin Lipidol 20(5):379–385. https://doi.org/10.1097/MOL.0b013e32832fa5c4
103. Shi H, Kokoeva MV, Inouye K, Tzameli I, Yin H, Flier JS (2006) TLR4 links innate immunity and fatty acid-induced insulin resistance. J Clin Invest 116(11):3015–3025. https://doi.org/10.1172/JCI28898

104. Howard JK, Flier JS (2006) Attenuation of leptin and insulin signaling by SOCS proteins. Trends Endocrinol Metab 17(9):365–371. https://doi.org/10.1016/j.tem.2006.09.007
105. Bjorbaek C, El-Haschimi K, Frantz JD, Flier JS (1999) The role of SOCS-3 in leptin signaling and leptin resistance. J Biol Chem 274(42):30059–30065. https://doi.org/10.1074/jbc.274.42.30059
106. Fresno M, Alvarez R, Cuesta N (2011) Toll-like receptors, inflammation, metabolism and obesity. Arch Physiol Biochem 117(3):151–164. https://doi.org/10.3109/13813455.2011.562514
107. Radaelli T, Varastehpour A, Catalano P, Hauguel-de Mouzon S (2003) Gestational diabetes induces placental genes for chronic stress and inflammatory pathways. Diabetes 52(12):2951–2958. https://doi.org/10.2337/diabetes.52.12.2951
108. Nelson SM, Sattar N, Freeman DJ, Walker JD, Lindsay RS (2007) Inflammation and endothelial activation is evident at birth in offspring of mothers with type 1 diabetes. Diabetes 56(11):2697–2704. https://doi.org/10.2337/db07-0662
109. McDonald EA, Wolfe MW (2011) The pro-inflammatory role of adiponectin at the maternal-fetal interface. Am J Reprod Immunol 66(2):128–136. https://doi.org/10.1111/j.1600-0897.2010.00971.x
110. von der Thusen JH, Kuiper J, van Berkel TJ, Biessen EA (2003) Interleukins in atherosclerosis: molecular pathways and therapeutic potential. Pharmacol Rev 55(1):133–166. https://doi.org/10.1124/pr.55.1.5
111. Ray PD, Huang BW, Tsuji Y (2012) Reactive oxygen species (ROS) homeostasis and redox regulation in cellular signaling. Cell Signal 24(5):981–990. https://doi.org/10.1016/j.cellsig.2012.01.008
112. Suematsu M, Katsuki A, Sumida Y, Gabazza EC, Murashima S, Matsumoto K, Kitagawa N, Akatsuka H, Hori Y, Nakatani K, Togashi K, Yano Y, Adachi Y (2005) Decreased circulating levels of active ghrelin are associated with increased oxidative stress in obese subjects. Eur J Endocrinol 153(3):403–407. https://doi.org/10.1530/eje.1.01977
113. Zhang X, Dong F, Ren J, Driscoll MJ, Culver B (2005) High dietary fat induces NADPH oxidase-associated oxidative stress and inflammation in rat cerebral cortex. Exp Neurol 191(2):318–325. https://doi.org/10.1016/j.expneurol.2004.10.011
114. Alcala M, Gutierrez-Vega S, Castro E, Guzman-Gutierrez E, Ramos-Alvarez MP, Viana M (2018) Antioxidants and oxidative stress: focus in obese pregnancies. Front Physiol 9:1569. https://doi.org/10.3389/fphys.2018.01569
115. Calder PC, Ahluwalia N, Brouns F, Buetler T, Clement K, Cunningham K, Esposito K, Jonsson LS, Kolb H, Lansink M, Marcos A, Margioris A, Matusheski N, Nordmann H, O'Brien J, Pugliese G, Rizkalla S, Schalkwijk C, Tuomilehto J, Warnberg J, Watzl B, Winklhofer-Roob BM (2011) Dietary factors and low-grade inflammation in relation to overweight and obesity. Br J Nutr 106(Suppl 3):S5-78. https://doi.org/10.1017/S0007114511005460
116. Knowles RG, Moncada S (1994) Nitric oxide synthases in mammals. Biochem J 298(Pt 2):249–258. https://doi.org/10.1042/bj2980249
117. Savini I, Catani MV, Evangelista D, Gasperi V, Avigliano L (2013) Obesity-associated oxidative stress: strategies finalized to improve redox state. Int J Mol Sci 14(5):10497–10538. https://doi.org/10.3390/ijms140510497
118. Hernandez-Trejo M, Montoya-Estrada A, Torres-Ramos Y, Espejel-Nunez A, Guzman-Grenfell A, Morales-Hernandez R, Tolentino-Dolores M, Laresgoiti-Servitje E (2017) Oxidative stress biomarkers and their relationship with cytokine concentrations in overweight/obese pregnant women and their neonates. BMC Immunol 18(1):3. https://doi.org/10.1186/s12865-016-0184-6
119. Qin LN, Feng AH, Yang CH, Xing HY (2005) Clinical values of superoxide dismutase and malondialdehyde detection in cord blood of newborns with fetal distress. Zhonghua Fu Chan Ke Za Zhi 40(5):312–314
120. Thangaratinam S, Rogozinska E, Jolly K, Glinkowski S, Roseboom T, Tomlinson JW, Kunz R, Mol BW, Coomarasamy A, Khan KS (2012) Effects of interventions in pregnancy on maternal weight and obstetric outcomes: meta-analysis of randomised evidence. BMJ 344:e2088. https://doi.org/10.1136/bmj.e2088

121. Dodd JM, Grivell RM, Deussen AR, Hague WM (2018) Metformin for women who are overweight or obese during pregnancy for improving maternal and infant outcomes. Cochrane Database Syst Rev 7:CD010564. https://doi.org/10.1002/14651858.CD010564.pub2
122. Ma RCW, Schmidt MI, Tam WH, McIntyre HD, Catalano PM (2016) Clinical management of pregnancy in the obese mother: before conception, during pregnancy, and post partum. Lancet Diabetes Endocrinol 4(12):1037–1049. https://doi.org/10.1016/S2213-8587(16)30278-9

Chapter 4
Role of Liver X Receptor in Cardiovascular Diseases

Tamhida Masi, Ramesh K. Goyal, and Bhoomika M. Patel

Abstract Cardiovascular diseases are the leading cause of death worldwide including various complications like atherosclerosis, myocardial infarction, diabetic cardiomyopathy, cardiac hypertrophy and cardiac fibrosis. Looking into the limitations and side effects of interventional and non-interventional treatment strategies, liver X receptors (LXRs) can be the novel targets as treatment strategy for cardiac complication. Nuclear receptors like liver X receptors (LXRs) are known to regulate various physiological functions like cholesterol and carbohydrate metabolism, energy expenditure and inflammation. Cholesterol derivatives, oxysterols were the first endogenous ligand found to activate LXRs whereas T0901317 and GW3965 were the potential synthetic LXR agonist reported. Various evidences have suggested that LXR may exert their beneficial role in heart disease. We reviewed recent data that shows a direct role of LXR agonist in various cardiovascular diseases like atherosclerosis, myocardial infarction, diabetic cardiomyopathy, cardiac hypertrophy, fibrosis. These accumulating evidences support that LXRs may represent a novel potential therapeutic target for various cardiovascular diseases.

Keywords Cardiovascular diseases · Liver X receptors · Atherosclerosis · Myocardial ischemia · Diabetic cardiomyopathy · Cardiac hypertrophy · Fibrosis · GW3965 · T0901317

T. Masi · B. M. Patel (✉)
Institute of Pharmacy, Nirma University, Sarkhej-Gandhinagar Highway, Ahmedabad 382 481, Gujarat, India
e-mail: drbhoomikampatel@gmail.com

R. K. Goyal (✉)
Delhi Pharmaceutical Sciences Research University, Delhi, India
e-mail: goyalrk@gmail.com

P. S. Tappia et al. (eds.), *Biochemistry of Cardiovascular Dysfunction in Obesity*, Advances in Biochemistry in Health and Disease 20,
https://doi.org/10.1007/978-3-030-47336-5_4

Introduction

Cardiovascular diseases (CVDs) are the leading cause of death worldwide with an estimated rate of around 17.9 million deaths in the year 2016, of which 85% of deaths are caused by heart attack and stroke [1]. According to American Heart Association, 840,678 deaths were reported in 2016 in US which was approximately 1 in 3 deaths and expected to reach around 23.6 million by 2030 [2]. Every year 11 million new cases of CVD are reported and 83.5 million population is living with CVD in European countries [3].

At present, smoking cessation therapy, antiplatelet agents, antithrombotic agents, lipid lowering agents, blood pressure lowering agents are used as non-interventional strategy for treatment of CVDs (Table 4.1) [4]. Chest pain, swelling in face, legs, hands, serious rashes, irregular heartbeats, painful erection in men, sudden weigh gain, stomach pain, dizziness, hyperuricemia, muscle cramps, frequent urination, dry mouth, cough, headache, vision problems and growth in body hairs are some of the side effects associated with the conventional antihypertensive drugs [5]. Statins (lipid lowering agents) are more likely to cause rhabdomyolysis, cognitive loss and pancreatic, hepatic and sexual dysfunction [6]. Bleeding events are major concern associated with antithrombotic agents [7]. In addition to drugs, surgery is often recommended for various cardiovascular ailment viz. Coronary artery bypass grafting (CABG), Transmyocardial laser revascularization (TMR), Heart valve replacement, Heart transplant, Open heart surgery, etc. are some types of surgery for cardiovascular complications. Bleeding, arrhythmias, damages to heart, kidneys, lungs and liver tissues are some of the complications linked with cardiac surgery and requires hospitalisation [8]. Hence, looking into the limitation and side-effects of interventional as well as non-interventional treatment strategies, new drugs acting on novel targets are needed of the hour.

Several novel targets like siRNA, renin angiotensin aldosterone system, opioids are identified for CVD [9–11]. Liver X receptors (LXRs) belong to the superfamily of nuclear receptors [12]. These receptors were originally cloned in 1990 as 'orphan receptors' during their discovery because no endogenous ligands were found [13, 14]. However, later in their discovery, oxysterols (a cholesterol derivatives) was found as first endogenous ligand to activate LXRs [15]. In addition to this, several potent non-steroidal synthetic LXR agonist have been reported which include T0901317 and GW3965 [16]. There are two different isoforms of LXRs are found namely LXRα and LXRβ which are encoded by nuclear receptor subfamily 1 group H member 3 (NR1H3) and nuclear receptor subfamily 1 group H member 2 (NR1H2) genes, respectively. Both LXRα and LXRβ are expressed at different locations in human body [17].

LXRs are termed as "cholesterol sensors" as they play significant role in maintaining the cholesterol and lipid homeostasis by regulating the expression of various genes [18]. Various studies proved LXRs to play important role in regulating the carbohydrate and energy metabolism through different metabolic pathways [19, 20].

Table 4.1 Current therapy used for preventing and treating CVDs

Class of agents	Indications	Side-effects
Anti-hypertensive agents		
Angiotensin-converting enzyme (ACE) inhibitors	• Acute myocardial infarction • Heart failure • Diabetic nephropathy • Hypertension • Progressive renal insufficiency	• Cough • Dizziness • Fast heart beat • Headache
Beta blockers	• Angina pectoris • Cardiac arrhythmia • Congestive heart failure • Glaucoma • Postural orthostatic tachycardia • Hypertension	• Tiredness • Upset stomach • Headache • Irregular heartbeats
Calcium channel blockers	• Hypertension • Cerebral vasospasm • Chest pain • Cardiac arrhythmia	• Feeling drowsy • Ankle swelling • Serious rashes • Fainting
Peripherally acting alpha-adrenergic blockers	• Hypertension • Pheochromocytoma • Congestive heart failure • Erectile dysfunction • Benign prostatic hyperplasia • Raynaud's disease • Post-traumatic stress disorder	• Vision problems • Decreased sexual ability • Painful erection in men
Vasodilators	• Hypertension • Pulmonary Hypertension	• Growth in body hairs • Dizziness • Problem breathing • Sudden weight gain
Angiotensin II antagonist	• Heart failure • Diabetic nephropathy • Hypertensive diabetic patient • Hypertension complicated by left ventricular hypertrophy	• Sore throat • Sinus problem • Heartburn • Backpain • Swelling face, throat, eyes, hands or ankles
Aldosterone inhibitor (Spironolactone)	• Hypertension • Heart failure • Hypokalemia	• Dehydration • Hyponatremia • Ataxia • Dizziness • Erectile dysfunction
Diuretics	• Hypertension • Ascites • Congestive cardiac failure • Edema of lungs, kidney and liver	• Frequent urination • Headache • Feeling thirsty • Muscle cramps • Hyperuricemia

(continued)

Table 4.1 (continued)

Class of agents	Indications	Side-effects
Centrally-acting alpha adrenergic	• Hypertension	• Dry mouth • Upset stomach • Feeling drowsy • Fever
Lipid lowering agents		
HMG-CoA inhibitors (Statins)	• Primary prevention of arterial disease • Secondary prevention of myocardial infarction Familial hypercholesterolemia	• Rhabdomyolysis • Cognitive loss • Pancreatic and hepatic dysfunction • Sexual dysfunction
Fibric acid derivatives (Fibrates)	• Atheromatous diseases • Primary triglyceridemic	• GI side effects • Urticaria • Hair loss • Anaemia
Bile acids-binding resins	• Hypercholesterolemia • Pruritis • Bile acid diarrhea	• Flushing • Palpitation • GI disturbance
Anti-thrombotic agents		
Antiplatelet drugs	• Primary prevention of artial thrombosis • Coronary heart disease • Stable and Unstable angina	• Bleeding events • Various GI bleeding
Anticoagulant drugs	• Atrial fibrillation • Coronary artery disease • Deep vein thrombosis • Ischemic stroke • Myocardial infarction • Pulmonary embolism	• Risk of bleeding • Alopecia • Osteoporosis
Thrombolytic/fibrinolytic drugs	• ST elevation MI • Stroke • Pulmonary embolism • Deep vein thrombosis Acute limb ischemia	• Hemorrhagic stroke • Low-grade fever • Hypotension

As they are involved in fat metabolism, LXRs can be one of the major potential therapeutic target for metabolic diseases like atherosclerosis and other cardiovascular disease [21]. Apart from these, LXRs is reported to be involved in treating cancer [22], chronic inflammation [23], Alzheimer's disease [24], skin diseases [25] and insulin sensitivity [26]. The current chapter shall give a descriptive understanding of role of LXRs in various CVDs.

Liver X Receptors

Liver X receptors (LXRs) belong to the subclass of nuclear hormone receptors and initially regarded as orphan receptors as there were no natural ligands discovered. But later on, several discoveries in LXRs suggested a cholesterol derivative, an oxysterols as potent activator of LXRs. LXRα and LXRβ are the two isoforms of LXRs which are expressed by NR1H3 and NR1H2 genes, respectively. LXRα is restricted to various tissues of liver, intestine, kidney, macrophages, adipose tissues, lungs and adrenal glands whereas LXRβ are found to expressed ubiquitously [17, 27]. Genes of human LXRα and LXRβ are located on chromosomal number 11p11.2 and 19p13.3 respectively [28]. LXRα consist of 447 amino acids sequences where as LXRβ consist of 460 amino acids sequences. Both these isomers show 77% similarity in their sequences and differ in one amino acid sequence at ligand binding domain. LXRα and LXRβ have four different domains in their structure which include: a N-terminal (AF1) activation domain, a DNA-binding domain, a ligand binding domain (LBD) and a C-terminal (AF2) domain. C-terminal interact with coactivator and corepressor which is responsible for the regulation of transcriptional activities [29, 30]. LXRs are ligand activated transcription factor which bind to retinoid X receptors (RXRs) and form permissive heterodimers. Hence, both RXR and LXR are activated by ligands of each other. RXR/LXR complex bind to LXR response element (LXRE) at DNA which contains direct repeats of AGGTCA which are separated by four nucleotides (DR-4). Silencing mediator for retinoic acid and thyroid hormone receptor (SMART) or nuclear receptor corepressor (NCOR) act as the corepressor for both LXRs and RXRs. Binding of these corepressor to LBD causes the inactivation of targeted genes where as RXR activator like retinoic acids and LXR activator like oxysterols are responsible for the transcriptional activity of targeted genes. Binding of ligand to the RXR/LXR heterodimers leads to conformational changes which results in release of corepressors and recruits the coactivators (Fig. 4.1) [31, 32].

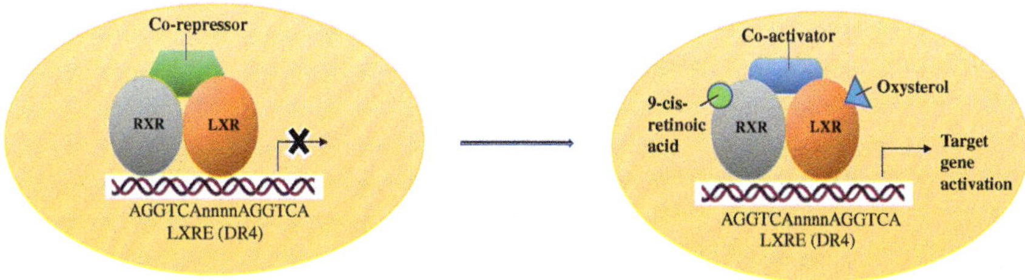

Fig. 4.1 Signal transduction of LXR/RXR. LXR forms heterodimer with RXR which binds to a LXRE. Signal transduction is inhibited when co-repressor binds to LXR/LXR. Following ligand binding to LXR or RXR, there is change in conformation of heterodimer, which leads to release of co-repressor and recruitment of co-activator takes place. This results in transcription of target genes. DR4—direct repeats separated by four nucleotides; LXR—liver X receptor; LXRE—LXR response element; RXR—retinoid X receptor

Physiological Functions of LXRs

Initial discovery pointed a significant role of LXRs in lipid metabolism. CYP7A1 is cytochrome enzyme in liver which is responsible for the conversion of cholesterol to bile acids. LXRα$^{-/-}$ and LXRβ$^{-/-}$ are LXRs knockout mice which were treated with high-cholesterol diet. LXRα$^{-/-}$ mice failed to convert cholesterol into bile acids and resulted in cholesterol accumulation whereas LXRβ$^{-/-}$ did not display this hepatic phenotype, suggesting LXRα plays prominent role in regulating CYP7A1 enzyme [33–35]. Another mechanism by which LXRs regulate cholesterol metabolism, is promoting the efflux of cholesterol in the faecal matter via elevated expression of intestinal ATP-binding cassette transporter (ABC) G5 and ABCG8 [36]. On the other hand, LXRs also promote the triglyceride (TG) and fatty acids (FA) synthesis in the liver by increasing the function of sterol regulatory element-binding protein 1C (SREBP1C), fatty acid synthase (FAS), acetyl CoA carboxylase (ACC) and stearoyl CoA denaturase-1 (SCD1) genes. Thus, the treatment with LXR agonist may result in lipogenesis in liver and serum [37]. LXRs also promote the process of reverse cholesterol transport by regulating the genes such as ATP-binding cassette transporter (ABC) A1 and ABCG1 and ADP-ribosylation factor like 7 (ARL7). These genes are responsible for the transport of the cholesterol from the peripheral tissues to the hepatic cells for excretion [38]. Mitochondrial oxidation of fatty acid can counteract the release of FA into the circulation. LXRs facilitate the oxidation of fatty acids in the mitochondria and reduces glucose oxidation in white adipose tissues (WAT) [39].

In addition to lipid metabolism, LXR agonists are also reported to regulate carbohydrate metabolism. In db/db mice, LXR agonist T0901317 inhibited phosphoenolpyruvate carboxykinase (PEPCK), glucose-6-phosphatase and fructose biphosphatase-1 in liver thereby producing suppression of gluconeogenesis, ultimately reducing the hepatic glucose output. Furthermore, in same study, treatment with T0901317 in fa/fa rats improved glucose tolerance and insulin sensitivity [26]. LXRs also maintain glucose homeostasis by increasing glucose uptake in WAT by inducing GLUT4, an insulin-dependent transporter [40].

Both LXRα and LXRβ shows their physiological action in regulating energy expenditure in brown adipose tissues (BAT). BAT regulate body temperature, prevent obesity and insulin resistance in the body. Uncoupling protein 1 (UCP-1) is found to be present prominently in BAT which is responsible for heat generation. Absence of both the LXR isoform in LXRα$^{-/-}$ and LXRβ$^{-/-}$ mice causes increases energy expenditure and high expression of UCP-1 [41]. Treatment of female C57BL/6 wild type mice with GW3965 produced a decreased energy expenditure (EE) and lower expression of UCP1. Thus, it can be concluded that LXRα and LXRβ agonist can be the therapeutic approach for obesity and maintaining energy homeostasis [41].

LXRs repress various pro-inflammatory mediators and reduces the inflammation. Treatment with LXR agonist GW3965 in rat Kupffer cells produced attenuation in expression of tumor necrosis factor alpha (TNFα) in dose-dependent manner [42]. In-vitro study in lipopolysaccharides (LPS)-induced inflammation in macrophages depicted repressed nuclear factor kappa-light-chain-enhancer of activated B cells

(NF-kB) and activator protein 1 (AP-1) when treated with LXR agonist GW3965 [43]. One mechanism by which LXR initiate its anti-inflammatory response is transrepression. In transrepression, in presence of inflammatory signals, release of corepressors like NCOR or SMART takes place via coronin-2A (CORO2A)-dependent manner. Additionally, NF-kB subunits p50 and p65 bind to the pro-inflammatory gene promoters. Moreover, inflammatory signal causes recruitment of coactivators. Activation of LXRs by its agonist leads to SUMOlyation of LXR by the SUMO2/3 protein. This LXR-SUMO complex binds to the corepressor complex. Binding of LXR-SUMO complex to corepressor like CORO2A mediate the interaction between LXR and CORO2A which impairs the inflammatory gene expression [44]. Physiological function of liver X receptor is mentioned in Fig. 4.2.

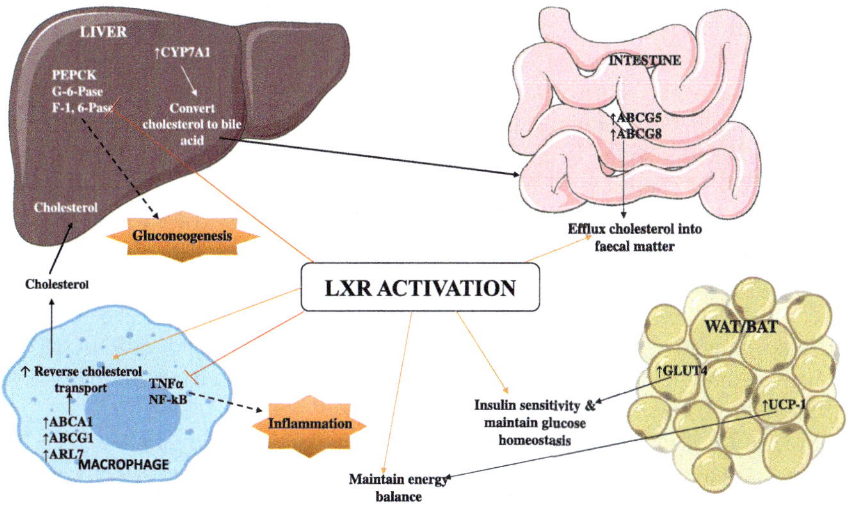

Fig. 4.2 Physiological function of LXRs. LXR activation leads to inhibition of gluconeogenesis in liver by downregulating PEPCK, G-6-Pase and F-1,6-Pase enzyme. LXR improves insulin sensitivity and maintain energy balance in body by increasing GLUT4 and UCP-1 expression in white and brown adipose tissues. LXR inhibit inflammation by blocking the recruitment of inflammatory cytokines like TNFα and NF-kB. LXRs also elevates reverse cholesterol transport in macrophages and other peripheral tissues by increasing the expression of ABCA1, ABCG1 and ARL7 and promotes the movement of cholesterol from peripheral tissues or macrophages to the liver. LXR activation increases the CYP7A1 expression and convert cholesterol to bile acids. This bile acid moves to intestine and excrete via faecal matter. Furthermore, LXR promotes efflux of cholesterol into the faecal matter by increasing the expression of ABCG5 and ABCG8. (black dotted line—inhibition of process; black and orange solid line—promotion of process; red line—inhibition of molecule expression); ABCA1—ATP binding cassette subfamily A member 1; ABCG1—ATP binding cassette subfamily G member 1; ABCG5—ATP-binding cassette transporter G5; ABCG8—ATP-binding cassette transporter G8; ARL7—ADP-ribosylation factor like 7; BAT—brown adipose tissue; CYP7A1—cytochrome P450 7A1; F-1,6-Pase—fructose biphosphatase-1; GLUT4—glucose transporter type-4; G-6-Pase—glucose-6-phosphatase; LXR—liver X receptor; NF-kB—nuclear factor kappa-light-chain-enhancer of activated B cells; PEPCK—phosphoenolpyruvate carboxykinase; TNFα—tumor necrosis factor; UCP-1 uncoupling protein 1; WAT—white adipose tissue

LXR in Cardiovascular Diseases

Atherosclerosis

Raised cholesterol level increases the risk of heart diseases and stroke. According to WHO report, 2.6 million deaths around the globe are estimated to cause due to hypercholesterolemia [45]. Atherosclerosis is the disease of arteries generally marked by chronic inflammation due to formation of fatty streak lesion and rupture of plaque and result in many cellular and molecular events. Increase in low-density lipoprotein (LDL) level in blood is one of the major causes of atherosclerosis initiation and development. Pathophysiology of atherosclerosis is marked by lipid accumulation in blood by macrophages leads to inflammation and formation of plaque. Various risk factors such as toxins, hyperglycemia, hypertension, infection, oxidized LDL and elevated homocysteine promotes atherogenesis by upregulating genes and activating macrophages and other inflammatory mediators resulting in lipid deposition and endothelial injury [46]. Proprotein convertase substilisin/Kexin type 9 (PCSK9) enzyme is produced and secreted mainly by liver. PCSK9 plays essential role in cholesterol metabolism by regulating plasma low-density lipoprotein cholesterol (LDL-C). PCSK9 binds to LDL receptor (LDLR) on the cell membrane of hepatocytes and causes lysosomal degradation of LDLR, which results in increased plasma LDL-C. Thus, inhibiting PCSK9 reduced the risk of atherosclerotic vascular disease [47]. miRNA-155 in C57BL/6J mice increased LDL accumulation and promoted inflammatory cytokines and miRNA-155-5p in human umbilical vein endothelial cells (HUVECs) caused atherosclerosis by TNFα-induced eNOS downregulation [48].

Role of LXR in lipid metabolism and inflammation is already discussed earlier. There are many preclinical evidences indicating LXR having anti-atherosclerotic property [49]. LXR agonist R211945 reduces the inflammatory response by decreasing number of macrophages, apolipoprotein B and oxidized phospholipids in plaque and is responsible for reversal cholesterol transport in New Zealand white rabbits [50]. In another study, treatment with LXR agonist T0901317 resulted in increased expression of chREBP mRNA expression and modulation of the carbohydrate-responsive element-binding protein (chREBP) and sterol regulatory element-binding protein 1C (SREBP-1C) activity, resulting into hepatic lipogenesis in chREBP knockout mice [51]. Heat shock proteins (HSPs) are responsible for the maintaining cellular homeostasis under stress condition. HSP70 is type of heat shock protein which activate the LXRα and its mRNA expression and facilitate the cholesterol efflux by inducing the level of ABCA1 and ABCG1 in zebrafish and also increased the cholesterol efflux from human macrophage foam cells [52]. Documented evidences suggested that natural substance quercetin interferes with the atherosclerosis development in ApoE$^{-/-}$ knockout mice by increasing the ABCA1, LXRα expression and by decreasing PCSK9, TNF-α, Il-6 and IL-10 expression [53]. Treatment with LXR agonist T0901317 in LDLR$^{-/-}$ male C57BL/6 mice reduced the atherosclerotic lesions by increasing the ABCA1 and ABCG1 mRNA expression and also revealed that macrophage LXRs are essential for antiatherogenic effect of T0901317 [54]. In another study, treatment with T0901717 in female apoE*3Leiden (E3L) mice decreased the atherosclerotic lesions number and lesions size in aortic valve area, reduced the expression of NF-kB, intercellular adhesion molecule (ICAM-1) and CD44 and increased the expression of ABCA1 and ABCG1 [55].

Myocardial Ischemia/Repurfusion Injury (MI/RI)

According to WHO, every year around 32.4 million population suffered from myocardial ischemia and stroke worldwide [56]. MI in layman term is known as heart attack, which is generally caused by reduced or stoppage of blood supply in the cardiomyocytes. In MI, balance between oxygen supply and demand is disturbed and leads to myocardial damage that is necrosis of heart muscles takes place [57]. Estrogen modulates heart and protects against MI [58]. The cardiac cell injury enhances the innate immunity response and thus inflammation and inflammatory cell infiltration are considered as hallmarks for myocardial ischemia and reperfusion injury. Various damage-associated molecular patterns (DAMPs) are released from the necrotic cells which further activates the toll-like receptors (TLRs) [59]. TLRs further activates the NF-kB, which release the pro-inflammatory mediators like IL-1α, IL-6, IL-18 and IL-1β and pro-inflammatory chemokines (C–C motif) ligand 2 (CCL2) and CCL5. These chemokines are found to be upregulated in ischemic heart [60]. Apoptosis is also mediated by caspase-12 and caspase-9 upregulation in the endoplasmic reticulum-mediated stress pathway and mitochondrial mediated stress pathway, respectively [61, 62]. Clinical study confirmed that level of circulating miRNA-30a-5p was significantly increased in patient who suffered left ventricular dysfunction and symptoms of heart failure six months after acute MI [63].

To investigate protective role of LRXs in heart, C57BL/6 male mice were subjected to left coronary artery occlusion to induce M/I injury. Treatment with LXR agonist GW3965 in mice showed a cardioprotective effect. It produced a decrease in infarct size and also shown an improvement in the left ventricular contractile function [64]. As mentioned earlier MI leads to apoptosis of the cardiac cells due to increased expression of caspase 3 (protein responsible for programmed cell death). Furthermore, in same study *in-vitro* treatment of cardiomyocytes with GW3965 resulted in attenuation of caspase 3 protein when exposed to hypoxia/reoxygenated condition [64]. In another study, cardioprotective role of LXRα/β dual agonist GW3965 was impaired in LXRα knockout mice but was not impaired in LXRβ knockout mice. Thus, it can be said that activating LXRα but not LXRβ isoform provides the protective role in MI. In same study, GW3963 and endogenous LXR agonist 22(R)-hydroxycholesterol [22(R)-HC] significantly inhibited caspase-12 and caspase-9 expression, attenuates the ROS formation and decreased myocardial apoptosis, infarct size and cardiac dysfunction in infarcted C57BL/6 mice [28]. Stem cell therapy can also be the strategy for the regeneration of the cardiac cells. Adipose derived mesenchymal stem cells (ADMSCs) are generally used for the clinical purpose for generation of cardiac cells. In Fluc$^+$-eGFP$^+$ transgenic C57BL/6a mice, LXR agonist T0901317 in combination with ADMSCs not only prevented MI injury but also improved the survival of the ADMSCs by downregulating TNFα, IL-6, ROS production, TLR4, MyD88, TRAF-6, IkBα and NF-kB-p65 and upregulating translocation of Nrf-2 from plasma to the nucleus. Thus, this can be concluded that treatment with LXR agonist result in downregulation of TLR4/NF-kB pathway and upregulation of Keap-1/Nrf-2 pathway and thus increases the viability of ADMSCs [65]. Deletion of LXRα in mice resulted in the decreased glucose uptake by downregulating GLUT1/4 and AMP-activated protein kinase (AMPK) phosphorylation, and increased the CD36 and thus worsens the MI injury in LXR$\alpha^{-/-}$ male mice compared to normal wild-type C57BL/6 mice. Thus, regulation of glucose metabolism by LXRs plays important role in improving the cardiac remodeling in MI [66].

Diabetic Cardiomyopathy (DCM)

Risk of developing CVD is 2–2.5 times higher in individual with type 2 diabetes mellitus (T2D) compared to nondiabetic individuals. Around 2.7 million patients with T2D shown to have heart failure in United Kingdom [67]. DCM is characterised by the ventricular dysfunction in the diabetic patient in absence of coronary artery disease like atherosclerosis and hypertension [68–70]. Clinical trials reported heart failure cases in diabetic patient were ranging from 19 to 26% [71]. Risk factors for the DCM are the hyperglycemia, hyperinsulinemia and insulin resistance in which liver plays a major role [72]. This results in insulin resistance in cardiac cells and metabolic disorders which further increases the mitochondrial dysfunction, inflammation, oxidative stress, AGEs production, ER-stress, cardiomyocyte death. Due to these, abnormalities like stiffness and hypertrophy takes place and resulting into diastolic and systolic dysfunction [73]. Several anti-hypertensives have been reported to play a role in diabetic cardiomyopathy [74–79]. Targets like serotonin [80], histone deacetylases [81] and estrogen receptors [82] are also reported to modulate diabetic cardiomyopathy. miRNA-1 found to have bidirectional role in pathogenesis of DCM. Upregulation of miRNA-1 produces apoptotic role in DCM whereas downregulation of miRNA-1 had anti-hypertrophic effect on cardiac cells. High glucose upregulates the miRNA-320 in cardiac cells and causes cardiac cells death [83].

As mentioned earlier, LXRs also play their significant role in glucose metabolism and improved various conditions like insulin resistance and hyperglycemia. PI3k/AKT/GLUT4 is insulin-stimulated pathway which is responsible for the glucose uptake in the cells. In this pathway, phosphorylation of AKT protein causes translocation of GLUT4 to the cellular membrane for entry of glucose into the cells [84]. Treatment with LXR agonist GW3965 in type 2 diabetes mellitus (T2D) db/db mice increased the phosphorylation of AKT protein and thus attenuates the insulin resistance and hyperglycemic condition in the mice, ultimately improved the cardiac function. Furthermore, GW3965 decreased ROS production, reduced apoptosis and improved LV dysfunction in db/db mice [85]. Diabetic heart losses its flexibility to use various kind of energy source due to lack of insulin and reduced uptake of glucose in cardiac cells. In T2D, lipolysis takes place due to less availability of insulin in adipose tissue. Thus, there is more availability of lipids and fatty acids to the cardiac cells. Cardiac cells utilise this freely available fatty acids which results in increased fatty acyl-CoA concentration. This fatty acyl-CoA are used for synthesis of diacylglycerol (DAG) which can be regarded as toxic. LXR agonist T0901317 reduced the accumulation of DAG in the heart of streptozotocin induced diabetic rats [86]. Role of LXRα in DCM is also reported to be mediated by the miRNA-1 in the cardiac cells. LXRα are the direct target for miRNA-1. Expression of miRNA-1 regulate the LXRα in the H9C2 cells. Interestingly, overexpression of miRNA-1 downregulated the expression of LXRα and aggravated the apoptotic condition in heart [87]. Confirmed role of miRNA in the DCM was given by Cheng et al. later in their discovery and suggested that overexpression of miRNA-1 is responsible for the glucose induced apoptosis in heart cells. He treated the rat H9C2 cardiac cells with anti-miRNA-1 and concluded that silencing of miRNA-1 provides the protection to the cardiac cells via inhibiting mitochondrial signaling pathway by increasing expression of LXRα [88]. In another study using H9C2 cells, GW3965 treatment inhibited hyperglycemia induced inducible nitric oxide synthase (iNOS), NF-kB, Caspase-3 and Cytochrome-C (myocardial injury marker) production thereby protecting cardiomyocyte against high glucose stress-induced injury [89].

Cardiac Hypertrophy

Cardiac hypertrophy is global disease with an estimated rate of 1 in 500 people [90]. It is common complication in around 30% of hypertensive individuals [91]. Cardiac hypertrophy is the compensatory mechanism to enhance the cardiac performance and lessen the ventricular wall tension and also oxygen consumption [92]. Thus, it results in the increase in the size of cardiomyocytes as a result of the arterial hypertension, myocardial infarction, inflammation and valvular cardiac disease [93]. Cardiac hypertrophy is enhanced by various mediators like fibrosis, over production of pro-inflammatory cytokines, autophagy suppression and hemodynamic stress [94]. Cardiac hypertrophy is reported to be modulated by various agents acting on targets like estrogen [95], histone deacetylases [96, 97] and MAP kinases [98]. miRNA-208a is cardiac specific miRNA which is upregulated in the cardiac hypertrophy by increasing the MCHβ expression in cardiomyocytes. Upregulation of miRNA-208a in reduces the Thrap-1 level and induces cardiac hypertrophy. Thus, provides the link between the thyroid hormone and cardiac hypertrophy [83].

In order to investigate role of LXRs in cardiac hypertrophy, LXRα knockout C57B6 mice were exposed to pressure overload induced hypertrophy. Treatment with the T0901317 in C57BL/6 inhibited angiotensin-II and liposaccharides (LPS) expression thereby producing suppression of NF-kB activity, ultimately producing the cardioprotective role against hypertrophy. Moreover, LXRs are also involved in the negative regulation of the cardiomyocyte growth [99]. T0901317 attenuates the cardiac hypertrophy by decreasing expression of endothelin-1 and atrial natriuretic factor (ANP) in cultured HL-1 cardiac cells. Treatment with T0901317 in C57BL/6J wild-type mice also showed a decrease in thickening of heart walls and left ventricular weight by reducing expression of endothelin-1 and ANP [100]. AZ876 is potent LXR agonist and prevents the cardiac hypertrophy without causing alteration in the lipid metabolism. AZ876 in C57Bl/6J mice reduces the transverse aortic constriction induced cardiac hypertrophy by reducing the heart weight and fibrosis without causing any effect to the blood pressure. This LXR agonist also reduces the overexpression of transforming growth factor beta (TGFβ) and angiotensin-II [101]. Increased LXRα expression prevented the development of left ventricular hypertrophy induced by high fat diet. Moreover, LXRα transgenic mice shows 1.5-fold increase in glucose uptake by increasing GLUT4 expression compared to non-transgenic wild-type mice [102].

Cardiac Fibrosis

Cardiac fibrosis is characterized by disturbance in balance between extracellular matrix production and degradation which finally lead to cardiac muscle dysfunction and reduction in overall heart function [103]. Extracellular matrix (ECM) is generally composed of the type I and type III collagen type. Collagen type I and type III in normal heart is found to be approximately 85% and 11%, respectively [104]. Collagen fibres are important for the normal contractile function of heart. Cardiac fibroblast regulates the synthesis and turnover of the collagen. Cardiac fibrosis leads to deposition of type I collagen, activation of fibroblast and myofibroblast differentiation [105]. Activating angiotensin II, atrial natriuretic peptide (ANP) and catecholamines leads to activation of fibroblast to myofibroblast. Activation of fibroblast to myofibroblast participate in elevation of collagen, cathedrin-11 and alpha-smooth muscle actin (α-SMA). Angiotensin-II also activate endothelin-1 and TGFβ which results in collagen and α-SMA synthesis [106]. Furthermore, downregulation of miRNA-29 in C57BL/6 mice provides a basis for mechanism of cardiac fibrosis [107].

Treatment with LXR agonist, AZ876 in the C57BL6/J mice shown marked decrease in collagen synthesis, reduced expression of TGFβ and α-SMA by reducing the expression of angiotensin-II. Moreover, AZ876 also downregulated connective tissue growth factor (ctgf) and collagen type I, alpha 1 (Col 1a 1) gene expression [101, 102]. Matrix metalloproteinases (MMPs) belongs to the family of endopeptidases containing Zn^{2+}. MMP-9 is responsible for the degradation of extracellular matrix (ECM) component during remodeling. In-vitro study revealed that LXR agonist GW3965 and T0901317 in macrophages obtained from mice represses the expression of MMP-9 by inhibiting the pro-inflammatory mediators such as NF-kB pathway [108]. The summary of key effects of liver X receptor is described in Table 4.2.

Table 4.2 Summary of key effects of liver X receptors in CVDs

Disease	Animal model	Key effects	References
Atherosclerosis	High cholesterol diet-induced atherosclerosis in New Zealand White Rabbits	• LXR agonist R211945 ↑ Reverse cholesterol transport ↓ Macrophages, apolipoprotein B and oxidized phospholipids	[50]
	High cholesterol diet fed Zebrafish	• HSP70 protein ↑ LXRα expression ↑ ABCA1 and ABCG1 gene	[52]
	ApoE$^{-/-}$ knockout mice	• Quercetin ↑ ABCA1, LXRα and PCSK9 gene expression	[53]
	chREBP knockout mice	• LXR agonist T0901317 ↑ chREBP and SREBP-1c mRNA expression	[51]
	LDLR$^{-/-}$ male C57BL/6 mice	• LXR agonist T0901317 ↑ ABCA1 and ABCG1 expression	[54]
	Female apoE*3Leiden (E3L) mice	• LXR agonist T0901317 ↓ Atherosclerotic lesion number and size ↓ NF-kB, ICAM-1 and CD44 ↑ ABCA1 and ABCG1 expression	[55]
Myocardial ischemia/ reperfusion injury [MI/RI]	Left coronary artery occlusion-induced C57BL/6 male mice	• LXR agonist GW3965 ↓ infarct size ↑ left ventricular contractile function	[64]
	In-vitro study of cardiomyocytes exposed to hypoxia/reoxygenated condition	• LXR agonist GW3965 ↓caspase-3 expression	[64]
	LXRα$^{-/-}$, LXRβ$^{-/-}$ and Infarcted wild-type C57BL/6 mice	• LXR agonist GW3965 ↓caspase-12 and caspase-9 expression ↓ ROS formation	[28]
	Fluc^{+}-eGFP^{+} transgenic C57BL/6a mice	• LXR agonist T0901317 in combination with ADMSCs ↓ TNFα, IL-6, ROS, TLR4, MyD88, TRAF-6, IkBα and NF-kB-p65 ↑ translocation of Nrf-2 from plasma to the nucleus	[65]

(continued)

Table 4.2 (continued)

Disease	Animal model	Key effects	References
	LXRα$^{-/-}$ male mice	• Decreases expression of LXRα ↓ GLUT1/4 expression compared to wild-type C57BL/6 mice ↑ AMPK phosphorylation	[66]
Diabetic cardiomyopathy	T2D db/db mice	• LXR agonist GW3965 ↑ phosphorylation of AKT protein ↓insulin resistance, apoptosis and hyperglycemia ↑ Left ventricular function	[85]
	Streptozotocin induced diabetic rats	• LXR agonist T0901317 ↓ accumulation of DAG in the heart	[86]
	Glucose induced apoptosis in H9C2 cells	• Anti-miRNA-1 treatment ↑ LXRα expression	[88]
	High glucose stress-induced injury in rat H9C2 cells	• LXR agonist GW3965 ↓ iNOS, NF-kB, Caspase-3 and Cytochrome-C	[89]
Cardiac hypertrophy	Pressure overload induced hypertrophy in LXRα knockout C57BL/6 mice	• LXR agonist T0901317 ↓ Angiotensin-II and liposaccharides (LPS) expression ↓ NF-kB activity ↓ Cardiomyocytes growth	[99]
	Cultured HL-1 cardiac cells	• LXR agonist T0901317 ↓ Endothelin-1 and ANP expression	[100]
	Pressure overload induced hypertrophy in LXRα knockout male C57BL/6J wild-type mice	• LXR agonist T0901317 ↓ Endothelin-1 and ANP expression ↓ Thickening of heart walls and left ventricular weight	[100]
	Transverse aortic constriction induced cardiac hypertrophy in C57BL/6J mice	• LXR agonist AZ876 ↓ Fibrosis and TGFβ expression	[101]

(continued)

Table 4.2 (continued)

Disease	Animal model	Key effects	References
Cardiac fibrosis	Transverse aortic constriction induced cardiac hypertrophy in C57BL/6J mice	• LXR agonist AZ876 ↓ Collagen synthesis ↓ Connective tissue growth factor (ctgf) and collagen type I, alpha 1 (Col 1a 1) gene expression	[101, 102]
	In-vitro study in macrophages obtained from mice	• LXR agonist GW3965 and T0901317 ↓ MMP9 expression and NF-kB	[108]

↑—increase and ↓—decrease

Conclusions

LXR agonists have long been established to play an important role in maintenance of lipid and glucose homeostasis, inflammation and energy balance in vertebrates. In addition to this, recent evidences have suggested that LXR exhibit wide range of beneficial effect in several cardiovascular diseases like atherosclerosis, myocardial infarction, diabetic cardiomyopathy, cardiac hypertrophy and cardiac fibrosis. Despite this, LXR agonists are not approved and have not been able to reach to the market. The future lies in designing and developing molecules, which are either selective agonists at LXRα or LXRβ, or are selective agonist of transrepression or tissue selective. Such focussed drug discovery process will be beneficial for patients suffering from CVDs.

References

1. https://www.who.int/en/news-room/fact-sheets/detail/cardiovascular-diseases-(cvds). Accessed 25 Nov 2019
2. https://healthmetrics.heart.org/wp-content/uploads/2019/02/At-A-Glance-Heart-Disease-and-Stroke-Statistics—2019.pdf. Accessed 27 Nov 2019
3. https://www.escardio.org/static_file/Escardio/About%20the%20ESC/Annual-Reports/ESC-Annual-Report-2019.pdf. Accessed 1 Dec 2019
4. Huffman MD, Bhatnagar D (2012) Novel treatments for cardiovascular disease prevention. Cardiovasc Ther 30:257–263
5. https://www.fda.gov/consumers/free-publications-women/high-blood-pressure-medicines-help-you. Accessed 1 Dec 2019
6. Golomb BA, Evans MA (2008) Statin adverse effects. Am J Cardiovasc Drugs 8:373–418
7. Weinberger J (2005) Adverse effects and drug interactions of antithrombotic agents used in prevention of ischaemic stroke. Drugs 65:461–471
8. https://www.nhlbi.nih.gov/health-topics/heart-surgery. Accessed 1 Dec 2019
9. Raghunathan S, Patel BM (2013) Therapeutic implications of small interfering RNA in cardiovascular diseases. Fundam Clin Pharmacol 27:1–20

10. Rawal H, Patel BM (2018) Opioids in cardiovascular disease: therapeutic options. J Cardiovas Pharmacol Ther 23:279–291
11. Patel BM, Mehta AA (2012) Aldosterone and angiotensin: role in diabetes and cardiovascular diseases. Eur J Pharmacol 697:1–2
12. Huang P, Chandra V, Rastinejad F (2010) Structural overview of the nuclear receptor superfamily: insights into physiology and therapeutics. Annu Rev Physiol 72:247–272
13. Apfel R, Benbrook D, Lernhardt E, Ortiz MA, Salbert G, Pfahl M (1994) A novel orphan receptor specific for a subset of thyroid hormone-responsive elements and its interaction with the retinoid/thyroid hormone receptor subfamily. Mol Cell Biol 14:7025–7035
14. Willy PJ, Umesono K, Ong ES, Evans RM, Heyman RA, Mangelsdorf DJ (1995) LXR, a nuclear receptor that defines a distinct retinoid response pathway. Genes Dev 9:1033–1045
15. Janowski BA, Willy PJ, Devi TR et al (1996) An oxysterol signalling pathway mediated by the nuclear receptor LXRα. Nature 383:728–731
16. Kick EK, Busch BB, Martin R, Stevens WC, Bollu V et al (2016) Discovery of Highly Potent Liver X Receptor β Agonists. ACS Medi Chem Lett 7:1207–1212
17. Repa JJ, Liang G, Ou J, Bashmakov Y, Lobaccaro JM et al (2000) Regulation of mouse sterol regulatory element-binding protein-1c gene (SREBP-1c) by oxysterol receptors, LXRα and LXRβ. Genes Dev 14:2819–2830
18. Wójcicka G, Jamroz-Wiśniewska A, Horoszewicz K, Bełtowski J (2007) Liver X receptors (LXRs). Part I: structure, function, regulation of activity, and role in lipid metabolism. Postepy Hig Med Dosw (Online) 61:736–759
19. Zhang Z, Chen H, Chen Z, Ding P, Ju Y et al (2019) Identify liver X receptor β modulator building blocks by developing a fluorescence polarization-based competition assay. Eur J Med Chem 178:458–467
20. Jakobsson T, Treuter E, Gustafsson JÅ, Steffensen KR (2012) Liver X receptor biology and pharmacology: new pathways, challenges and opportunities. Trends Pharmacol Sci 33:394–404
21. Tontonoz P, Mangelsdorf DJ (2003) Liver X receptor signaling pathways in cardiovascular disease. Mol Endocrinol 17:985–993
22. Ju X, Huang P, Chen M, Wang Q (2017) Liver X receptors as potential targets for cancer therapeutics. Oncol Lett 14:7676–7680
23. Steffensen KR, Jakobsson T, Gustafsson JÅ (2013) Targeting liver X receptors in inflammation. Expert Opin Ther Targ 17:977–990
24. Sandoval-Hernandez AG, Buitrago L, Moreno H, Cardona-Gómez GP, Arboleda G (2015) Role of liver X receptor in AD pathophysiology. PLoS ONE 10:e0145467
25. Ouedraogo ZG, Fouache A, Trousson A, Baron S, Lobaccaro JM (2017) Role of the liver X receptors in skin physiology: putative pharmacological targets in human diseases. Chem Phys Lipid 207:59–68
26. Cao G, Liang Y, Broderick CL, Oldham BA, Beyer TP et al (2003) Antidiabetic action of a liver x receptor agonist mediated by inhibition of hepatic gluconeogenesis. J Biol Chem 278:1131–1136
27. Tobin KA, Ulven SM, Schuster GU, Steineger HH, Andresen SM et al (2002) Liver X receptors as insulin-mediating factors in fatty acid and cholesterol biosynthesis. J Biol Chem 277:10691–10697
28. He Q, Pu J, Yuan A, Lau WB, Gao E et al (2014) Activation of liver-X-receptor α but not liver-X-receptor β protects against myocardial ischemia/reperfusion injury. Circul Heart Fail 7:1032–1041
29. Ni M, Zhang B, Zhao J, Feng Q, Peng J et al (2019) Biological mechanisms and related natural modulators of liver X receptor in nonalcoholic fatty liver disease. Biomed Pharmacother 113:108778
30. Färnegårdh M, Bonn T, Sun S, Ljunggren J, Ahola H et al (2003) The three-dimensional structure of the liver X receptor β reveals a flexible ligand-binding pocket that can accommodate fundamentally different ligands. J Biol Chem 278:38821–38828

31. Zelcer N, Tontonoz P (2006) Liver X receptors as integrators of metabolic and inflammatory signaling. J Clin Investig 116:607–614
32. Calkin AC, Tontonoz P (2012) Transcriptional integration of metabolism by the nuclear sterol-activated receptors LXR and FXR. Nat Rev Mol Cell Biol 13:213
33. Steffensen KR, Gustafsson JÅ (2004) Putative metabolic effects of the liver X receptor (LXR). Diabetes 53:S36–S42
34. Peet DJ, Turley SD, Ma W, Janowski BA, Lobaccaro JM, Hammer RE, Mangelsdorf DJ (1998) Cholesterol and bile acid metabolism are impaired in mice lacking the nuclear oxysterol receptor LXRα. Cell 93:693–704
35. Alberti S, Schuster G, Parini P, Feltkamp D, Diczfalusy U et al (2001) Hepatic cholesterol metabolism and resistance to dietary cholesterol in LXRβ-deficient mice. J Clin Investig 107:565–573
36. Yasuda T, Grillot D, Billheimer JT, Briand F, Delerive P et al (2010) Tissue-specific liver X receptor activation promotes macrophage reverse cholesterol transport in vivo. Arterioscler Thromb Vasc Biol 30:781–786
37. Schultz JR, Tu H, Luk A, Repa JJ, Medina JC, Li L, Schwendner S, Wang S, Thoolen M, Mangelsdorf DJ, Lustig KD (2000) Role of LXRs in control of lipogenesis. Genes Dev 14:2831–2838
38. Delvecchio CJ, Bilan P, Nair P, Capone JP (2008) LXR-induced reverse cholesterol transport in human airway smooth muscle is mediated exclusively by ABCA1. Am J Physiol-Lung Cell Mole Physiol 295:L949–L957
39. Stenson BM, Ryden M, Steffensen KR, Wåhlén K, Pettersson AT et al (2009) Activation of liver X receptor regulates substrate oxidation in white adipocytes. Endocrinology 150:4104–4113
40. Gabbi C, Warner M, Gustafsson JA (2009) Minireview: liver X receptor β: emerging roles in physiology and diseases. Mol Endocrinol 23:129–136
41. Korach-André M, Archer A, Barros RP et al (2011) Both liver-X receptor (LXR) isoforms control energy expenditure by regulating brown adipose tissue activity. Proc Natl Acad Sci 108:403–408
42. Wang YY, Dahle MK, Steffensen KR et al (2009) Liver X receptor agonist GW3965 dose-dependently regulates lps-mediated liver injury and modulates posttranscriptional TNF-α production and p38 mitogen-activated protein kinase activation in liver macrophages. Shock 32:548–553
43. Ogawa S, Lozach J, Benner C et al (2005) Molecular determinants of crosstalk between nuclear receptors and toll-like receptors. Cell 122:707–721
44. Schulman IG (2017) Liver X receptors link lipid metabolism and inflammation. FEBS Lett 591:2978–2991
45. https://www.who.int/gho/ncd/risk_factors/cholesterol_text/en/. Accessed 1 Dec 2019
46. Jamkhande PG, Chandak PG, Dhawale SC et al (2014) Therapeutic approaches to drug targets in atherosclerosis. Saudi Pharma J 22:179–190
47. Pott J, Schlegel V, Teren A et al (2018) Genetic regulation of PCSK9 (proprotein convertase subtilisin/kexin type 9) plasma levels and its impact on atherosclerotic vascular disease phenotypes. Circul Geno Precis Medi 11:e001992
48. Jackson AO, Regine MA, Subrata C, Long S (2018) Molecular mechanisms and genetic regulation in atherosclerosis. IJC Heart Vascul 21:36–44
49. Terasaka N, Hiroshima A, Koieyama T et al (2003) T-0901317, a synthetic liver X receptor ligand, inhibits development of atherosclerosis in LDL receptor-deficient mice. FEBS Lett 536:6–11
50. Vucic E, Calcagno C, Dickson SD et al (2012) Regression of inflammation in atherosclerosis by the LXR agonist R211945: a noninvasive assessment and comparison with atorvastatin. JACC Cardiovasc Imag 5:819–828
51. Cha JY, Repa JJ (2007) The liver X receptor (LXR) and hepatic lipogenesis the carbohydrate-response element-binding protein is a target gene of LXR. J Biol Chem 282:743–751

52. Gungor B, Vanharanta L, Hölttä-Vuori M et al (2019) HSP70 induces liver X receptor pathway activation and cholesterol reduction in vitro and in vivo. Molecul Metabol 28:135–143
53. Li SS, Cao H, Shen DZ et al (2019) Effect of quercetin on atherosclerosis based on expressions of ABCA1, LXR-α and PCSK9 in ApoE-/-mice. Chin J Integrat Med 30:1–8
54. Levin N, Bischoff ED, Daige CL et al (2005) Macrophage liver X receptor is required for antiatherogenic activity of LXR agonists. Arterioscler Thromb Vasc Biol 25:135–142
55. Verschuren L, de Vries-van der Weij J, Zadelaar S, et al (2009) LXR agonist suppresses atherosclerotic lesion growth and promotes lesion regression in apoE* 3Leiden mice: time course and mechanisms. J Lipid Res 50:301–311
56. https://www.who.int/cardiovascular_diseases/priorities/secondary_prevention/country/en/index1.html. Accessed 1 Dec 2019
57. Lu L, Liu M, Sun R et al (2015) Myocardial infarction: symptoms and treatments. Cell Biochem Biophys 72:865–867
58. Rayabarapu N, Patel BM (2014) Beneficial role of tamoxifen in isoproterenol-induced myocardial infarction. Can J Physiol Pharmacol 92:849–857
59. Liu J, Wang H, Li J (2016) Inflammation and inflammatory cells in myocardial infarction and reperfusion injury: a double-edged sword. Clini Medi Insights Cardiol 10:79–84
60. Ong SB, Hernández-Reséndiz S, Crespo-Avilan GE et al (2018) Inflammation following acute myocardial infarction: multiple players, dynamic roles, and novel therapeutic opportunities. Pharmacol Ther 186:73–87
61. Szegezdi EV, Fitzgerald UN, Samali A (2003) Caspase-12 and ER-stress-mediated apoptosis: the story so far. Ann N Y Acad Sci 1010:186–194
62. Li P, Zhou L, Zhao T et al (2017) Caspase-9: structure, mechanisms and clinical application. Oncotarget 8:23996–24008
63. Maciejak A, Kostarska-Srokosz E, Gierlak W, Dluzniewski M, Kuch M, Marchel M, Opolski G, Kiliszek M, Matlak K, Dobrzycki S, Lukasik A (2018) Circulating miR-30a-5p as a prognostic biomarker of left ventricular dysfunction after acute myocardial infarction. Scienti Rep 8:9883
64. Lei P, Baysa A, Nebb HI et al (2013) Activation of Liver X receptors in the heart leads to accumulation of intracellular lipids and attenuation of ischemia–reperfusion injury. Basic Res Cardiol 108:323
65. Wang Y, Li C, Cheng K et al (2014) Activation of liver X receptor improves viability of adipose-derived mesenchymal stem cells to attenuate myocardial ischemia injury through TLR4/NF-κB and Keap-1/Nrf-2 signaling pathways. Antioxid Redox Signal 21:2543–2557
66. Ji Q, Zhao Y, Yuan A et al (2017) Deficiency of liver-X-receptor-α reduces glucose uptake and worsens post-myocardial infarction remodeling. Biochem Biophys Res Commun 488:489–495
67. Gulsin GS, Athithan L, McCann GP (2019) Diabetic cardiomyopathy: prevalence, determinants and potential treatments. Thera Adv Endocrinol Metabol 10:2042018819834869
68. Patel BM, Mehta AA (2013) Choice of anti-hypertensive agents in diabetic subjects. Diabetes Vascu Dis Res 10:385–396
69. Goyal BR, Mehta AA (2013) Diabetic cardiomyopathy: pathophysiological mechanisms and cardiac dysfuntion. Hum Exp Toxicol 32:571–590
70. Patel BM (2019) Histone deacetylase and oxidative stress: role in diabetic cardiomyopathy. In: Chakraborti S, Dhallan N, Ganguly N, Dikshit M (eds) Oxidative stress in heart disease. Springer, Singapore, pp 413–425
71. Jia G, Hill MA, Sowers JR (2018) Diabetic cardiomyopathy: an update of mechanisms contributing to this clinical entity. Circ Res 122:624–638
72. Patel BM, Goyal RK (2019) Liver and insulin resistance: new wine in old bottle!!! Eur J Pharmacol 862:172657
73. Pan J, Guleria RS, Zhu S, Baker KM (2014) Molecular mechanisms of retinoid receptors in diabetes-induced cardiac remodeling. J Clini Medi 3:566–594
74. Goyal BR, Mesariya P, Goyal RK, Mehta AA (2008) Effect of telmisartan on cardiovascular complications associated with streptozotocin diabetic rats. Mol Cell Biol 1;314(1–2):123–131

75. Goyal BR, Solanki N, Goyal RK, Mehta AA (2009) Investigation into the cardiac effects of spironolactone in the experimental model of type 1 diabetes. J Cardiovasc Pharmacol 54:502–509
76. Goyal BR, Parmar K, Goyal RK, Mehta AA (2011) Beneficial role of telmisartan on cardiovascular complications associated with STZ-induced type 2 diabetes in rats. Pharmacol Rep 63:956–966
77. Patel BM, Kakadiya J, Goyal RK, Mehta AA (2013) Effect of spironolactone on cardiovascular complications associated with type-2 diabetes in rats. Exp Clin Endocrinol Diabetes 121:441–447
78. Goyal BR, Patel MM, Bhadada SV (2011) Comparative evaluation of spironolactone, atenolol, metoprolol, ramipril and perindopril on diabetes induced cardiovascular complications in type 1 diabetes in rats. Int J Diabetes Metabol 19:11–18
79. Patel BM, Bhadada SV (2014) Type 2 diabetes-induced cardiovascular complications: comparative evaluation of spironolactone, atenolol, metoprolol, ramipril and perindopril. Clin Exp Hypertens 36:340–347
80. Raghunathan S, Tank P, Bhadada S, Patel B (2014) Evaluation of buspirone on streptozotocin induced type 1 diabetes and its associated complications. BioMed Res Int 2014: 2014.
81. Patel BM, Raghunathan S, Porwal U (2014) Cardioprotective effects of magnesium valproate in type 2 diabetes mellitus. Eur J Pharmacol 728:128–134
82. Rabadiya S, Bhadada S, Dudhrejiya A, Vaishnav D, Patel B (2018) Magnesium valproate ameliorates type 1 diabetes and cardiomyopathy in diabetic rats through estrogen receptors. Biomed Pharmacother 97:919–927
83. Ghosh N, Katare R (2018) Molecular mechanism of diabetic cardiomyopathy and modulation of microRNA function by synthetic oligonucleotides. Cardiovasc Diabetol 17:43
84. Hou N, Mai Y, Qiu X et al (2019) Carvacrol attenuates diabetic cardiomyopathy by modulating the PI3K/AKT/GLUT4 pathway in diabetic mice. Front Pharmacol 10:998
85. He Q, Pu J, Yuan A et al (2014) Liver X receptor agonist treatment attenuates cardiac dysfunction in type 2 diabetic db/db mice. Cardiovasc Diabetol 13:149
86. Harasiuk D, Baranowski M, Zabielski P et al (2016) Liver x receptor agonist to901317 prevents diacylglycerols accumulation in the heart of streptozotocin-diabetic rats. Cell Physiol Biochem 39:350–359
87. Cheng Y, Zhang D, Zhu M et al (2017) Liver X receptor α is targeted by microRNA-1 to inhibit cardiomyocyte apoptosis through a ROS-mediated mitochondrial pathway. Biochem Cell Biol 96:11–18
88. Cheng Y, Zhao W, Zhang X et al (2018) Downregulation of microRNA-1 attenuates glucose-induced apoptosis by regulating the liver X receptor α in cardiomyocytes. Experi Thera Medi 16:1814–1824
89. He Q, Wang F, Fan Y et al (2018) Differential effects of and mechanisms underlying the protection of cardiomyocytes by liver-X-receptor subtypes against high glucose stress-induced injury. Biochem Biophys Res Commun 503:1372
90. Semsarian C, Ingles J, Maron MS, Maron BJ (2015) New perspectives on the prevalence of hypertrophic cardiomyopathy. J Am Coll Cardiol 65:1249–1254
91. Cramariuc D, Gerdts E (2016) Epidemiology of left ventricular hypertrophy in hypertension: implications for the clinic. Expert Rev Cardiovasc Ther 14:915–926
92. Samak M, Fatullayev J, Sabashnikov A et al (2016) Cardiac hypertrophy: an introduction to molecular and cellular basis. Medi Sci Monitor Basic Res 22:75
93. Ma Z, Deng C, Hu W et al (2017) Liver X receptors and their agonists: targeting for cholesterol homeostasis and cardiovascular diseases. Curr Issues Mol Biol 22:41–64
94. Shimizu I, Minamino T (2016) Physiological and pathological cardiac hypertrophy. J Mol Cell Cardiol 97:245–262
95. Patel BM, Desai VJ (2014) Beneficial role of tamoxifen in experimentally induced cardiac hypertrophy. Pharmacol Rep 66:264–272
96. Raghunathan S, Goyal RK, Patel BM (2016) Selective inhibition of HDAC2 by magnesium valproate attenuates cardiac hypertrophy. Can J Physiol Pharmacol 95:260–267

97. Patel BM (2018) Sodium butyrate controls cardiac hypertrophy in experimental models of rats. Cardiovasc Toxicol 18:1–8
98. Sharma B, Chaube U, Patel BM (2019) Beneficial effect of Silymarin in pressure overload induced experimental cardiac hypertrophy. Cardiovasc Toxicol 19:23–35
99. Wu S, Yin R, Ernest R et al (2009) Liver X receptors are negative regulators of cardiac hypertrophy via suppressing NF-κB signalling. Cardiovasc Res 84:119–126
100. Kuipers I, Li J, Vreeswijk-Baudoin I et al (2010) Activation of liver X receptors with T0901317 attenuates cardiac hypertrophy in vivo. Eur J Heart Fail 12:1042–1050
101. Cannon MV, Yu H, Candido WM et al (2015) The liver X receptor agonist AZ876 protects against pathological cardiac hypertrophy and fibrosis without lipogenic side effects. Eur J Heart Fail 17:273–282
102. Cannon MV, Silljé HH, Sijbesma JW, Khan MA, Steffensen KR, van Gilst WH, de Boer RA (2016) LXRα improves myocardial glucose tolerance and reduces cardiac hypertrophy in a mouse model of obesity-induced type 2 diabetes. Diabetologia 59:634–643
103. Ma ZG, Yuan YP, Wu HM, Zhang X, Tang QZ (2018) Cardiac fibrosis: new insights into the pathogenesis. Int J Biolog Sci 14:1645
104. Bashey RI, Martinez-Hernandez A, Jimenez SA (1992) Isolation, characterization, and localization of cardiac collagen type VI. Associations with other extracellular matrix components. Circul Res 70:1006–1017
105. Hinderer S, Schenke-Layland K (2019) Cardiac fibrosis–a short review of causes and therapeutic strategies. Adv Drug Deliv Rev 146:77–82
106. Tian J, An X, Niu L (2017) Myocardial fibrosis in congenital and pediatric heart disease. Experi Ther Medi 13:1660–1664
107. Van Rooij E, Sutherland LB, Thatcher JE et al (2008) Dysregulation of microRNAs after myocardial infarction reveals a role of miR-29 in cardiac fibrosis. Proc Natl Acad Sci 105:13027–13032
108. Castrillo A, Joseph SB, Marathe C et al (2003) Liver X receptor-dependent repression of matrix metalloproteinase-9 expression in macrophages. J Biol Chem 278:10443–10449

Chapter 5
Role of NLRP3 Inflammasomes in Obesity-Induced Cardiovascular Diseases

Krishna M. Boini, Pin-Lan Li, and Saisudha Koka

Abstract The prevalence of obesity is increasing at an alarming rate in many countries across the world. This is a significant concern given that obesity is associated with several metabolic complications including cardiovascular diseases such as myocardial infarction, hypertension, atherosclerosis, dyslipidemia, chronic kidney disease, insulin resistance and type 2 diabetes mellitus. The discovery of the NLRP3 (NLR family, pyrin domain containing 3) inflammasome as an intracellular machinery responsible for the activation of inflammation in variety of tissues or organs opened new avenues for treatment of a host of obesity-induced cardiovascular disorders. Here, we summarize our current understanding on how the NLRP3 inflammasome is involved in obesity and associated cardiovascular complications. The modulation of NLRP3 inflammasomes may have a great impact in the development of novel therapeutic modalities in obesity induced cardiovascular diseases. We review various NLRP3 inflammasome-targeted strategies and the evidence supporting the role of the NLRP3 inflammasome in obesity induced cardiovascular complications such as atherosclerosis, hypertension, myocardial infarction and adverse cardiac remodeling.

Keywords Obesity · NLRP3 inflammasome · Cardiovascular diseases · Atherosclerosis · Hypertension · Myocardial infarction

---

K. M. Boini · S. Koka (✉)
Department of Pharmacological and Pharmaceutical Sciences, College of Pharmacy, University of Houston, 4849 Calhoun Rd, Room 4042, Houston, TX 77204, USA
e-mail: sskoka@uh.edu

P.-L. Li
Department of Pharmacology and Toxicology, Virginia Commonwealth University, Richmond, VA 23298, USA

P. S. Tappia et al. (eds.), *Biochemistry of Cardiovascular Dysfunction in Obesity*, Advances in Biochemistry in Health and Disease 20,
https://doi.org/10.1007/978-3-030-47336-5_5

Introduction

Obesity is a multi-factorial disorder, which is often associated with several comorbid disease conditions such as insulin resistance, type 2 diabetes mellitus, dyslipidemia, atherosclerosis myocardial infarction, hypertension, and chronic kidney diseases [1–4]. Its prevalence is on continuous rise in all age groups across many developed countries in the world [5]. Epidemiological data reveals that 30% of the world population is obese or overweight [6]. Obesity which was once viewed as a lifestyle "choice"—the choice to overeat and under exercise, is now being considered more appropriately as a chronic disease by the modern world [7]. Chronic inflammation is the hallmark of obesity. Adipose tissue acts as an active metabolic tissue which secretes multiple metabolically important proteins known as 'adipokines'. These adipokines play a major role in the development of insulin resistance and cardio-vascular complications associated with obesity [8–11]. Recent studies recognized the role of NLRP3 (nucleotide oligomerization domain (NOD)-like receptor protein with pyrin domain containing 3) inflammasome in obesity and related cardiovascular complications. This review discusses the molecular mechanisms of NLRP3 inflammasome activation involved in chronic inflammation associated with the progression of obesity and obesity-induced cardiovascular diseases (CVDs) like atherosclerosis, hypertension, myocardial infarction and highlights various therapeutic approaches to inhibit inflammasome formation which could play a key role in the treatment of several chronic inflammatory disease conditions.

NLRP3 Inflammasome as a Sensor of Inflammation

The inflammasomes have been identified as an intracellular machinery responsible for the activation of inflammatory responses in a variety of tissues or organs [12, 13]. They are made of multiprotein complexes consisting a NOD-like receptor (NLR) protein, the pro-protein, procaspase-1 and adaptor molecule ASC (Apoptosis-associated speck-like protein containing a CARD) [13]. Pattern recognition receptors (PRRs) present in cytosol initiate inflammasome complex formation by recognizing pathogen-associated molecular patterns (PAMPs) and danger-associated molecular patterns (DAMPs). PAMPs and DAMPs are also detected by various immune cells and macrophages which initiate a series of inflammatory reactions [14–16]. NLRP1, NLRP3, NLRP6, NLRP7, NLRP12, NLRC4 and the HIN-200 family member AIM2 (absent in melanoma-2) which constitute the nucleotide binding domain and leucine-rich (NLR) protein family, toll-like receptors (TLRs), retinoic acid-inducible gene I-like helicases (RLHs) can act as PRRs [17]. The inflammasomes are formed upon exposure to pathogens, reactive oxygen species (ROS), and environmental irritants [18]. Among the NLR family, the NLRP3 inflammasome (also referred to as cryopyrin and NLRP3) is the best characterized and most extensively studied inflammasome. NLRP3 has been found to detect endogenous stress-associated danger signals including PAMPs (viruses, intracellular bacteria, cell-wall components), bacterial toxins and extracellular components such as adenosine triphosphate (ATP), uric acid, β-amyloid, cholesterol crystals, calcium pyrophosphate dehydrate, sphingosine, homocysteine and DAMPs to produce local tissue sterile inflammation [19–21].

Several diseases like acute myocardial infarction, diabetes mellitus, liver toxicity, gout etc. are known to be associated with NLRP3 inflammasome formation [8, 13, 21, 22]. NLRP3 has leucine-rich repeats (LRRs) at the C-terminal and pyrin domain (PYD) at the N-terminal along with a nucleotide-binding and oligomerization domain (NBD) which forms the core of the NLRP3 inflammasome. An adapter protein ASC which contains a PYD domain and a caspase recruitment domain (CARD) is required for pro-caspase-1 recruitment and the activation of NLRP3 [23–25]. Procaspase-1 is activated into capsapse-1 which further activates cytokines IL-1 β and IL-18 initiating the inflammatory process [23, 25].

Activation of Inflammasome in Obesity-Induced Cardiovascular Diseases

Obesity is a major risk factor for cardiovascular disorders and obese patients have significantly elevated morbidity and mortality due to CVDs [8, 9]. Activation of the NLRP3 inflammasome plays a central role in obesity-induced chronic inflammation and CVDs [26]. Accordingly, several studies have reported an association of the NLRP3 inflammasome with obesity, insulin resistance and type 2 diabetes mellitus [26–28]. Over-nutrition associated with obesity is known to activate NLRP3 inflammasome [29]. Mounting evidence from the studies in animal models of obesity and obese human subjects indicate that obesity is associated with increased NLRP3 expression in adipose tissue [28, 30–33]. Activation of NLRP3 is a two-step process which includes priming and triggering [34]. The first or "priming" signal is carried by DAMPs/PAMPs like glucose, palmitate, ceramide, uric acid or lipopolysaccharide (LPS) which are identified by toll-like receptors (TLRs) or cytokines such as tumor necrosis factor α (TNF-α), activating mainly nuclear factor-κB (NF-κB)-dependent signaling which results in the expression and of NLRP3 components and inactive precursors of IL-1β namely pro-IL-1β, pro-IL-18, and transcriptional expression of inflammasome such as NLRP3 deubiquitination and ASC phosphorylation [35]. The second or "triggering" signal involve the oligomerization of inactive NLRP3, ASC and procaspase-1, resulting in proteolytic cleavage of caspase-1 instigating caspase-1 activation [25]. Activated caspase-1 then processes pro-IL-1β and pro-IL-18 to their mature forms that are rapidly secreted from the cell which results in the maturation of IL-1β and leading to final inflammasome formation.

There are three important mechanisms underlying NLRP3 inflammasome activation including reactive oxygen species (ROS) activation, lysosome rupture and ion channel gating. The regulation of NLRP3 inflammasome activity by several triggers can be explained by these three mechanisms. NLRP3 is a sensor for changes in cellular oxidative stress, where increases in ROS activate this inflammasome. Nicotinamide adenine dinucleotide phosphate oxidase (NOX)-derived ROS activate NLRP3 inflammasomes in the pathogenesis of homocystine-induced glomerular sclerosis or adipokine visfatin or cholesterol crystals or 7-keto-induced atherosclerosis [36]. More recent studies have also implicated mitochondrial ROS in inflammasome activation [37, 38]. One recent study found that thioredoxin-interacting protein (TXNIP), a binding partner of the antioxidant protein thioredoxin, can directly bind NLRP3 and regulate inflammasome activation in a ROS-inducible manner [22, 39]. Another mechanism of inflammasome activation is potassium (K^+) efflux through endogenous ion channels. Intracellular potassium is an indicator of membrane integrity and loss of potassium leads to NLRP3 inflammasome activation by disruption of cell membrane integrity [40]. In addition, crystals like monosodium urate, amyloid-β, alum silica and asbestos are phagocytized by cells leading to the release of proteolytic lysosomal contents into the cytosol, resulting in NLRP3 inflammasome activation through lysosomal destabilization [20, 41]. The formation and activation of NLRP3 inflammasome is shown in Fig. 5.1.

The NLRP3 Inflammasome and Obesity

Chronic low-grade inflammation is an important feature of obesity. Obesity-induced inflammatory responses involve both innate and adaptive immune processes which indicate the potential roles for NLRP3 inflammasome activity. NLRP3 inflammasome components are expressed in adipose tissue macrophages. During obesity, adipocyte secretion of proinflammatory or pathogenic adipokines, is increased markedly. The role of NLRP3 inflammasome in the pathogenesis of obesity was supported by data showing that $NLRP3^{-/-}$ and $ASC^{-/-}$ knockout mice are protected against high fat diet (HFD)-induced obesity [28, 42, 43]. IL-1β and NLRP3 mRNA expression in visceral adipose is markedly decreased upon caloric restriction and increased in obese diabetic and HFD fed mice relative to their lean controls on standard chow diet [28, 44]. Adipocyte differentiation is also regulated by NLRP3 inflammasome activation [44]. NLRP3 and IL-1β gene expressions in the adipose tissue are reduced in type 2 diabetics following exercise and calorie restriction [28]. Similarly, serum adipokines are decreased upon fat mass reduction and increased upon excess calorie intake [45–47]. HFD-fed mice have exhibited increased Caspase-1 mRNA activity and enhanced adipose tissue IL-18 protein levels [28, 44, 48]

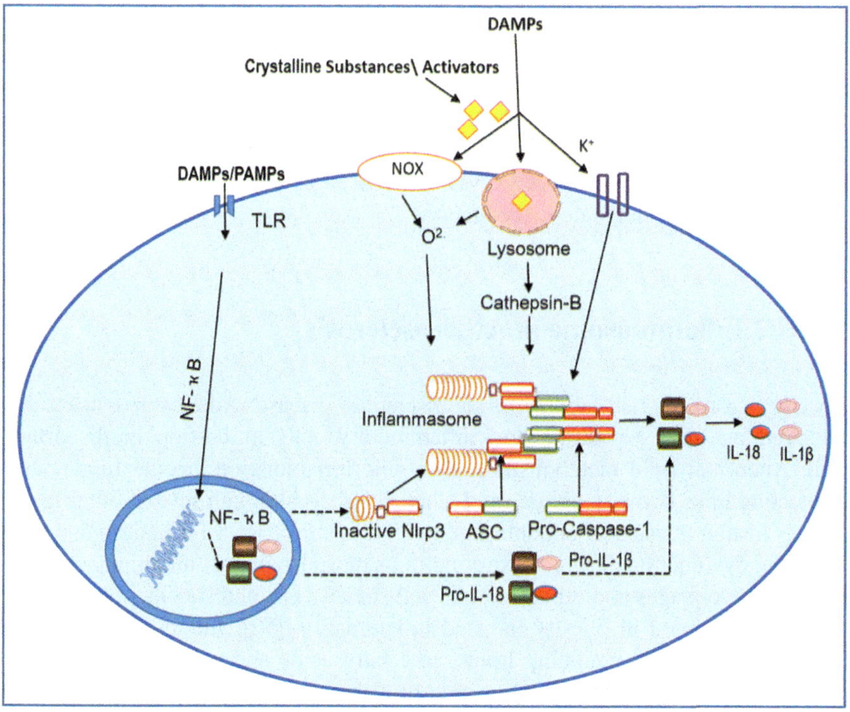

Fig. 5.1 NLRP3 inflammasome activation and formation. In obese conditions, damage-associated molecular patterns (DAMPs) and pathogen-associated molecular patterns (PAMPs) activate macrophages in adipose tissue which act via toll-like receptors (TLRs) which activate nuclear factor-κB (NF-κB) resulting in the expression of NLRP3 components and inactive precursors of IL-1β namely pro-IL-1β, pro-IL-18 which remain in cytoplasm. In obesity, DAMPs also directly leads to NLRP3 inflammasome formation via three main pathways including reactive oxygen species (ROS) generation, lysosomal destabilization and rupture and K^+ ion channel gating, all of which lead to the activation of NLRP3 inflammasome through the oligomerization of the inactive NLRP3, the apoptosis-associated speck-like protein (ASC), and the procaspase-1 components. The activated inflammasome produce active interleukin-1β or -18 (IL-1β and IL-18) from pro-IL-1β and pro-IL-18 which initiate inflammatory responses in obesity induced cardiovascular diseases

The NLRP3 Inflammasome in Obesity Induced Cardiovascular Diseases

Obesity is associated with the dysregulation of multiple metabolic factors that increase inflammasome activation resulting in an increased incidence of CVDs. In particular, abdominal obesity is a known risk factor for CVDs worldwide. CVD mortality and morbidity is significantly increased in obese individuals [3]. CVD risk is associated with the inflammatory status of an obese individual. Obese individuals with CVD risk often exhibit increased metabolic biomarkers such as adiponectin, cytokines which lead to vascular endothelial dysfunction and atherosclerosis [49].

Several cardiometabolic disease states display enhanced inflammatory responses and cytokines like IL-1β and IL-18 [50, 51]. Cardiovascular injury in obesity is produced by stimuli which induce NLRP3 inflammasome activation. Hence it has been recently suggested that both NLRP3 and pro-inflammatory cytokines are biomarkers of cardiovascular risk in obese patients. Molecular mechanisms of NLRP3 inflammasome in various obesity induced CVDs such as atherosclerosis, hypertension and myocardial infarction are described below.

NLRP3 Inflammasome in Atherosclerosis

Obesity is a major risk factor for atherosclerotic disease. Atherosclerosis is characterized by lipid deposition and inflammatory cells infiltration in the arterial wall. Abnormal lipid metabolism and chronic inflammation are the fundamental connecting links between obesity and atherosclerosis although several other factors such as insulin resistance, endoplasmic reticulum stress, ROS formation, and mitochondrial dysfunction during obese conditions may also impact inflammasome activation. Macrophages and other immune cells like T cells and B cells in the adipose tissue are activated in obesity induced inflammation [52]. Inflammasome activation is triggered by circulating lipids, free fatty acids and crystalline cholesterol (ChC) [53]. The NLRP3 inflammasome was shown to participate in the development of atherosclerosis making it a common mechanism and potential target of increased cardiometabolic risk in obesity and atherosclerosis [19]. ChCs are engulfed by macrophages which results in lysosomal destabilization and cathepsin B release leading to Nlrp3 inflammasome activation and accumulation of inflammatory cells [19]. Low-density lipoprotein (LDL) cholesterol in the vasculature is phagocytized by macrophages resulting in lipid rich foam cell formation and inflammation leading to plaque destabilization. Lysosomal damage by ChCs leads to caspase-1-dependent macrophage IL-1β secretion [19]. The progression of atherosclerotic lesions is associated with Pro-IL-1β [54]. Yajima et al. demonstrated for the first time that NLRP3 inflammasome contributes to atherogenesis. Studies in $ASC^{-/-}$ mice showed that ASC deficiency reduced the expression of IL-1β and IL-18 and attenuated atherosclerotic lesion formation [55]. Deficiency of NLRP3, ASC, or IL 1α/β is shown to decrease inflammasome- dependent IL-18 levels and attenuated atherosclerosis [19]. Studies in both human and murine macrophages have shown that ChCs induce NLRP3 inflammasome dependent IL-1β production which leads to both the development and destabilization of atherosclerotic lesions [56]. Bone marrow transplantation from NLRP3 or IL-1α/IL-1β-deficient mice significantly reduced atherosclerosis in high-cholesterol fed LDL receptor-deficient ($Ldlr^{-/-}$) mice indicating that macrophage-derived NLRP3 and IL-1 are important determining factors for cholesterol-driven atherosclerosis [19]. Studies in apolipoprotein E ($ApoE^{-/-}$) mice showed that nuclear factor erythroid 2- related factor (Nrf2) plays an essential role in the NRLP3 activation and atherosclerosis progression [57]. Endogenous cholesterol crystals act as pro-atherogenic danger signals. They initiate inflammation in the vasculature via

Nrf2 pathway. It has been identified that Nrf2 is required to induce inflammatory effects of IL-1β and IL-1α which occur both in a NLRP3/caspase-1-dependent and independent manner [57]. NLRP3 inflammasomes also cause atherogenesis via hyperhomocysteinemia-induced inflammation. A study done in $ApoE^{-/-}$ mice demonstrated that activation of NLRP3 inflammasome contributes to atherogenesis and hyperhomocysteinemia (HHcy) and that deletion of NLRP3 inflammasome ameliorates HHcy-induced inflammation and atherosclerosis [58]. Studies in human macrophages and aorta of patients with atherosclerosis also confirmed the role NLRP3 in atherosclerosis [59]. Paramel et al. recently reported that atherosclerotic plaques have high expression of NLRP3 inflammasome-related genes which were found to be activated upon exposure to ATP and cholesterol crystals [60]. Lectin-like oxidized low-density lipoprotein receptor-1 (LOX-1) expression, mitochondrial DNA damage and ROS generation were induced in human macrophages exposed to lipopolysaccharide followed by NLRP3 inflammasome activation. It has been demonstrated that in cardiovascular diseases such as atherosclerosis and myocardial ischemia, LOX-1-mediated autophagy and mtDNA damage induce NLRP3 inflammasome activation. Inhibitors of ROS are known to reduce NLRP3 inflammasome formation while autophagy inhibitors activate NLRP3 inflammasome [61].

Taken together, all these findings indicate that the NLRP3 inflammasome-mediated signaling is involved in atherosclerosis. Hence, NLRP3-inflammasome is an essential step in the inflammatory process involved in atherosclerosis and inflammasome inhibitors have a therapeutic role in the prevention of cardiovascular damage induced by obesogenic diets. In fact, cholesterol crystal-induced NLRP3 inflammasome activation is inhibited by arglabin which is an inflammasome inhibitor has shown to reduce atherosclerotic lesions in apolipoprotein mice [62].

NLRP3 Inflammasome in Hypertension

Obesity has been shown to increase the risk of high blood pressure [63]. Several studies indicated a clear association between increased weight gain and obesity with elevated blood pressure [7]. Obese people have a 3.5 fold increased risk of developing hypertension compared to the non-obese counterparts. Inflammasome activity and resulting IL-1β production was shown to play a key role in the progression of hypertension. Hypertensive patients were found to have elevated levels of circulating IL-1β [64]. Salt-sensitive hypertension is also characterized by chronic inflammation and high-salt-induced inflammation and oxidative stress activate NF-κB which is an effective activator of NLRP3 and contributes to the pathophysiology of hypertension [65]. Inhibition of NF-κB attenuates oxidative stress, NLRP3 inflammasome formation and activation followed by caspase-1 activation thus resulting in the reduction of pro-inflammatory cytokines in the paraventricular nucleus of salt-sensitive hypertensive rats [66]. NLRP3 inflammasome activation leads to the production of IL-1β which is an essential pro-inflammatory cytokine with pleiotropic effects. IL-1β inhibitor, gevokizumab, attenuated oxidative stress and release of pro-inflammatory cytokines leading to suppression of sympathoexitation in the paraventricular nucleus of salt-sensitive hypertensive Dahl rats, indicating that inhibition of IL-1β centrally reduces oxidative stress and renin-angiotensin system activation and delays hypertension-induced cardiovascular damage [67]. In hypertension, Angiotensin II (Ang II) is an essential vasoconstrictive peptide produced upon activation of renin-angiotensin system. Ang II is an important regulator of the inflammatory processes in hypertension. It has been demonstrated that Ang II infusion in murine hearts leads to NLRP3 inflammasome activation and increased expression of cytokines like IL-1β [68]. It has been demonstrated that NLRP3 inflammasome leads to cardiac remodeling independent of hypertension, since NLRP3 inflammasome inhibition attenuated Ang II-induced cardiac fibrosis without affecting blood pressure [68]. Thus, it can be concluded that in hypertensive cardiac diseases NLRP3 inflammasome/IL-1β nexus could be a potential therapeutic target for intervention.

NLRP3 Inflammasome in Myocardial Infarction and Cardiac Remodeling

Epidemiological studies suggest that both overweight and obesity increases the incidence of acute myocardial infarction. Myocardial infarction is commonly caused by the rupture of atherosclerotic plaques in the coronary arteries, narrowing of arteries and insufficient blood supply to the heart. Myocardial infarction is accompanied by inflammatory process leading to pathological changes in the myocardium and tissue injury. Inflammation is also involved in the tissue repair and recovery after myocardial infarction [69]. Studies in HFD diet-induced obese mice demonstrated large infarct sizes compared to mice on standard diet. In addition HFD mice also had increased NLRP3 inflammasome formation and activation [70]. NLRP3$^{-/-}$ mice were protected against the vascular hyper permeability and damage induced by HFD. Myocardial ischemia–reperfusion injury was also attenuated upon NLRP3 inhibition [71]. In myocardial ischemia by coronary artery ligation model, it was found that NLRP3 was upregulated in the cardiac fibroblasts of the ischemic hearts [72]. NLRP3 expression, capase-1 activity and IL-1β were increased in mouse models of ischemia-reperfusion injury [73]. Cytokine expression and inflammatory cell infiltration is significantly diminished in ASC$^{-/-}$or caspase-1$^{-/-}$ mice. ASC$^{-/-}$ or caspase-1$^{-/-}$ mice also exhibit reduced left ventricular dysfunction after myocardial infarction coupled with decreased infarct size and myocardial fibrosis. Taken together it indicates that inflammasome components, such as NLRP3 and ASC play a role in the myocardial damage following permanent coronary ligation in HFD mice.

Obesity is very well associated with left ventricular hypertrophy and cardiac remodeling. Several pathologies such as valvular heart diseases, myocarditis, myocardial infarction and dilated cardiomyopathy which gradually lead to progressive decompensation are accompanied by cardiac remodeling [74]. NF-κB and p38MAPK signaling are regulated by caspase activation and recruitment domain 3

(CARD3) which acts as a positive modulator of ventricular remodeling and dysfunction after myocardial infarction [75]. NLRP3 inflammasome activation is partially involved in the inflammatory responses following myocardial remodeling [76]. CARD3 and caspase-1 being important components of NLRP3 inflammasome activation play an integral role in cardiac remodeling following NLRP3 inflammasome activation [75].

Therapeutic Strategies Targeting Inflammasome

Effective inhibition of the NLRP3 inflammasome would be very useful in the treatment of multiple inflammatory and metabolic disease conditions. Therefore, there is an active search for therapeutic modulators of NLRP3 inflammasome pathways to address the unmet clinical needs. So far some of the major therapeutic strategies developed in this line include inhibition of NLRP3 inflammasome, inhibition of IL-1β and caspase-1 inhibition. The potential inflammasome inhibitors and their mechanism of action is described in Table 5.1.

Summary

In conclusion, NLRP3 inflammasome plays an important role in the pathogenesis of obesity and associated CVDs like atherosclerosis, hypertension, and myocardial infarction all of which lead to the increased mortality and morbidity in obese patients. NLRP3 inflammasome and IL-1β activity seem to regulate PAMPs and DAMPs during injury, inflammation, infection, or stress and contributes to a number of inflammatory disorders as well as obesity induced cardiovascular pathologies. Understanding the molecular components involved in the NLRP3 inflammasome formation is important to the development of safe therapeutic strategies. Experimental evidence from both in vitro and in vivo studies suggest that pharmacological interference with inhibitors targeting various mechanistic pathways leading NLRP3 inflammasome formation may offer novel approaches for treatment and prevention of obesity and associated CVD complications. Although there are no selective NLRP3 inhibitors available clinically at present, results from several undergoing translational studies with selective NLRP3 inhibitors are eagerly awaited in our fight against inflammatory diseases.

Table 5.1 List of the NLRP3 inflammasome inhibitors

NLRP3 inflammasome inhibitor	Mechanism of action
Direct/Indirect inhibition of NLRP3inflammasome	
MCC950	Inhibits NLRP3-induced ASC speck formation and oligomerization
BAY 11-7082	Alkylates cysteine residues within the ATPase region of NLRP3 and inhibits NLRP3 ATPase activity
OLT1177	Inhibits NLRP3 oligomerization
Glyburide	Prevents ASC oligomerization independent of K_{ATP} channels
BHB	Inhibits K^+ efflux and blocks ASC oligomerization (β-hydroxybutyrate)
Colchicine	Lysosomal destabilization, inhibition of the polymerization of ASC
IL-1 β antagonists	
Anakinra	IL-1β receptor antagonist which blocks 1β activity
Canakimumab	Monoclonal antibody which blocks 1β activity
Caspase-1 inhibitors	
VX-740 & VX765	Blocks pro-IL-1β cleavage
Autophagy inducers	
Metformin	Activate AMPK/autophagy and inhibit NLRP3 inflammasome
Resveratrol	Induces autophagy and attenuates mitochondrial damage
Arglabin	Induces autophagy process and inhibit NLRP3 inflammasome

Acknowledgements This work was supported by grants from American Heart Association [19AIREA34380223], National Institutes of Health [1R56HL143809-01A1] to S.K and National Institutes of Health [DK104031] to K.B.

Conflicts of Interests The authors of this manuscript declare that they have no conflicts of interests.

References

1. Czernichow S, Kengne AP, Stamatakis E et al (2011) Body mass index, waist circumference and waist-hip ratio: which is the better discriminator of cardiovascular disease mortality risk?: evidence from an individual-participant meta-analysis of 82864 participants from nine cohort studies. Obes Rev 12:680–687
2. Singh GM, Danaei G, Farzadfar F et al (2013) Global Burden of Metabolic Risk Factors of Chronic Diseases Collaborating G, Asia-Pacific Cohort Studies C, Diabetes Epidemiology: Collaborative analysis of Diagnostic criteria in E, Emerging Risk Factor C, Prospective Studies C: The age-specific quantitative effects of metabolic risk factors on cardiovascular diseases and diabetes: a pooled analysis. PLoS ONE 8:e65174

3. Van Gaal LF, Mertens IL, De Block CE (2006) Mechanisms linking obesity with cardiovascular disease. Nature 444:875–880
4. Yusuf S, Hawken S, Ounpuu S et al (2005) Obesity and the risk of myocardial infarction in 27,000 participants from 52 countries: a case-control study. Lancet 366:1640–1649
5. Boini KM, Hussain T, Li PL et al (2017) Trimethylamine-N-oxide instigates NLRP3 inflammasome activation and endothelial dysfunction. Cell Physiol Biochem 44:152–162
6. Chooi YC, Ding C, Magkos F (2019) The epidemiology of obesity. Metabolism 92:6–10
7. Nammi S, Koka S, Chinnala KM, Boini KM (2004) Obesity: an overview on its current perspectives and treatment options. Nutr J 3:3
8. Hubert HB, Feinleib M, McNamara PM, Castelli WP (1983) Obesity as an independent risk factor for cardiovascular disease: a 26-year follow-up of participants in the Framingham heart study. Circulation 67:968–977
9. Ritchie SA, Connell JM (2007) The link between abdominal obesity, metabolic syndrome and cardiovascular disease. Nutr Metab Cardiovasc Dis 17:319–326
10. Weiss R, Dziura J, Burgert TS et al (2004) Obesity and the metabolic syndrome in children and adolescents. N Engl J Med 350:2362–2374
11. Yanovski SZ, Yanovski JA (2002) Obesity. N Engl J Med 346:591–602
12. Mariathasan S, Newton K, Monack DM et al (2004) Differential activation of the inflammasome by caspase-1 adaptors ASC and Ipaf. Nature 430:213–218
13. Strowig T, Henao-Mejia J, Elinav E, Flavell R (2012) Inflammasomes in health and disease. Nature 481:278–286
14. Meylan E, Tschopp J, Karin M (2006) Intracellular pattern recognition receptors in the host response. Nature 442:39–44
15. Pedra JH, Cassel SL, Sutterwala FS (2009) Sensing pathogens and danger signals by the inflammasome. Curr Opin Immunol 21:10–16
16. Zhong Y, Kinio A, Saleh M (2013) Functions of NOD-like receptors in human diseases. Front Immunol 4:333
17. Davis BK, Wen H, Ting JP (2011) The inflammasome NLRs in immunity, inflammation, and associated diseases. Annu Rev Immunol 29:707–735
18. Broz P, Dixit VM (2016) Inflammasomes: mechanism of assembly, regulation and signalling. Nat Rev Immunol 16:407–420
19. Duewell P, Kono H, Rayner KJ et al (2010) NLRP3 inflammasomes are required for atherogenesis and activated by cholesterol crystals. Nature 464:1357–1361
20. Halle A, Hornung V, Petzold GC et al (2008) The NALP3 inflammasome is involved in the innate immune response to amyloid-beta. Nat Immunol 9:857–865
21. Martinon F, Petrilli V, Mayor AT et al (2006) Gout-associated uric acid crystals activate the NALP3 inflammasome. Nature 440:237–241
22. Zhou R, Tardivel A, Thorens B et al (2010) Thioredoxin-interacting protein links oxidative stress to inflammasome activation. Nat Immunol 11:136–140
23. Latz E, Xiao TS, Stutz A (2013) Activation and regulation of the inflammasomes. Nat Rev Immunol 13:397–411
24. Schroder K, Tschopp J (2010) The inflammasomes 140:821–832
25. Schroder K, Zhou R, Tschopp J (2010) The NLRP3 inflammasome: a sensor for metabolic danger? Science 327:296–300
26. Stienstra R, Tack CJ, Kanneganti TD et al (2012) The inflammasome puts obesity in the danger zone. Cell Metab 15:10–18
27. Lee HM, Kim JJ, Kim HJ et al (2013) Upregulated NLRP3 inflammasome activation in patients with type 2 diabetes. Diabetes 62:194–204
28. Vandanmagsar B, Youm YH, Ravussin A et al (2011) The NLRP3 inflammasome instigates obesity-induced inflammation and insulin resistance. Nat Med 17:179–188
29. Traba J, Sack MN (2017) The role of caloric load and mitochondrial homeostasis in the regulation of the NLRP3 inflammasome. Cell Mol Life Sci 74:1777–1791
30. Kursawe R, Dixit VD, Scherer PE et al (2016) A role of the inflammasome in the low storage capacity of the abdominal subcutaneous adipose tissue in obese adolescents. Diabetes 65:610–618

31. Mocanu AO, Mulya A, Huang H et al (2015) Effect of Roux-en-Y gastric bypass on the NLRP3 inflammasome in adipose tissue from obese rats. PLoS ONE 10:e0139764
32. Nagareddy PR, Kraakman M, Masters SL et al (2014) Adipose tissue macrophages promote myelopoiesis and monocytosis in obesity. Cell Metab 19:821–835
33. Esser N, L'Homme L, De Roover A et al (2013) Obesity phenotype is related to NLRP3 inflammasome activity and immunological profile of visceral adipose tissue. Diabetologia 56:2487–2497
34. Elliott EI, Sutterwala FS (2015) Initiation and perpetuation of NLRP3 inflammasome activation and assembly. Immunol Rev 265:35–52
35. Stutz A, Kolbe CC, Stahl R et al (2017) NLRP3 inflammasome assembly is regulated by phosphorylation of the pyrin domain. J Exp Med 214:1725–1736
36. Martinon F (2010) Signaling by ROS drives inflammasome activation. Eur J Immunol 40:616–619
37. Won JH, Park S, Hong S et al (2015) Rotenone-induced impairment of mitochondrial electron transport chain confers a selective priming signal for NLRP3 inflammasome activation. J Biol Chem 290:27425–27437
38. Zhou R, Yazdi AS, Menu P, Tschopp J (2011) A role for mitochondria in NLRP3 inflammasome activation. Nature 469:221–225
39. Abais JM, Xia M, Li G et al (2014) Nod-like receptor protein 3 (NLRP3) inflammasome activation and podocyte injury via thioredoxin-interacting protein (TXNIP) during hyperhomocysteinemia. J Biol Chem 289:27159–27168
40. Petrilli V, Papin S, Dostert C et al (2007) Activation of the NALP3 inflammasome is triggered by low intracellular potassium concentration. Cell Death Differ 14:1583–1589
41. Hornung V, Bauernfeind F, Halle A et al (2008) Silica crystals and aluminum salts activate the NALP3 inflammasome through phagosomal destabilization. Nat Immunol 9:847–856
42. Stienstra R, van Diepen JA, Tack CJ et al (2011) Inflammasome is a central player in the induction of obesity and insulin resistance. Proc Natl Acad Sci U S A 108:15324–15329
43. Wen H, Gris D, Lei Y et al (2011) Fatty acid-induced NLRP3-ASC inflammasome activation interferes with insulin signaling. Nat Immunol 12:408–415
44. Stienstra R, Joosten LA, Koenen T et al (2010) The inflammasome-mediated caspase-1 activation controls adipocyte differentiation and insulin sensitivity. Cell Metab 12:593–605
45. Boini KM, Xia M, Koka S et al (2017) Sphingolipids in obesity and related complications. Front Biosci (Landmark Ed) 22:96–116
46. Trayhurn P, Wood IS (2005) Signalling role of adipose tissue: adipokines and inflammation in obesity. Biochem Soc Trans 33:1078–1081
47. Lyon CJ, Law RE, Hsueh WA (2003) Minireview: adiposity, inflammation, and atherogenesis. Endocrinology 144:2195–2200
48. Membrez M, Ammon-Zufferey C, Philippe D et al (2009) Interleukin-18 protein level is upregulated in adipose tissue of obese mice. Obesity (Silver Spring) 17:393–395
49. Kizer JR (2014) Adiponectin, cardiovascular disease, and mortality: parsing the dual prognostic implications of a complex adipokine. Metabolism 63:1079–1083
50. Dinarello CA (2009) Immunological and inflammatory functions of the interleukin-1 family. Annu Rev Immunol 27:519–550
51. Feve B, Bastard JP (2009) The role of interleukins in insulin resistance and type 2 diabetes mellitus. Nat Rev Endocrinol 5:305–311
52. Odegaard JI, Chawla A (2008) Mechanisms of macrophage activation in obesity-induced insulin resistance. Nat Clin Pract Endocrinol Metab 4:619–626
53. De Nardo D, Latz E (2011) NLRP3 inflammasomes link inflammation and metabolic disease. Trends Immunol 32:373–379
54. Lu X, Kakkar V (2014) Inflammasome and atherogenesis. Curr Pharm Des 20:108–124
55. Yajima N, Takahashi M, Morimoto H et al (2008) Critical role of bone marrow apoptosis-associated speck-like protein, an inflammasome adaptor molecule, in neointimal formation after vascular injury in mice. Circulation 117:3079–3087

56. Rajamaki K, Lappalainen J, Oorni K et al (2010) Cholesterol crystals activate the NLRP3 inflammasome in human macrophages: a novel link between cholesterol metabolism and inflammation. PLoS ONE 5:e11765
57. Freigang S, Ampenberger F, Spohn G et al (2011) Nrf2 is essential for cholesterol crystal-induced inflammasome activation and exacerbation of atherosclerosis. Eur J Immunol 41:2040–2051
58. Wang R, Wang Y, Mu N et al (2017) Activation of NLRP3 inflammasomes contributes to hyperhomocysteinemia-aggravated inflammation and atherosclerosis in apoE-deficient mice. Lab Invest 97:922–934
59. Zheng F, Xing S, Gong Z, Xing Q (2013) NLRP3 inflammasomes show high expression in aorta of patients with atherosclerosis. Heart Lung Circ 22:746–750
60. Paramel Varghese G, Folkersen L et al (2016) NLRP3 inflammasome expression and activation in human atherosclerosis. J Am Heart Assoc 5:e003031
61. Ding Z, Liu S, Wang X, Dai Y et al (2014) LOX-1, mtDNA damage, and NLRP3 inflammasome activation in macrophages: implications in atherogenesis. Cardiovasc Res 103:619–628
62. Abderrazak A, Syrovets T, Couchie D et al (2015) NLRP3 inflammasome: from a danger signal sensor to a regulatory node of oxidative stress and inflammatory diseases. Redox Biol 4:296–307
63. Din-Dzietham R, Liu Y, Bielo MV, Shamsa F (2007) High blood pressure trends in children and adolescents in national surveys, 1963 to 2002. Circulation 116:1488–1496
64. Krishnan SM, Sobey CG, Latz E et al (2014) IL-1beta and IL-18: inflammatory markers or mediators of hypertension? Br J Pharmacol 171:5589–5602
65. Yu XJ, Zhang DM, Jia LL et al (2015) Inhibition of NF-kappaB activity in the hypothalamic paraventricular nucleus attenuates hypertension and cardiac hypertrophy by modulating cytokines and attenuating oxidative stress. Toxicol Appl Pharmacol 284:315–322
66. Qi J, Yu XJ, Shi XL et al (2016) NF-kappaB blockade in hypothalamic paraventricular nucleus inhibits high-salt-induced hypertension through NLRP3 and caspase-1. Cardiovasc Toxicol 16:345–354
67. Qi J, Zhao XF, Yu XJ et al (2016) Targeting interleukin-1 beta to suppress sympathoexcitation in hypothalamic paraventricular nucleus in Dahl salt-sensitive hypertensive rats. Cardiovasc Toxicol 16:298–306
68. Gan W, Ren J, Li T et al (1864) The SGK1 inhibitor EMD638683, prevents Angiotensin II-induced cardiac inflammation and fibrosis by blocking NLRP3 inflammasome activation. Biochim Biophys Acta Mol Basis Dis 1–10:2018
69. Frangogiannis NG (2014) The inflammatory response in myocardial injury, repair, and remodelling. Nat Rev Cardiol 11:255–265
70. Mastrocola R, Collino M, Penna C et al (2016) Maladaptive modulations of NLRP3 inflammasome and cardioprotective pathways are involved in diet-induced exacerbation of myocardial ischemia/reperfusion injury in mice. Oxid Med Cell Longev 2016:3480637
71. Marchetti C, Chojnacki J, Toldo S et al (2014) A novel pharmacologic inhibitor of the NLRP3 inflammasome limits myocardial injury after ischemia-reperfusion in the mouse. J Cardiovasc Pharmacol 63:316–322
72. Sandanger O, Ranheim T, Vinge LE et al (2013) The NLRP3 inflammasome is up-regulated in cardiac fibroblasts and mediates myocardial ischaemia-reperfusion injury. Cardiovasc Res 99:164–174
73. Liu Y, Lian K, Zhang L et al (2014) TXNIP mediates NLRP3 inflammasome activation in cardiac microvascular endothelial cells as a novel mechanism in myocardial ischemia/reperfusion injury. Basic Res Cardiol 109:415
74. Sutton MG, Sharpe N (2000) Left ventricular remodeling after myocardial infarction: pathophysiology and therapy. Circulation 101:2981–2988
75. Li L, Wang X, Chen W et al (2015) Regulatory role of CARD3 in left ventricular remodelling and dysfunction after myocardial infarction. Basic Res Cardiol 110:56
76. Cassel SL, Sutterwala FS (2010) Sterile inflammatory responses mediated by the NLRP3 inflammasome. Eur J Immunol 40:607–611

Chapter 6
Vicious Link of Obesity with Cardiometabolic and Renal Diseases

Amrit Pal Singh, Tajpreet Kaur, and Harpal Singh Buttar

Abstract Obesity is escalating all over the world and prevails among 13% of adult population. World Health Organization (WHO) has estimated that excessive body weight and obesity related incidences of type 2 diabetes mellitus (T2D) and cardiovascular diseases (CVDs) has increased nearly fourfold over the last 25 years. Excessive deposition of peripheral and visceral fat also causes metabolic syndrome and renal complications. In obese subjects, the risk of non-communicable diseases (NCDs) such as musculoskeletal and neurodegenerative disorders, infertility, and breast cancer is relatively higher than lean persons. The white adipocytes secrete a wide variety of bioactive chemicals such as adipokines, resistin, leptin, interleukins (IL-1β, IL-6), tumor necrosis factor-α (TNF-α), interferon-γ (IFN-γ), monocyte chemoattractant protein-1 (MCP-1), free fatty acid, macrophages infiltration, mast cell degranulation, plasminogen activator inhibitor-1 (PAI-1), endothelial adherence molecules and oxidative stress. These bioactive chemicals play crucial role in the pathogenesis of obesity-induced disorders like insulin resistance, dyslipidemia, metabolic syndrome, atherosclerosis, thrombosis, vasculopathy, high blood pressure, glomerulopathy and glomerulosclerosis. Well planned health care strategies are needed to reduce the risk of nongenetic factors associated with obesity, and their links with T2D, CVDs and renal diseases. The health-care burden related to NCDs such as obesity, T2D, and CVDs and neurodegenerative disorders, renal diseases and cancer is escalating worldwide. People need to think about the cost-effective measures such as lifestyle modifications, unhealthy dietary habits, physical activity, and consumption of healthful foods containing green vegetables, fruits, and Mediterranean-type diet consisting of olive oil, poultry and fish, dairy products, fiber rich foods, and low amount of red meat. The focus of this review is to highlight the

A. P. Singh (✉)
Department of Pharmaceutical Sciences, Guru Nanak Dev University, Amritsar, Punjab, India
e-mail: apsingh82@gmail.com

T. Kaur
Department of Pharmacology, Khalsa College of Pharmacy, Amritsar, India

H. S. Buttar
Department of Pathology & Laboratory Medicine, Faculty of Medicine, University of Ottawa, Ottawa, ON, Canada

P. S. Tappia et al. (eds.), *Biochemistry of Cardiovascular Dysfunction in Obesity*, Advances in Biochemistry in Health and Disease 20,
https://doi.org/10.1007/978-3-030-47336-5_6

relationships of obesity-induced production of inflammatory cytokines, adipokines, distortion of carbohydrate and lipid metabolism, kidney malfunction, activation of the renin-angiotensin-aldosterone system, arterial hypertension, and heart attack and stroke.

Keywords Obesity · Cardiometabolic disorders · Obesity-associated kidney dysfunction · Activation of renin-angiotensin-aldosterone system · Arterial hypertension · CVDs

Introduction

Obesity is characterized by excessive accumulation of body fat and is recognized as a global epidemic health problem. The incidence of obesity has doubled between 1980 and 2014, and is increasing among children and middle age-populations [1, 2]. It is estimated that about 8.5% of world population is inflicted with obesity that includes around 78 million people worldwide [3]. Industrialization and urbanization in developing countries have not only changed the living standards, but have also modified the lifestyles and dietary habits of children and adults. Excessive intake of carbohydrate diets, fatty foods, sugar loaded drinks, salty foods and inactivity promote rapid weight gain in all populations [4–7]. It has been reported that lower energy expenditure due to lack of physical activity promotes weight gain, and life expectancy is reduced by almost seven years in obese persons [8–10]. Generally, accumulation of excessive fat occurs in skeletal muscles, liver, peritoneum, gastro-intestinal tract, abdomen, buttocks and mammary tissues [11–13]. Obesity causes deterioration of metabolic functions, and consequently produces pathological changes in the cardiovascular and renal systems [14]. Overwhelming number of studies have indicated that obesity is a major risk factor for neuro-degenerative disorders, cardiovascular diseases (CVDs), type-2 diabetes mellitus (T2D), fatty liver, respiratory problems, renal disorders and cancer [15–22]. There is a higher prevalence of metabolic syndrome, osteoarthritis and pulmonary hypoventilation in obese patients [15, 23–25].

During the last 25 years, obesity related T2D and CVDs have become a major public health hazard in developed and developing countries [26, 27]. T2D and CVD rates have risen dramatically in Asia, Africa, and Middle Eastern countries [28], and premature mortality and morbidity due to heart disease and stroke are escalating in the relatively younger population under the age of 50 years [29, 30]. It has been reported that in Asian countries, obesity related disorders such as T2D, hypertension and CVDs often occur in younger age groups than western countries [31, 32]. CVD-related death rate is generally higher among low and middle income populations [33–35]. Heart attack and stroke impose very high burden on the health care system due to prolonged rehabilitation, hospital and drug costs as well as employee absenteeism. It has been estimated that obesity related disorders cost around 4 to 8% of total health care budget in several countries [35–39]. Thus, prevention and management of obesity, T2D, CVDs and kidney diseases are targeted as the most important public health issues worldwide. Currently, cost-effective public health policies are being directed to lay out appropriate anti-obesity measures in children and adults, including healthy dietary habits, consumption of fresh vegetables and fruits, reduction of sugar loaded beverages, and intake of Mediterranean style diets and encourage exercise as useful non-pharmacologic therapies. Several types of dietary formulations and herbal remedies are also being promoted in lay press to reduce weight gain [25, 40–42].

Most commonly used index of obesity is body mass index (BMI) which measures body weight in kilograms divided by height in meters square (kg/m^2) [1]. According to World Health Organization (WHO), individuals with BMI ranging from 18.5 to 25.0 kg/m^2 are considered normal healthy. On the other hand, people with BMI < 18.5 kg/m^2 are underweight. Those with BMI 25–30 kg/m^2 are regarded over-weight, whereas individuals with BMI > 30–35.5 kg/m^2 are considered moderately obese (Class I), and BMI > 35.5–40 kg/m^2 severely obese (Class II), and BMI > 40 kg/m^2 are categorized as extremely obese (Class III). Apart from BMI, MRI, anthropometry and computed tomography densitometry are other approaches to measure obesity [1].

There are 2 types of adipose tissues, white adipose tissue (WAT) and brown adipose tissue (BAT). The total body composition of lean adult men and women consists of about 20 % white adipose tissue (WAT), however, in obese humans WAT can increase far upto >40% [43–46]. The WAT serves as passive depot for energy storage in the form of lipids, and release of free fatty acids and adipokines. Adipocytes of white adipose tissue have endocrine and paracrine functions and secrete wide range of adipokines which regulate appetite, insulin sensitivity and angiogenesis [47–49]. The WAT produces pro-inflammatory cytokines such as tumor necrosis factor TNF-α, interleukins like IL-1, IL-6, IL-8, IL-18, monocyte chemoattractant protein-1 (MCP-1), transforming growth factor-β (TGF-β), adrenomedullin and calcitonin gene-related peptide, which are responsible for causing pathological conditions like endothelial inflammation and atherosclerosis, CVDs, T2D and metabolic syndrome [50–54]. Obesity promotes pathological conditions such as metabolic syndrome which is characterized by hyperglycemia, insulin resistance, dyslipidemia, thrombosis, atherosclerosis, hypertension, and systemic inflammation [47, 55]. WAT produced adipokines like leptin, resistin, adiponectin, TNF-α, and IL-6 govern food intake, energy balance and insulin sensitivity by impinging upon the hypothalamus and vagal inputs [56–58]. Augmentation of WAT and deficiency of BAT produces increased releases of adipokines, pro-inflammatory markers, angiotensin converting enzyme (ACE) and reduction of lipoprotein lipase [59, 60].

In neonates, BAT make upto 5% of total body weight which helps to prevent hypothermia due to higher ratio of body-surface area to body-volume and lack of thermal insulators like subcutaneous fat and body hairs. BAT serves as alternative to regulate body heat [61, 62]. Mitochondrial rich BAT activates peroxisome proliferator activated receptor-γ (PPAR-γ) and uncoupling protein-1 (UCP-1) which improves metabolic rate and reduces body weight gain through thermogenesis of lipids present in WAT and increases utilization of energy [63, 64]. The decreased BAT abates expression of UCP-1 which declines utilization of fat stored in white adipose tissue [47, 63–65].

It has been hypothesized that altered gut microbiota (bacteriodes, staphylococcus, bi-fido bacterium, lactobacillus) increase the risk of childhood overweight [66]. Intake of probiotics like yogurt and cheese improve growth of beneficial bacteria in the gut and reduce the incidence of childhood obesity [67, 68]. Prebiotics like indigestible oligosaccharides also promote growth of beneficial bacteria in the gastrointestinal tract and reduce the risk of irritable bowel syndrome [69].

Obesity-Induced Cardiovascular Disorders

The obesity related cardiometabolic alterations are illustrated in Fig. 6.1. Obesity produces cardio-metabolic syndrome which is characterized by high blood pressure, hyperglycemia, hyperlipidemia, reduced high density lipoproteins (HDL), increased low density lipoprotein (LDL), atherosclerosis and vascular diseases. WAT causes the development of systemic inflammatory state that provokes high incidences of cardiometabolic disorders in children and young adults [70–74]. In obesity, the circulating blood volume is increased that puts high demand on function of left ventricle and consequently causes left ventricular remodeling. Generally, these effects result in left ventricular dilation with eccentric left ventricular hypertrophy, increased cardiac output and cardiac overload. The cardiac hypertrophy increases cardiac mass due to

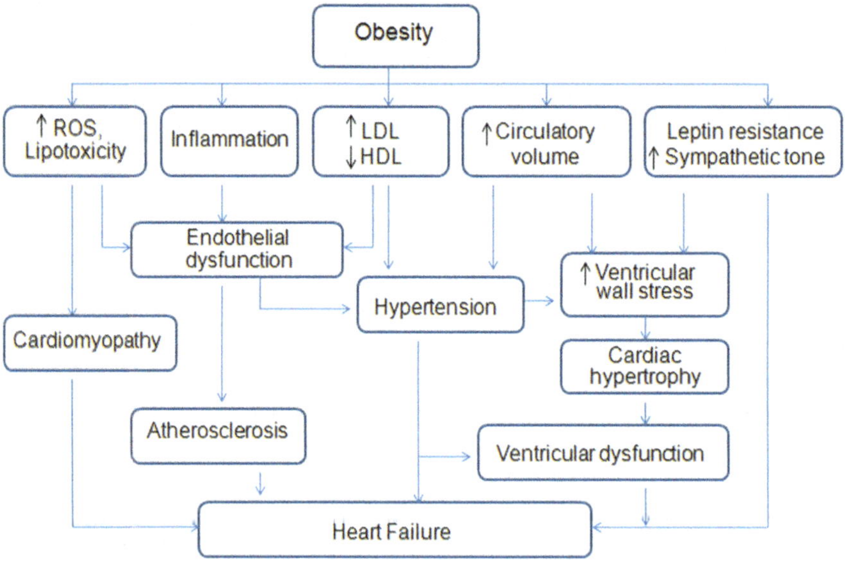

Fig. 6.1 Diagrammatic representation of various mechanisms involved in obesity-induced ROS lipotoxicity, inflammation, high density HDL and low density LDL lipoproteins, endothelial dysfunction, atherosclerosis, and cardiovascular events and heart failure. The upward arrows indicate increase and downward arrows indicate decrease

compensatory mechanism to manage pressure overload or volume overload, thereby producing excessive stress on the heart function [2, 75]. In obese patients, there is increase in chances of developing heart failure is enhanced by 5–7% for every 1 BMI (kg/m^2) increase and contributes upto 11–14% of all heart failure cases [76].

The Frank-Starling curve is often shifted to the left because of increase in filling pressure and volume, thus increasing cardiovascular work. Obesity is responsible for alteration in various cardiovascular function i.e. alterations in volume and pressure overload [30, 77]. The activation of renin angiotensin aldosterone system (RAAS) causes cardiac overload through an increase in blood volume in obese individuals [78]. Obesity also leads to left atrial enlargement, both by increased circulating blood volume as well as abnormal left ventricular diastolic filling [78, 79]. These abnormalities not only increase the risk of heart failure, but also left atrial enlargement, increase of atrial flutter and cardiovascular complications [80].

Obese patients are more likely to be hypertensive than lean patients and their high blood pressure is primarily due to increase in systolic arterial pressure [81]. Nearly 60% of obese persons are hypertensive owing to the activation of RAAS, increased sensitivity of the sympathetic nervous system, leptin resistance, elevated cardiac output, expanded vascular volume, and reduction in cardiopulmonary function [82–84]. In obese patients, there is also escalation of inflammatory proteins in the systemic circulation, intracellular adhesion molecule-1 (ICAM-1), vascular cell adhesion molecule-1 (VCAM-1), and P-selectin as well as increase in C-reactive protein (CRP). CRP also activates endothelial nuclear factor-κB (NF-κB), IL-1β, plasminogen activator inhibitor-1 (PAI-1), IL-6, TNF-α, MCP-1, endothelin-1 and inhibition of endothelial nitric oxide synthase (eNOS) [85]. PAI-1 serves as a regulatory cascade of coagulation, which is primarily derived from platelets and gets elevated in case of inflammatory disorders during obesity. It also promotes hypercoagulation and contributes to cause atherosclerosis through deposition of platelets and fibrin molecules in blood vessels [85, 86]. TNF-α and IL-6 contributes towards vasculopathy. IL-6 declines lipoprotein lipase activity, which increases lipid uptake by vessel walls, resulting in atherosclerosis [87, 88]. The WAT secretes serum amyloid A that also contribute towards atherogenesis [48, 85]. The vascular inflammation is amplified by CRP and increased expression of adhesion proteins and cytokines that cause adherence of leukocytes in blood vessels. The inhibition of eNOS causes vasoconstriction, induction of endothelin-1 and P-selectin, pro-aggregatory effects, and consequently increases the risk of CVDs [89, 90]. The production of angiotensinogen and ACE from the adipocytes causes inflammation of blood vessels through stimulation of MCP-1, VCAM-1, and ICAM-1, increase in blood volume due to the Na^+ and body water expansion. Obesity also causes hardening of blood vessels by promoting the secretion of aldosterone and stimulating β-1 receptors present in juxtaglomerular cells of the kidney [91]. Escalation of leptin resistance in obesity causes insulin resistance, depresses parasympathetic activity and stimulates sympathetic activity, thereby causing vasculopathy [92, 93].

The subepicardial adipose tissue (SEAT) releases FFA that provides energy to epimyocardium and coronary arteries. The nerve growth factor (NGF) and mast cell infiltration are also expressed in atherosclerotic SEAT. The SEAT produces atherosclerotic lesions in left anterior descending coronary artery resulting in myocardial infarction [94].

Obesity-Induced T2D and Associated Complications

Figure 6.2 summarizes obesity induced T2D and associated complications. Insulin deficiency and hyperglycemia are the key features of T2D. Obesity down-regulates the GLUT-4 expression and decrease in glucose uptake by tissues and organs, thereby causing hyperglycaemia [95–97]. On the other hand, obesity-induced up-regulation of resistin and decreased adiponectin levels contributes to peripheral insulin resistance. The enhanced levels of blood glucose and microalbuminuria lead to diabetic nephropathy and cardiovascular complications [48, 57, 98–101]. Sustained hyperglycaemia-induced oxidative stress also causes retinopathy and neuropathy [102, 103].

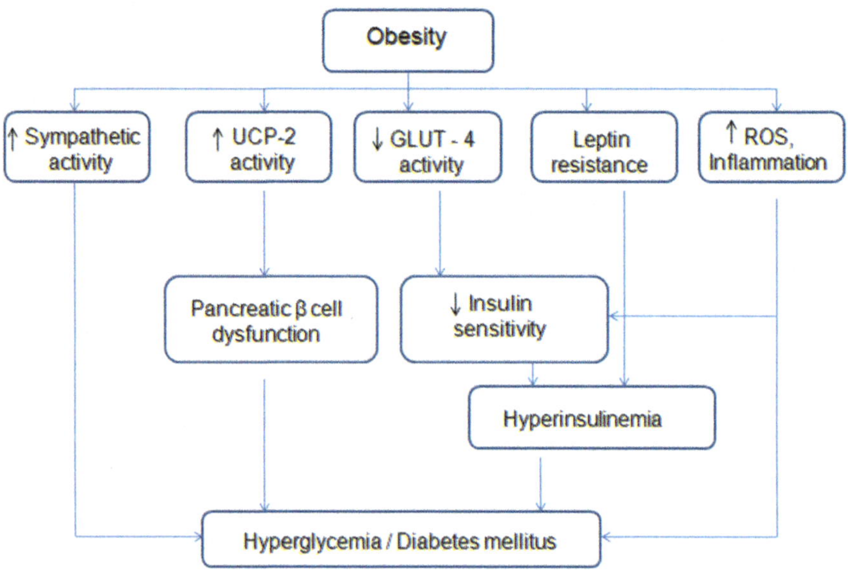

Fig. 6.2 Diagrammatic representation of various mechanisms involved in obesity-induced metabolic syndrome: increased sympathetic activity, high UCP-2 and low GLUT-4 activity, leptin resistance, increased production of ROS, pancreatic B-cell dysfunction, reduced insulin sensitivity and hyperinsulinemia, hyperglycemia and diabetes mellitus. The upward arrows indicate increase and downward arrows indicate decrease

Oxidative stress induced by increased glucose level results in the pathogenesis of various chronic complications of diabetes mellitus [104, 105]. Various metabolic alterations are induced by hyperglycemia and results in remarkable increase in oxidative stress to cells. Redox imbalance or oxidative stress is the outcome of intense generation of reactive oxygen species (ROS) and of reactive nitrogen species (RNS) [106, 107]. When not counteracted by endogenous pro-oxidants and anti-oxidant molecules, oxidative stress can cause DNA lesions and loss of cell membrane integrity due to lipid peroxidation as well as protein and carbohydrate structural changes [107–109]. Oxidative damage is linked to many types of pathologies such as cardiometabolic diseases (e.g. atherosclerosis, hypertension, T2D, CVDs), neurodegenerative disorders, auto-immune and rheumatic diseases or cancer [107, 110, 111]. Endogenous anti-oxidants that can scavenge free radicals include glutathione and enzymes (catalase, superoxide dismutase) produced internally, and dietary antioxidants like vitamin A, C and E. Elements like selenium, zinc, copper and iron-binding proteins such as ferritin and transferrin contribute to anti-oxidant defense by quenching free radicals and inhibiting lipid peroxidation [112, 113]. Mediterranean style of diet containing fruits, nuts, vegetables and foods rich in anti-oxidants reduce the risk of CVDs, T2D, and non-communicable diseases by lowering oxidative stress in cells [25, 40–42].

Obesity-Induced Renal Malfunction

Figure 6.3 depicts obesity-induced changes in renal architecture and its functions. In obese persons, the kidney weight is increased by upto 40%. Obesity causes podocyte injury, expansion of mesangial cells, glomerulosclerosis, and increase the risk of renal cell carcinoma [16, 114]. As mentioned above, the WAT increases production of RAAS and inflammatory cytokines, consequently resulting in lipotoxicity. The leptin resistance suppresses parasympathetic activity and increases renal sympathetic excitation that alters baroreflex control in obese persons [19, 100]. The increased production of pro-inflammatory molecules and oxidative stress leads to renal malfunction [115].

Increase in ACE and angiotensin activities in obesity enhance tubular reabsorption of Na^+ and water by the kidney tubules and lead to efferent arteriolar constriction of glomerulus thereby causing hypertensive nephrosclerosis [116]. Elevated levels of leptin resistance in obesity may predispose the individual to glomerulosclerosis through up-regulation of transforming growth factor-β (TGF-β). Hyperinsulinemia stimulates the production of insulin like growth factors (IGF), that promotes glomerular hypertrophy, whereas hyperlipidemia promotes glomerulosclerosis with the engagement of low density lipoprotein receptors on mesangial cells and fibrogenic chemokines [19]. In nephropathy, there is increase in intimal thickening and narrowing of lumen of renal arteries and arterioles. The dysregulated adipokines leads to hemodynamic and structural changes in kidneys and renal malfunction [116, 117]. Diabetes-induced nephropathy and immunoglobulin A induced glomerulonephritis

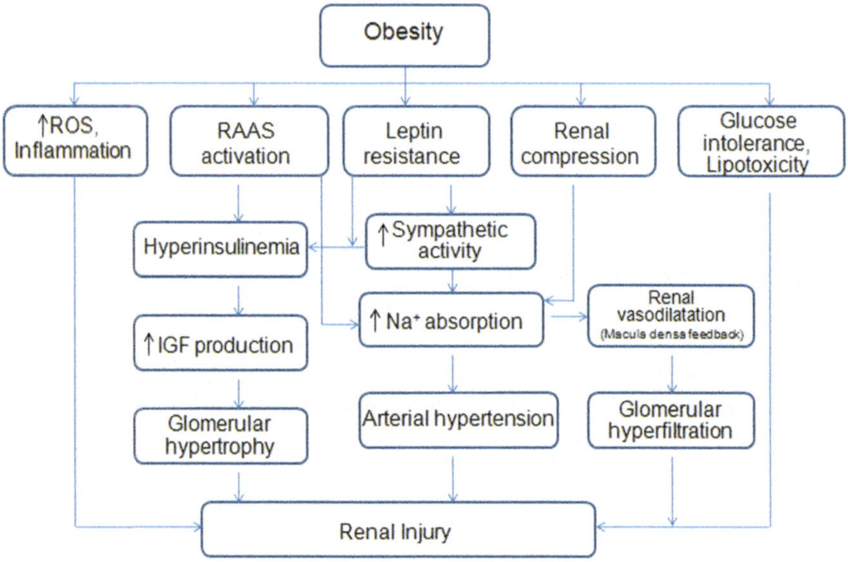

Fig. 6.3 Diagrammatic representation of various mechanisms involved in obesity-induced kidney malfunction and its sequela: renal compression and vasodilation, glomerular hypertrophy, increased activities of the autonomic sympathetic system and renin-angiotensin-aldosterone system (RAAS), increased absorption of Na^+ and arterial hypertension

are aggravated during obesity [118]. Figure 6.3 summarises obesity induced changes in renal architecture and its functions.

Obesity-Induced Inflammatory Disorders

Besides secreting beneficial adipokines, the white adipocytes produce pro-inflammatory cytokines which increase the risk of non-communicable diseases. Obesity also promotes circulating levels of inflammatory proteins like C-reactive protein (CRP-1), IL-6, vascular cell adhesion molecule-1 (VCAM-1), serum amyloid A3 (SAA3), fibrinogen, IFN-γ, MCP-1, PAI-1 and mast cell degranulation [85, 119]. The bioactive products released from macrophages and mast cells cause inflammatory diseases, namely atherosclerosis, osteoarthritis and auto-immune problems. Obesity-induced collagen deposition results in systemic inflammatory complications such as thrombosis, vasculopathy, fatty liver, respiratory disorders, glomerulosclerosis, glomerulopathy, decreased in insulin sensitivity, decreased glucose transporter-4 (GLUT-4) expression, tissue fibrosis and some cancers [14, 85, 119, 120]. Reduced levels of adiponectin have been noted to suppression of gluconeogenesis and causing inflammatory disorders [48, 57, 58].

Conclusion

The occurrence of obesity has escalated globally especially among children and middle age populations. Increased intake of saturated fat and carbohydrate rich diets and lower energy expenditure due to inactivity result in excessive accumulation of WAT in the body. In obese men and women (BMI > 30 kg/m^2), WAT plays a crucial endocrine/paracrine role and secrete a wide variety of adipokines and pro-inflammatory cytokines. Obesity-induced up-regulation of inflammatory cytokines are linked with an array of pathological conditions such as hyperglycaemia, insulin resistance, atherosclerosis, hyperlipidemia, hypertension, and increased risk of T2D, CVDs, renal dysfunction, and increased risk of breast cancer. The high occurrence of premature mortality and morbidity associated with obesity, T2D and CVDs create unusual economic pressure on the family and society as well as national economy due to prolonged treatment, rehabilitation, hospitalization, drug costs and employee absenteeism. To decrease the incidences of obesity related complications described in this review, holistic approaches are needed to educate people about the significant health risks associated with obesity and to promote healthy eating habits and benefits of exercise. There is overwhelming evidence that intake of Mediterranean-type diet containing vegetables, fruits, omega-3-fatty acids, fish and poultry diet, low fat dairy products and olive oil help to reduce the incidence of obesity as well as obesity linked disorders discussed in this review.

Conflict of Interest The authors declare no conflict of interest.

References

1. Rubenstein AH (2005) Obesity: a modern epidemic. Trans Am Clin Climatol Assoc 116:103–113
2. Lavie CJ, Alpert MA, Arena R et al (2013) Impact of obesity and the obesity paradox on prevalence and prognosis in heart failure. JACC: Heart Fail 1:93–102
3. World Health Organization (2015) Obesity and overweight. Fact sheet number 311, January
4. Mohan V, Shanthirani S, Deepa R et al (2001) Intra-urban differences in the prevalence of the metabolic syndrome in southern India—the Chennai urban population study (CUPS No. 4). Diabet Med 18:280–287
5. Deshmukh PR, Gupta SS, Dongre AR et al (2006) Relationship of anthropometric indicators with blood pressure levels in rural Wardha. Indian J Med Res 123:657–664
6. Poobalan A, Aucott L (2016) Obesity among young adults in developing countries: a systematic overview. Curr Obes Rep 5:2–13
7. Popkin BM, Adair LS, Wen NS (2012) Now and then: the global nutrition transition: the pandemic of obesity in developing countries. Nutr Rev 70:3–21
8. Müller MJ, Enderle J, Bosy-Westphal A (2016) Changes in energy expenditure with weight gain and weight loss in humans. Curr Obes Rep 5:413–423
9. Peeters A, Barendregt JJ, Willekens F et al (2003) Obesity in adulthood and its consequences for life expectancy: a life-table analysis. Ann Intern Med 138:24–32
10. Noeman SA, Hamooda HE, Baalash AA (2011) Biochemical study of oxidative stress markers in the liver, kidney and heart of high fat diet induced obesity in rats. Diabetol Metab Syndr 3:17

11. Friedman JM, Halaas JL (1998) Leptin and the regulation of body weight in mammals. Nature 395:763–770
12. Griffin N, Burke C, Grant LA (2011) Common primary tumours of the abdomen and pelvis and their patterns of tumour spread as seen on multi-detector computed tomography. Insights Imag 2:205–214
13. Hovey RC, Aimo L (2010) Diverse and active roles for adipocytes during mammary gland growth and function. J Mammary Gland Biol Neoplasia 15:279–290
14. Hall ME, do Carmo JM, da Silva AA et al (2014) Obesity, hypertension, and chronic kidney disease. Int J Nephrol Renovasc Dis 7:75–88
15. Marceau P, Biron S, Hould FS et al (1999) Liver pathology and the metabolic syndrome X in severe obesity. J Clin Endocrinol Metab 84:1513–1517
16. Calle E, Rodriguez C, Walker-Thurmond K, Thun M (2003) Overweight, obesity, and mortality from cancer in a prospectively studied cohort of US adults. N Engl J Med 48:1625–1638
17. Orlić L, Mikolasevic I, Bagic Z et al (2014) Chronic kidney disease and nonalcoholic fatty liver disease-is there a link? Gastroenterolo Res Pract. Article ID 847539
18. Hamilton MT, Hamilton DG, Zderic TW (2007) Role of low energy expenditure and sitting in obesity, metabolic syndrome, type 2 diabetes, and cardiovascular disease. Diabetes 56:2655–2667
19. Naumnik B, Myśliwiec M (2010) Renal consequences of obesity. Med Sci Monit 16:RA163–170
20. Felizardo RJ, da Silva MB, Aguiar CF, Câmara NO (2014) Obesity in kidney disease: a heavy weight opponent. World J Nephrol 3:50–63
21. Kyrou I, Randeva HS, Weickert MO, Clinical problems caused by obesity. Endotext [Internet]
22. Paschos P, Paletas K (2009) Non alcoholic fatty liver disease and metabolic syndrome. Hippokratia 13:9–19
23. Villareal DT, Apovian CM, Kushner RF, Klein S (2005) Obesity in older adults: technical review and position statement of the American Society for Nutrition and NAASO, The Obesity Society. Am J Clin Nutr 82:923–934
24. King LK, March L, Anandacoomarasamy A (2013) Obesity & osteoarthritis. Indian J Med Res 138:185–193
25. Asif M (2014) The prevention and control the type-2 diabetes by changing lifestyle and dietary pattern. J Educ Health Promot 3:1
26. World Health Organization (2016) Cardiovascular diseases (CVDs). Fact sheet number, June
27. Seidell JC (2000) Obesity, insulin resistance and diabetes-a worldwide epidemic. Br J Nutr Suppl 1:S5-8
28. Hossain P, Kawar B, El Nahas M (2007) Obesity and diabetes in the developing world-a growing challenge. N Engl J Med 356:213–215
29. Eckel RH (1997) Obesity and heart disease. A statement for healthcare professionals from the nutrition committee, American heart association. Circulation 96:3248–3250
30. Poirier P, Giles TD, Bray GA et al (2006) Obesity and cardiovascular disease: pathophysiology, evaluation, and effect of weight loss. An update of the 1997 American Heart Association scientific statement on obesity and heart disease from the obesity committee of the council on nutrition, physical activity, and metabolism. Circulation 113:898–918
31. Ramachandran A, Snehalatha C (2010) Rising burden of obesity in Asia. J Obes pii 868573
32. Mungreiphy NK, Dhall M, Tyagi R et al (2009) Ethnicity, obesity and health pattern among Indian population. J Nat Sci Biol Med 3:52–59
33. Li N, Katzmarzyk PT, Horswell R et al (2014) BMI and coronary heart disease risk among low-income and underinsured diabetic patients. Diabetes Care 37:3204–3212
34. Levine JA (2011) Poverty and obesity in the U.S. Diabetes 60:2667–2668
35. Laslett LJ, Alagona P Jr, Clark BA 3rd et al (2012) The worldwide environment of cardiovascular disease: prevalence, diagnosis, therapy, and policy issues: a report from the American College of Cardiology. J Am Coll Cardiol 60(25 Suppl):S1-49

36. Seidell JC (1995) The impact of obesity on health status: some implications for health care costs. Int J Obes Relat Metab Disord 19(Suppl 6):S13-16
37. Wolf AM, Colditz GA (1998) Current estimates of the economic cost of obesity in the United States. Obes Res 6:97–106
38. Counterweight Project Team (2005) The impact of obesity on drug prescribing in primary care. Br J Gen Pract 55:743–749
39. World Health Organization (2016) Obesity and overweight. Fact sheet number, June
40. Georgoulis M, Kontogianni MD, Yiannakouris N (2014) Mediterranean diet and diabetes: prevention and treatment. Nutrients 6:1406–1423
41. Yumuk VA, Tsigos CB, Fried MC et al (2015) European guidelines for obesity management in adults. Obes Facts 8:402–424
42. Swinburn BA, Caterson I, Seidell JC, James WPT (2004) Diet, nutrition and the prevention of excess weight gain and obesity. Public Health Nutr 7:123–146
43. Rinzler CA (2009) The encyclopedia of cosmetic and plastic surgery, chapter 1, p 6
44. Dietz H (2005) Clinical obesity in adults and children, chapter 2, p 103
45. Palmer AK, Kirkland JL (2016) Aging and adipose tissue: potential interventions for diabetes and regenerative medicine. Exp Gerontol 86:97–105
46. World Health Organization (WHO) (2000) Obesity: preventing and managing the global epidemic. Report of a WHO consultation. World Health Organization, Geneva, Switzerland Technical Report Series 894
47. Coelho M, Oliveira T, Fernandes R (2013) Biochemistry of adipose tissue: an endocrine organ. Arch Med Sci 9:191–200
48. Trayhurn P, Bing C (2006) Appetite and energy balance signals from adipocytes. Philos Trans R Soc Lond B Biol Sci 361:1237–1249
49. Ronti T, Lupattelli G, Mannarino E (2006) The endocrine function of adipose tissue: an update. Clin Endocrinol 64:355–365
50. Johnson AR, Milner JJ, Makowski L (2012) The inflammation highway: metabolism accelerates inflammatory traffic in obesity. Immunol Rev 249:218–238
51. González-Muniesa P, Garcia-Gerique L, Quintero P et al (2016) Effects of hyperoxia on oxygen-related inflammation with a focus on obesity. Oxid Med Cell Longev 8957827
52. Ramos-Nino ME (2013) The role of chronic inflammation in obesity-associated cancers. ISRN Oncology. Article ID 697521
53. Trayhurn P (2013) Hypoxia and adipose tissue function and dysfunction in obesity. Physiol Rev 93:1–21
54. Lv Y, Zhang SY, Liang X et al (2016) Adrenomedullin 2 enhances beiging in white adipose tissue directly in adipocyte-autonomous manner and indirectly through activation of M2 macrophages. J Biol Chem 291:23390–23402
55. Prasad GVR (2014) Metabolic syndrome and chronic kidney disease: current status and future directions. World J Nephrol 3:210–219
56. Rabe K, Lehrke M, Parhofer KG, Broedl UC (2008) Adipokines and insulin resistance. Mol Med 14:741–751
57. Kwon H, Pessin JE (2013) Adipokines mediate inflammation and insulin resistance. Front Endocrinol (Lausanne) 4:71
58. Rosen ED, Spiegelman BM (2006) Adipocytes as regulators of energy balance and glucose homeostasis. Nature 444:847–853
59. de Kloet AD, Krause EG, Woods SC (2010) The renin angiotensin system and the metabolic syndrome. Physiol Behav 100:525–534
60. Boscaro M, Giacchetti G, Ronconi V (2012) Visceral adipose tissue: emerging role of gluco- and mineralocorticoid hormones in the setting of cardiometabolic alterations. Ann N Y Acad Sci 1264:87–102
61. Aherne W, Hull D (1966) Brown adipose tissue and heat production in the newborn infant. J Pathol Bacteriol 91:223–234
62. Carter BW, Schucany WG (2008) Brown adipose tissue in a newborn. Proc (Bayl Univ Med Cent) 21:328–330

63. Hageman RS, Wagener A, Hantschel C et al (2010) High-fat diet leads to tissue-specific changes reflecting risk factors for diseases in DBA/2J mice. Physiol Genomics 42:55–66
64. Commins SP, Watson PM, Padgett MA et al (1999) Induction of uncoupling protein expression in brown and white adipose tissue by leptin. Endocrinology 140:292–300
65. Fromme T, Klingenspor M (2011) Uncoupling protein 1 expression and high-fat diets. Am J Physiol Regul Integr Comp Physiol 300:R1–R8
66. Koleva PT, Bridgman SL, Kozyrskyj AL (2015) The infant gut microbiome: evidence for obesity risk and dietary intervention. Nutrients 7:2237–2260
67. Sanchez M, Panahi S, Tremblay A (2015) Childhood obesity: a role for gut microbiota? Int J Environ Res Public Health 12:162–175
68. Reid G (2015) The growth potential for dairy probiotics. Inter Dairy J 49:16–22
69. Yoo JY, Kim SS (2016) Probiotics and prebiotics: present status and future perspectives on metabolic disorders. Nutrients 8:173
70. Ashen MD (2008) Management of cardiometabolic syndrome in the primary and secondary prevention of cardiovascular disease. J Nurse Pract 4:673–680
71. Aronne LJ, Isoldi KK (2007) Overweight and obesity: key components of cardiometabolic risk. Clin Cornerstone 8:29–37
72. Skinner AC, Perrin EM, Moss LA, Skelton JA (2015) Cardiometabolic risks and severity of obesity in children and young adults. N Engl J Med 373:1307–1317
73. Peterson LR, Waggoner AD, Schechtman KB et al (2004) Alterations in left ventricular structure and function in young healthy obese women: assessment by echocardiography and tissue Doppler imaging. J Am Coll Cardiol 43:1399–1404
74. Wong CY, O'Moore-Sullivan T, Leano R et al (2001) Alterations of left ventricular myocardial characteristics associated with obesity. Circulation 110:3081–3087
75. Lavie CJ, Milani RV, Ventura HO et al (2007) Disparate effects of left ventricular geometry and obesity on mortality in patients with preserved left ventricular ejection fraction. Am J Cardiol 100:1460–1464
76. Kenchaiah S, Evans JC, Levy D et al (2002) Obesity and the risk of heart failure. N Engl J Med 347:305–313
77. Lavie CJ, Milani RV, Ventura HO (2009) Obesity and cardiovascular disease: risk factor, paradox, and impact of weight loss. J Am Coll Cardiol 53:1925–1932
78. Ebong IA, Goff DC, Rodriguez CJ et al (2014) Mechanisms of heart failure in obesity. Obes Res Clin Pract 8:e540–e548
79. Lavie CJ, Amodeo C, Ventura HO, Messerli FH (1987) Left atrial abnormalities indicating diastolic ventricular dysfunction in cardiopathy of obesity. Chest 92:1042–1046
80. Wang TJ, Parise H, Levy D et al (2004) Obesity and the risk of new-onset atrial fibrillation. JAMA 292:2471–2477
81. Messerli FH, Ventura HO, Reisin E et al (1982) Borderline hypertension and obesity: two prehypertensive states with elevated cardiac output. Circulation 66:55–60
82. Hall JE, da Silva AA, do Carmo JM et al (2010) Obesity-induced hypertension: role of sympathetic nervous system, leptin, and melanocortins. J Biol Chem 285:17271–17276
83. Bravo PE, Morse S, Borne DM et al (2006) Leptin and hypertension in obesity. Vasc Health Risk Manag 2:163–169
84. da Silva AA, do Carmo J, Dubinion J, Hall JE (2009) Role of sympathetic nervous system in obesity related hypertension. Curr Hypertens Rep 11:206
85. Skurk T, Havner H (2004) Obesity and impaired fibrinolysis: role of adipose production of plasminogen activator inhibitor-1. Int J Obes Relat Metab Disord 28:1357–1361
86. De Pergola G, Pannacciulli N (2002) Coagulation and fibrinolysis abnormalities in obesity. J Endocrinol Invest 25:899–904
87. Xu H, Uysal KT, Becherer JD et al (2002) Altered tumor necrosis factor-alpha (TNF-alpha) processing in adipocytes and increased expression of transmembrane TNF-alpha in obesity. Diabetes 51:1876–1883
88. Yudkin JS, Kumari M, Humphries SE, Mohamed-Ali V (2000) Inflammation, obesity, stress and coronary heart disease: is interleukin-6 the link? Atherosclerosis 148:209–214

89. Ziccardi P, Nappo F, Giugliano G et al (2002) Reduction of inflammatory cytokine concentrations and improvement of endothelial functions in obese women after weight loss over one year. Circulation 105:804–809
90. Lagrand WK, Visser CA, Hermens WT et al (1999) C-reactive protein as a cardiovascular risk factor more than an epiphenomenon? Circulation 100:96–102
91. Tham DM, Martin-McNulty B, Wang YX et al (2002) Angiotensin II is associated with activation of NF-kappaB mediated genes and downregulation of PPARs. Physiol Genomics 11:21–30
92. Reilly MP, Igbal N, Schutta M et al (2004) Plasma leptin levels are associated with coronary atherosclerosis in type 2 diabetes. J Clin Endocrinol Metab 89:3872–3878
93. Correia ML, Haynes WG (2004) Obesity-related hypertension: is there a role for selective leptin resistance? Curr Hypertens Rep 6:230–235
94. Chaldakov GN, Stankulov IS, Fiore M et al (2001) Adipoendocrinology and adipoparacrinology: emerging fields of the study on the adipose tissue. Biomed Rev 12:31–39
95. McArdle MA, Finucane OM, Connaughton RM et al (2013) Mechanisms of obesity-induced inflammation and insulin resistance: insights into the emerging role of nutritional strategies. Front Endocrinol (Lausanne) 4:52
96. Herman MA, Kahn BB (2006) Glucose transport and sensing in the maintenance of glucose homeostasis and metabolic harmony. J Clin Invest 116:1767–1775
97. Wilcox G (2005) Insulin and insulin resistance. Clin Biochem Rev 26:19–39
98. Lim AKH (2014) Diabetic nephropathy-complications and treatment. Int J Nephrol Renovasc Dis 7:361–381
99. Satchell SC, Tooke JE (2008) What is the mechanism of microalbuminuria in diabetes: a role for the glomerular endothelium? Diabetologia 51:714–725
100. Sandhu JS, Singla M, Ahuja A et al (2004) Renal risks of obesity. JIACM 5:335–338
101. Picard F, dos Santos P, Catargi B (2013) Diabetes, obesity and heart complications. Rev Prat 63:759–764
102. Wu Y, Tang L, Chen B (2014) Oxidative stress: implications for the development of diabetic retinopathy and antioxidant therapeutic perspectives. Oxid Med Cell Longev 752387
103. Kowluru RA, Chan PS (2007) Oxidative stress and diabetic retinopathy. Exp Diabetes Res 43603
104. Brownlee M (2001) Biochemistry and molecular cell biology of diabetic complications. Nature 14:813–820
105. Ceriello A (2003) New insights on oxidative stress and diabetic complications may lead to a 'causal' anti-oxidant therapy. Diabetes Care 26:1589–1596
106. Matough FA, Budin SB, Hamid ZA et al (2012) The role of oxidative stress and antioxidants in diabetic complications. Sultan Qaboos Univ Med J 12:5–18
107. Rahman T, Hosen I, Islam MMT, Shekhar HU (2012) Oxidative stress and human health. Adv Biosci Biotechnol 3:997–1019
108. Rahal A, Kumar A, Singh V et al (2014) Oxidative stress, prooxidants, and antioxidants: the interplay. Biomed Res Int 761264
109. Bhattacharyya A, Chattopadhyay R, Mitra S, Crowe SE (2014) Oxidative stress: an essential factor in the pathogenesis of gastrointestinal mucosal diseases. Physiol Rev 94:329–354
110. Kabel AM (2014) Free radicals and antioxidants: role of enzymes and nutrition. World J Nutrition and Health 2:35–38
111. Rahman K (2007) Studies on free radicals, antioxidants, and co-factors. Clin Interv Aging 2:219–236
112. Lobo V, Patil A, Phatak A, Chandra N (2010) Free radicals, antioxidants and functional foods: impact on human health. Pharmacogn Rev 4:118–126
113. Young IS, Woodside JV (2001) Antioxidants in health and disease. J Clin Pathol 54:176–186
114. Eknoyan G (2011) Obesity and chronic kidney disease. Nefrologia 31:397–403
115. Kim SY, Lim AY, Jeon SK et al (2011) Effects of dietary protein and fat contents on renal function and inflammatory cytokines in rats with adriamycin-induced nephrotic syndrome. Mediators Inflamm 945123

116. Chauhan V, Vaid M, Chauhan N, Parashar A (2012) The obesity epidemic and kidney disease: a literature review. In: Sahay M (ed) Diseases of renal parenchyma, pp 45–56. InTech Publisher
117. Main J (2005) Atherosclerotic renal artery stenosis, ACE inhibitors, and avoiding cardiovascular death. Heart 91:548–552
118. Sawanobori E, Terada N, Fujii Y et al (2008) Ultrastructural study of human glomerular capillary loops with IgA nephropathy using quick-freezing and deep-etching method. Histol Histopathol 3:297–307
119. Rocha VZ, Folco EJ (2011) Inflammatory concepts of obesity. Int J Inflam 2011:529061
120. Reena KT, Kaur A et al (2016) Mast cell stabilizers obviate high fat diet-induced renal dysfunction in rats. Eur J Pharmacol 777:96–103

Chapter 7
Role of Sodium-Glucose Co-transporters on Cardiac Function in Metabolic Syndrome Mammalians

Belma Turan

Abstract Metabolic syndrome (MetS) is increasingly common among humans all over the world and a combination of serious pathological conditions occurred together. MetS induces increasing of risks for several organ dysfunctions including heart and type 2 diabetes (T2DM) since it is closely linked to overweight or obesity and inactivity among humans, currently, also linking to insulin resistance. To prevent MetS, it is needed first to have a healthy lifestyle, however, several therapeutic approaches also are in used to lighten its risky effects. Sodium-glucose co-transporter 2 (SGLT2) and 1 (SGLT1) inhibitors are relatively new glucose-lowering agents that work by increasing urinary glucose excretion through the kidneys, exerting their action independently of insulin. However, there are a number of side effects of these agents in humans with MetS. Nevertheless, different research teams, recently, demonstrated that SGLT2 inhibitors (SGLT2is) exert important cardioprotective effects in patients with MetS and T2DM via lowering the high risks for cardiovascular morbidity and mortality. Furthermore, it has been also emphasized that SGLT2is-associated cardioprotection in insulin-resistant overweights rats includes prevention of prolongation in ventricular-repolarization via marked augmentation of mitochondrial function together with normalization of oxidative stress followed by improvement of fusion-fission proteins, without its glucose-lowering effect. Moreover, two recent clinical studies announced that SGLT2is, electrophysiologically, could provide marked protective effects on electrocardiographic parameters in T2DM patients. Therefore, in the present review article, it has been documented the recent data related to SGLT2is on both experimental and clinical studies and their outcomes in terms of either adverse, beneficial, or both effects.

Keywords High carbohydrate diet · Overweight body · Insulin resistance · Cardiac electrophysiology · Sodium-glucose co-transporters · Mitochondria

B. Turan (✉)
Department of Biophysics, Faculty of Medicine, Ankara University, Ankara, Turkey
e-mail: belma.turan@medicine.ankara.edu.tr

P. S. Tappia et al. (eds.), *Biochemistry of Cardiovascular Dysfunction in Obesity*, Advances in Biochemistry in Health and Disease 20,
https://doi.org/10.1007/978-3-030-47336-5_7

Introduction

Patients with metabolic syndrome (MetS), at most, characterized with insulin resistance and overweight and/or obese body, most of the cells cannot respond to insulin, and, thereby glucose cannot influx into the cells normally. As a result, a serious number of pathological stimuli arise into the tissues and related functional changes in organs [1, 2]. Increased oxidative stress parallel to depressed antioxidant defense in the heart is one most common factor in cardiovascular disorders in MetS mammalians with insulin resistance [1]. In this regard, studies demonstrated elevated oxidative damage, at most, due to increased reactive oxygen species, ROS, together with depressed antioxidant protection in patients with the MetS [3–5].

Metabolic syndrome, characterized by insulin resistance and generally further developed type 2 diabetes (T2DM), is one of the cardiovascular risk factors for humans, including high blood pressure, atherosclerotic alterations (including endothelial dysfunction), left ventricular dysfunction (including long-QT syndrome), and hypertrophy besides others [1, 2, 6]. All those alterations underline serious cardiovascular diseases. Therefore, it can be clearly understandable why cardiovascular diseases arise through more that one reason in MetS humans as well as experimental animals. Furthermore, studies emphasized that more than one organs are affected by MetS, together and/or individually [3, 7–10].

One common organ system affected in the MetS is the cardiovascular system. Among this system, the heart in the MetS is affected by organ-specific insulin resistance and increased oxidative stress [9–11]. A close relationship between heart dysfunction and increased oxidative stress (due to both increases in ROS and RNS), alone and/or together with depressed antioxidant defense in MetS mammalians have been shown with the results of many experimental and clinical studies [3, 10, 12, 13]. More importantly, a cross-correlation between systemic insulin resistance, oxidative stress, and development of T2DM diabetes has been demonstrated with a widespread clinical study [14]. Moreover, the findings of several experimental studies on MetS emphasized that cardiomyocardium has specific own oxidative stress and insulin resistance besides systemic ones, which underline the MetS associated heart dysfunction in mammalians [9, 11, 15–17]. The basic mechanisms responsible for alterations in the heart in MetS, further leading to cardiac disease, are highlighted in Fig. 7.1.

Metabolic Syndrome and Heart Function

As mentioned in the introduction and due to already published data, the percentage of morbidity and mortality is seriously high among humans with MetS [18–20]. According to criteria announced by WHO, patients with MetS have impaired glucose tolerance, increased levels of insulin triglycerides, and/or low HDL cholesterol in their sera, parallel to hypertension and high body mass index. These patients usually

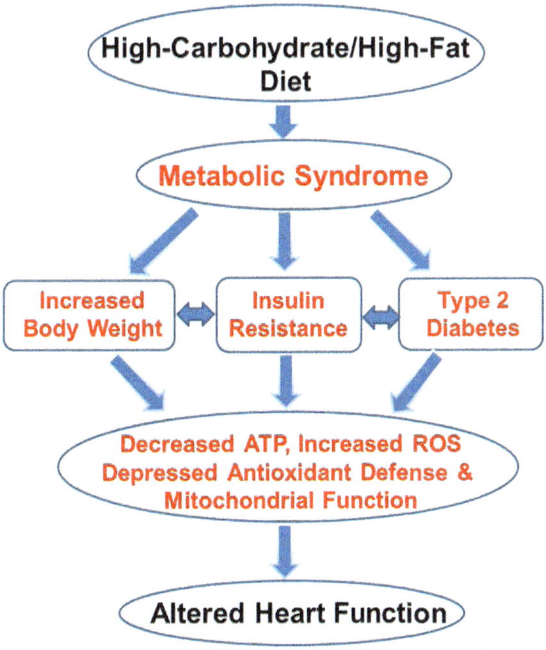

Fig. 7.1 Potential mechanisms are responsible for cardiac dysfunction in metabolic syndrome mammalians via either high-carbohydrate or high-fat diet. Metabolic syndrome generally is characterized with over body weight, hyperinsulinemia with slightly (but significantly) increased blood glucose and insulin resistance which followed with type 2 diabetes (T2DM) in humans. All those changes can basically promote the severe alterations many parameters in cardiomyocytes, such as decreases in cellular ATP level and mitochondrial functions, decreases in antioxidant defense system together with increases in oxidative stress. All those changes underline the development of serious cardiac dysfunction in mammalians

have insulin resistance and then followed with T2DM. In these regards, Tenerz A, and co-workers demonstrated nicely the cross-correlation between diabetes and insulin resistance in patients with MetS who have an acute myocardial infarction (using a data from a total of 145 patients) without previously known T2DM [20]. A population-based study performed among 106,470 residents of Olmsted County who have abnormal cardiac structure and function in the MetS is given by Aijaz and co-workers [21]. Interestingly, their data demonstrated that women with MetS had an incidence of early left ventricular dysfunction characterized by increased mass index and diastolic dysfunction in the left ventricle. Moreover, the relationship between oxidative stress and cardiovascular dysfunction in the MetS and diabetes has been documented, previously [1, 22]. As an example, authors have shown that there is a close relationship between increased oxidative stress with depressed antioxidant defenses definite with decreased superoxide dismutase activity in a patient with the MetS [5, 12].

Microscopic analysis of the heart tissue documented that there were marked disorganization of cardiac myofibrils parallel to the loose of their integrity, smaller diameters in myofibrils, increases in the content of connective tissues localized to myofibrils and vessels in the heart and vessels isolated from MetS rats [11]. There were also significantly seen intracellular vacuolization, intracellular lipid inclusion, and alterations in both composition and function of cardiomyocyte organelles in left ventricular cardiomyocytes from MetS rats. Notably, most of all those changes were supporting the hypothesis on MetS-associated direct targeting of heart with increased oxidative stress. In addition, similar findings in MetS heart were already shown such as significantly pale staining in tissue sections, eosinophilia loss in some myofibrils, and marked heterogeneity in the cytoplasm [23, 24]. Indeed, all those changes are linking to intracellular vacuolization and defects of organelles, ones similar to those of observed streptozotocin-diabetic rat heart [25]. More importantly, those structural alterations have been demonstrated in other animal models with increased oxidative stress status [23]. Of note, those changes observed in the heart of animals either MetS and/or increased oxidative stress conditions are fitting to the changes observed in individuals having obesity [24].

Experimental data provided further information related to the demonstration of the role of oxidative stress on heart dysfunction in MetS mammalians. In the concept of this information, research data showed seriously production of hydrogen peroxide, increases in lipid and protein oxidation in the heart of obesity-related diabetic mice [26]. More importantly, development of insulin resistance in mice markedly increased a direct ROS production the heart, being independent of hyperglycemia and hyperinsulinemia [15], while and the cardiac high superoxide production and depressed antioxidant enzymes were also observed in the high fat-diet fed rats [16, 17].

In the subcellular levels, both experimental and clinical data emphasized that mitochondrial dysfunction plays an important role in the pathogenesis of MetS-associated heart dysfunction. Indeed, even early studies have pointed out not only hyperlipidemia or hyperinsulinemia but also every component of the risk factors associated with MetS can modulate independently the mitochondrial function [17, 27, 28]. Besides animal model studies [29], clinical data in obese or T2D patients showed the changes in cardiac oxygen consumption, the ratio of phosphocreatine/ATP and atrium mitochondrial oxygen consumption [30–32]. As summary, the mechanisms for impaired cardiac mitochondrial function in the MetS include the insulin resistance, mitochondrial uncoupling, increased mitochondrial oxidative stress [33], and impaired mitochondrial Ca^{2+}-handling [9, 34, 35].

Of note, although obesity is one of the well-characterized risk factors for heart failure in MetS individuals, some authors, depending on the outcomes of studies, discussed "the obesity paradox" for heart diseases in humans whether as a true protective effect or not [36]. Despite the close correlation between MetS and cardiac dysfunction, numerous studies have documented that individuals with overweight and/or obese and diagnosed with cardiovascular disease, have a better prognosis than the none of ones with overweight or obese patients [37]. In this regard, some studies also reported that patients with higher BMI have more convenient lifespan than those of lower BMI as well as high rates of hypertension and T2DM [38–40].

In those studies, authors even showed that cardiovascular mortality was significantly lower in both overweight and obese patients compared with that in slim ones. Notably, these observations were also observed in patients regardless of their age and gender differences, as well as regardless of their systolic or diastolic heart failure [40]. More importantly, some studies also reported that MetS patients with heart failure without T2DM had the best survival compared to those of heart failure patients without T2DM, whereas heart failure patients with T2DM (non MetS) showed the worst survival compared to the other groups [41].

As a summary, both experimental and clinical investigations have shown that the structure of the left ventricle is significantly impaired in patients with MetS, together with left ventricular both systolic and diastolic dysfunctions. Thus it can be concluded that the impaired global left ventricular function is actually the result of impairment of several factors, including increased oxidative stress in MetS individuals. The degree of structural and functional damage increased with the number of risk factors for MetS. Further studies are necessary for complementing the influence of MetS on the left ventricular structure and function. Findings show that not only structure but also a function of the heart are affected by every component of the MetS as either alone or multi-factorial ways as well as either independently or combined manner. It is well-accepted that their combination effects affect the heart more seriously [13, 42].

Role of SGLTs in Cardiac Function

One of the glucose transporter families is membrane specific carrier proteins, named as the sodium-glucose cotransporters (SGLTs). Two members of this family are SGLT1 and SGLT2 which are functioning as sugar transporters across the cell membrane. T2DM, with importantly increasing rate among humans all over the world, and heart failure coincide, usually, and, in turn, it can contribute to development of heart failure, while an important percentage of heart failure patients have T2DM [43, 44]. This serious event faced highlights our needs to develop novel therapeutic agents, which will not only improve the altered glycemic system but also protect the cardiovascular system against hyperglycemia and hyperinsulinemia associated damages.

Many anti-diabetic agents drugs were used for different trials as parallel to insulin therapy, including metformin groups [45, 46] and incretin therapies, including dipeptidyl peptidase-4 (DPP-4) inhibitors [47]. However, there are no consistent findings associated with the well-controlled progress in the disease. SGLT2 inhibitors are oral antidiabetic agents currently approved for the treatment of T2D [48]. It has been demonstrated SGLT2 inhibitors (SGLT2i) could reduce alterations in cardiovascular events under cardiovascular outcome trials [49–53]. SGLT2is, despite these compensatory mechanisms, due to the positive effects on lowering the hyperglycemia, they exert clinically relevant promoting the weight loss of obese and/or overweight people. Furthermore, the combination of SGLT2 inhibitors with other drugs seems to have more efficient for protection and management of T2DM [54–58].

Although it has been documented that SGLT1 expression was in the small intestine, liver, lung, kidney, and heart, we, recently, demonstrated that SGLT2 expression is not only in the kidney [59] but also in the heart, as well [34]. Several theories have been put forward to explain the profound beneficial effects of SGLT2is on cardiovascular disorders [53, 60], Turan's team demonstrated that their cardioprotective action was due to their existence in the heart thereby directly affecting the cardiac system under hyperinsulinemia and hyperglycemia [34]. Due to both our findings and literature data, SGLT2is, including empagliflozin, provides marked cardioprotection through reducing the hyperglycemia in T2DM patients via increasing glucose excretion via renal system, thereby reducing the occurrence of nonfatal myocardial infarction, or nonfatal stroke and finally decreasing the cardiovascular mortality [61–63] whereas some of them demonstrated SGLT2is empagliflozin did not affect the rates of myocardial infarction or stroke but reduced the rate of cardiovascular mortality, admission for heart failure, and all-cause mortality [49]. Besides cardioprotective actions, SGLT2 inhibition reduces inflammation and attenuates nephropathy in T2DM patients [64]. More importantly, some studies emphasized the beneficial effects of SGLT2is as their action on the body weight lose [63, 64].

Consequently, it can be summarized that cardiovascular outcome trials have shown why SGLT2 inhibitors are effective anti-diabetic agents to reduce cardiovascular alterations, particularly heart failure in diabetic patients. However, there is an important mechanistic discussion on the context of completed and ongoing trials of SGLT2is in the prevention and treatment of heart failure in individuals with and without diabetes. Therefore, one can suggest putative mechanisms associated with the underlying events of SGLT2is-related cardioprotection: SGLT2i can (1) recover the depressed cardiac metabolism and bioenergetics [65, 66] and myocardial Na^+/H^+ exchange [56, 67], (2) improve the ventricular loading via a reduction in increased both preload and afterload, due to, at most, reduction in high blood pressure [53, 60, 68], (3) recovery in stuructural alterations in both kidney and heart [69], and (4) improvement in adipokines and cytokines and adipose tissue accumulation into epidardium [70, 71].

SGLT2 Inhibitors and Cardiovascular Protection: Experimental and Clinical Trial Outcomes of Diabetics

Generally, it had been preferred the use of insulin in T2DM patients for a long time, the high mortality rates and uncontrolled glycemic level with serious cardiac problems could not be prevented. Therefore, besides insulin and insulin-sensitizing therapy, in the recent decade, some clinical trials outcomes demonstrated the important benefits with non-insulin therapies on glycemic control and heart dysfunction [72]. Among them, the first experimental studies tried to asses their in vitro and in vivo pharmacology in either hyperglycemic cells or T2DM animals [73]. Han et al. performed the experimental studies with SGLT2is (i.e. dapagliflozin) and used both acute and multi-dose studies in normal and diabetic rats to improve fed and fasting plasma glucose levels as well as to improve glucose utilization after multi-dose treatment [73]. Their data strongly emphasized that SGLT2is can be accepted as an efficacious treatment for T2DM. Latter experimental studies with SGLT2is provided important novel mechanisms associated with their benefits, particularly in cardiovascular disorders. Hansen HH et al. treated Zucker diabetic fatty rats with empagliflozin and showed that this inhibitör preserved β-cell mass through the restoration of glucose homeostasis, at most, through insulin-independent pathways [74]. In another study, using dapagliflozin, Lee T-M et al. also provided a novel mechanism of SGLT2is-associated cardiac benefits including attenuation of cardiac fibrosis, which is through the regulation of the macrophage polarization via STAT3 signaling in infarcted rat hearts [69]. More importantly, we also demonstrated, for the first time, that dapagliflozin-treatment of MetS rats, but not insulin-treatment, provided important cardioprotective action, at most, though not only ECG (prevention of QT-prolongation) but also improvements in the depressed left ventricular developed pressure, heart rate and relaxation activity of vessel system [34].

Most of the studies on the cardioprotective role of SGLT2 inhibitors are concentrated in the field of their important cardioprotection in T2DM patients. Although diabetes mellitus is still very complex and a set of syndromes and requires medications usually insulin for T1DM and oral antidiabetics for T2DM, SGLTis are new classes of antidiabetic drugs, working via kidney system to extrude the urinary glucose [38–40, 75].

Comparison to that SGLT1, SGLT2 is a low affinity but high capacity transporter and works to absorb the most of glucose in the proximal tubule in humans [76]. Although it was believed that SGLT2 exists in the kidney while SGLT1 is found in the skeletal muscles and heart, recently the presence of SGLT2 in the mammalian cardiomyocytes has been demonstrated [34]. Among others, SGLT2is did bring an imposing strategy for the treatment of T2DM patients, particularly against cardiovascular disorders. More importantly, besides their insulin-independent effects, clinical outcomes led to re-consider their protective actions when used in multi-drug treatment approaches [57, 58]. With this consideration, the recent studies emphasized the benefits of using an SGLT2i in dual combination with metformin and triple combination with a glucagon-like peptide 1 receptor agonist, dipeptidyl peptidase 4 inhibitor, or other glucose-lowering agent to treat T2DM patients [57, 77]. However, a number of questions appeared on the reduction of cardiovascular disease outcomes trials with SGLT2is.

If one looks at the literature data on SGLT2is-associated outcomes, Vasilakou et al. performed a meta-analysis and examined the glucose-lowering efficacy of SGLT2is in T2DM patients through analysis of renal function [78]. Their data demonstrated that SGLT2is increased plasma glucagon levels and stimulated hepatic glucose production, via being independent of insulin resistance and β-cell damage. Similarly, another study performed in T2DM patients, Zinman et al. performed a clinical study by using a SGLT2 inhibitor, in addition to standard care, on cardiovascular morbidity and mortality at high cardiovascular risk (EMPA-REG OUTCOME Trial) while their data provided important novel action of SGLT2is in T2DM patients through inducing lower rate of cardiovascular complications [49]. Following these important clinical trials reports, SGLT2is is advanced to use primarily by clinicians for their benefits in the prevention of cardiovascular disease, rather than focusing on their glucose-control in T2DM patients. Another cardiovascular trial outcome study with SGLT2is was CANVAS program [52]. They performed the analysis in over 10,000 T2DM patients with high risk for cardiovascular disease. They used canagliflozin and demonstrated that this inhibitor provided also important benefits in renal pathologies. In both clinical trials, the collected data showed that the promise findings associated with low hospitalization period for cardiac diseased patients. In addition, the results of Han JH et al. provided new information related with SGLT2is benefits such as antiatherosclerotic action and lowering of the levels of C-reactive protein, tumor necrosis factor-α, interleukin-6 and monocyte chemoattractant protein-1 in sera [79]. Similar to our experimental findings, Sato T et al. examined the efficiency of different SGLT2is on the reversibility of ventricular repolarization heterogeneity in T2DM patients [80]. Their analysis demonstrated the normalization of QTc dispersion, independently of their effects on glycaemic control in those patients.

Taken into consideration the data with clinical trials outcomes, authors and the publication of DECLARE–TIMI 58 (https://www.jwatch.org/na47925/2018/11/10/cardiovascular-effects-sglt2-inhibitors), in very important review articles, emphasized that SGLT2is have unique actions in T2DM patients such as lowering high blood glucose being independent of insulin level with acceptable risk factor profiles as well as their benefits in cardiovascular endpoints and sympathetic overactivity [53, 81–84]. However, the mechanisms by which SGLT2 inhibition improves cardiovascular outcomes are not fully understood.

Effects of SGLT2 Inhibitors on the Morphology of the Heart

Studies have shown that demonstration of the effects of SGLT1is on the heart is a relatively easy compared to those of SGLT2is because there are importantly high expression levels of SGLT1s in mammalian hearts, while SGLT2s are expressed in the kidney [85]. Furthermore, we recently have shown the important amount of SGLT2 expression in left ventricular cardiomyocytes from male rats, while its expression was found to be increased significantly in the cardiomyocytes from MetS rats [34].

In experimental studies, dapagliflozin treatment of infarcted rats provided important attenuation in cardiac fibrosis as well as increases in the collagen formation in the left ventricle [69]. In that study, light microscopy analysis demonstrated that the infarcted area of the left ventricle in dapagliflozin-treated rats was very thin. In addition, there was differentiated scar tissue four weeks after infarction. Those benefits in those hearts were confirmed with the increased maximal rate of left ventricular contractive and relaxation activities favorable remodeling in the left ventricle.

Histologic investigations in the heart from diabetic, the heart failure modeled, and also the myocardial ischemic modeled experimental animals treated with SGLT2is have shown the marked improvements in the morphology of those samples [86–88].

Those reports demonstrated that SGLT2is could significantly decrease the weight of left ventricle, at most, reduction in cardiomyocyte size, interstitial fibrosis, and infiltration of interstitial macrophage. In addition, those treatments induced a significant improvement in both tissue and cardiomyocyte levels, such as decreasing in the cross-sectional area of cardiomyocytes and inhibition in collagen I and III depositions [55]. Furthermore, the data with Lin et al. reported that empagliflozin could decrease the fibrosis and thickening in pericoronary arteria [88]. Moreover, Hammoudi et al. studied the effect of empagliflozin in genetic diabetic mouse [89]. Authors have demonstrated the prevention of the cardiac hypertrophy and remodeling markers (such as extracellular signal-regulated kinases, c-Jun NH2-terminal kinases, and p38) [90], as well as significant attenuation in disordered cell arrays and focal necrosis with that treatment, markedly [91]. All those reports clearly suggest that SGLT2is are potentially important agents to improve the morphology of the diabetic heart. More importantly, the reports on the investigation of SGLT2is treatment (i.e. dapagliflozin) in different pathological heart models (i.e. myocardial ischemia, heart failure), showed that those drugs could attenuate myocardial infarct size [92], myofibroblast infiltration and cardiac fibrosis [69] while empagliflozin pretreatment improved the cardiac edema and deformed cardiac chambers [53, 93].

There are also important reports related to the effects of SGLT2is and human heart tissue. Lin et al. treated T2DM patients with empagliflozin treatment of T2DM patients and reported that that treatment improved cardiac interstitial fibrosis, coronary arterial thicking and remodeling, cardiac interstitial macrophage infiltration and cardiac antioxidant enzyme levels as well as vascular dysfunction [88]. In a recent clinical trial, Januzzi et al. also tested the effect of canagliflozin treatment for 2-year inelderly T2DM patients and their data demonstrated that the SGLT2i treatment prevented the alterations not completely but delayed the development of heart failure compared to nontreated group [94]. Overall, taken into consideration the outcomes of the clinical trials documented widely in recent review articles, SGLT2is have cardioprotective effects in not only T2DM but also other types of the pathological heart [58, 82]. Of note, it seems even great need to substantiate their safety and efficacy, the recent reports emphasized the important valuable therapeutic side of SGLT2is for reduction of cardiovascular risk in diabetic patients as well as for acceptable preventive agents in nondiabetic patients.

Effects of SGLT2 Inhibitors on Cardiac Oxidative Stress

There is a number of proposals on the potential mechanisms responsible for cardioprotection with the treatment of SGLT2is in different types of cardiac pathologies. Considering the current findings, it can be proposed that the direct cardiac-targeting effect of SGLT2is includes their effects mediated through reduction of oxidative stress and recovery in both ionic dyshomeostasis and mitochondrial dysfunction in cardiomyocytes, besides their systemic effects by hemodynamic and metabolic actions [95, 96].

The cardiac effects of SGLT2is are demonstrated by different research teams through their effects on cardiac oxidative stress. Indeed, the important role of increased oxidative stress in the pathogenesis of cardiac pathologies including hypertrophy and remodeling as well as diabetic cardiomyopathy [97–100]. The contribution of increased oxidative and nitrosative stress to hyperglycemia associated cardiomyocyte dysfunction was widely investigated by Turan's team using different animal models [25, 97, 98, 101–104]. In those studies, antioxidants, and agents having antioxidant-like actions provided important cardioprotection against desired increases in oxidative stress at both system and cell levels. Recent investigations, at both experimental level and clinical outcomes, have shown that SGLT2is can present effects through their antioxidant actions, independent from their glucose-lowering effects. In these regards, we and others treated genetically prediabetic and/or MetS rats with SGLT2is and demonstrated their action on the reduction of increased oxidative stress at tissue, cellular and systemic levels, significantly [34, 69, 87, 88, 105]. These recoveries in oxidative stress are parallel to the augmentation in both structure and function of the samples. Consistent with previously published findings, we, recently, determined the levels of ROS and RNS in isolated left ventricular cardiomyocytes from MetS rats, using specific fluorescence probes for the production of these oxidant agents. Dapaliflozin treatment of the MetS rats protected significantly the cells against the production of these oxidant agents and the levels of both ROS and RNS were found to be a similar level to those of control cells [34]. As a summary, one can propose that all above cardioprotective effects of SGLT2is might be attributed to their direct cardiac-targeting effect, at least, due to lowering of cardiac oxidative stress, being independent of their antidiabetic actions.

Effects of SGLT2 Inhibitors on Ultrastructure and Function of Mitochondria in Hyperglycemic Cardiomyocytes

It is well known that there is a close relationship between the increased amount of ROS production and mitochondrial dysfunction in cells. Supporting this statement, several reports demonstrated how mitochondria are crucial in different intracellular signal transduction pathways, in part, through changes in mitochondrial Ca^{2+} ($[Ca^{2+}]_{Mit}$)-homeostasis in cardiovascular diseases [106–108]. Indeed, even early studies emphasized the mitochondrial role in energy production, and thereby, a well-controlled modulation of $[Ca^{2+}]_{Mit}$-homeostasis also in cardiomyocytes among others [109–111]. More importantly, studies also have shown the association between mitochondrial dysfunction and decreased ATP production [112], while a reduction of ROS via mitochondria-sensitive protein modifications prevented chronic heart failure associated remodeling [113, 114] as well as the role of this relation in development of glucose tolerance, insulin resistance, and cardiac diastolic dysfunction in mammalians [108].

We and others have shown the association between mitochondrial dysfunction and the development of diabetic cardiomyopathy [34, 115, 116]. Furthermore, our recent data, as well as other reports, showed the benefits of SGLT2is treatment on mitochondrial dysfunction in rat diabetes models [34, 86, 92]. SGLT2is exerted important benefits on the morphology of mitochondria in diabetic animals, such as attenuation in abnormal inter-myofibrillar mitochondria, normalization of disorganized sarcomeres, recovery in reduced matrix electron density, loss of cristae and mitochondrial fragmentation [86]. Consisting of those findings, there were irregularly partitioned and clustered inter-myofibrillar mitochondria, together with numerous lysosomes and irregularly arranged mitochondrial crysta in isolated ventricular cardiomyocytes from MetS rats (Fig. 7.2). Furthermore, we observed marked recovery in inter-myofibrillar mitochondria, normal appearance of sarcomere organizations and very little amount fragmented mitochondria in isolated cardiomyocytes from MetS rats treated with dapagliflozin (Fig. 7.3). More importantly, when we treated the MetS rats with insulin, we observed more fragmented mitochondria and numerous lysosomes together with a reduced but still significant amount of irregularly arranged mitochondrial crysta in those cardiomyocytes (Fig. 7.4).

SGLT2is were used for the recovery of mitochondria in other types of cardiac pathologies such as ischemia–reperfusion, cardiac arrest, and diabetes [92, 117–119]. In most studies, dapagliflozin treatment prevented the depolarization and mitochondrial swelling, markedly improved the mitochondrial morphology through attenuating mitochondrial fragmentation, loss of cristae and fusion of cristae. Furthermore, dapagliflozin treatment increased the expressions of some proteins essential for the regulation of cardiac mitochondrial fatty acid oxidation and some others such as

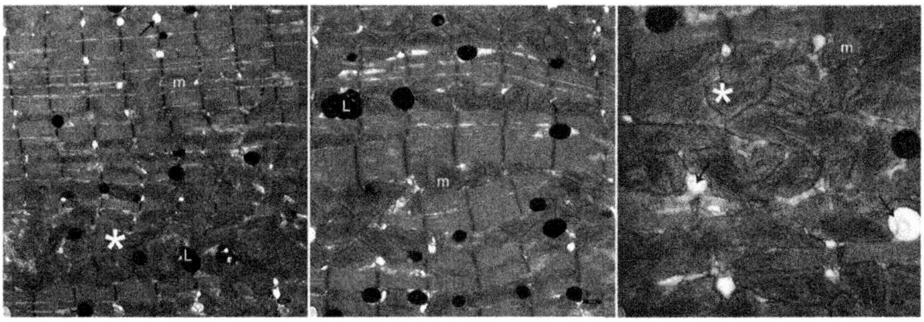

Fig. 7.2 Electron microscopic examination of isolated cardiomyocytes from the MetS rats. Rats were received 32% sucrose into their drinking water for 30 weeks and then the development of metabolic syndrome (MetS) was confirmed by determination of body weight, fasting blood glucose level, insulin level, oral glucose tolerance test, and insulin resistance, as described elsewhere [10, 11]. Transmission electron microscopy analysis (LEO Electronenmikroskopie, Oberkochen, Germany) showed that there are markedly observed irregularly partitioned and clustered intermyofibrillar mitochondria, numerous lysosomes, irregularly arranged mitochondrial crysta in isolated left ventricular cardiomyocytes. Shorten symbols are *m*; mitochondrion, *L*; lysosome, *arrow;* Z-line, *thin arrow;* T tubules, *asterisk;* partitioned mitochondria. Magnifications: ×7,750 (left), ×10,000 (middle), × 21,560 (right)

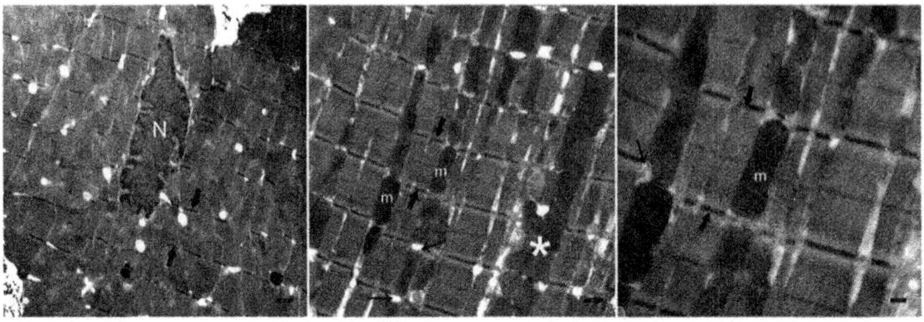

Fig. 7.3 Electron microscopic examination of isolated cardiomyocytes from dapagliflozin treated MetS rats. MetS rats were treated with dapagliflozin (5 mg/kg, Bristo-myers Squibb Manufacturing Company, Humacao, Porto Riko) for 2 weeks. The freshly isolated left ventricular cardiomyocytes were examined with transmission electron microscopy analysis (LEO Electronenmikroskopie, Oberkochen, Germany). In these samples, there are the well-organized intermyofibrillar mitochondria along sarcomere and there are a very little number of partitioned mitochondria can be detected. Shorten symbols: *N;* nucleus, *m*; mitochondrion, *L*; lysosome, *arrow*; Z-line, *thin arrow;* T-tubules, *asterisk;* mitochondria. Magnifications: ×7,750 (left), ×10,000 (middle), ×21,560 (right)

Fig. 7.4 Electron microscopic examination of isolated cardiomyocytes from insulin-treated MetS rats. For comparison to dapagliflozin effect, MetS rats were treated with insulin (0.15 mg/kg, Humalog Mix25 Kwikpen, Lilly) for 2 weeks, and then the isolated left ventricular cardiomyocytes were examined with transmission electron microscopy (LEO Electronenmikroskopie, Oberkochen, Germany). There are relatively less but significantly seen lysosomes and partitioned mitochondria with relatively organized intermyofibrillar mitochondria along sarcomere in these samples. Shorten symbols: *m*; mitochondrion, *L*; lysosome, *arrow*; Z-line, *thin arrow;* T-tubules, *asterisk;* partitioned mitochondria. Magnifications: ×7,750 (left), ×10,000 (middle), ×21,560 (right)

DRP1, MFN1, MFN2 and OPA1 responsible from proper mitochondria functioning [92, 115, 117–120].

Overall, it can be concluded that a strategy with SGLT2Is to improve mitochondrial dynamics, mitochondrial function, as well as mitochondrial morphology, ROS production, biogenesis and protein expressions would help to attenuate the development of risk factors for cardiovascular dysfunction under not only hyperglycemia but also other pathologies in the body of mammalians. However, these benefits are still unclear and need further investigations.

Beneficial Effects of SGLT2 Inhibitors on Ionic Mechanisms in Hyperglycemic Cardiomyocytes

Metabolic syndrome is metabolic syndrome and closely associated with overweightness and/or obesity in humans, which is generally followed with T2DM. More importantly, this syndrome has also serious risk factor for cardiovascular disorders, characterized with insulin resistance and long-QT interval in their ECGs [121–125]. Since the very long period, it is well known that there are severe alterations in the heart from diabetics, basically left ventricular dysfunction, decreases in the heart rate, and increases both systolic and diastolic pressures. In a short statement, there are marked depressions in both electrical and mechanical properties of the myocardium from diabetics, at both body and organ as well as cardiomyocyte levels [126–131]. Besides above studies, we investigated the contribution of altered sarcoplasmic reticulum (SR) function to altered intracellular Ca^{2+}-cycling in isolated left ventricular cardiomyocytes from MetS rats with depressed left ventricular function [9]. Our single cell level examinations demonstrated that there were markedly increased basal level of Ca^{2+} consisting with depressed SR Ca^{2+}-loading and SERCA2a activity, and leaky-ryanodine receptor (RyR2) function and inhibited sodium-calcium exchanger (NCX) [9]. Furthermore, we have also shown that there were markedly prolonged action potentials, at most, through inhibited voltage-dependent K^+-channel currents [34]. In that study, we also determined significantly increased voltage-dependent Na^+-channel currents with no change in voltage-dependent Ca^{2+}-channel currents as well as increased intracellular pH level. Importantly, dapagliflozin treatment of either MetS rats of cardiomyocytes isolated from MetS rats exhibited marked protection against all above changes [34]. Our investigations with SGLT2i in MetS rats, overall, confirmed the previous clinical outcomes through a mechanism of insulin-independent pathways, in part, mediated with well-controlled oxidative stress in cardiomyocytes.

Among well-known events contributing to maintenance of cardiac function under physiological condition, intracellular homeostasis of both Ca^{2+} and Na^+ are important players in cardiomyocytes and both can increase in hyperglycemic heart, further leading to heart failure [132–135]. Experimental studies with SGLT2is have shown, depending on either SGLT2 activity or not, the important benefits via improvement of intracellular Ca^{2+} and Na^+ levels in cardiomyocytes [67]. This inhibitor treatment also exhibited marked normalization in mitochondrial Ca^{2+} level, either directly or indirectly [34]. Moreover, the authors mentioned that those effects with SGLT2i were similar to the effect of Na^+/H^+-exchange inhibitor [135]. Moreover, it has been emphasized that NHE inhibitors directly inhibited cardiac NHE inducing a markedly controlled intracellular level of Na^+ and mitochondrial Ca^{2+} level, which can further prevent sudden cardiac death in mammalians [136, 137]. In addition, other actors, played an important contribution to cytosolic ionic levels, can also be under controlled with SGLT2is in diabetics [138]. In that concent, we, recently, have examined the effect of SGLT2i, dapagliflozin, on cardiac Zn^{2+}-transporters and cellular Zn^{2+}-level as well as oxidative stress and matrix metalloproteinase (MMP) status in MetS rat cardiomyocytes [139]. In that examination, we demonstrated that dapagliflozin treatment of MetS rats presented an important anti-oxidant like action and therefore provided marked protective effect on cellular Zn^{2+}-homeostasis affecting protein expression levels of Zn^{2+}-transporters, MMPs. And increased oxidative stress. Overall, all above reports strongly imply the important roles of SGLT2is in cardiovascular disorders, under pathological conditions, at most under hyperglycemic and hyperinsulinemic conditions in mammalians.

Conclusions

Following a demonstration of antidiabetic agents, SGLT2is were generally under use for diabetics as a novel approach, at most through their inhibitory action on renal glucose reabsorption. After the discovery of their preventive actions on the development of risk factors for heart dysfunction, independent of their glucose-lowering effects, SGLT2is are currently used as cardioprotective agents, at least, to reverse the ventricular repolarization heterogeneity, not only in T2DM patients but also other patients with heart failure, independent of their glycemic control action [53, 56, 58, 80, 95, 140–142]. Those cardioprotective effects of SGLT2is include their mediation to reduce inflammation, oxidative stress, apoptosis, mitochondrial dysfunction and ionic dyshomeostasis at the cellular level in the heart.

Taken as a whole, the results of recent cardiovascular protection with SGLT2is, in different pathological conditions, support the opening of a new therapeutic approach for the prevention of cardiovascular risk factors. However, more studies associated with the efficacy and safety of those inhibitors are needed for a better understanding of their meditative actions in the cardiovascular system. Overall, the reports on beneficial effects of SGLT2is, a new class of anti-diabetic agents, can confer significant cardiovascular protection in patients with different origins.

Acknowledgements All experimental studies mentioned in this text were supported by a grant from TUBITAK SBAG-214S254.

Competing Interests The author declares no competing interests.

References

1. Ilkun O, Boudina S (2013) Cardiac dysfunction and oxidative stress in the metabolic syndrome: an update on antioxidant therapies. Curr Pharm Des 19(27):4806–4817
2. Bugger H, Abel ED (2008) Molecular mechanisms for myocardial mitochondrial dysfunction in the metabolic syndrome. Clin Sci (Lond) 114(3):195–210
3. Palmieri VO et al (2006) Systemic oxidative alterations are associated with visceral adiposity and liver steatosis in patients with metabolic syndrome. J Nutr 136(12):3022–3026
4. Grundy SM (2007) Metabolic syndrome: a multiplex cardiovascular risk factor. J Clin Endocrinol Metab 92(2):399–404

5. Armutcu F et al (2008) Oxidative stress markers, C-reactive protein and heat shock protein 70 levels in subjects with metabolic syndrome. Clin Chem Lab Med 46(6):785–790
6. Grundy SM et al (2005) Diagnosis and management of the metabolic syndrome: an American Heart Association/National Heart, Lung, and Blood Institute scientific statement: executive summary. Crit Pathw Cardiol 4(4):198–203
7. Nicolson GL (2007) Metabolic syndrome and mitochondrial function: molecular replacement and antioxidant supplements to prevent membrane peroxidation and restore mitochondrial function. J Cell Biochem 100(6):1352–1369
8. Voulgari C et al (2010) The impact of metabolic syndrome on left ventricular myocardial performance. Diabetes Metab Res Rev 26(2):121–127
9. Okatan EN, Durak AT, Turan B (2016) Electrophysiological basis of metabolic-syndrome-induced cardiac dysfunction. Can J Physiol Pharmacol 94(10):1064–1073
10. Durak A et al (2017) Onset of decreased heart work is correlated with increased heart rate and shortened QT interval in high-carbohydrate fed overweight rats. Can J Physiol Pharmacol 95(11):1335–1342
11. Okatan EN et al (2015) Profiling of cardiac beta-adrenoceptor subtypes in the cardiac left ventricle of rats with metabolic syndrome: comparison with streptozotocin-induced diabetic rats. Can J Physiol Pharmacol 93(7):517–525
12. Ford ES et al (2003) The metabolic syndrome and antioxidant concentrations: findings from the Third National Health and Nutrition Examination Survey. Diabetes 52(9):2346–2352
13. Chinali M et al (2008) Cardiac markers of pre-clinical disease in adolescents with the metabolic syndrome: the strong heart study. J Am Coll Cardiol 52(11):932–938
14. Meigs JB et al (2007) Association of oxidative stress, insulin resistance, and diabetes risk phenotypes: the Framingham Offspring Study. Diabetes Care 30(10):2529–2535
15. Boudina S et al (2009) Contribution of impaired myocardial insulin signaling to mitochondrial dysfunction and oxidative stress in the heart. Circulation 119(9):1272–1283
16. Saraiva RM et al (2007) Reduced neuronal nitric oxide synthase expression contributes to cardiac oxidative stress and nitroso-redox imbalance in ob/ob mice. Nitric Oxide 16(3):331–338
17. Serpillon S et al (2009) Superoxide production by NAD(P)H oxidase and mitochondria is increased in genetically obese and hyperglycemic rat heart and aorta before the development of cardiac dysfunction. The role of glucose-6-phosphate dehydrogenase-derived NADPH. Am J Physiol Heart Circ Physiol 297(1):H153–62
18. Isomaa B et al (2001) Cardiovascular morbidity and mortality associated with the metabolic syndrome. Diabetes Care 24(4):683–689
19. Ford ES, Giles WH (2003) A comparison of the prevalence of the metabolic syndrome using two proposed definitions. Diabetes Care 26(3):575–581
20. Tenerz A et al (2003) Diabetes, insulin resistance, and the metabolic syndrome in patients with acute myocardial infarction without previously known diabetes. Diabetes Care 26(10):2770–2776
21. Aijaz B et al (2008) Abnormal cardiac structure and function in the metabolic syndrome: a population-based study. Mayo Clin Proc 83(12):1350–1357
22. Huang PL (2009) eNOS, metabolic syndrome and cardiovascular disease. Trends Endocrinol Metab 20(6):295–302
23. Marfella R et al (2009) Myocardial lipid accumulation in patients with pressure-overloaded heart and metabolic syndrome. J Lipid Res 50(11):2314–2323
24. Kishi S et al (2014) Association of obesity in early adulthood and middle age with incipient left ventricular dysfunction and structural remodeling: the CARDIA study (Coronary Artery Risk Development in Young Adults). JACC Heart Fail 2(5):500–508
25. Ayaz M et al (2002) Protective effect of selenium treatment on diabetes-induced myocardial structural alterations. Biol Trace Elem Res 89(3):215–226
26. Boudina S et al (2007) Mitochondrial energetics in the heart in obesity-related diabetes: direct evidence for increased uncoupled respiration and activation of uncoupling proteins. Diabetes 56(10):2457–2466

27. Boveris A (1984) Determination of the production of superoxide radicals and hydrogen peroxide in mitochondria. Methods Enzymol 105:429–435
28. St-Pierre J et al (2002) Topology of superoxide production from different sites in the mitochondrial electron transport chain. J Biol Chem 277(47):44784–44790
29. Duncan JG et al (2007) Insulin-resistant heart exhibits a mitochondrial biogenic response driven by the peroxisome proliferator-activated receptor-alpha/PGC-1alpha gene regulatory pathway. Circulation 115(7):909–917
30. Neubauer S et al (1997) Myocardial phosphocreatine-to-ATP ratio is a predictor of mortality in patients with dilated cardiomyopathy. Circulation 96(7):2190–2196
31. Peterson LR et al (2004) Effect of obesity and insulin resistance on myocardial substrate metabolism and efficiency in young women. Circulation 109(18):2191–2196
32. Anderson EJ et al (2009) Substrate-specific derangements in mitochondrial metabolism and redox balance in the atrium of the type 2 diabetic human heart. J Am Coll Cardiol 54(20):1891–1898
33. Nakamura H et al (2012) p53 promotes cardiac dysfunction in diabetic mellitus caused by excessive mitochondrial respiration-mediated reactive oxygen species generation and lipid accumulation. Circ Heart Fail 5(1):106–115
34. Durak A et al (2018) A SGLT2 inhibitor dapagliflozin suppresses prolonged ventricular-repolarization through augmentation of mitochondrial function in insulin-resistant metabolic syndrome rats. Cardiovasc Diabetol 17(1):144
35. Oliveira PJ et al (2003) Enhanced permeability transition explains the reduced calcium uptake in cardiac mitochondria from streptozotocin-induced diabetic rats. FEBS Lett 554(3):511–514
36. Perrone-Filardi P et al (2015) The role of metabolic syndrome in heart failure. Eur Heart J 36(39):2630–2634
37. Gaddam KK, Ventura HO, Lavie CJ (2011) Metabolic syndrome and heart failure–the risk, paradox, and treatment. Curr Hypertens Rep 13(2):142–148
38. Gupta PP, Fonarow GC, Horwich TB (2015) Obesity and the obesity paradox in heart failure. Can J Cardiol 31(2):195–202
39. Lavie CJ et al (2013) Impact of obesity and the obesity paradox on prevalence and prognosis in heart failure. JACC Heart Fail 1(2):93–102
40. Oreopoulos A et al (2008) Body mass index and mortality in heart failure: a meta-analysis. Am Heart J 156(1):13–22
41. Kalantar-Zadeh K et al (2004) Reverse epidemiology of conventional cardiovascular risk factors in patients with chronic heart failure. J Am Coll Cardiol 43(8):1439–1444
42. Kenchaiah S et al (2002) Obesity and the risk of heart failure. N Engl J Med 347(5):305–313
43. Ather S et al (2012) Impact of noncardiac comorbidities on morbidity and mortality in a predominantly male population with heart failure and preserved versus reduced ejection fraction. J Am Coll Cardiol 59(11):998–1005
44. Mentz RJ et al (2014) Noncardiac comorbidities in heart failure with reduced versus preserved ejection fraction. J Am Coll Cardiol 64(21):2281–2293
45. Vella S et al (2010) The use of metformin in type 1 diabetes: a systematic review of efficacy. Diabetologia 53(5):809–820
46. George P, McCrimmon RJ (2013) Potential role of non-insulin adjunct therapy in Type 1 diabetes. Diabet Med 30(2):179–188
47. Garg SK et al (2013) Effect of sitagliptin on post-prandial glucagon and GLP-1 levels in patients with type 1 diabetes: investigator-initiated, double-blind, randomized, placebo-controlled trial. Endocr Pract 19(1):19–28
48. Washburn WN, Poucher SM (2013) Differentiating sodium-glucose co-transporter-2 inhibitors in development for the treatment of type 2 diabetes mellitus. Expert Opin Investig Drugs 22(4):463–486
49. Zinman B et al (2015) Empagliflozin, cardiovascular outcomes, and mortality in type 2 diabetes. N Engl J Med 373(22):2117–2128
50. Verma S et al (2018) Empagliflozin reduces cardiovascular events, mortality and renal events in participants with type 2 diabetes after coronary artery bypass graft surgery: subanalysis of the EMPA-REG OUTCOME(R) randomised trial. Diabetologia 61(8):1712–1723

51. Fitchett D et al (2018) Effects of empagliflozin on risk for cardiovascular death and heart failure hospitalization across the spectrum of heart failure risk in the EMPA-REG OUTCOME(R) trial. Eur Heart J 39(5):363–370
52. Neal B et al (2017) Canagliflozin and cardiovascular and renal events in type 2 diabetes. N Engl J Med 377(7):644–657
53. Verma S, McMurray JJV (2019) The serendipitous story of SGLT2 inhibitors in heart failure. Circulation 139(22):2537–2541
54. Kosiborod M et al (2017) Lower risk of heart failure and death in patients initiated on sodium-glucose cotransporter-2 inhibitors versus other glucose-lowering drugs: the CVD-REAL study (Comparative effectiveness of cardiovascular outcomes in new users of sodium-glucose cotransporter-2 inhibitors). Circulation 136(3):249–259
55. Ye Y et al (2017) SGLT-2 inhibition with dapagliflozin reduces the activation of the Nlrp3/ASC inflammasome and attenuates the development of diabetic cardiomyopathy in mice with type 2 diabetes. Further augmentation of the effects with saxagliptin, a DPP4 inhibitor. Cardiovasc Drugs Ther 31(2):119–132
56. Packer M et al (2017) Effects of sodium-glucose cotransporter 2 inhibitors for the treatment of patients with heart failure: proposal of a novel mechanism of action. JAMA Cardiol 2(9):1025–1029
57. van Baar MJB et al (2018) SGLT2 inhibitors in combination therapy: from mechanisms to clinical considerations in type 2 diabetes management. Diabetes Care 41(8):1543–1556
58. Lahnwong S, Chattipakorn SC, Chattipakorn N (2018) Potential mechanisms responsible for cardioprotective effects of sodium-glucose co-transporter 2 inhibitors. Cardiovasc Diabetol 17(1):101
59. Vrhovac I et al (2015) Localizations of Na(+)-D-glucose cotransporters SGLT1 and SGLT2 in human kidney and of SGLT1 in human small intestine, liver, lung, and heart. Pflugers Arch 467(9):1881–1898
60. Sattar N et al (2016) SGLT2 inhibition and cardiovascular events: why did EMPA-REG outcomes surprise and what were the likely mechanisms? Diabetologia 59(7):1333–1339
61. Plosker GL (2012) Dapagliflozin: a review of its use in type 2 diabetes mellitus. Drugs 72(17):2289–2312
62. Lamos EM, Younk LM, Davis SN (2013) Canagliflozin, an inhibitor of sodium-glucose cotransporter 2, for the treatment of type 2 diabetes mellitus. Expert Opin Drug Metab Toxicol 9(6):763–775
63. Saeed MA, Narendran P (2014) Dapagliflozin for the treatment of type 2 diabetes: a review of the literature. Drug Des Devel Ther 8:2493–2505
64. De Nicola L et al (2014) Sodium/glucose cotransporter 2 inhibitors and prevention of diabetic nephropathy: targeting the renal tubule in diabetes. Am J Kidney Dis 64(1):16–24
65. Ferrannini E, Mark M, Mayoux E (2016) CV protection in the EMPA-REG OUTCOME trial: a "thrifty substrate" hypothesis. Diabetes Care 39(7):1108–1114
66. Lopaschuk GD et al (2010) Myocardial fatty acid metabolism in health and disease. Physiol Rev 90(1):207–258
67. Baartscheer A et al (2017) Empagliflozin decreases myocardial cytoplasmic Na(+) through inhibition of the cardiac Na(+)/H(+) exchanger in rats and rabbits. Diabetologia 60(3):568–573
68. Chilton R et al (2015) Effects of empagliflozin on blood pressure and markers of arterial stiffness and vascular resistance in patients with type 2 diabetes. Diabetes Obes Metab 17(12):1180–1193
69. Lee TM, Chang NC, Lin SZ (2017) Dapagliflozin, a selective SGLT2 Inhibitor, attenuated cardiac fibrosis by regulating the macrophage polarization via STAT3 signaling in infarcted rat hearts. Free Radic Biol Med 104:298–310
70. Garvey WT et al (2018) Effects of canagliflozin versus glimepiride on adipokines and inflammatory biomarkers in type 2 diabetes. Metabolism 85:32–37
71. Sato T et al (2018) The effect of dapagliflozin treatment on epicardial adipose tissue volume. Cardiovasc Diabetol 17(1):6

72. Khalaf KI, Taegtmeyer H (2012) After avandia: the use of antidiabetic drugs in patients with heart failure. Tex Heart Inst J 39(2):174–178
73. Han S et al (2008) Dapagliflozin, a selective SGLT2 inhibitor, improves glucose homeostasis in normal and diabetic rats. Diabetes 57(6):1723–1729
74. Hansen HH et al (2014) The sodium glucose cotransporter type 2 inhibitor empagliflozin preserves beta-cell mass and restores glucose homeostasis in the male zucker diabetic fatty rat. J Pharmacol Exp Ther 350(3):657–664
75. Hinnen D (2013) The role of the kidney in hyperglycemia: a new therapeutic target in type 2 diabetes mellitus. J Cardiovasc Nurs 28(2):157–165
76. Novak LM, Kruger DF (2017) Bolstering your armamentarium with SGLT2 inhibitors. Nurse Pract 42(10):28–34
77. Lim S, Eckel RH, Koh KK (2018) Clinical implications of current cardiovascular outcome trials with sodium glucose cotransporter-2 (SGLT2) inhibitors. Atherosclerosis 272:33–40
78. Vasilakou D et al (2013) Sodium-glucose cotransporter 2 inhibitors for type 2 diabetes: a systematic review and meta-analysis. Ann Intern Med 159(4):262–274
79. Han JH et al (2017) The beneficial effects of empagliflozin, an SGLT2 inhibitor, on atherosclerosis in ApoE (-/-) mice fed a western diet. Diabetologia 60(2):364–376
80. Sato T et al (2017) Effect of sodium-glucose co-transporter-2 inhibitors on impaired ventricular repolarization in people with Type 2 diabetes. Diabet Med 34(10):1367–1371
81. Hsia DS, Heymsfield SB (2018) Reducing diabetes risk at an early age. Nat Med 24(6):708–710
82. Jia X et al (2018) SGLT2 inhibitors and cardiovascular outcomes: current perspectives and future potentials. Curr Diab Rep 18(9):63
83. Zhang XL et al (2018) Cardiovascular safety, long-term noncardiovascular safety, and efficacy of sodium-glucose cotransporter 2 inhibitors in patients with type 2 diabetes mellitus: a systemic review and meta-analysis with trial sequential analysis. J Am Heart Assoc 7(2)
84. Verma S, McMurray JJV (2018) SGLT2 inhibitors and mechanisms of cardiovascular benefit: a state-of-the-art review. Diabetologia 61(10):2108–2117
85. Kashiwagi Y et al (2015) Expression of SGLT1 in human hearts and impairment of cardiac glucose uptake by phlorizin during ischemia-reperfusion injury in mice. PLoS ONE 10(6):e0130605
86. Habibi J et al (2017) Sodium glucose transporter 2 (SGLT2) inhibition with empagliflozin improves cardiac diastolic function in a female rodent model of diabetes. Cardiovasc Diabetol 16(1):9
87. Kusaka H et al (2016) Empagliflozin lessened cardiac injury and reduced visceral adipocyte hypertrophy in prediabetic rats with metabolic syndrome. Cardiovasc Diabetol 15(1):157
88. Lin B et al (2014) Glycemic control with empagliflozin, a novel selective SGLT2 inhibitor, ameliorates cardiovascular injury and cognitive dysfunction in obese and type 2 diabetic mice. Cardiovasc Diabetol 13:148
89. Hammoudi N et al (2017) Empagliflozin improves left ventricular diastolic dysfunction in a genetic model of type 2 diabetes. Cardiovasc Drugs Ther 31(3):233–246
90. Liang L, Jiang J, Frank SJ (2000) Insulin receptor substrate-1-mediated enhancement of growth hormone-induced mitogen-activated protein kinase activation. Endocrinology 141(9):3328–3336
91. Zhou Y, Wu W (2017) The sodium-glucose co-transporter 2 inhibitor, empagliflozin, protects against diabetic cardiomyopathy by inhibition of the endoplasmic reticulum stress pathway. Cell Physiol Biochem 41(6):2503–2512
92. Tanajak P et al (2018) Cardioprotection of dapagliflozin and vildagliptin in rats with cardiac ischemia-reperfusion injury. J Endocrinol 236(2):69–84
93. Shi X et al (2017) Effect of empagliflozin on cardiac biomarkers in a zebrafish model of heart failure: clues to the EMPA-REG OUTCOME trial? Mol Cell Biochem 433(1–2):97–102
94. Januzzi JL Jr et al (2017) Effects of canagliflozin on cardiovascular biomarkers in older adults with type 2 diabetes. J Am Coll Cardiol 70(6):704–712

95. Marx N, McGuire DK (2016) Sodium-glucose cotransporter-2 inhibition for the reduction of cardiovascular events in high-risk patients with diabetes mellitus. Eur Heart J 37(42):3192–3200
96. Tanaka A, Node K (2017) Emerging roles of sodium-glucose cotransporter 2 inhibitors in cardiology. J Cardiol 69(3):501–507
97. Aydemir-Koksoy A et al (2010) Antioxidant treatment protects diabetic rats from cardiac dysfunction by preserving contractile protein targets of oxidative stress. J Nutr Biochem 21(9):827–833
98. Ayaz M et al (2004) Effects of selenium on altered mechanical and electrical cardiac activities of diabetic rat. Arch Biochem Biophys 426(1):83–90
99. Yamamoto E et al (2006) Enhancement of cardiac oxidative stress by tachycardia and its critical role in cardiac hypertrophy and fibrosis. J Hypertens 24(10):2057–2069
100. Griendling KK, Sorescu D, Ushio-Fukai M (2000) NAD(P)H oxidase: role in cardiovascular biology and disease. Circ Res 86(5):494–501
101. Turan B (2010) Role of antioxidants in redox regulation of diabetic cardiovascular complications. Curr Pharm Biotechnol 11(8):819–836
102. Turan B, Vassort G (2011) Ryanodine receptor: a new therapeutic target to control diabetic cardiomyopathy. Antioxid Redox Signal 15(7):1847–1861
103. Vassort G, Turan B (2010) Protective role of antioxidants in diabetes-induced cardiac dysfunction. Cardiovasc Toxicol 10(2):73–86
104. Okatan EN, Tuncay E, Turan B (2013) Cardioprotective effect of selenium via modulation of cardiac ryanodine receptor calcium release channels in diabetic rat cardiomyocytes through thioredoxin system. J Nutr Biochem 24(12):2110–2118
105. Oelze M et al (2014) The sodium-glucose co-transporter 2 inhibitor empagliflozin improves diabetes-induced vascular dysfunction in the streptozotocin diabetes rat model by interfering with oxidative stress and glucotoxicity. PLoS ONE 9(11):e112394
106. Giorgi C et al (2012) Mitochondrial Ca(2+) and apoptosis. Cell Calcium 52(1):36–43
107. Nazarewicz RR et al (2013) Nox2 as a potential target of mitochondrial superoxide and its role in endothelial oxidative stress. Am J Physiol Heart Circ Physiol 305(8):H1131–H1140
108. Jeong EM et al (2016) Role of mitochondrial oxidative stress in glucose tolerance, insulin resistance, and cardiac diastolic dysfunction. J Am Heart Assoc 5(5)
109. Carafoli E, Lehninger AL (1971) A survey of the interaction of calcium ions with mitochondria from different tissues and species. Biochem J 122(5):681–690
110. Denton RM, McCormack JG (1980) The role of calcium in the regulation of mitochondrial metabolism. Biochem Soc Trans 8(3):266–268
111. Rizzuto R et al (1992) Rapid changes of mitochondrial Ca^{2+} revealed by specifically targeted recombinant aequorin. Nature 358(6384):325–327
112. Mansouri A et al (2006) Alterations in mitochondrial function, hydrogen peroxide release and oxidative damage in mouse hind-limb skeletal muscle during aging. Mech Ageing Dev 127(3):298–306
113. Dey S et al (2018) Mitochondrial ROS drive sudden cardiac death and chronic proteome remodeling in heart failure. Circ Res 123(3):356–371
114. Owada T et al (2017) Resolution of mitochondrial oxidant stress improves aged-cardiovascular performance. Coron Artery Dis 28(1):33–43
115. Galloway CA, Yoon Y (2015) Mitochondrial dynamics in diabetic cardiomyopathy. Antioxid Redox Signal 22(17):1545–1562
116. Montaigne D et al (2014) Myocardial contractile dysfunction is associated with impaired mitochondrial function and dynamics in type 2 diabetic but not in obese patients. Circulation 130(7):554–564
117. Sharp WW et al (2015) Inhibition of the mitochondrial fission protein dynamin-related protein 1 improves survival in a murine cardiac arrest model. Crit Care Med 43(2):e38-47
118. Duncan JG (2011) Mitochondrial dysfunction in diabetic cardiomyopathy. Biochim Biophys Acta 1813(7):1351–1359

119. Ong SB et al (2010) Inhibiting mitochondrial fission protects the heart against ischemia/reperfusion injury. Circulation 121(18):2012–2022
120. Lucas E et al (2016) Obesity-induced cardiac lipid accumulation in adult mice is modulated by G protein-coupled receptor kinase 2 levels. Cardiovasc Diabetol 15(1):155
121. Yang XH et al (2017) The relationship between insulin sensitivity and heart rate-corrected QT interval in patients with type 2 diabetes. Diabetol Metab Syndr 9:69
122. Torekov SS et al (2014) KCNQ1 long QT syndrome patients have hyperinsulinemia and symptomatic hypoglycemia. Diabetes 63(4):1315–1325
123. Alpert MA (2001) Obesity cardiomyopathy: pathophysiology and evolution of the clinical syndrome. Am J Med Sci 321(4):225–236
124. Gastaldelli A et al (2000) Insulin prolongs the QTc interval in humans. Am J Physiol Regul Integr Comp Physiol 279(6):R2022–R2025
125. de Simone G et al (1994) Relation of obesity and gender to left ventricular hypertrophy in normotensive and hypertensive adults. Hypertension 23(5):600–606
126. Yaras N et al (2005) Effects of diabetes on ryanodine receptor Ca release channel (RyR2) and Ca^{2+} homeostasis in rat heart. Diabetes 54(11):3082–3088
127. Wehrens XH, Marks AR (2003) Altered function and regulation of cardiac ryanodine receptors in cardiac disease. Trends Biochem Sci 28(12):671–678
128. Choi KM et al (2002) Defective intracellular Ca(2+) signaling contributes to cardiomyopathy in Type 1 diabetic rats. Am J Physiol Heart Circ Physiol 283(4):H1398–H1408
129. Pierce GN, Kutryk MJ, Dhalla NS (1983) Alterations in Ca^{2+} binding by and composition of the cardiac sarcolemmal membrane in chronic diabetes. Proc Natl Acad Sci USA 80(17):5412–5416
130. Fein FS, Sonnenblick EH (1994) Diabetic cardiomyopathy. Cardiovasc Drugs Ther 8(1):65–73
131. Rubler S et al (1972) New type of cardiomyopathy associated with diabetic glomerulosclerosis. Am J Cardiol 30(6):595–602
132. Priest BT, McDermott JS (2015) Cardiac ion channels. Channels (Austin) 9(6):352–359
133. Lambert R et al (2015) Intracellular Na^+ concentration ($[Na^+]_i$) is elevated in diabetic hearts due to enhanced Na^+-glucose cotransport. J Am Heart Assoc 4(9):e002183
134. Baartscheer A et al (2005) Chronic inhibition of Na^+/H^+-exchanger attenuates cardiac hypertrophy and prevents cellular remodeling in heart failure. Cardiovasc Res 65(1):83–92
135. Baartscheer A et al (2003) Increased Na^+/H^+-exchange activity is the cause of increased $[Na^+]$i and underlies disturbed calcium handling in the rabbit pressure and volume overload heart failure model. Cardiovasc Res 57(4):1015–1024
136. Liu T, O'Rourke B (2008) Enhancing mitochondrial Ca^{2+} uptake in myocytes from failing hearts restores energy supply and demand matching. Circ Res 103(3):279–288
137. Liu T et al (2014) Inhibiting mitochondrial Na^+/Ca^{2+} exchange prevents sudden death in a Guinea pig model of heart failure. Circ Res 115(1):44–54
138. Joubert M et al (2017) The sodium-glucose cotransporter 2 inhibitor dapagliflozin prevents cardiomyopathy in a diabetic lipodystrophic mouse model. Diabetes 66(4):1030–1040
139. Olgar Y, Turan B (2019) A sodium-glucose cotransporter 2 (SGLT2) inhibitor dapagliflozin comparison with insulin shows important effects on Zn(2+)-transporters in cardiomyocytes from insulin-resistant metabolic syndrome rats through inhibition of oxidative stress (1). Can J Physiol Pharmacol 97(6):528–535
140. Lim VG et al (2019) SGLT2 inhibitor, canagliflozin, attenuates myocardial infarction in the diabetic and nondiabetic heart. JACC Basic Transl Sci 4(1):15–26
141. Vardeny O, Vaduganathan M (2019) Practical guide to prescribing sodium-glucose cotransporter 2 inhibitors for cardiologists. JACC Heart Fail 7(2):169–172
142. Garcia-Ropero A et al (2019) Metabolism of the failing heart and the impact of SGLT2 inhibitors. Expert Opin Drug Metab Toxicol 15(4):275–285

Chapter 8
The Roles of HIFs in the Complications of Diabetes

Nuray Yazihan and Mehtap Kacar

Abstract Diabetes is mainly defined as disturbances of glucose and lipid metabolism, and characterized by hyperglycemia. The progress of diabetes affects most of the tissues and organs such as endothelium, retina, heart, kidney, and brain. Glycolysis and metabolism are controlled by oxygen (O_2) which is essential for the maintenance of life of all aerobic organisms. Disturbances of O_2 level results in disturbed mitochondrial respiration, metabolic and oxidative stress. Generally adipose tissue O_2 level is lower than the alveolar and vascular area and brown and white adipose tissues have different O_2 concentrations. With increased nutrient intake and expansion of adipose tissue, O_2 need will be increased and induce hypoxia-inducible factors (HIFs). Insufficient vasculature and blood flow to adipose tissue increase expression of HIF-1 dependent genes to induce angioneogenesis. Hypoxia in adipose tissue and adipocytes was shown to inhibit insulin-responsive pathways such as IRS and protein kinase B, receptor autophosphorylation and insulin-dependent glucose transport and have the capacity to control inflammation and immune cell polarity. These mechanisms and cascades take part in the pathogenesis of metabolic syndrome and diabetes and subsequent complications. In this chapter, we will discuss the role of HIFs in the pathogenesis of diabetes-associated vascular and renal complications.

Keywords Hypoxia inducible factors (HIF) · Diabetes · Adipocyte · Mitochondrial stress · Cardiovascular complications

N. Yazihan (✉)
Interdisciplinary Food, Metabolism and Clinical Nutrition Department, Institute of Health Sciences, Ankara University, Ankara, Turkey
e-mail: nurayyazihan@yahoo.com

Faculty of Medicine, Department of Pathophysiology, Ankara University, Ankara, Turkey

M. Kacar
Faculty of Medicine, Department of Physiology, Yeditepe University, Istanbul, Turkey

P. S. Tappia et al. (eds.), *Biochemistry of Cardiovascular Dysfunction in Obesity*, Advances in Biochemistry in Health and Disease 20,
https://doi.org/10.1007/978-3-030-47336-5_8

Oxygen (O_2) is essential for the maintenance of life of all aerobic organisms. O_2 is needed for aerobic glycolysis. Decreased O_2 level disturbs mitochondrial respiration and oxidative phosphorylation, induces mitochondrial stress and starts anaerobic metabolism. Tissue O_2 level, especially adipose tissue O_2 level is lower than alveolar and vascular area. O_2 levels are lower in white adipose tissue compared to brown adipose tissue. Both acute and chronic excessive intake of nutrients results in adipose tissue expansion especially white tissue expansion by a combination of increase intracellular lipid deposition, hypertrophy and hyperplasia of adipocytes. Increased fat volume needs more O_2 and induce hypoxia inducible factors (HIFs). Insufficient vasculature and blood flow to adipose tissue increases expression of HIF-1 dependent genes to support angioneogenesis. Hypoxia in adipose tissue and adipocytes was shown to inhibit insulin responsive pathways such as IRS and protein kinase B, receptor autophosphorylation and insulin dependent glucose transport. Both induction of hypoxia with cobalt, and overexpression of HIF-1 in adipocytes were shown to decrease insulin responses [20, 42, 51, 52].

Pancreatic islets receive almost 1/6 of pancreatic total blood flow, but they only account for 1–2% of the volume. Blood supply and O_2 content is important for regulation of endocrine function of the pancreas. During hyperglycemia, islet cells need more oxygen and in case of O_2 deficiency pseudohypoxia occurs. HIFs are critical for switching from aerobic to anaerobic glycolysis by induction of GLUT1, multiple glycolytic enzymes and most importantly control of excess mitochondrial oxygen consumption. Chronic intermittent hypoxia interferes pancreatic β-cell function and impairs basal and glucose induced insulin secretions. Chronic hypoxia disturbs mitochondrial functions and increases reactive oxygen radical formation which further increases the pancreatic damage [15, 53, 54].

It is well known that hypoxia has a significant role on adipocyte functions, control of metabolism and development of inflammation in obesity-related diseases especially in the pathogenesis of metabolic syndrome, diabetes and subsequent complications. In this chapter we will discuss the role of HIFs in pathogenesis of diabetes associated vascular and renal complications.

Hypoxia Inducible Factors

Hypoxia inducible factors (HIFs) play important roles in cellular adaptation to hypoxia inflammation, stress conditions and nutrient deprivation. There are three members of the family: HIF-1α, HIF-2α, HIF-3α. HIF-1α and HIF-2α are the main factors that regulate hypoxic responses. HIF-1α is associated with glycolytic gene expression, whereas HIF-2α is associated with lipid metabolism. Expression of erythropoietin is under control of HIF-2α. HIF-1α is found decreased during preadipocyte to adipocyte differentiation. HIF-2α and HIF-3α expressions are increased in mature adipocytes. HIF-2α-specific target genes are involved in the regulation of function and/or differentiation of stem cell, cell cycle progression of renal carcinoma cells. ATP dependent K channels are main regulators of metabolism, insulin secretion and glucose transport are also under control of HIF-1α. Although limited information is found about HIF-3α; its isoforms HIF-3α2 and 3α3 inhibit HIF-1α and HIF-2α gen expressions; and hypoxia activated glycolytic genes [3, 11, 22, 23, 34, 46].

In hypoxic conditions, HIF-1 pathway regulates the responses of organism. There are two different subunits, HIF-1α and HIF-1β, and they are encoded by the HIF1α and HIF1β genes, respectively. Both of subunits are expressed constitutively, but hypoxia affects only HIF-1α expression. HIF-1α undergo degradation by oxygen dependent reactions in normoxia, the changes in the transcription of hypoxia related genes are not induced. The prolyl hydroxylase domain proteins (PHDs) of HIF-1α continues with hydroxylation reaction with oxygen and convert prolines. The hydroxylated prolines noticed by the von Hippel-Lindau (VHL) protein, then ubiquitination occurs and is degraded in the 26S proteasome. In hypoxic conditions, degradation of HIF-1α is inhibited. Then, HIF-1α and HIF-1β subunits generate a heterodimer and translocate to the nucleus. HIF-1α binds to hypoxia-responsive elements (HREs) on DNA, and affects several hundred hypoxia-responsive genes. As response to hypoxia, HIF-1 signaling triggers many different adaptive and responsive signaling pathways such as the switch from oxidative phosphorylation to anaerobic glycolysis, angiogenesis, erythropoiesis, and cell survival. The responses to physiological hypoxia can be observed during development and growth process of human life. The pathological hypoxia presents during pathophysiological events [9].

HIFs are oxygen sensitive response factors that regulate metabolic adaptation. Now, it is accepted that HIFs are master regulators of all hypoxic responses and cellular survival. HIFs have a central role in control of crosstalk between inflammatory cells and tissue microenvironment. HIFs are oxygen sensitive response factors that regulate metabolic adaptation. Now, it is accepted that HIFs are master regulators of all hypoxic responses and cellular survival. HIFs have a central role in control of crosstalk between inflammatory cells and tissue microenvironment. While adipose tissue increases in size, vascular supply need will be increase and in the case of inadequate O_2 supply local hypoxia will occur. Local hypoxia will induce HIF-1 related genes that will activate local inflammation, fibrosis, neoangiogenesis, smooth muscle proliferation, and anaerobic glycolysis. Both HIF-1α and HIF-2α are hypoxia responsive genes but it has been recently shown that HIF-1α and HIF-2α antagonize each other in the regulation of macrophage polarization and inflammatory responses of the tissues. M1 and M2 polarizations are important for fate of the system. M2 polarization is regulated by HIF-2 activation and NO production, resulting in different effects on the vascular responses and cardiovascular complications of hyperglycemia induced pseudohypoxia. It is demonstrated that M2 polarization with HIF-2α decreases the adipose tissue inflammation and insulin resistance. HIF-2α gene modification could modulate proinflammatory responses of adipose tissue macrophages in high fat diet. HIF-2α is important for protection of the body against obesity induced inflammation and insulin resistance. Understanding mechanisms that regulate hyper nutrition induced expansion of adipose tissue is critical for identify obesity related metabolic syndrome and diabetes pathophysiology. These are also important for determining of therapeutic strategies for minimizing the complications of obesity and/or metabolic

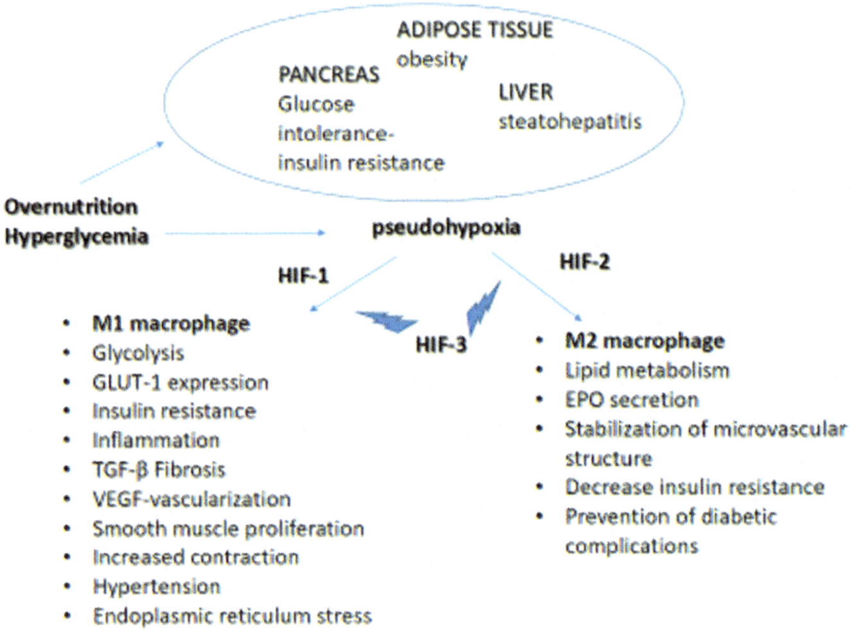

Fig. 8.1 Hypoxia and hyperglycemia

syndromes. It is interesting that intestinal HIF-2α signaling was positively correlated with body-mass index and hepatic steatosis. It is shown that overexpression of both HIF-1α and HIF-2α could induce hepatic steatosis. Hepatic HIF-2α is critical for regulation of lipid metabolism more than HIF-1α [2, 10, 16, 36, 57] (Fig. 8.1).

Pathophysiology of Diabetic Macro and Microvascular Complications

Diabetes is mainly defined as disturbances of glucose and lipid metabolism, and characterized by hyperglycemia. It affects most of the tissues and organs such as endothelium, retina, heart, kidney, and brain. It leads to different complications including cardiovascular disease, stroke, nephropathy, and retinopathy. In humans, metabolic reactions are mostly oxygen dependent (aerobic metabolism). Hyperglycemia affects HIF-1 signaling. Hyperglycemia that is typical sign of diabetes leads to pseudohypoxia because of increased glucose concentration in the tissues. Glucose is used by alternative pathways such as polyol pathway in the patients with diabetes. Increased activation of polyol pathway causes increased ratio of NADH/NAD+ which resembles pyruvate/lactate ratio. Pseudohypoxia with NADH/NAD+ stimulates the production of reactive oxygen species ROS. Increased ROS and pseudohypoxia stimulate transcription of HIF-1α mRNA. Diabetic or hyperglycemic pseudo hypoxia disturbs HIF-1 signaling. This conflict generates inappropriate response against hypoxia in the cells. The dysregulation of HIF-1 signaling plays main role in the development of diabetic complications [9].

Hyperglycemia blocks the hypoxia-induced HIF-1α stabilization. Hypoxia and hyperglycemia are two important factors in development of diabetic complication such as retinopathy, nephropathy, neuropathy, atherosclerosis and foot ulcers [7].

HIF-1α regulated genes are linked to cell adaptation and survival mechanisms which are angiogenesis, anaerobic glycolytic pathway, erythropoiesis, wound healing, cell growth, proliferation, differentiation, survival and apoptosis. Some of these genes are vascular endothelial growth factor (VEGF), haem oxygenase 1 (HO-1), nitric oxide synthase (NOS), endothelin, erythropoietin (EPO), lactate dehydrogenase A (LDH-A), Glucose transporter 1 (GLUT-1), Glucose transporter 3 (GLUT-3), C-X-C chemokine receptor type 4 (CXCR4) and stromal cell-derived factor-1 (SDF-1) and p53. The other subunit of HIF, HIF-1β is the aryl hydrocarbon receptor nuclear translocator (ARNT) plays an important role in activation of pancreatic β cells for secreting insulin by stimulation of glucose. Hyperglycemia leads suppression of HIF-1β; and disturbs insulin secretion from pancreatic islet cells. HIF-1α is an essential factor for expression of HIF-1β hence for maintaining beta cell function. In normoxic condition, glucose alone is not enough to activate the HIF-1α signaling; but normal concentration of glucose is needed for HIF-1α protein expression and activation as a response to hypoxia. The dysfunction of HIF-1α transactivation is one of the reason of inhibition of angiogenesis and related target gene expressions. It may be linked with impaired wound healing in diabetic patients. Hyperglycemia increases oxidative stress and production of ROS which affect HIF-1α signaling. Especially superoxide radicals (O_2-) degrades HIF-1α by activating hydroxylation of prolines and also increasing ubiquitin-proteasome activity. In addition to degradation, $O2^-$ also causes suppression of HIF-1α formation [56].

Vascular lesions and other associated problems are important complications of diabetes. Diabetic vasculopathy is characterized by loss of hypoxia-dependent angiogenesis and, actually, dysfunction of HIF pathway in vascular beds of many tissues such as skin, nerves, brain, skeletal muscle, heart, kidney [50].

There is a strong correlation between diabetes and cardiovascular diseases such as atherosclerosis, hypertension, peripheral vascular diseases, stroke, delayed wound healing and also cardiomyopathy [9, 50]. These complications are mostly linked with defects in responses against hypoxia in the vascular tissues. The production of VEGF that is an angiogenic growth factor decreases in diabetes. Impaired HIF-1α transactivation is responsible for decreased VEGF expression in hypoxic diabetic tissues [50].

Although VEGF synthesis is mainly promoted by hypoxia, it can be stimulated by several factors such as gender, smoking, hyper- and hypoglycemia, hypercholesterolemia, hypoxia, and stress. In diabetics, chronic hyperglycemia leads to dysregulation of HIF-1α and VEGF expression in the micro-vascular endothelial cells (ECs). In addition of oxygen, insulin, insulin-like growth factor-I (IGF-I), IGF-II and AGEs can affect HIF pathway at different steps of this pathway. The patients with diabetes have increased levels of IGF-1. IGF-I promotes HIF-1α expression, then VEGF production increases. It has been reported that IGF-1 stimulated HIF-1α activation leads to increased VEGF expression. VEGF regulates angiogenesis, it is important in development of vascular complication of diabetes [25].

The other important microvascular complication of diabetes is diabetic retinopathy. HIF-1 target genes are necessary for vascular stability, healthy retinal function, and development. Retina has physiologically very low levels of oxygen. Hyperglycemia leads to decrease oxygen levels in the retina in early stage of diabetes. Hypoxia causes increased oxidative stress, inflammation, loss of pericytes in the retina. On the other hand, HIF-1α function is suppressed in diabetes. The inactivity of HIF-1α signaling provides augmented hypoxic damage [13].

In addition to hypoxia, IGF-1 induced inappropriate activation of HIF-1α can provide pathophysiology of diabetic retinopathy. The inappropriate and exaggerated activation of HIF-1α leads to upregulation of VEGF in the retinal cells [25].

Development of diabetic cardiomyopathy is independent to the vascular complications of diabetes. Diabetic cardiomyopathy is characterized by inhibition of glycolysis in myocardial cells, loss of capillaries, increased lipid accumulation, fibrotic tissue changes, advanced glycation end products (AGEs) formation, apoptosis, and mitochondrial dysfunction in the myocardium. In diabetic condition, myocardial cells use glucose fatty acids rather than glucose for their metabolism. It is resulted from the decreased GLUT-1 transporter expression and decreased hexokinase II enzyme. Their genes are regulated by HIF-1 signaling [9].

Diabetic cardiomyopathy is associated with high glucose content of myocardial cells, increased oxidative stress, abnormal stimulation of neuroendocrine system, chronic inflammatory status, and myocardial apoptosis. HIF-1α affects thymic cells and hence T cells. Researcher argued that HIF-1α in the T-cells play a central role for protection from myocardial cell damage. They showed that T cell specific deficiency of HIF-1α caused severe damage the myocardium of diabetic mice. HIF-1α is expressed in lymphocytes under the hypoxic and hyperglycemic conditions. Hyperglycemia decreases HIF-1α expression and function and changes the expression profile of HIF-1α. Hyperglycemia promotes expression of cellular adhesion molecules of monocytes and leads to increased adherence of monocytes on the endothelial cells. Hyperglycemia also increases the endothelial permeability. Increased endothelial permeability generates to accelerated lipid deposition and macrophage recruitment in the endothelium [32].

Diabetic individuals have two-to-fourfold increased risk of cardiovascular disease. The cardiovascular diseases are usually associated with development of atherosclerosis. Atherosclerosis is a chronic inflammatory disease. The increased inflammation and accelerated atherosclerosis are observed in diabetic patients. Hyperglycemia, chronic inflammatory status, and hypoxia affect vascular tissue, especially the endothelium. All of these factors contribute to endothelial damage and dysfunction. Endothelial dysfunction is the first step of pathophysiological mechanism of atherosclerosis. Endothelial dysfunction is characterized by the loss of the physiological functions and properties of endothelium. Endothelial tissues have a tendency for vasodilatation, its surface exhibits antiadhesive and antiaggregant properties. Endothelial cells synthesis and secretes many substances such as endothelin-1, tromboxan A2, nitric oxide (NO), prostacyclin etc. Hyperglycemia induces apoptotic signal pathways including β-1integrin signaling, p38 MAPK, JNK (cJun-N-terminal protein kinase) in the endothelial cells. Hence, hyperglycemia generates apoptotic death of endothelial cells. Diabetic endothelium tissues have a tendency for vasoconstriction and thrombus formation. Vasoconstriction occurs disturbances of blood flow, then disturbed blood flow stress leads to endoplasmic reticulum stress in the endothelial cells [4].

EPC are derived bone marrow; they are localized bone marrow and blood. They can differentiate to mature endothelial cell and promote both the new vessels formation and the endothelial repair. Hypoxia is the strongest stimulus for EPC mobilization and differentiation. In addition of hypoxia, release of proangiogenic factors such as VEGF, SDF-1α, angiopoietin 1, hepatocyte growth factor (HGF), platelet derived growth factor (PDGF), monocyte chemotactic protein-(MCP-) 1, and macrophage inflammatory protein-(MIP-) 1 promote EPC migration and differentiation [27].

The capacity of circulating EPC pool is a new indicator for evaluation of cardiovascular health. Diabetic individuals have decreased capacity of this pool compared with healthy subjects. On the other hand, their EPCs also have functional disruption that is characterized by decreased proliferation capacity, shortened survival time, decreased adhesion and migration. For this reason; diabetic patients have disturbed hypoxic response and limited neovascularization in ischemic conditions. The dysfunction and decreased level of EPC are associated with macro- and microvascular complications, cardiomyopathy, nephropathy, neuropathy of diabetes [4].

The dysregulation of HIF under the hyperglycemic conditions seems to be responsible for reduced collateral vessels formation induced by coronary ischemia in diabetics. VEGF and its receptor level decrease in diabetics. Increased ROS levels are linked with deficiency of angiogenesis. No is the most important endothelial mediator, and endothelial dysfuntion is mostly linked with deficiency of NO by reduced eNOS, eNOS coupling and increased oxidative stress. The tetrahydrobiopterin (BH4) I a precursor for synthesis of NO. Hyperglycemia causes reduced BH4 level by oxidation. Thereby; eNOS activity and NO production decrease in diabetic conditions. Endothelial cells have a tendency for vasoconstriction by increased endothelin-1 and Angiotensin II expression in the patients with diabetes [21].

Hyperglycemia is associated with increased ROS production. Especially O_2^- leads to reduced NO bioavailability which causes inhibition of HIF-1α. Hyperglycemia generates increased endothelial permeability, increased expression of adhesion molecules, decreased NO synthesis, tendency for thrombosis and vasoconstriction in the endothelium. This process is mediated by increased ROS production, formation of AGEs, activation of protein kinase C pathway in diabetes [14].

Hyperglycemia is associated with decreased HIF-1α expression in the myocardial cells. Researchers showed that impaired HIF-1α signaling leads to loss of protective mechanisms in the myocardium. They demonstrated that upregulation of HIF-1α by treatment caused high level of VEGF and eNOS, then NO increased and the infarction area size decreased [55].

HIF-1β (ARNT) mainly mediates beta-cells function in the pancreas. Gunton et al. in 2005 firstly showed that decreased ARNT levels in the pancreatic beta cells from diabetic patients [18].

On the other hand, HIF-1β plays a crucial role together with HIF-1α in the response to ischemia of myocardial cells. HIF-1β is essential for glucose metabolism in the many tissues including endothelium. The disruption of endothelial ARNT affects negatively blood vessel formation. ARNT is also an important cardiac metabolic mediator, it plays roles in regulation of cardiac function. It is observed that decreased ARNT levels in myocardial cells is linked with heart failure and increased mortality in diabetes. Interestingly the liver cells have decreased ARNT levels in diabetic patients. Decreased ARNT levels are associated with impaired insulin-stimulated glucose uptake in the endothelial cells and skeletal muscle cells. Experimental studies were demonstrated that ablation of liver specific ARNT caused exhibition of properties of type 2 diabetes such as increased lipogenic gene expression, increased hepatic gluconeogenesis, and decreased serum beta-hydroxybutyrate levels. Insulin resistance may be linked with downregulation of ARNT in the hepatocytes of diabetic individuals [48].

The inflammatory response in the pathogenesis of atherosclerosis is generated by interactions between plasma lipoproteins, monocytes/macrophages, T lymphocytes, endothelial cells, and smooth muscle cells as well as the extracellular matrix of the arteries. Macrophages are the most important cells in pathogenesis of atherosclerosis and play crucial roles in the generation of foam cells which produce inflammatory mediators. M1 and M2 macrophages present in the atherosclerotic plaques. M1 macrophages play an important role in the development of plaque, on the other hand, M2 macrophages help to regression of inflammation. Hyperglycemia and advanced glycation end products (AGEs) effect macrophages polarization [12, 26, 29, 30, 33, 39].

Hypoxia in Diabetic Nephropathy Pathogenesis

Diabetic nephropathy is one of the most common complication of diabetes. It generally progresses to end stage renal failure. HIF-1α dysregulation is one of the important mechanisms in pathophysiology of diabetic nephropathy. Kidneys are important for the regulation of body fluid composition. Although the blood supply of kidney is less than 1% of the total blood supply, oxygen consumption by the kidneys account for more than 10% of total oxygen intake. Na-K ATP ases in the cortical proximal tubules depletes 80% of renal oxygen and function as the oxygen sensors of the body. Na-K ATP ase are also important for regulation of erythropoietin synthesis. Hyperglycemia with glomerular hyperfiltration, osmotic diuresis, and increased glucose reabsorption via sodium glucose transporters increase oxygen needs of kidney tissue.

Researchers shown that cobalt chloride ($CoCl_2$) can be useful for prevention of hypoxia related complications in diabetes. Cobalt (Co) is a transition metal that is able to activate HIF-1α signaling similar to hypoxia. $CoCl_2$ leads to inhibition of

Fe^{+2} dependent prolyl hydroxylase domain proteins and also blocks the binding of HIF-1α to von Hippel-Lindau protein (pVHL), so the degradation of HIF-1α is inhibited via Co. Some studies shown that $CoCl_2$ might be play protective role in diabetic nephropathy by blocking degradation of HIF. There are two main important mechanisms in pathophysiology of diabetic nephropathy: first; hyperglycemia leads to renal hypoxia by increasing oxygen consumption and mitochondrial dysfunction, second: submaximal HIF activation. It is demonstrated that activation of HIF signaling prevents diabetic nephropathy [1, 19, 38].

Chronic hyperglycemia activates inflammatory and hypoxic signaling pathways in the endothelial cells of glomeruli. The activation of HIF-1α leads to VEGF gene activation, and then VEGF levels increase in this area. VEGF binds to its receptors that are localized on the endothelial cell membrane, and increases permeability of glomerular capillaries and glomerular filtration membranes. This process causes to loss of protein and development of proteinuria. The result will be increased inflammatory activity, migration of mononuclear phagocytes and other immune cells into glomeruli. They produce many inflammatory substances such as cytokines, TGF-β. Increased TGF-β induces collagen synthesis, increases thickening of glomerular capillary basement membrane and promotes accumulation and synthesis of extracellular matrix. All of these events lead to development of glomerulosclerosis and diabetic nephropathy [58].

The about 40% of diabetic patients have diabetic nephropathy, but only less than 10% of these patients develop chronic renal failure. The patients with diabetes cannot use glucose for their cellular metabolism an ATP production by oxidation in mitochondria. They generate a shift in the glucose metabolism. Generally, glucose is metabolized by five different ways: pentose phosphate pathways, AGEs pathway, sorbitol pathway, polyol pathway, and hexosamine pathway. At the end of these pathways toxic glucose metabolites such as lactate, sorbitol, diacylglycerol (DAG) and methylglyoxal (MG) are generated and accumulated. These toxic end products can contribute to diabetic nephropathy pathogenesis. Otto Warburg firstly described "Warburg effect" in tumor cells. Warburg observed that tumor cells produced ATP by oxygen independent way, and their intracellular lactate levels increased after this process. The both mitochondrial dysfunction and the Warburg effect play pivotal roles in the development of diabetic nephropathy [63].

Diabetes leads to hypoxia in the kidneys. The different cells in the kidneys give different responses to hypoxia. In the mesangial cells hyperglycemia stimulates HIF activity by different mechanisms such as ADAM 17 [40]. ADAM 17 upregulation is mediated by HIF-1α and also epidermal growth factor receptor (EGFR)/ADAM 17 signaling. EGF induce upregulation of TGF-β [31].

Glomerulosclerosis and proteinuria are common findings of the diabetic nephropathy. Tubular cells have high metabolic activity with numerous mitochondria. Hyperglycemia contributes to oxidative stress. Increased oxidative stress affects mitochondria, oxygen consumption increases and decreases mitochondrial respiration in the tubular cells. Normal hypoxia-induced HIF activation cannot occur in these cells. Hypoxic damage starts in tubular cells, then tubulointerstitial fibrosis, albuminuria generate further damage. In glomerular cells exaggerated HIF activation causes glomerulosclerosis [40].

Hyperglycemia upregulates sodium glucose cotransporters (SGLTs) and increases reabsorption of sodium and glucose in the tubular cells. Increased activity of SGLT enhances oxygen consumption with quabain sensitive oxygen utilization. This mechanism increases the oxygen demand. On the other hand, hyperglycemia promotes glomerular hyperfiltration. The Pasteur effect refers to the adaptive responses to hypoxic conditions and is characterized by decreased oxidative phosphorylation and increased anaerobic glycolytic pathways activation for production of ATP. The kidneys prefer anaerobic glycolysis accomplished by the Pasteur effect under normoxic and hyperglycemic conditions and kidneys use other alternative glycolytic pathways such as AGEs, hexosamine pathways. HIF pathways mediates the glycolytic switch by regulation of glycolytic transporters and enzymes [49]. AGEs activate HIF-1α mainly with a lesser extent for HIF-2. Upregulation and normal function of HIF-1α are critical for prevention of renal complications in diabetes [5, 38].

Hypoxia begins the process of endoplasmic reticulum (ER) stress and dysfunction. ER stress causes disturbances of protein synthesis and accumulation of unfolded proteins in the cells. Hypoxia-induced ER stress plays important roles in the pathophysiology of diabetic retinopathy which causes damage of podocytes and apoptosis of tubular cells [35].

Hyperglycemia, accumulation of AGEs, oxidative stress and cytokines all mediate kidney injury in the diabetic patients. Hyperglycemia, Angiotensinogen II, protein kinase C, TGF-β, ROS and inflammation are key mediators of the activation of HIF signaling by oxygen-independent way. HIF-1 also regulates genes of proteins that play important roles in oxidative stress, regulation of glucose and matrix metabolisms. Researchers demonstrated that HIF-1 mediates matrix accumulation, renal hypertrophy, and AGEs formation by increased glucose uptake in the glomerular mesangial cells. Under the hyperglycemic conditions renal medullar and tubular cells upregulate GLUT-1 expression [37].

Increased inflammatory status also affects microvascular endothelial cells and stimulates procoagulation cascade. Hyperglycemia generates increased expression of HIF-1α and PTEN (phosphatase and tensin homolog). The microangiopathy is characterized by basement membrane thickness and thrombosis in the capillaries. The microangiopathy that is resulted from hyperglycemia and hypoxia causes nerve degeneration in diabetes. The endoneurinal capillaries and epineurinal vessels are disrupted by hyperglycemia and hypoxia. The impaired blood supply of nerve tissues occurs development of neuropathy [28].

Induction of ATP dependent potassium (K-ATP) channels are key regulators of the cellular metabolism. Blockage of K-ATP channels regulate glucose uptake, insulin secretion and sensitivity and are important regulators of the hypoxic responses. We found that activation of K-ATP is protective in ischemic injury models. Blockage of K-ATP channels diminishes HIF-1α mediated cytoprotection and inflammation in kidney tissue [59, 62]. Similar to K-ATP channels, *N*-methyl-D-aspartate (NMDA) receptors are shown to be therapeutic target to decrease diabetic nephropathy and other complications of diabetes. They are a class of cation-selective ionotropic receptors with a high intrinsic Ca^{2+} permeability with multiple subunits multiple subunits (NR1, NR2A-B-C-D, NR3A-3B etc.). These receptors modulate renal blood flow, glomerular and tubular functions. Depending on the site of action, agonist potencies of effector endogenous diacidic molecules show diversity. For example, L-glutamate, L-aspartate function in the tubular site but they are relatively weak agonists for podocyte NMDA receptor. NMDA receptors are important for regulation of erythropoietin secretion and functions in kidney. In addition to conventional drug therapies, new promising drug treatments are defined as ACE inhibitors, Angiotensinogen receptor blockers, dipeptidyl peptidase (DPP-4), sodium-glucose co-transporter (SGLT)-2 inhibitors, includes the blockage of NMDA receptors and different inflammatory pathways and markers [41, 44, 60]. Diversity of functions of NMDA receptors could limit their usage in diabetic nephropathy. Blockage of NMDA receptors inhibits erythropoietin mediated protection in spinal cord trauma [61]. The mechanisms that involved in the pathogenesis of diabetic nephropathy are summarized in Fig. 8.2.

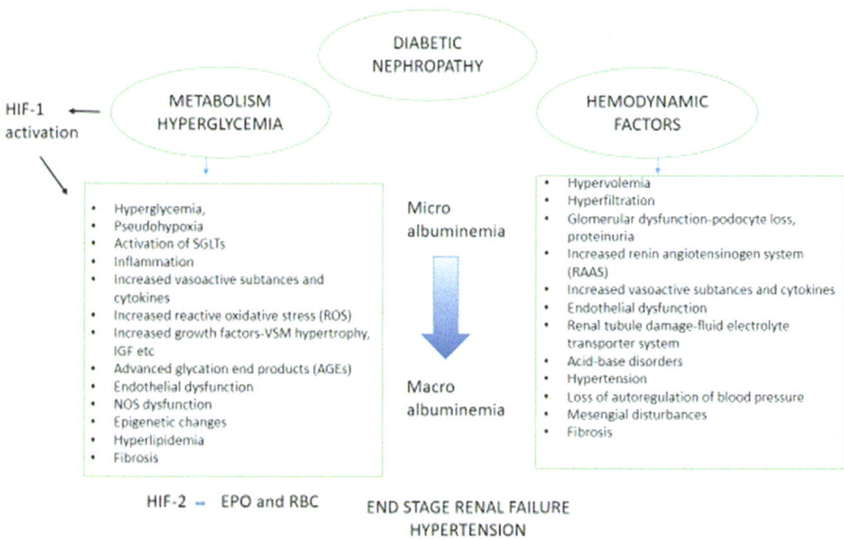

Fig. 8.2 Pathophysiology of diabetic nephropathy

Role of Vascular Insufficiency and Hypoxia in Delayed Wound Healing Process of Diabetic Patients

Impaired wound healing is one of the important problems in diabetic patients. It usually presents as diabetic foot ulceration. Underlying pathogenic mechanism of delayed healing is still unclear. In physiological conditions; the wound healing process is characterized by inflammation, migration and proliferation of fibroblasts and keratinocytes, new vessel formation, matrix synthesis and deposition, epithelization and remodeling. In diabetic conditions; this process is disturbed, new vessel formation decreases, proliferation and migration of cells reduce, and epithelization cannot be generated. Hypoxia is occurred by decreases in blood supply and increases of oxygen consumption during the wound healing period in the affected area. HIF-1 stimulates new vessel formation (angiogenesis) via activating the transcription of angiogenic factors such as VEGF, angiopoietin 2, fibroblast growth factor (FGF) 2, and increased recruitment of endothelial progenitor cells (EPC) in the damaged area. Especially EPC recruitment in wound area is mostly important for neovascularization. HIF-1α stimulates keratinocytes migration, type I collagen and fibronectin synthesis in addition to EPC accumulation. HIF-1α affects myeloid cells; then it leads to increased bactericidal activity in them. Myeloid cells produce antimicrobial peptides, proteases, TNF-α, NO and other defensive molecules at the high levels under the HIF effect. HIF-1α causes increased expression of glucose transporters 1 and 3 (GLUT-1, GLUT-3) and activation of glycolytic enzyme that is lactate dehydrogenase. Researchers shown that severe hypoxia and low levels of HIF-1α in the diabetic wound. HIF-1α signaling is dysregulated by hyperglycemia. Hyperglycemia leads to accumulation of dicarbonyl metabolite methylglyoxal (MGO) in the cells. MGO inhibits HIF-1α dependent gene activation. Diabetes leads to increased chronic inflammatory status and levels of ROS. Increased ROS dysregulates HIF-1 signaling. Researchers shown that applying of antioxidant treatment such as α-tocopherol restores decreased HIF-1 activity [8].

Hyperglycemia induced HIF dysregulation plays main role in development of diabetic foot ulcers and delayed wound healing. There are multilevel interactions between HIF signaling pathway and hyperglycemia. Hyperglycemia promotes pVHL-dependent ubiquitination of HIF-1; hence it leads to HIF destabilization. During the wound-healing process HIF stabilization is very important factor for regulation of repair and healing process [6].

Under the increased blood glucose levels or oxidative stress; glucose and its degradation products such as glyoxal, methylglyoxal, 3-deoxyglucosone react non-enzymatically with amino group of proteins to produce a Schiff base [43].

Schiff base is labile, then it converts to the more stable Amadori products. Only a small part of Amadori-products forms AGEs via irreversible chemical reactions. AGEs are different than other Amadori-products because of their irreversible nature [45].

AGEs are generated by the Maillard reaction. The high levels of AGEs stimulate their receptor expression in the cells. Increased AGEs cause increased oxidative stress and inflammation in the vascular beds. Endothelial dysfunction, procoagulant state, atherosclerosis formation are generated by increased AGEs that promote target gene expression. Serum levels of AGEs are an important marker for evaluation of endothelial dysfunction in the patients with diabetes. The high levels of AGEs are associated with left ventricular diastolic dysfunction and vascular stiffness in Type I diabetes. On the other hand, AGE intake with foods is the other important source of AGE. If meats or fatty foods are heated or baked at a high temperature, the browning reaction generates in the foods. The browning reaction is the same AGE reaction. Finally, the AGE content of these foods increases. There is a strong correlation between amount of AGE intake by diet and serum AGE levels [43]. For this reason, diet management is very important point for prevention and treatment of vascular complication in diabetics.

Hyperglycemia stimulates glycation of various structural and functional proteins including plasma proteins, extracellular matrix proteins (ECM), albumin, LDL, fibrinogen, immunoglobulins, complements and collagen. The non-enzymatic modification of these proteins may lead to many pathophysiological changes such as acceleration of atherosclerosis, glomerular dysfunction, decreased nitric oxide synthesis, reduced fibrinolysis, activation of platelet adhesion, increased oxidative stress, disturbances of immune system regulation, alteration of extracellular matrix composition, and endothelial dysfunction. AGEs play very important roles in the pathogenesis of diabetic complications such as atherosclerosis, retinopathy, cataract, neuropathy, nephropathy and cardiomyopathy. Glycation of eye lens protein is responsible for development of diabetic cataract, which is a cause of blindness. The accumulation of ECM protein in the glomerular mesangial and tubulointerstitial area is specific pathological features of diabetic nephropathy. The increased AGEs formation lead to imbalance between the synthesis and degradation of ECM components, thereby; causes the pathologic accumulation of collagens, fibronectins, and laminins [47].

The plasma levels of ischemia-modified albumin (Wu et al.), glycated albumin (GA), fructosamine, and AGEs are important indicators for vascular event prediction in diabetic patients [17].

Chronic subclinical inflammation is an important risk factors for development of many complications of diabetes, including neuropathy, nephropathy, cardiomyopathy, macro-vasculopathy, micro-vasculopathy and foot ulcers. RAGE is localized in many cells and tissues including lung, liver, vascular endothelium, monocytes, dendritic cells, and neurons. The diabetic patient tends to have increased serum AGEs that activate pro-inflammatory cells. However increased activation of inflammatory cells stimulate chronic subclinical inflammation. Many tissues undergo inflammatory damage because of AGE-induced diffuse chronic subclinical inflammation. AGEs accrue extensive tissue damage and also organ dysfunction. In addition to the damage that they cause, AGEs lead to inhibition of the repair process after these damages have occured. There is a vicious cycle between chronic subclinical inflammation and oxidative stress; each of them stimulates other. The immune system regulation disrupts in diabetic patients, they cannot generate the appropriate response against diabetic wound or infections [24].

References

1. Allison SJ (2014) HIF activation in prevention of diabetic nephropathy. Nat Rev Nephrol 10(11):612–613
2. Aouadi M (2014) HIF-2α blows out the flames of adipose tissue macrophages to keep obesity in a safe zone. Diabetes 63(10):3169–3171
3. Augstein A, Poitz DM, Braun-Dullaeus RC, Strasser RH, Schmeisser A (2011) Cell-specific and hypoxia-dependent regulation of human HIF-3α: inhibition of the expression of HIF target genes in vascular cells. Cell Mol Life Sci 68(15):2627–2642
4. Avogaro A, Albiero M, Menegazzo L, de Kreutzenberg S, Fadini GP (2011) Endothelial dysfunction in diabetes: the role of reparatory mechanisms. Diabetes Care 34(Supplement 2):S285–S290
5. Bondeva T, Heinzig J, Ruhe C, Wolf G (2013) Advanced glycated end-products affect HIF-transcriptional activity in renal cells. Mol Endocrinol 27(11):1918–1933
6. Botusan IR, Sunkari VG, Savu O, Catrina AI, Grünler J, Lindberg S, Pereira T, Ylä-Herttuala S, Poellinger L, Brismar K (2008) Stabilization of HIF-1α is critical to improve wound healing in diabetic mice. Proc Natl Acad Sci 105(49):19426–19431
7. Catrina S-B, Okamoto K, Pereira T, Brismar K, Poellinger L (2004) Hyperglycemia regulates hypoxia-inducible factor-1α protein stability and function. Diabetes 53(12):3226–3232
8. Catrina SB, Zheng X (2016) Disturbed hypoxic responses as a pathogenic mechanism of diabetic foot ulcers. Diabetes/Metab Res Rev 32:179–185
9. Cerychova R, Pavlinkova G (2018) HIF-1, metabolism, and diabetes in the embryonic and adult heart. Front Endocrinol 9
10. Choe SS, Shin KC, Ka S, Lee YK, Chun J-S, Kim JB (2014) Macrophage HIF-2α ameliorates adipose tissue inflammation and insulin resistance in obesity. Diabetes 63(10):3359–3371
11. Covello KL, Kehler J, Yu H, Gordan JD, Arsham AM, Hu C-J, Labosky PA, Simon MC, Keith B (2006) HIF-2α regulates Oct-4: effects of hypoxia on stem cell function, embryonic development, and tumor growth. Genes Dev 20(5):557–570
12. Dai J, Zhang X, Li L, Chen H, Chai Y (2017) Autophagy inhibition contributes to ROS-producing NLRP3-dependent inflammasome activation and cytokine secretion in high glucose-induced macrophages. Cell Phys Biochem 43(1):247–256
13. Ekberg NR, Eliasson S, Li YW, Zheng X, Chatzidionysiou K, Falhammar H, Gu HF, Catrina S-B (2019) Protective effect of the HIF-1A Pro582Ser polymorphism on severe diabetic retinopathy. J Diabetes Res 2019
14. Gao W, Ferguson G, Connell P, Walshe T, O'brien C, Redmond EM, Cahill PA (2014) Glucose attenuates hypoxia-induced changes in endothelial cell growth by inhibiting HIF-1α expression. Diabetes Vasc Dis Res 11(4):270–280
15. Gerber PA, Rutter GA (2017) The role of oxidative stress and hypoxia in pancreatic beta-cell dysfunction in diabetes mellitus. Antioxid Redox Signal 26(10):501–518
16. Gonzalez FJ, Xie C, Jiang C (2018) The role of hypoxia-inducible factors in metabolic diseases. Nat Rev Endocrinol 1
17. Guerin-Dubourg A, Cournot M, Planesse C, Debussche X, Meilhac O, Rondeau P, Bourdon E (2017) Association between fluorescent advanced glycation end-products and vascular complications in type 2 diabetic patients. BioMed Res Int 2017
18. Gunton JE, Kulkarni RN, Yim S, Okada T, Hawthorne WJ, Tseng Y-H, Roberson RS, Ricordi C, O'Connell PJ, Gonzalez FJ (2005) Loss of ARNT/HIF1β mediates altered gene expression and pancreatic-islet dysfunction in human type 2 diabetes. Cell 122(3):337–349
19. Haase VH (2015) A breath of fresh air for diabetic nephropathy. Am Soc Nephrol
20. Hoevenaars FP, Keijer J, van der Stelt I, Duivenvoorde LP, Herreman L, van Nes R, Friedecký D, Hegeman MA, van Schothorst EM (2019) White adipose tissue response of obese mice to ambient oxygen restriction at thermoneutrality: response markers identified, but no WAT inflammation. Genes 10(5):359

21. Howangyin KY, Silvestre J-S (2014) Diabetes mellitus and ischemic diseases: molecular mechanisms of vascular repair dysfunction. Arteriosclerosis Thrombosis Vasc Biol 34(6):1126–1135
22. Hu C-J, Iyer S, Sataur A, Covello KL, Chodosh LA, Simon MC (2006) Differential regulation of the transcriptional activities of hypoxia-inducible factor 1 alpha (HIF-1α) and HIF-2α in stem cells. Mol Cell Biol 26(9):3514–3526
23. Hu C-J, Wang L-Y, Chodosh LA, Keith B, Simon MC (2003) Differential roles of hypoxia-inducible factor 1α (HIF-1α) and HIF-2α in hypoxic gene regulation. Mol Cell Biol 23(24):9361–9374
24. Hu H, Jiang H, Ren H, Hu X, Wang X, Han C (2015) AGEs and chronic subclinical inflammation in diabetes: disorders of immune system. Diabetes/Metab Res Rev 31(2):127–137
25. Jiang F, Tang Y-T, Guo L, Jiao X-Y (2013) The role of insulin-like growth factor I and hypoxia inducible factor 1α in vascular endothelial growth factor expression in type 2 diabetes. Ann Clin Lab Sci 43(1):37–44
26. Jin X, Yao T, Zhou Z, Zhu J, Zhang S, Hu W, Shen C (2015) Advanced glycation end products enhance macrophages polarization into M1 phenotype through activating RAGE/NF-κB pathway. BioMed Res Int 2015
27. Kamei N, Atesok K, Ochi M (2017) The use of endothelial progenitor cells for the regeneration of musculoskeletal and neural tissues. Stem Cells Int 2017
28. Kan H-W, Hsieh J-H, Chien H-F, Lin Y-H, Yeh T-Y, Chao C-C, Hsieh S-T (2018) CD40-mediated HIF-1α expression underlying microangiopathy in diabetic nerve pathology. Dis Models Mech 11(4):dmm033647
29. Kantera JE, Kramera F, Barnharta S, Averilla MM, Vivekanandan-Girib A, Vickeryc T, Lid LO, Beckere L, Yuane W, Chaite A (2012) Diabetes promotes an inflammatory macrophage phenotype and atherosclerosis through acyl-CoA synthetase. Blood 15(20):25
30. Kraakman MJ, Murphy AJ, Jandeleit-Dahm K, Kammoun HL (2014) Macrophage polarization in obesity and type 2 diabetes: weighing down our understanding of macrophage function? Front Immunol 5:470
31. Li R, Uttarwar L, Gao B, Charbonneau M, Shi Y, Chan JS, Dubois CM, Krepinsky JC (2015) High glucose up-regulates ADAM17 through HIF-1α in mesangial cells. J Biol Chem 290(35):21603–21614
32. Lin Y, Tang Y, Wang F (2016) The protective effect of HIF-1α in T lymphocytes on cardiac damage in diabetic mice. Ann Clin Lab Sci 46(1):32–43
33. Liu Y-C, Zou X-B, Chai Y-F, Yao Y-M (2014) Macrophage polarization in inflammatory diseases. Int J Biol Sci 10(5):520
34. Lum JJ, Bui T, Gruber M, Gordan JD, DeBerardinis RJ, Covello KL, Simon MC, Thompson CB (2007) The transcription factor HIF-1α plays a critical role in the growth factor-dependent regulation of both aerobic and anaerobic glycolysis. Genes Dev 21(9):1037–1049
35. Maekawa H, Inagi R (2017) Stress signal network between hypoxia and ER stress in chronic kidney disease. Front Physiol 8:74
36. Michailidou Z, Morton NM, Navarrete JMM, West CC, Stewart KJ, Fernández-Real JM, Schofield CJ, Seckl JR, Ratcliffe PJ (2015) Adipocyte pseudohypoxia suppresses lipolysis and facilitates benign adipose tissue expansion. Diabetes 64(3):733–745
37. Nayak BK, Shanmugasundaram K, Friedrichs WE, Cavaglierii RC, Patel M, Barnes J, Block K (2016) HIF-1 mediates renal fibrosis in OVE26 type 1 diabetic mice. Diabetes 65(5):1387–1397
38. Nordquist L, Friederich-Persson M, Fasching A, Liss P, Shoji K, Nangaku M, Hansell P, Palm F (2015) Activation of hypoxia-inducible factors prevents diabetic nephropathy. J Am Soc Nephrol 26(2):328–338
39. Peled M, Fisher EA (2014) Dynamic aspects of macrophage polarization during atherosclerosis progression and regression. Front Immunol 5:579
40. Persson P, Palm F (2017) Hypoxia-inducible factor activation in diabetic kidney disease. Curr Opin Nephrol Hypertens 26(5):345–350
41. Rao V, Rao LV, Tan SH, Candasamy M, Bhattamisra SK (2018) Diabetic nephropathy: an update on pathogenesis and drug development. Diabetes Metab Syndr Clin Res Rev

42. Regazzetti C, Peraldi P, Grémeaux T, Najem-Lendom R, Ben-Sahra I, Cormont M, Bost F, Le Marchand-Brustel Y, Tanti J-F, Giorgetti-Peraldi S (2009) Hypoxia decreases insulin signaling pathways in adipocytes. Diabetes 58(1):95–103
43. Rhee SY, Kim YS (2018) The role of advanced glycation end products in diabetic vascular complications. Diabetes Metab J 42(3):188–195
44. Roshanravan H, Kim EY, Dryer SE (2016) NMDA receptors as potential therapeutic targets in diabetic nephropathy: increased renal NMDA receptor subunit expression in Akita mice and reduced nephropathy following sustained treatment with memantine or MK-801. Diabetes 65(10):3139–3150
45. Schalkwijk CG, Miyata T (2012) Early-and advanced non-enzymatic glycation in diabetic vascular complications: the search for therapeutics. Amino Acids 42(4):1193–1204
46. Shay JE, Simon MC (2012) Hypoxia-inducible factors: crosstalk between inflammation and metabolism. In: Seminars in cell & developmental biology. Elsevier, pp 389–394
47. Singh VP, Bali A, Singh N, Jaggi AS (2014) Advanced glycation end products and diabetic complications. Korean J Physiol Pharmacol 18(1):1–14
48. Su L, Knapp M, Wu R (2017) Possible role for HIF-1β/ARNT in the regulation of vascular function in diabetes and cardiovascular disease. Curr Trends Cardiol 1(1):3–4
49. Takiyama Y, Haneda M (2014) Hypoxia in diabetic kidneys. BioMed Res Int 2014
50. Thangarajah H, Vial IN, Grogan RH, Yao D, Shi Y, Januszyk M, Galiano RD, Chang EI, Galvez MG, Glotzbach JP (2010) HIF-1α dysfunction in diabetes. Cell Cycle 9(1):75–79
51. Trayhurn P (2014) Hypoxia and adipocyte physiology: implications for adipose tissue dysfunction in obesity. Annu Rev Nutr 34:207–236
52. Trayhurn P (2019) Oxygen—a critical, but overlooked, nutrient. Front Nutr 6
53. Wang N, Khan SA, Prabhakar NR, Nanduri J (2013) Impairment of pancreatic β-cell function by chronic intermittent hypoxia. Exp Physiol 98(9):1376–1385
54. Wang Y, Hai B, Niu X, Ai L, Cao Y, Li R, Li Y (2016) Chronic intermittent hypoxia disturbs insulin secretion and causes pancreatic injury via the MAPK signaling pathway. Biochem Cell Biol 95(3):415–420
55. Wu J, Yang L, Xie P, Yu J, Yu T, Wang H, Maimaitili Y, Wang J, Ma H, Yang Y (2017) Cobalt chloride upregulates impaired HIF-1α expression to restore sevoflurane post-conditioning-dependent myocardial protection in diabetic rats. Front Physiol 8:395
56. Xiao H, Gu Z, Wang G, Zhao T (2013) The possible mechanisms underlying the impairment of HIF-1α pathway signaling in hyperglycemia and the beneficial effects of certain therapies. Int J Med Sci 10(10):1412
57. Xie C, Yagai T, Luo Y, Liang X, Chen T, Wang Q, Sun D, Zhao J, Ramakrishnan SK, Sun L (2017) Activation of intestinal hypoxia-inducible factor 2α during obesity contributes to hepatic steatosis. Nat Med 23(11):1298
58. Xu X, Chen P, Zheng Q, Wang Y, Chen W (2011) Effect of pioglitazone on diabetic nephropathy and expression of HIF-1α and VEGF in the renal tissues of type 2 diabetic rats. Diabetes Res Clin Pract 93(1):63–69
59. Yazihan N, Ataoglu H, Kavas GO, Akyurek N, Yener B, Aydm C (2008) The effect of K-ATP channel blockage during erythropoietin treatment in renal ischemia-reperfusion injury. J Invest Surg 21(6):340–347
60. Yazihan N, Uzuner K (2005) A new aspect to erythropoietin; Renal NMDA receptors. In: FASEB J. Federation Amer Soc Exp Biol 9650 Rockville Pike, Bethesda, MD 20814-3998 USA, pp A1152–A1152
61. Yazihan N, Uzuner K, Salman B, Vural M, Koken T, Arslantas A (2008) Erythropoietin improves oxidative stress following spinal cord trauma in rats. Injury 39(12):1408–1413
62. Yilmaz TU, Yazihan N, Dalgic A, Kaya EE, Salman B, Kocak M, Akcil E (2015) Role of ATP-dependent K channels in the effects of erythropoietin in renal ischaemia injury. Indian J Med Res 141(6):807
63. Zhang G, Darshi M, Sharma K (2018) The warburg effect in diabetic kidney disease. In: Seminars in nephrology. Elsevier, pp 111–120

Chapter 9
Cellular and Molecular Mechanisms Contributing to Cardiac Hypertrophy in Obesity and Obesity-Related Hypertension

Norma L. Gómez-Viquez, Patrick Mailloux-Salinas, Tzindilú Molina-Muñoz, Rodrigo Guevara, Guadalupe Bravo, and Karla Carvajal

Abstract Obesity, defined as a harmful accumulation of body fat, increases the risk of cardiac hypertrophy, which leads to heart failure (HF). Obesity often coexists with hypertension, a condition that aggravates cardiac hypertrophy and accelerates the progression of HF. The cellular and molecular mechanisms involved in cardiac hypertrophy during obesity, and those that explain synergic effect of obesity and hypertension on cardiac hypertrophy are still an issue of investigation. So far, evidence suggests that obesity promotes cellular events related to insulin resistance, neurohormonal over-activation, oxidative stress, chronic inflammation and perturbation of cellular signaling, which are some of the processes involved in the onset of cardiac hypertrophy and hypertension, and indeed the exacerbation of these events appears to explain, in part, the worsening of cardiac hypertrophy when obesity is accompanied by hypertension. In this chapter, we analyze data that may help to clarify the participation of the complex interconnecting mechanisms that are evoked by obesity and hypertension, when allied together to induce cardiac hypertrophy. Better understanding of these mechanisms will allow us to have an improved management of obesity and obesity-related hypertension, as well as to achieve more effective prophylactic therapies and opportune diagnosis of these clinical conditions.

Keywords Cardiac hypertrophy · Hypertension · Obesity · Signaling pathways

---

N. L. Gómez-Viquez (✉) · P. Mailloux-Salinas · T. Molina-Muñoz · R. Guevara · G. Bravo
Departamento de Farmacobiología, Centro de Investigación y de Estudios Avanzados, Instituto Politécnico Nacional, Mexico City, Mexico
e-mail: letyviquez@hotmail.com

K. Carvajal
Laboratorio de Nutrición Experimental, Instituto Nacional de Pediatría, Mexico City, Mexico

R. Guevara
Maestría en Ciencias Biológicas, Centro Tlaxcala de Biología de la Conducta, Universidad Autónoma de Tlaxcala, Tlaxcala, Mexico

P. S. Tappia et al. (eds.), *Biochemistry of Cardiovascular Dysfunction in Obesity*, Advances in Biochemistry in Health and Disease 20,
https://doi.org/10.1007/978-3-030-47336-5_9

Introduction

Over the last three decades, the prevalence of obesity has increased dramatically worldwide, leading to a parallel increase in the rates of its comorbidities, such as insulin resistance (IR), diabetes mellitus type 2 (DMT2), hypertension and dyslipidemia, which are important risk factors for cardiovascular disease (CVD), the main cause of death globally. Accordingly, obesity is considered by the World Health Organization (WHO) as one of the "major public health problems of the 21st century" [1, 2].

Obesity causes hemodynamic and non-hemodynamic abnormalities that contribute to the development of a cardiomyopathy independently of hypertension. Obesity cardiomyopathy is commonly characterized by an early subclinical left ventricular (LV) dysfunction that is followed by a late LV hypertrophy (LVH), both of which are important risks factors for heart failure (HF). Nonetheless, obesity often coexist with hypertension, an issue that is clinically relevant because the concomitant presence of hypertension and obesity enhances the prevalence and severity of LVH and accelerates its progression to HF [2–7].

The mechanisms underlying LVH in obesity and those involved in the worsening of LVH in obesity-related hypertension are still being unraveled. Clinical and experimental evidence suggest that increased fat accumulation favors a pathologic environment associated to increased neurohormonal stimulation, IR, oxidative stress (OS), and inflammation, conditions that underwrite the onset of hypertension and LVH, implying that obesity, hypertension and LVH share the same pathogenic mechanisms [2–6, 8, 9]. On the other hand, current studies suggest that LVH in obesity is linked to the stimulation of a web of signaling pathways that initiate the genetic program of cardiac hypertrophy at nucleus, and the exacerbation of these signaling cascades have been recently associated with the aggravation of cardiac hypertrophy in obesity-related hypertension [2, 10–12]. In this chapter we will discuss the cellular and molecular mechanisms that could contribute to the development of LVH and hypertension in obesity, and the way they could interact to intensify cardiac illness.

Cardiac Disease in Obesity and Obesity-Related Hypertension

Obesity is a non-communicable disease, characterized by an increase in the body weight due to excessive fat accumulation. According to the WHO, the prevalence of obesity in world´s adult population has nearly tripled since 1975. In 2016, 39% of adults aged 18 years and over (39% of men and 40% of women) were overweight and 13% (11% of men and 15% of women) were obese. At present, obesity is recognized as one of the major public health problems worldwide due to its negative impact in people´s health, the many comorbidities that accompanies it, and the economic costs linked to prevention and treatment of the disease [1]. Clinical evidence has demonstrated that obesity leads to alterations in the geometry and function of the heart, and predisposes to HF, the leading global cause of death and disability in obese and diabetic patients. Moreover, it has been established that obesity contributes to the onset of several chronic diseases, among them, systemic hypertension, which increases morbidity and mortality by worsening cardiac remodeling and accelerating the progression of HF and thus, amplifying CV risk [2–6]. This issue is clinically relevant given the high prevalence of hypertension in obese subjects. Epidemiological data suggest that 60–70% of hypertension in adults is directly attributable to adiposity. Indeed, the high prevalence of hypertension among patients with obesity accounts for 78% of incident hypertension in men and 64% of incident hypertension in women, and the prevalence increase with severity of obesity. Accordingly, obese people have a 3.5 fold increased likelihood of having hypertension and it has been estimated that the increased risk of developing hypertension is 20–30% for every 5% increment in weight gain [13]. The synergistic effect between obesity and hypertension on cardiac disease has been established by studies showing that prevalence of LVH in normotensive obese subjects reaches 13% whereas in hypertensive individuals with morbid obesity ranges over 75%. Similarly, in a cohort of hypertensive patients the prevalence of LVH was 12% for normal weight individuals, 25% for overweight individuals and 48% for obese individuals [14]. In addition, there is evidence that hypertension or higher systolic blood pressures, even if they are not in the hypertensive range, are associated with a greater extent of LVH and cardiac dysfunction [2, 5]. Notably, it has been reported in population studies, that future weight gain is significantly higher in patients with hypertension than in normotensive individuals, suggesting that hypertension per se contributes to increase obesity, implying a further link between obesity and hypertension [15]. Thus, considering the role of obesity and hypertension in the development of LVH and HF, as well as the frequent overlapping of these diseases and their synergistic effects on cardiac illness, it is important to establish the molecular mechanisms underlying the synergic relation of the two conditions.

Pathogenesis of Hypertension in Obesity

The mechanisms linking obesity with hypertension and cardiac disease have not yet been conclusively defined. As mentioned above, obesity, and in particular the excessive visceral fat distribution, predispose to hemodynamic and non-hemodynamic conditions that lead to changes in the structure and function of the heart, and contribute to the onset of hypertension, which in turn exacerbates cardiac damage and increases the risk of HF [2–6, 9]. Some of the non-hemodynamic factors involved in the development of cardiac disease and hypertension in obesity are dyslipidemia, OS, increased production of pro-inflammatory cytokines, macrophage infiltration, lipotoxicity, increased epicardial fat deposition, and increased activation of renin-angiotensin-aldosterone system (RAAS) and sympathetic nervous system (SNS) (Fig. 9.1) [2–6]. All these factors are strongly associated to IR, an early event thought

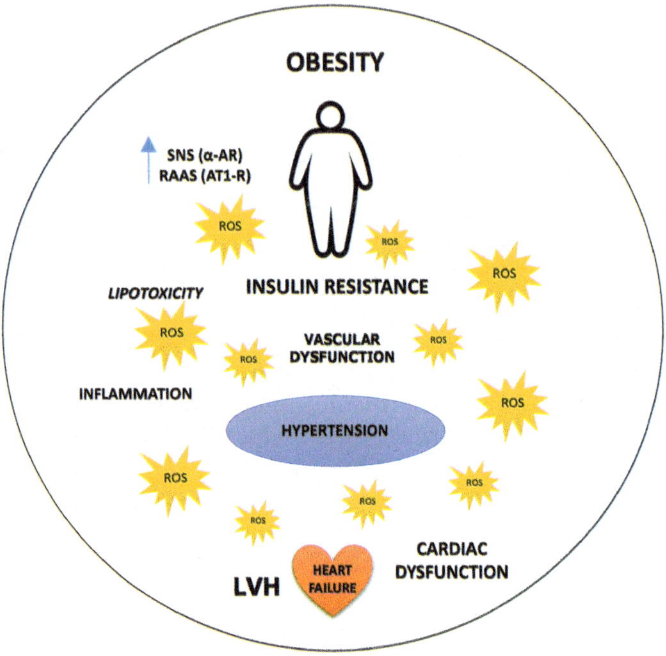

Fig. 9.1 Pathogenic mechanisms leading to hypertension, LVH and HF in obesity. The diagram shows that obesity results in a constellation of inter-related pathologic events that lead to hypertension, LVH, cardiac dysfunction and thus to HF. These cellular events include hemodynamic factors associated to increased volume and pressure load, and non-hemodynamic alterations linked to neurohormonal over-activation (mediated by SNS and RAAS), insulin resistance, oxidative stress, lipotoxicity and inflammation. LVH = left ventricular hypertrophy. HF = heart failure. SNS = sympathetic nervous system. RAAS = renin-angiotensin-aldosterone system

to be caused by obesity. Indeed, accumulated clinical and experimental evidence indicates that IR plays an important role in the development of obesity-related hypertension [16], and it has been reported that IR in vascular beds precedes the development of hypertension in spontaneous hypertensive rats [17, 18]. These studies point to IR as a key early event in the development of hypertension.

It has been established that metabolic alterations derived from IR, and overactivation of RAAS and SNS in vascular system lead to activation of signaling pathways, which underlay the development of an endothelial dysfunction and arterial stiffness that precede the onset of hypertension [4, 6]. In addition, overactivation of RAAS and SNS are also involved in the obesity-related hemodynamic changes that contribute to hypertension development. In this sense, obesity has been associated with an increase in central and total blood volume and a mild systemic vascular resistance, conditions that favor an augmentation of stroke volume and thus a high cardiac output. This high cardiac output has been attributed to the amplified metabolic demand that results from increased fat mass, and has been involved in the raise of systemic blood pressure. The effects that overactivation of RAAS and SNS have on heart, kidneys and vascular system, result in increased cardiac output, exacerbation of volume expansion and raised vascular tone, enhancing hemodynamic changes that lead to elevation of blood pressure, and thus to the onset of hypertension. This hemodynamic profile is different from that occurring in lean individuals with essential hypertension, which is characterized by high peripheral resistance and low circulating intravascular volume [3, 5, 6, 19]. In addition, the systemic vascular resistance tends to be higher in hypertensive than in normotensive obese individuals, but it is lower than in lean hypertensive patients with equivalent blood pressure values [5, 19, 20]. Therefore, it has been considered that hypertensive obese patients exhibit an inappropriate lower systemic vascular resistance face to augmented cardiac output.

Pathogenesis of Cardiac Disease in Obesity and Obesity-Related Hypertension

Similarly to vascular tissue, in the heart, obesity causes a complex and interconnected network of cellular events that share IR as a core and include OS, inflammation, apoptosis, fibrosis, abnormal remodeling, impaired Ca^{2+} handling and metabolic unbalance (Fig. 9.1). These events contribute to the onset of a subclinical early cardiac dysfunction followed by a late LVH, leading over time towards HF [2]. Additionally, the increase in central and total blood volume caused by obesity leads the heart to operate at unnecessarily high filling pressures, and produces a form of volume overload, implicated in the development of an eccentric LVH. Yet, obesity is commonly accompanied by a mixed eccentric/concentric LVH, suggesting that besides LV preload, there are other factors contributing to LVH. In this sense, it is well stablished that concentric LVH is linked to an increase in LV afterload due to pressure overload. Therefore, it has been proposed that concentric LVH in obesity may be a consequence of the elevation of systemic blood pressure that results from the increase in blood volume. Similarly, the lack of the normal decrease in systolic and diastolic blood pressure during sleep and obstructive sleep apnea, both associated with an increase in SNS, contribute to hypertension and concentric LVH in obesity. Notably, high systolic blood pressure values in obesity, even if they are not at the hypertensive range, have been strongly associated with the extent of LVH, underscoring the importance of afterload as a determinant of cardiac hypertrophy severity. Consequently, there are compiling reasons to believe that concomitant presence of hypertension and obesity, affects the heart structure more adversely than either condition alone, by increasing LV afterload and LV preload [2, 3, 5, 8, 9, 19–21]. Furthermore, it has been shown that IR and sympathetic overdrive are higher in hypertensive obese patients than in lean hypertensive or normotensive obese subjects [21], suggesting that exacerbation of these factors may also contribute to the greater heart damage induced by the combined presence of obesity and hypertension.

Molecular Mechanisms Contributing to Cardiac Hypertrophy in Obesity and Obesity-Related Hypertension

The cellular and molecular mechanisms involved in LVH in obesity and those contributing to LVH exacerbation when hypertension is present, are still poorly understood. It is well stablished that LVH, defined as an abnormal increase in LV mass, initiates in response to detrimental stimuli that evoke several pathological events, which in turn stimulate an intricate network of canonical unfavorable signaling cascades. These signaling pathways enhance protein synthesis, induce re-expression of fetal genes, and change the expression and function of several proteins, leading to progressive alterations in the size, geometry, composition and function of heart. At the cellular level these alterations are manifested by abnormalities in growth, contraction, Ca^{2+} handling, ionic flux, energy metabolism, extracellular matrix arrangement, cytoskeletal structure, collagen deposition, and so on. Most research done to date has used experimental models of different cardiomyopathies and in vitro studies to identify the molecular components of the signaling cascades involved in onset and progression of LVH [2, 22, 23]. In the case of obesity and obesity-related hypertension, these studies are just beginning, but experimental evidence supports the pivotal role of increased neurohormonal stimulation, IR, inflammation, macrophage infiltration, and OS on signaling cascades of cardiac hypertrophy [2, 9, 24]. Moreover, recent evidence in experimental models of obesity with hypertension suggest that these signaling cascades are exacerbated when both diseases coexist [10–12].

Neurohormonal Stimulation

Obesity is normally accompanied by IR, a condition that favors activation of SNS. Obesity also leads to increased activation of RAAS, which contributes to IR and SNS activation, implying a cross-talking among these processes [4]. Activation of SNS and RAAS promote a concentric LVH by exerting a direct effect on LV myocardium through specific $G_{q/11}$-protein coupled receptors (GqPCRs) activated by norepinephrine (NE) and angiotensin II (AngII), respectively. NE binds to alpha-adrenergic receptors (αAR), while AngII binds to angiotensin receptor type 1 (AT1). Activation of these receptors generates a cascade of intracellular signals that lead to activation and auto-phosphorylation of protein kinase activated by the complex Ca^{2+}-calmodulin (CaMKII). CaMKII mediates phosphorylation of class II histone deacetylases (HDACs) in the nucleus, inducing their translocation out of the nucleus and relieving repression of transcription factor MEF2 to initiate the transcription of hypertrophy genes. Stimulation of GqPCRs also results in activation of calcineurin (CaN), a phosphatase that dephosphorylates NFAT driving its translocation to the nucleus where it interacts with the transcription factor GATA4 to initiate the genetic program of hypertrophy [25]. AngII also exacerbates CaMKII activity by increasing its oxidation, via stimulation of NADPH oxidase (Nox) that lead to an increase in generation of reactive species of oxygen (ROS) [26]. On the other hand, AT1R can interact with non-receptor type tyrosine kinases and receptor-type tyrosine kinases leading to activation of MAPK (ERK/JNK) cascades involved in cardiac hypertrophy [27]. AT1R can also lead to IR by inducing protein degradation and/or inhibition of insulin receptor substrate (IRS) and PI3K messenger [28], or to inflammation by increasing TNFα synthesis [29]. It has been reported that ATIR inhibition decreases cardiac remodeling and IR in obese rats by normalizing ERK expression and increasing IP3K phosphorylation [30]. Additionally, AT1R antagonist decreases cardiac remodeling, susceptibility to ischaemic/reperfusion and TNFα synthesis in obese rats [31]. On the other hand, auto-phosphorylation and oxidation of CaMKII has been linked to OS exacerbation, apoptosis, inflammation, fibrosis and cardiac hypertrophy in cardiac H9C2 cells treated with palmitate and in mice fed with a high fat diet (HFD) [32]. Nevertheless, in this study activation of CaMKII was related to stimulation of Toll-like receptor 4 (TLR4), a metabolic sensor of saturated fatty acids, suggesting that CaMKII mediates a cross-talk between TLR4 and GqPCRs. Similarly, enhanced activity of CaN and NFAT3 have been implicated in the development of LVH in obese Zucker rats [33].

On the other hand, in vitro studies in neonatal cardiac cells show that enhanced expression of G proteins-coupled receptor kinase (GRK2), by activation of AT1R and αAR, results in cardiac hypertrophy by increasing the binding of GRK2 to PI3K, leading to phosphorylation of protein kinase B (PKB/AKT) and subsequent inactivation of glycogen synthase kinase 3 beta (GSK3β), and thus promoting activation and nuclear translocation of NFAT. The relevance of GRK2 in cardiac hypertrophy was also demonstrated in GRK2 knockout mice that exhibited attenuated hypertrophy in response to pressure overload [34]. In addition, humans and mice with HF display a lineal correlation between GRK2 expression and cardiac hypertrophy severity [35]. Notably, increased cardiac expression of GRK2 was found in obese ob/ob and HFD-fed mice [36, 37], and $GRK2^{+/-}$ mice fed with a HFD show an attenuated obese and IR phenotype, as well as decreased cardiac hypertrophy and fibrosis [37], indicating that GRK2 play an important role in modulating obesity-induced IR and LVH.

Insulin Resistance, Inflammation and Oxidative Stress

Obesity results in increased plasma free fatty acid (FFA) levels and enhanced production of pro-inflammatory cytokines that favor the development of IR in the heart, which in turn increases the production of local cytokines such as IL6 and IL1β, leading to exacerbation of IR. Increased FFA metabolism in the heart mediated by IR contributes to lipotoxicity and mitochondrial damage leading to increased ROS production and thus to OS leading to enhanced IR [7]. Thus, IR, OS and inflammation form an interactive network of cellular signals that aggravate cardiac damage.

In the heart, insulin controls energy metabolism and growth by interacting with its receptor. This binding promotes tyrosine auto-phosphorylation in the receptor and in insulin receptor substrate (IRS), thereby initiating two main signal branches: one mediated by the IP3K/AKT that regulates glucose and lipid metabolism; and another one mediated by interaction of IP3K/AKT and RAS/MAPK to regulate transcription of genes involved in cell growth and differentiation [38]. Insulin pathway also imbricates with inflammatory pathways by the activation of serine kinases IKKβ and JNK, which inactivate IRS promoting IR. Besides, insulin activates the cytokines transcription factors NFKβ and AP1 (FOS/JUN) resulting in a depraved loop that exacerbates inflammation and IR. Cardiac IR in obesity has been associated to blunted activation of the insulin signaling cascade via IP3K/AKT, which seems to favor the growth effects of insulin through MAPK pathway. Decrease action of IP3K/AKT pathway has been linked to increased serine phosphorylation of IRS promoted by OS, inflammation or AT1R activation [7, 28, 38].

Growing evidence supports that mitochondrial dysfunction is associated to cardiac diseases. Indeed, in obesity the vascular tissue and heart are characterized by augmented mitochondrial ROS production promoting oxidative damage in proteins which results in cardiac and endothelial cell dysfunction, leading to hypertension and HF. Under excessive ROS production, nitric oxide (NO) produced by endothelial cells and activated macrophages, serves as precursor of reactive nitrogen species (RNS) such as the very toxic peroxynitrite ($ONOO^-$) [39]. Under increased ROS and RNS production, proteins, including ionic channels, transporters and components of signaling pathways, undergo oxidative post-translational modifications (OPTMs), including S-nitrosylation, S-glutathionilation, N-sulfonylation, nitration and disulfide bonds [40], which may contribute to cardiac dysfunction and LVH [41]. The participation of OS in obesity-related cardiac hypertrophy was recently evidenced in mice fed with a high fat-high sucrose diet, by showing that over-expression of catalase activity can revert diastolic dysfunction and LVH [42]. These authors also found that mitochondrial complexes I and II underwent OPTM associated with a decreased activity, which in turn drove a 3-fold increase in H_2O_2 production, whereas ATP synthesis dropped. Cys100 or Cys103 from complex II subunit B were proposed as targets of reversible OPTM [42]. Thus, as the main source of ROS production, mitochondria are also a prey of their self-activity which conduces to further oxidative damage during obesity.

Obesity-related OS seems to modulate key proteins in cardiac signaling that may conduce to hypertrophy. In this sense, CaMKII may undergo oxidation at Met281/283 and S-nytrosilation at Cys290 that promote activation of the enzyme in the same way than auto-phosphorylation [43]. During obesity, oxidation of CaMKII has emerged as an important factor for cardiac hypertrophy, apoptosis, fibrosis, and inflammation [32]. Another important kinase that is affected by OPTM, is cardiac liver kinase 1 (LKB1). Cardiac LKB1 is one of the main upstream kinases in the AMP-dependent kinase (AMPK) pathway. AMPK is the major energy gauge controlling energy production/utilization in the cell. Recently, Calamares et al. demonstrated that in mice fed a HFD, cardiac LKB1 activity is inhibited, blunting the downstream phosphorylation cascaded mediated by AMPK. They suggested that inactivation of the enzyme may be mediated by OPTM at Lys96-97 [45]. Moreover, this group and others have proposed that diminished cardiac LKB1 activity contributes to the development of hypertrophy [44, 45].

Hypertrophied adipose cells undergo macrophage infiltration and increase adipokine secretion leading to a chronic inflammatory stage, characterized by the presence of TNF-α, IL-1β and IL-6, which in turn promotes ROS and RNS production [46] that target blood vessels structures, causing endothelial dysfunction and thus hypertension [39]. On the other hand, exacerbation of the inflammatory stage in the myocardium by OS has been reported. Evidence of this was provided from obese mice lacking fibronectin type III domain containing 5 (FNDC5), the precursor protein of the obesity protecting hormone irisin, showing that in the heart of the mutant animals, obesity induced by a HFD causes more severe hypertrophy, as well as a major inflammatory state and OS. The lack of FNDC5 also promoted an increase in inflammatory signaling cascade JAK2/STAT3 [47]. These results further clarify the synergic participation of OS and inflammation in obesity-related cardiac hypertrophy. This interconnection may be explained, in part, by augmentation of the JAK2/STAT3 cascade that control cytokines synthesis and, in consequence, OS. Actually, in obese Zucker rats, Chen et al. found that activation and expression of STAT3 was elevated, along with augmented IL6 production. They also proved that IL6 stimulates MERK5/ERK5 pathway that mediate hypertrophy gene expression [33].

Exacerbation of Cardiac Hypertrophy in Obesity-Related Hypertension

Given the clinical relevance of hypertension in obesity over the last years, investigations focused on the molecular mechanisms that lead to LVH aggravation in obesity-related hypertension are emerging. Phillip-Couderc et al. performed a cardiac transcriptome analysis in obese dogs fed with a HFD that developed hypertension before the onset of LVH [48]. In this study the authors found that, in the absence of LVH, obesity-related hypertension is already accompanied by cardiac changes in expression of genes related to extracellular matrix remodeling, energy metabolism, ion flux, cell proliferation, stress response, signal transduction, hormones and, cytoskeletal, nuclear and sarcolemma structure, suggesting that obesity-related hypertension produces early functional alterations in the heart that precede the structural changes associated with remodeling. Unfortunately, this study does not allow to elucidate the genetic changes that are specifically induced by obesity. Holzem et al. used a mice fed with a HFD to evaluate the effects of obesity and pressure overload induced by aortic constriction, on cardiac hypertrophy. After 10 weeks on HFD the control animals did not develop hypertension neither LVH, whereas those with aortic constriction had an increase in systolic blood pressure, LV mass and cell size, that were accompanied by over-activation of the JNK cascade, indicating the importance of hypertension to prompt hypertrophy in obesity [10]. Reedy et al. observed that, after 20 weeks of a HFD, mice exhibited LVH with only a slightly increase in systolic blood pressure. The treatment of HFD-fed mice with AngII caused a significant increase in systolic blood pressure that was associated with a greater LVH. Hypertensive HDF-fed animals also showed the highest level of inflammatory cytokines, IL6, ILβ1, macrophage CD68 marker, and fibrosis marker TGFβ, among others. Similarly, hypertensive HFD-fed mice exhibited a major increase in AKT activation and expression of glycolytic enzymes [11]. These results support the synergic effect of obesity and hypertension on cardiac metabolic and structure remodeling.

Conclusions

Obesity and hypertension converge in the disruption of key cellular and metabolic processes, such as mitochondrial dysfunction, OS, inflammation and cellular signaling, hence disturbances originated during obesity may be the critical point that directs the development of hypertension and then, both conditions favor the creation of a complex network of cellular events that trigger signaling cascades for hypertrophy gene transcription. In particular, OS derived from obesity, appears as a critical point in the control and prevention of further oxidative damage, which could consequently prevent the formation of a metabolic environment that facilitates further cardiac injury. Attention should be paid on deeply unraveling the molecular causes involved in this synergy adjustment between obesity and hypertension in order to improve the prevention, early diagnosis and treatment of obese patients.

References

1. World Health Organization. Obesity and Overweight. 2018
2. Abel ED, Litwin SE, Sweeney G (2008) Cardiac remodeling in obesity. Physiol Rev 88(2):389–419
3. Good D, Morse SA, Ventura HO, Reisin E (2008) Obesity, hypertension, and the heart. J Cardiometab Syndr 3(3):168–172
4. DeMarco VG, Aroor AR, Sowers JR (2014) The pathophysiology of hypertension in patients with obesity. Nat Rev Endocrinol 10(6):364–376
5. Alpert MA, Omran J, Bostick BP (2016) Effects of obesity on cardiovascular hemodynamics, cardiac morphology, and ventricular function. Curr Obes Rep 5(4):424–434
6. Seravalle G, Grassi G (2017) Obesity and hypertension. Pharmacol Res 122:1–7
7. Carvajal K, Balderas-Villalobos J, Bello-Sanchez MD et al (2014) Ca^{2+} mishandling and cardiac dysfunction in obesity and insulin resistance: role of oxidative stress. Cell Calcium 56(5):408–415
8. Peterson LR (2006) Obesity and insulin resistance: effects on cardiac structure, function, and substrate metabolism. Curr Hypertens Rep 8(6):451–456
9. Alpert MA, Karthikeyan K, Abdullah O, Ghadban R (2018) Obesity and cardiac remodeling in adults: mechanisms and clinical implications. Prog Cardiovasc Dis 61(2):114–123
10. Holzem KM, Marmerstein JT, Madden EJ, Efimov IR (2015) Diet-induced obesity promotes altered remodeling and exacerbated cardiac hypertrophy following pressure overload. Physiol Rep 3(8):e12489

11. Reddy SS, Agarwal H, Barthwal MK (2018) Cilostazol ameliorates heart failure with preserved ejection fraction and diastolic dysfunction in obese and non-obese hypertensive mice. J Mol Cell Cardiol 123:46–57
12. Linz D, Hohl M, Mahfoud F, Reil JC et al (2012) Cardiac remodeling and myocardial dysfunction in obese spontaneously hypertensive rats. J Transl Med 10:187
13. Kotchen TA (2010) Obesity-related hypertension: epidemiology, pathophysiology, and clinical management. Am J Hypertensionolume 23(11):1170–1178
14. de Simone G, Izzo R, De Luca N, Gerdts E (2013) Left ventricular geometry in obesity: is it what we expect? Nutr Metab Cardiovasc Dis 23(10):905–912
15. Kannel WB, Brand N, Skinner JJ Jr, Dawber TR et al (1967) The relation of adiposity to blood pressure and development of hypertension: the framingham study. Ann Intern Med 67(1):48–59
16. Zhou MS, Wang A, Yu H (2014) Link between insulin resistance and hypertension: what is the evidence from evolutionary biology? Diabetol Metab Syndr 6:12
17. Shimamoto K, Ura N (2006) Mechanisms of insulin resistance in hypertensive rats. Clin Exp Hypertens 28(6):543–552
18. Li R, Zhang H, Wang W, Wang X et al (2010) Vascular insulin resistance in prehypertensive rats: role of PI3-kinase/Akt/eNOS signaling. Eur J Pharmacol 628(1–3):140–147
19. Lavie CJ, Messerli FH (1986) Cardiovascular adaptation to obesity and hypertension. Chest 90(2):275–279
20. Messerli FH (1982) Cardiovascular effects of obesity and hypertension. Lancet 22:1165–1168
21. Weidmann P, de Courten M, Boehlen L, Shaw S (1993) The pathogenesis of hypertension in obese subjects. Drugs 2:197–208
22. Kehat I, Molkentin JD (2010) Molecular pathways underlying cardiac remodeling during pathophysiologic stimulation. Circulation 122(25):2727–2735
23. Frey N, Olson EN (2003) Cardiac hypertrophy: the good, the bad, and the ugly. Annu Rev Physiol 65:45–79
24. Guzzardi MA, Iozzo P (2011) Fatty heart, cardiac damage, and inflammation. Rev Diabet Stud 8(3):403–417
25. Salazar NC, Chen J, Rockman HA (2007) Cardiac GPCRs: GPCR signaling in healthy and failing hearts. Biochim Biophys Acta 1768:1006–1018
26. Wagner S, Dantz C, Flebbe H, Azizian A et al (2014) NADPH oxidase 2 mediates angiotensin II-dependent cellular arrhythmias via PKA and CaMKII. J Mol Cell Cardiol 75:206–215
27. Rose BA, Force T, Wang Y (2010) Mitogen-activated protein kinase signaling in the heart: angels versus demons in a heart-breaking tale. Physiol Rev 90(4):1507–1546
28. Olivares-Reyes J, Arellano-Plancarte A, Castillo-Hernandez J (2009) Angiotensin II and the development of insulin resistance: Implications for diabetes. Mol Cell Endocrinol 302(2):128–139
29. Kalra D, Sivasubramanian N, Mann DL (2002) Angiotensin II induces TNFA biosynthesis in the adult mammalian heart through a protein kinase-C dependent pathway. Circulation 105:2198–2205
30. Oliveira-Junior SA, Martinez PF, Guizoni DM, Campos DH et al (2014) AT1 receptor blockade attenuates insulin resistance and myocardial remodeling in rats with diet-induced obesity. PLoS ONE 9(1):e86447
31. du Toit EF, Nabben M, Lochner A (2005) A potential role for angiotensin II in obesity induced cardiac hypertrophy and ischaemic/reperfusion injury. Basic Res Cardiol 100:346–354
32. Zhong P, Quan D, Peng J, Xiong X et al (2017) Role of CaMKII in free fatty acid/hyperlipidemia-induced cardiac remodeling both in vitro and in vivo. J Mol Cell Cardiol 109:1–16
33. Chen WK, Yeh YL, Lin YM, Lin JY et al (2014) Cardiac hypertrophy-related pathways in obesity. Chin J Physiol 57(3):111–120
34. Schlegel P, Reinkober J, Meinhardt E, Tscheschner H et al (2017) G protein-coupled receptor kinase 2 promotes cardiac hypertrophy. PLoS ONE 12(7):e0182110
35. Dorn GW (2009) GRK mythology: G-protein receptor kinases in cardiovascular disease. J Mol Med (Berl) 87(5):455–463

36. Lucas E, Jurado-Pueyo M, Fortuño MA, Fernández-Veledo S et al (1842) Downregulation of G protein-coupled receptor kinase 2 levels enhances cardiac insulin sensitivity and switches on cardioprotective gene expression patterns. Biochim Biophys Acta 2448–2456:2014
37. Lucas E, Vila-Bedmar R, Arcones AC, Cruces-Sande M et al (2016) Obesity-induced cardiac lipid accumulation in adult mice is modulated by G protein-coupled receptor kinase 2 levels. Cardiovasc Diabetol 15(1):155
38. Jia G, DeMarco VG, Sowers JR (2016) Insulin resistance and hyperinsulinaemia in diabetic cardiomyopathy. Nat Rev Endocrinol. 12(3):144–153
39. Palomo I, Alarcón M, Moore-Carrasco R, Argilés JM (2006) Hemostasis alterations in metabolic syndrome. Int J Mol Med 18(5):969–974
40. Zima AV, Blatter LA (2006) Redox regulation of cardiac calcium channels and transporters. Cardiovasc Res 71:310–321
41. Kimberley M, Ritchie RH, Delbridge L (2009) Reactive oxygen species and insulin resistant cardiomyopathy. Aust Physiol Soc 40:27–35
42. Sverdlov AL, Elezaby A, Qin F, Behring JB et al (2016) Mitochondrial reactive oxygen species mediate cardiac structural, functional, and mitochondrial consequences of diet-induced metabolic heart disease. J Am Heart Assoc 5:e002555
43. Mattiazzi A, Bassani RA, Escobar AL, Palomeque J et al (2015) Chasing cardiac physiology and pathology down the CaMKII cascade. Am J Physiol Heart Circ Physiol 308(10):H1177–H1191
44. Calamaras TD, Lee C, Lan F, Ido Y, Siwik DA et al (2012) Post-translational modification of serine/threonine kinase LKB1 via adduction of the reactive lipid species 4-Hydroxy-trans-2-nonenal (HNE) at lysine residue 97 directly inhibits kinase activity. J Biol Chem 287(50):42400–42406
45. Ikeda Y, Sato K, Pimentel DR, Sam F et al (2009) Cardiac-specific deletion of LKB1 Leads to Hypertrophy and Dysfunction. J Biol Chem 284:35839
46. Marseglia L, Manti S, D'Angelo G, Nicotera A et al (2015) Oxidative stress in obesity: a critical component in human diseases. Int J Mol Sci 16(1):378–400
47. Geng Z, Fan WY, Zhou B, Ye C et al (2019) FNDC5 attenuates obesity-induced cardiac hypertrophy by inactivating JAK2/STAT3-associated inflammation and oxidative stress. J Transl Med 17(1):107
48. Philip-Couderc P, Smih F, Pelat M, Vidal C et al (2003) Cardiac transcriptome analysis in obesity-related hypertension. Hypertension 41(3):414–421

Chapter 10
Role of Epicardial Adipose Tissue in Heart Failure: From Basic to Clinical Perspectives

Hao Zhang, Mahmoud Gheblawi, Jiu-Chang Zhong, and Gavin Y. Oudit

Abstract Obesity, which is highly associated with insulin resistance (IR), systemic inflammation, metabolic disorders and cardiovascular diseases (CVD) including hypertension, hyperlipidemia, coronary artery disease (CAD) and ultimately heart failure (HF), is an epidemic problem with growing population. Fatty acid metabolism consisting of anabolism and catabolism is crucial in ensuring constant energy supply to almost all vital organs in physiological milieu, and it is well maintained under exquisite regulatory mechanisms dependent on dynamic external and intrinsic factors. More importantly, localized adipose tissues, such as epicardial adipose tissue (EAT) and perivascular adipose tissue (PVAT), also exert regulatory roles as an endocrine and paracrine gland on the heart and whole-body vasculatures via the proximal secretion of hormones, adipokines, cytokines and microRNAs. Notably, emerging evidences confirming the flip-flop protective effects of EAT on failing hearts and PVAT on vascular tone have questioned the empirically-believed negative relation between regional adiposity (mainly visceral) and cardiovascular events, though the mechanistic underpinning remains largely unknown. This chapter provides an overview of adipose tissue physiology and EAT-mediated pathophysiological progression to heart failure with preserved ejection fraction (HFpEF) in obesity, interprets the current findings of possible interplays from pre-clinical and clinical models. Furthermore, we highlight the heuristic translational insights on early diagnosis, intervention and therapeutic options to balance the physiological

H. Zhang · G. Y. Oudit (✉)
Division of Cardiology, Faculty of Medicine and Dentistry, Department of Medicine, University of Alberta, Edmonton, AB, Canada
e-mail: gavin.oudit@ualberta.ca

M. Gheblawi · G. Y. Oudit
Faculty of Medicine and Dentistry, Department of Physiology, University of Alberta, Edmonton, AB, Canada

H. Zhang · M. Gheblawi · G. Y. Oudit
Mazankowski Alberta Heart Institute, University of Alberta, Edmonton, AB, Canada

J.-C. Zhong
Heart Center and Beijing Key Laboratory of Hypertension, Beijing Chaoyang Hospital, Capital Medical University, Beijing 100020, China

P. S. Tappia et al. (eds.), *Biochemistry of Cardiovascular Dysfunction in Obesity*, Advances in Biochemistry in Health and Disease 20,
https://doi.org/10.1007/978-3-030-47336-5_10

and pathological equilibrium of EAT averting unfavorable cardiovascular insults in the context of obesity.

Keywords Heart failure · Epicardial adipose tissue · Obesity · Metabolic syndrome · Insulin resistance · Inflammation · Renin-angiotensin system · Adipocytokine · Apelin · Human explanted hearts

Introduction

In contemporary society, obesity strikingly impinges on the public health with over 500 M patients diagnosed as of 2013, which projected to rise in the forthcoming decades imposing enormous socio-economic burden including medical costs and poor labour resource [1]. Meanwhile, cardiovascular disease is another epidemiology attributed to coronary artery diseases, stroke, hypertension and heart failure, ranked as the leading cause of deaths globally [1–3]. Obesity is a eutrophic genetic disease characterized by pathologically excessive adipose tissue accumulation in the body which is chronically induced by heterogeneous factors, such as over-rich calorie diet and sedentary lifestyle, and complicated by the systemic metabolic syndrome, namely, type 2 diabetes mellitus, insulin resistance, dyslipidemia [4]. Recently, the site-specific properties beyond simply being an energy reservoir of adipose tissue have drawn much interest, and given the close anatomic proximity to the omnivorous heart, any perturbation of EAT or PVAT profile would promptly trigger pathogenic adipokines secretion adversely jeopardizing coronary arteries and myocardium. Not surprisingly, paralleled correlations between the regional adipose distribution (e.g. volume and thickness) and detrimental CVD events (i.e. atrial fibrillation, atherosclerosis and left ventricular diastolic dysfunction), possibly due to enhanced inflammation, disturbed adipocytokines and lipotoxicity, have been demonstrated by copious literatures [5–9]. Specifically, obesity confers stronger prediction to heart failure with preserved ejection fraction (HFpEF, EF $\geq$ 50%) versus heart failure with reduced ejection fraction (HFrEF, EF < 50%) especially in females [10–12]. However, controversial evidences exist in a few studies claiming the cardio-protective effects like plaque stability; thus improving prognosis from obese individuals with CVD, which is the so-called "obesity paradox" awaiting further elucidation underlying the epigenetic shift [13–15]. Despite the emerging appreciation of adipose tissue as a cardiometabolic risk factor and a modifier routinely assessed by modern imaging techniques (i.e. 2D transthoracic echocardiography and cardiac magnetic resonance imaging), why, how and to what extent should we tackle this intractable illness still remain poorly understood. Moreover, the pursuit of investigation is partially held back by the disparate adiposity pattern between murine models and human beings. Therefore, in this review, we set to cover the biological role of adipose tissue in relation to obesity, to illustrate the central importance of maintaining normal epicardial adipic profiles, to emphasize the ambivalent EAT-derived effects on failing hearts and to explore any possible mechanisms (e.g. ACE2/Apelin axis) that could potentially illuminate the pathway searching for therapeutic hope.

Obesity and Developmental Physiology of Adipose Tissue

Obesity, or adiposity, is an ailing condition in which exceptional high fats or lipids accumulate within body. It is advisedly diagnosed using the age-independent body mass index (BMI) with a combination of descriptive anthropometry (i.e. waist circumference and waist-to-hip ratio) in clinical practices (Fig. 10.1A), which avoids the gender-related difference of fatness distribution [14, 16, 17]. To state, the splice of the beige tissue, mainly composed of subcutaneous adipose tissue (SAT), tends to deposit primarily around the gluteal-femoral part in female vis a vis the archetypical accumulation of visceral fats (VAT) in male's upper body (Fig. 10.1B) [18, 19]. In addition, the favorable side effects on female verse harmfulness on male also help to partition the sexual dimorphism of fat proportions, presumably resulted from dividing lipolytic actions as well as distinct pro-inflammatory profiles [14, 20]. Biologically, our bodies have evolved two heterotypes of adipose tissue—white (WAT) and brown adipose tissue (BAT)—to ensure functional diversities as we grow. For example, WAT is responsible for energy storage, thermal insulation, internal organ protection and excessive-metabolite buffering via expansion and proliferation despite just one large lipid droplet and sparse mitochondria are encircled within the parenchymal cells, whilst BAT is indispensably providing non-shievering heat to control body temperature by interacting with mitochondrial uncoupling protein-1 (UCP1) [21, 22]. Both SAT and VAT fall into the broad categorization of WAT being more metabolically active. Intriguingly, multiple lines of evidences indicate the WAT-to-BAT turnover can reverse from WAT-dominant throughout development to BAT-differentiation under physiological (cold exposure) and pharmacological (e.g. activating β adrenergic receptor and peroxisome proliferator-activated receptor λ) stimulations [22–25]. Inviting though bariatric insights are perceived to fine-tune the adipose tissue phenotypic plasticity, minimal is yet unraveled in human trails. In terms of the adipotic constituents, adipocyte is literally recognized as the featured cellular type storing fatty acids (FFAs) in the form of triglyceride (TG), which only accounts for a small compositional proportion compared with the sizeable remnants including pre-adipocytes, macrophages, neutrophils and stem cells [26, 27]. Simultaneously, adipocytes are capable of secreting over fifty cytokines and peptides collectively known as adipocytokines in (patho-)physiological status [14, 20, 28]. It is the cellulous and promiscuous nature of adipose tissue that dynamically orchestrates energy supply, hormonal secretosome (via endocrine, paracrine and autocrine) and inflammatory regulations (i.e. superseding M1 macrophagic phenotype in times of morbidly fluctuating lipids) [29, 30].

As for cardiovascular system, WAT is present as three distinct subtypes: the EAT, the PVAT and the pericardial adipose tissue (PAT). Lately, the serendipitous role of ectopic fat depot like EAT has drawn much attention especially in cardiometabolism,

A

Guideline from WHO	
1. BMI (kg/m²)	Disease Stratification
< 18.5	Underweight
18.5 – 24.9	Normal weight
25.0 – 29.9	Overweight
30.0 – 34.9	Obesity class I
35.0 – 39.9	Obesity class II
> 39.9	Obesity class III (morbidly)
2. Waist circumference (cm)	
Male	> 102
Female	> 88
3. Waist-to-hip ratio	
Male	> 0.9
Female	> 0.85

B Gender-dimorphism of fat distribution

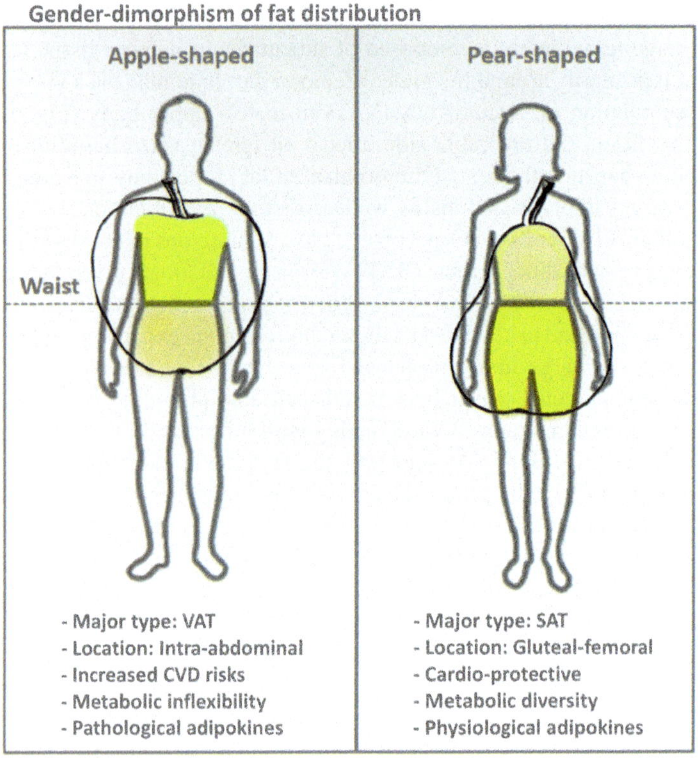

Fig. 10.1 **A** Obesity classification for adults from WHO guidelines, based on BMI and additional physical parameters. WHO: world health organization; BMI: body mass index. **B** "Apple-shaped" versus "Pear-shaped" adipose tissue distribution between men and women. In females, adipose tissue appears to accumulate more in the lower body, mainly subcutaneous adipose tissue around the gluteal-femoral area, while in males fat accumulates in the visceral area. The gender-specific fatness distribution is believed to correlate with different systemic and cardiovascular effects. Females are benefiting from the lower-body white adipose tissue with metabolic cardioprotection and salubrious secretosome, while intra-abdominal adipogenesis brings myriads of fateful cardiovascular consequences on males

given the constantly high-energy demand of heart which is met dominantly (~60%) by β-oxidation of the esterified fatty acids. This process is facilitated via EAT by virtue of its neighboring anatomic location and seamless contact with the myocardium [30, 31]. However, higher reliance on fatty acid metabolism in the setting of preceding cardiac pathologies, statistically about 80% in CAD and 90% in diabetes, sensitizes the failing heart and exacerbates the progression to diastolic HF [31]. During remodeling, the EAT could potentially alleviate lipotoxicity by buffering the surplus metabolite. On the other hand, contradictory findings as to the counteractive effects of epicardial adipogenesis on the myocardial dysfunctionality, presumably triggered by impaired EAT quality or limited buffering capacity under chronic overnutrition, complicate the whole edifice of understanding towards the localized interplay [32, 33]. Accordingly, future studies on balancing the seesaw of EAT are certainly warranted to ameliorate or even recuperate the cardiac health.

Physiological Role of EAT

Anatomical Biology

The first of the three WATs present in the cardiovasculature is the EAT constituting the upper layer of the myocardium's epithelium, the epicardium, which is located between the visceral pericardium and the myocardium (Fig. 10.2A) [34]. Furthermore, the PAT is located above the serous pericardium on the surface of the fibrous pericardium (Fig. 10.2A) [34]. Lastly, it is the PVAT that encompasses many of the greater and lesser vessels within the body such as the aorta and efferent and afferent renal arteries [35–37].

EAT Versus PAT: Distinct Distribution and Function

The EAT constitutes 20% of the weight of healthy adult human hearts and arises from the same splanchnopleuric mesodermal BAT lineage as mesenteric and omental fat cells [35–38]. PAT, on the other hand, arises from the primitive thoracic mesenchyme and consists of the adipose tissue enveloping the pericardium which may extend over 80% of the pericardial surface and compose of up to 20–50% of the cardiac mass [36]. The EAT is supplied nutrients by the coronary arteries and their vasa vasorum comprising a part of both of their adventitia indicating EAT's contiguity with both the myocardium and the coronary arteries while the PAT is supplied by the internal thoracic artery and is not in contiguity with the myocardium [35, 36, 38–41]. The EAT begins at the branching of the coronary arteries from the aortic root and traces around the atrioventricular grooves and then down following interventricular grooves and coronary branches, both on and within the myocardium until reaching

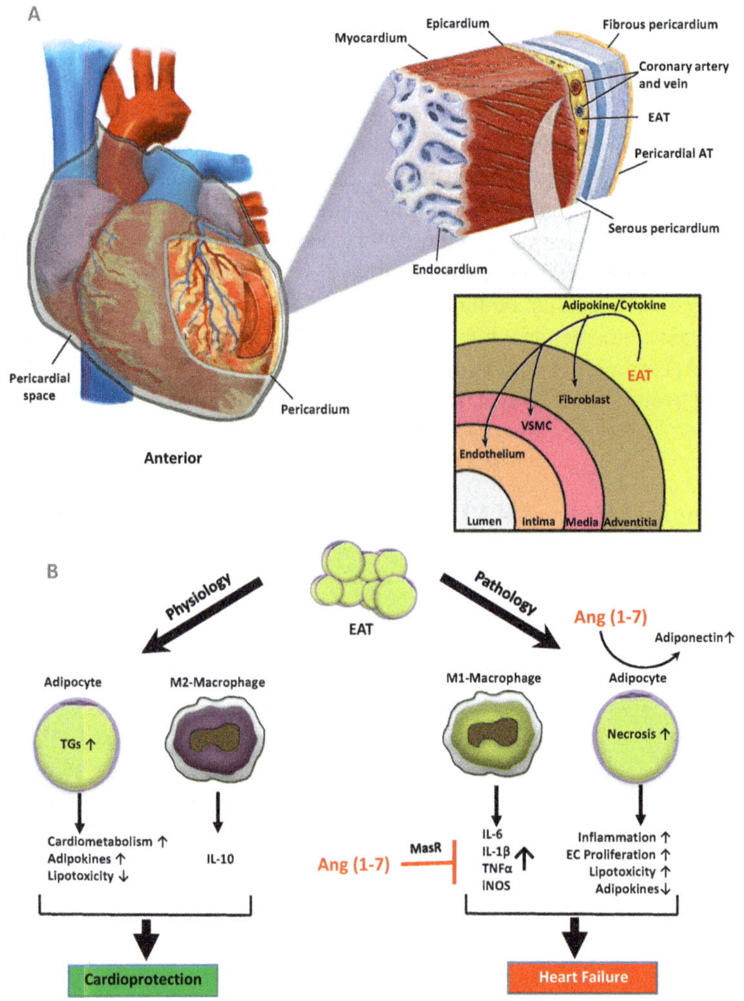

Fig. 10.2 The role of epicardial adipose tissue in cardiac (patho-)physiology. **A** In the human heart, EAT is located in the atrioventricular and interventricular grooves, surrounding the significant branches of coronary arteries, both left and right ventricles, atria and the apex (a). The blow-up above in (a) delineates the relative anatomic position between epicardial and pericardial adipose tissues, each covering epicardium and fibrous pericardium respectively; meanwhile, the sketch below in (a) depicts the cross-sectional perivascular contour beneath EAT. **B** Due to the anatomical closeness between EAT/PVAT and myocardium, they may function more than a local fatty acids reservoir, which readily ensures adequate energy supply to cardiomyocytes in stress conditions, such as CAD and ischemia. The physiological nature of EAT/PVAT could turn into pathological upon adipose tissue inflammation. It is macrophage polarizing to proinflammatory CD 11c^{+} M1-macrophages in EAT/PVAT that contributes to the pathogenesis of CAD, localized inflammation and lipotoxicity, which inevitably leads to HF. AT: adipose tissue; EAT: epicardial adipose tissue; EC: endothelial cell; VSMC: vascular smooth muscle cell; Ang (1–7): angiotensin 1–7; IL: interleukin; iNOS: inducible nitric oxide synthase; CD 11c^{+} M1-macrophages: classically activated inflammatory phenotype of macrophages; MasR: Mas receptor; TGs: triacylglycerols; TNF-α: tumor necrosis factor-α. Reproduced with permission from Patel et al. [29]

the apex of the heart [37, 41, 42]. The EAT's mechanical function is the reduction of torsional strain along interventricular grooves and at the base of the coronary arteries in addition to limiting frictional strain between the exterior surface of the coronaries with the serous pericardium [38, 43]. Additionally, the EAT plays an imperative role in maintaining physiological metabolism and homeostasis of the heart [38], because it consists of adipocytes, preadipocytes, fibroblasts, macrophages, endothelial cells, microvasculature, and ganglia [27, 38, 44, 45]. Most adipocytes in EAT are preadipocytes and are generally smaller in cell size compared to visceral fat elsewhere in the body as there is limited room to expand within the pericardial sac before impairing cardiac function [29, 37, 38, 41, 46]. The constant demand for energy by the heart keeps lipid storage, and thereby cell size, at a minimum in EAT [37, 38, 41, 45, 46].

Metabolic Profile of EAT

The heart's primary energy source comes from β-oxidation of esterified fatty acids which accounts for 50–70% of its energy consumption [29, 31, 47]. FFAs are transported from the blood into the EAT employing the cluster of differentiation 36 (CD36), a plasma membrane-associated protein which is responsible for the FFAs uptake by the heart [38, 47, 48]. Here, FFAs are deposited as myocardial TG until later being degraded back into FFAs and released as cardiac metabolic requirements demand [38, 47, 48].

Myocardial metabolism further favours FFAs utilization over glucose in times of diabetes, exercise, limited food intake, and obesity [38, 41]. Likewise, the EAT does not readily uptake glucose innately as other adipose tissues do, as observed in its reduced amount of both insulin receptors and glucose transporters [49]. A probable cause for EAT's lack of insulin bioactivity could be the latter's anti-lipolytic effects mediated primarily through downstream AMPK activation which runs contrary to EAT's role as an energy depot for the myocardium, despite the lack of clear elucidation as to the communication means between the EAT and myocardium [29, 36, 50]. However, we can infer that the energy for cardiac metabolism primarily originates from this source because of the fact that WAT lipolysis in addition to FFA synthesis and release is highest in EAT [41]. Furthermore, the EAT not only serves as a temporary TG depot but exhibits cardioprotective buffering mechanisms during hyperlipidemia preventing ectopic dyslipidemia within the myocardium by storing FFAs in addition to its hypothesized action in secreting FFAs back into the circulation against their concentration gradients [30, 38, 41, 47, 51].

Secretosome of EAT

The EAT containing miscellaneous components is a large secretosome all releasing an array of adipokines, cytokines and hormones eventuating in autocrine along with myocardial endocrine and paracrine signaling due to the absence of an anatomical barrier between them [30, 36, 38, 52]. Secretions may also result in paracrine and endocrine effects upon the coronary arteries either through direct diffusion across their adventitia, media, and intima or that of their vasa vorsum which are equally engulfed in EAT [38, 53].

The adipocytes from the EAT release adipokines such as adiponectin, apelin, and leptin in addition to cytokines such as tumor growth factor-β (TGF-β) and monocyte chemoattractant protein-1 (MCP-1) [52]. Among them, the anti-oxidant adipokine adiponectin is responsible for promoting glucose uptake, insulin sensitivity and lipid catabolism [46, 52, 54]. The protein hormone also inhibits eNOS-related vasodilation, platelet aggregation, thrombosis and macrophage activation [46, 52]. While it can be produced by EAT adipocytes, cardiomyocytes as well as vascular stromal cells [52], the apelin secreted from the adipocytes leads to the vasodilation of arteries, improved cardiac contractility, output, and recovery from ischemia while decreasing oxidative stress, cardiac infarct size and fibrosis [55–57]. As a circulating hormone, leptin affects body weight, food intake, fat mass, and metabolism either through direct action on tissues or through neuroendocrine signaling and is found in levels proportional to insulin [27, 52]. As for cytokines, MCP-1 is a cytokine released by EAT adipocytes at a much higher rate than SAT usually in response to oxidative stress, cytokines or hormonal factors, and functions in attracting monocytes to infiltrate the EAT and mature into resident macrophages in addition to recruitment of T-lymphocytes [30, 52, 58]. The cytokines released by EAT such as IL-10, IL-6, TGF-β, tumor necrosis factor-α (TNF-α) are quite potent to the extent that only 10% released need binding to a receptor to elicit a response and serve functions as both inflammatory and anti-inflammatory mediators of the immune system [45, 52, 59]. Likewise, these cytokines and adipokines secreted by the EAT are capable of crossing the tunicas of the coronary arteries and their vasa vasorum leading to endothelial dysfunction, smooth muscle proliferation, and destabilization of atherosclerotic plaques [30, 43].

Pathological Role of EAT in HF with Obesity

Two Subtypes of HF: HFpEF Versus HFrEF

HFpEF is a sign of diastolic impairment in which the LV can no longer fill appropriately due to either physical restriction from a tightening pericardial space to both hypertrophic and fibrotic factors which limit the LV's relaxation and filling capabilities [60]. Patients express preserved LV ejection fractions (≥50%) with decreased cardiac output and impaired LV global longitudinal strain (GLS) [10, 61]. HFrEF, on the other hand, is a marker of systolic dysfunction (with underlying diastolic dysfunction) wherein LV pumping (EF < 50%) is diminished primarily due to cardiac dilation arising from morbidities such as aortic stenosis, hypertension, chronic hypertrophy, CAD or myocardial infarction (MI) [10, 30, 60]. The synopsis is that an overworked LV weakens over time and ceases to expel sufficient amounts of the end-diastolic volume resulting in reduced ejection fractions and pulmonary tension [62]. In cases of MI, blood supply is cut off to parts of the ventricular (mainly LV) which leads to the spread of necrosis factors as well as deposition of fibrotic tissue reducing the contractility capabilities of the myocardium [30, 63].

Association Between HFpEF and Obesity

HFpEF is the primary form of HF associated with obesity and is marked by diastolic dysfunction of the LV and constriction of available space within the pericardial sac limiting the extent of cardiac remodeling [29, 30, 36]. The expansion of the EAT preadipocytes limits the amount of space within the pericardial sac available for the LV to relax into during diastole leading to restrictive cardiomyopathy [29, 37, 38, 46]. Concurrently the burden of the increased weight of the expanded EAT connected to the myocardium forces the LV to work harder in ejecting blood during systole triggering in hypertrophic remodeling [36, 38]. Obesity-related EAT expansion, therefore, results in diastolic dysfunction marked by cardiac hypertrophy and restrictive cardiomyopathy, indicative of HFpEF [30, 38, 64]. The EAT is hypothesized to be capable of transporting FFAs across concentration gradients whereby exerting cardioprotective effects during instances of hyperlipidemia [38, 51]. This function, however, is compromised with the onset of obesity and HFpEF in which there is a marked expansion of EAT (Fig. 10.3) losing the cardioprotective ability in regulating FFAs levels [38, 39, 51].

Metabolic and Secretosome Alterations of EAT in Obese Patients with HF

Excess adiposity can induce inflammation at both organ and systemic levels, and this is no exception for the interplay between EAT expansion and the myocardial injury [52]. Increased adipokines and cytokines from the EAT, which may release more than 50 kinds adipocytokines when in a pathological state, and invading macrophages dysregulate the myocardium's naturally flexible metabolic energy source preference of 50–70% from FFAs to sole FFAs utilization through induced IR [14, 27–29, 47]. The increased usage of FFAs mediated by the EAT also leads to the buildup of reactive

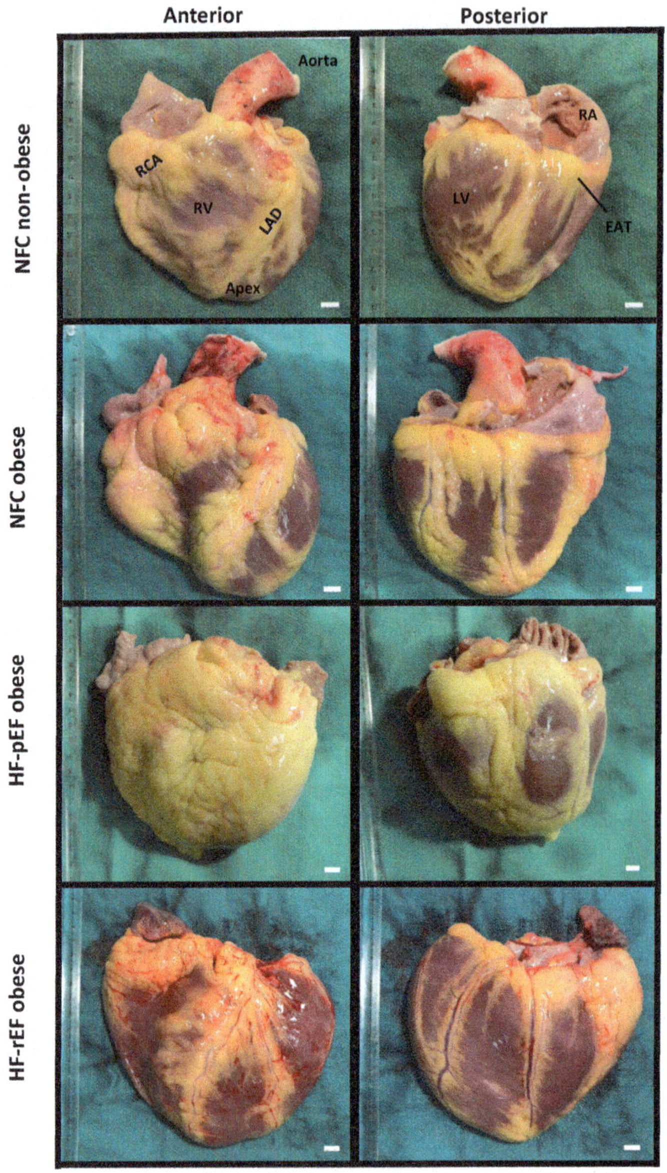

Fig. 10.3 Extensive epicardial adipose tissue in human hearts. Representative anterior and posterior views of human explanted hearts from non-failing controls (NFC) without and with obesity/overweight (BMI < 25.0 kg/m^2, heart weight = 438 g, ejection fraction = 65%; BMI > 30.0 kg/m^2, heart weight = 316 g, ejection fraction = 55%), heart failure with preserved ejection fraction (HF-pEF) and obesity (BMI = 30.82 kg/m^2, heart weight = 546 g; EF = 50%), and heart failure with reduced ejection fraction (HF-rEF) and obesity/overweight (BMI = 31.31 kg/m^2, heart weight = 480 g; EF = 24.9%) depicting the areal distribution of epicardial adipose tissue. Scale bar = 1 cm. LAD indicates left anterior descending coronary artery, RCA indicates right coronary artery

oxygen species (ROS) and dyslipidemia in the myocardium resulting in fibrosis and MI [49].

Obesity leads to elevated levels of leptin and resistin which are typically low in non-obese persons who have to counteract higher levels of adiponectin instead [52]. The anti-oxidant adipokine adiponectin which confers insulin sensitivity, declines with the development of obesity as hypoxia-induced apoptosis and mitochondrial ROS production elevate [39, 46, 52, 65]. Elevated EAT secretion of leptin and resistin increases IL-6, TNF-α, and interferon gamma (IFNγ) cytokines as well as chemokine MCP-1 levels which stimulates inflammation in obese patients by promoting monocyte invasion and maturation in addition to attracting T-leukocytes. Both in turn release further pro-inflammatory factors elevating myocardial IR, inflammation and lipotoxicity (Fig. 10.2B) [38, 52].

EAT-Mediated Inflammation in Obese Patients with HF

Macrophages stimulated by EAT's adipokine and chemokine factors undergo polarization into the M1 classic pro-inflammatory phenotype [30, 66]. These macrophages express $CD11c^{+}$ and release IL-6, IL-1B, TNF-α and MCP-1 which further recruit and retain M1 macrophages capable of nitric oxide (NO) production, through the iNOS pathway, further inflaming the organ (Fig. 10.2B) [66]. M1 macrophage self-recruitment resembles a feed-forward regulation mechanism by which inflamed EAT falsely identifies inflammation due to elevated resistin and leptin levels, which leads to macrophage recruitment and maturation which go on to release further pro-inflammatory factors as well as hormones such as the cardiac remodeling angiotensin II (Ang II) [38, 54, 66, 67]. The alternative non-inflammatory M2 phenotype is stimulated by IL-10, amongst other factors, and marked by the CD206+ antigen in extracellular matrix (ECM) deposition which has overall beneficial and therapeutic effects in the EAT (Fig. 10.2B) [54, 66]. The M2 phenotype likewise increases insulin sensitivity through TGF-β as well as sequestering the proliferation of EAT preadipocytes [29, 54, 68].

Translational Approaches: From Basic to Clinical Insights

Limitations of Model Organisms

The spread and eventual encompassing of the myocardium with EAT is not seen in other mammals as compared to human hearts [29, 37, 41]. Marchington et al. conducted significant histological and enzymatic activity studies on EAT from various mammals, both model organisms and not [41]. They discovered that none of their experimental specimens had enough EAT which resulted in pooling samples from different specimens for each species to have an adequate amount to perform analysis on. While the human myocardium and coronary arteries are enveloped and impregnated by the EAT, other mammals including mice, hardly have any EAT, to begin with [37, 41]. However, the mouse model is extensively used in cardiac research despite the EAT is primarily focused around the atrioventricular groves and defined more proximally to the aorta, whereas in humans the EAT spreads as a continuous body structure from the point of coronary branching from the aortic root to the apex of the heart [30, 38, 41, 42].

A possible reason for the reduced EAT in mice is their elevated metabolic rate. The normal murine heartbeat is between 500–700 beats per minute while the human heart beats in comparison at a significantly lower rate [69, 70]. As by the square-cube law, where the volume increases at a faster rate to that of the surface area, a mouse's body size demands more cardiac work to sustain metabolic levels in order to regulate body temperature, whereas larger animals do not need to sustain elevated rates of metabolism to thermoregulate once they reach their ideal body temperature [71]. This decreased cardiac output thereby allows for EAT build-up for times of elevated activity or stress when sudden energy expenditure would be required [38]. Similarly, it has been observed that EAT adipocytes contain UCP-1 signifying that when larger mammals are in need to thermoregulate their entire bodies in cold temperatures, thermogenesis of the EAT can occur to maintain high cardiac function while the peripheries catch up [35, 38]. Exposure to the cold causes activation of cardiac natriuretic peptides which function in converting white adipocytes into a beige phenotype whereby they increase in mass through replication with smaller size and exhibit thermogenesis through the UCP-1 pathway [38, 72]. This observation is synonymous with Marchington et al.'s that EAT seems to be most prevalent in larger mammals and specifically amongst the carnivorous like obese monkey [41].

Human Explanted Hearts: An Optimal Alternative

Taking this all into account the use of recipient explanted human heart tissue is a sound alternative allowing the interpretation of laboratory observations in a more clinically relevant and translational setting where the ideal animal models (primates) would be too costly and unethical to run for every experimentation. The usage of human explanted heart tissue would allow for the comparison of effects seen in pre-clinical models to human tissue, helping elucidate what mechanisms are conserved between species. In the case of obesity and HF, usage of explanted hearts allows us to compare the extent of EAT between different HF subtypes. The EAT appears to be quite enlarged in obese-HFpEF patients as compared to obese-HFrEF, and the latter also seem to have less EAT than the non-failing control hearts from obese donors (NFC-obese) but similar to the ones from NFC non-obese donors (Fig. 10.3). It is not evidentially clear as to why the EAT expansion is markedly increased in HFpEF cases; however, it should be noted that HFrEF depicts a more advanced form of HF where cardiac remodeling can no longer counter myocardial insult resulting in dilation [29]. Another noteworthy point is that despite the EAT in HFrEF not appearing to be as enlarged as in HFpEF, it is just as inflamed and insulin resistant (Fig. 10.3) [29].

Therapeutic Targets: ACE2/Apelin Pathway

A more predictive marker of HF incident occurring is chronic inflammation in conjunction with metabolic syndrome (MetS) as Suzuki et al. demonstrated [73]. Patients with elevated levels of both IL-6 and C-reactive protein as well as elevated markers of MetS (IR, BMI, waist circumference) had worse outcomes and a higher incidence of HF [73]. IR and inflammation are critical markers of the MetS relating to HF, which coincidentally occur almost simultaneously in the EAT [36, 45]. The prevention of HF incidents in obese patients should start through the reduction of EAT's chronic inflammation and IR which can be treated with therapeutics targeted at alterations in either the renin-angiotensin (RAS) or the apelin system [45, 54, 73].

The RAS is the primary hormonal system involved in the pathogenesis of obesity with relationship to the heart through cardiac remodeling [29]. The RAS has both endocrine and paracrine effects and exists as an independent system at the local level in various tissues [74, 75]. Here, angiotensinogen is cleaved by renin into Ang I which is further degraded by angiotensin-converting-enzyme (ACE) into Ang II acting primarily upon the Ang II type 1 Receptor (AT_1R), and this is referred to as the ACE/Ang II/AT_1R axis [74, 75]. Ang II is also cleaved by ACE2 into Ang 1–7 which acts upon the Mas receptor (MasR), as a counter-regulatory arm of the RAS [74, 75]. In addition, ACE2 can also convert Ang I into Ang 1–9 which is further cleaved by either the neutral endopeptidase Neprilysin (NEP) or ACE, and both of them are capable of converting it into Ang 1–7 [74, 75]. The two regulatory axes of the RAS confer different effects on the myocardial tissue. The ACE/Ang II/AT_1R axis is primarily functional in short term recovery of cardiac output through increased contractility and vasoconstriction and is a potential therapeutic for vasodilatory shock [74–76]. HF arising from obesity is demarked by increased IR and chronic inflammation of both the myocardium and the EAT in addition to maladaptive and pathological cardiac remodeling of the ventricles [75]. Upregulated cardiac Ang II levels attempts to remodel in order to maintain regular heart function despite the restrictive nature of HFpEF [29, 63, 77]. Genetic deletion of ACE2 further worsens an obese model as a result of the vasodilatory, anti-inflammatory and blunted insulin-positive effects [54, 75]. Addition of ACE2 to elevate local Ang 1–7 levels in the heart has been shown to ameliorate the effects of chronic Ang II stimulation and is a potential therapeutic target for IR and HF [63, 78].

Apelin is a 77-amino prepropeptide adipokine secreted in the heart and its knockout leads to IR and HF [57, 79]. Apelin is degraded into three main isoforms: Pyr-apelin-13, apelin-17, and apelin-36. Each has a conserved 12-member C-terminus which confers receptor binding activity, with Pyr-apelin-13 and apelin-17 being the most biologically active [57]. Apelin has been shown to ameliorate hypoxia-induced apoptosis, mitochondrial ROS generation as well as improve insulin sensitivity and glucose utilization in a diet-induced-obesity model [65, 80]. The naturally occurring apelin analogues are short-lived in the body (<4 min) as their C-terminal phenylalanine is cleaved off by ACE2 while NEP acts on the conserved "RPRL" motif resulting in conformational changes that can alter binding affinity to the apelin receptor (APJ receptor) [56, 57].

In obese animal models, apelin has been shown to ameliorate ischemia/reperfusion (I/R) injury through ROS mediation, resulting in the reduction of myocardial apoptosis and injury as well as reduced mitochondrial damage [65]. Mice fed a high-fat diet displayed signs of cardiac dysfunction exhibiting reduced ejection fractions and fractional shortenings as well as elevated levels of serum insulin and apelin levels in adipose tissue [65, 81]. Aged apelin mutant mice developed progressive impairment of myocardial contractility along with systolic dysfunction, and loss of apelin contributed to HF in response to both pressure-overload and Ang II infusion [82, 83]. Conversely, mice hearts on a regular diet and given an apelin-13 infusion fared much better off than regular diet only hearts, and a similar observation was noted between high-fat diet hearts and high-fat diet hearts with apelin-13 infusion [65]. Moreover, apelin has direct effects on the propagation of action potential and contractility in cardiomyocytes, and the mechanisms involved in the inotropic effects may be associated with increased myofilament sensitivity to Ca^{2+} [84]. Infusion with apelin leads to reduced caspase-3 activity, a biomarker of myocardial infarction, and increased anti-apoptotic protein Bcl-2 expression, which were secondarily verified by staining for infarcted areas [65]. This concludes that the ACE2/Apelin signaling is a potential therapeutic avenue requiring more research. Current work is looking at developing analogues of apelin limiting its susceptibility to biodegradation by ACE2 and NEP [55, 56, 78, 85].

Conclusion

Elucidating the specific subtypes of HFpEF can further our understanding as to why EAT expands in this phenotype as compared to HFrEF. Iacobellis et al. have demonstrated that EAT increases proportionally to LV hypertrophy and independently to BMI while others have shown that elevated levels of abdominal VAT (as measured by waist circumference) indicate increased EAT in addition to increased LV size (Table 10.1) [86–88]. The Paulus and Tschöpe's hypothesis outlines a course of pathology for HFpEF in which comorbidities such as obesity, MetS and IR lead to elevated adiposity which in turn leads to systemic inflammation. And this results in coronary microvasculature endothelial dysfunction spreading into adjacent myocardium and causing elevated hypertrophy and fibrosis due to decreased levels of protein kinase G (PKG), cyclic guanosine monophosphate content (cGMP) and NO bioavailability [89]. The inflammation although does not have to be systemically occurring as it is present at the locale of the expanding EAT which favorably

Table 10.1 Summary of clinical studies relating obesity with heart failure

	Cohort	Findings
Brouwers et al. [98]	PREVEND 8592 participants	Elevated BMI signified similarly increased prognosis of both HFrEF and HFpEF for primarily Dutch participants
Dalos et al. [94]	Division of Cardiology of the Medical University of Vienna 193 participants	Amongst HFpEF patients, the higher the BMI score, and age of participants correlated to increased exercise intolerance NYHA classifications
Eaton et al. [96]	WHI 42,170	Diabetes increased the risk factor for both HF subtypes amongst postmenopausal women. Greater BMI associated with the development of HFpEF but not HFrEF, especially amongst African-American women
Ho et al. [62]	CHS, FHS & PREVEND 28,820 (FHS 9496, CHS 5277, PREVEND 7369, MESA 6678) participants	BMI generally was a strong predictor of HF incidence with overlap between HFpEF and HFrEF diagnosis
Ingelsson et al. [99]	ULSAM 1187 participants	Abdominal and total adiposity as measured by waist circumference and BMI, respectively, both independently predicted incidence of HF as did insulin resistance
Pandey et al. [5]	JHS 2602 participants (some belonging to ARIC)	BMI among African-Americans is more strongly correlated to risk of HF than VAT levels as measured by CT-scan without deducing HF subtypes

(continued)

Table 10.1 (continued)

	Cohort	Findings
Savji et al. [10]	CHS, FHS, MESA & PREVEND 22,681 participants	BMI in addition to insulin resistance were independent indicators of HFpEF than HFrEF, with BMI more strongly associated with HFpEF than HFrEF in women than men
Tsujimoto and Kajio [100]	TOPCAT 3310 participants	Patients with HFpEF and abdominal obesity were at a higher risk of cardiovascular-related mortality than patients with HFpEF and no abdominal obesity among propensity-score matched ($n = 1058$) and total subject comparisons when grouped based on waist circumference but not when grouped based on BMI
Vardeny et al. [101]	ARIC 12,606 participants	Insulin resistance and higher BMI independently associated with an increased incidence of HF

Adapted from Savji et al. [10]. BMI: Body Mass Index; HF: Heart Failure; HFpEF: Heart Failure Preserved Ejection Fraction; HFrEF: Heart Failure Reduced Ejection Fraction; VAT: Visceral Adipose Tissue allows inflammatory and paracrine factors to traverse past the smooth muscle cells of the media and onto the endothelium [54, 89]. Kremen et al. were able to show that an expanded EAT in obese patients undergoing cardiac surgery was able to confer systemic IR through elevation of resistin and leptin [90]. Upon operation, MCP-1 levels elevated first in the EAT indicating recruitment of monocytes and other immune cells followed by elevations of TNF-α and IL-6. Resistin and leptin levels increased significantly enough in patients despite adiponectin levels remaining relatively stable [52, 90]. The elevation of leptin, resistin, IL-6, and TNF-α along with the hyperglycemia indicates that the EAT conferred some IR [90]. In addition, the elevated hypertrophy and fibrosis by both myocytes and fibroblasts demands increased energy from the EAT, furthering its expansion and eventual engulfment of the heart's epicardium (Fig. 10.3).

Meanwhile, confirmation of HFrEF is through the dilation of the LV, atrophy of myocytes and fibrosis [91]. Wherein there is less energy demanded by the myocardium leading to the lack of EAT expansion even with the comorbidity of obesity. As the current mechanisms of communication between the EAT and the myocardium have not been elucidated, as previously mentioned, we cannot discern this expansion of the original hypothesis with absolute certainty [38]. A seminal study exploring the relationship between EAT adipocyte and cardiomyocyte rates of apoptosis and proliferation in combination with endothelial dysfunction in NFC, HFpEF, and HFrEF would be required.

With reduced cardiovascular mortality risks, patients diagnosed with HF have a better prognosis if they are obese as compared to their lean counterparts, which is known as the "obesity paradox" [92, 93]. The obesity and HFpEF paradox is a U-shaped relationship where extremely obese, and those underweight both experience worse outcomes than patients who are overweight to mildly obese indicating that the underlying subtypes of HFpEF may differ in their pathology [94]. Differences in HF outcomes are primarily from variations in BMI measurements in addition to varied definitions of obesity based on ethnicity [5, 92, 95]. For example, African-American women and men seem to be more susceptible to HFpEF then their counterparts as measured by BMI (Table 10.1) [5, 96]. However, those in the PREVEND cohort primarily consisting of Dutch participants had BMI associated with higher risk of HFpEF and HFrEF (Table 10.1) [5, 96]. Different ethnicities and different cohorts of people vary in their susceptibility to different HF subtypes as does the relationship between the anthropometric measurement and HF subtypes. These variations have prompted the application of lower cut-off points to ascribe both overweightness and obesity in Asian populations that differ from both traditional and the WHO's guidelines (Fig. 10.1A) [95, 97]. The general trend, however, is that obesity, diabetes mellitus and IR all tend to correlate with the prognosis of HFpEF over HFrEF (Table 10.1).

Acknowledgements We acknowledge funding support from the Canadian Institutes of Health Research (CIHR) and Heart and Stroke Foundation to GYO. HZ is supported by a China Scholarship Council (CSC) Award.

References

1. Benjamin EJ, Virani SS, Callaway CW et al (2018) Heart disease and stroke statistics-2018 update: a report from the American Heart Association. Circulation 137:e67–e492
2. Heart Disease (2017) Center for Disease Control and Prevention, National Center for Health Statistics website. http://www.cdc.gov/nchs/fastats/heart-disease.htm
3. Jackson SL, Tong X, King RJ et al (2018) National burden of heart failure events in the United States, 2006 to 2014. Circul Heart Fail 11:e004873
4. Khan A, Raza S, Khan Y et al (2012) Current updates in the medical management of obesity. Recent Pat Endocr Metab Immune Drug Discov 6:117–128
5. Pandey A, Kondamudi N, Patel KV et al (2018) Association between regional adipose tissue distribution and risk of heart failure among blacks. Circul Heart Fail 11:e005629
6. Picard FA, Gueret P, Laissy JP et al (2014) Epicardial adipose tissue thickness correlates with the presence and severity of angiographic coronary artery disease in stable patients with chest pain. PLoS ONE 9:e110005
7. Park HE, Choi SY, Kim M (2014) Association of epicardial fat with left ventricular diastolic function in subjects with metabolic syndrome: assessment using 2-dimensional echocardiography. BMC Cardiovasc Disord 14:3
8. Morales-Portano JD, Peraza-Zaldivar JA, Suarez-Cuenca JA et al (2018) Echocardiographic measurements of epicardial adipose tissue and comparative ability to predict adverse cardiovascular outcomes in patients with coronary artery disease. Int J Cardiovasc Imaging 34:1429–1437
9. Nerlekar N, Muthalaly RG, Wong N et al (2018) Association of volumetric epicardial adipose tissue quantification and cardiac structure and function. J Am Heart Assoc 7:e009975
10. Savji N, Meijers WC, Bartz TM et al (2018) The association of obesity and cardiometabolic traits with incident HFpEF and HFrEF. JACC Heart Fail 6:701–709
11. Wang YC, Liang CS, Gopal DM et al (2015) Preclinical systolic and diastolic dysfunctions in metabolically healthy and unhealthy obese individuals. Circul Heart Fail 8:897–904
12. Russo C, Jin Z, Homma S et al (2011) Effect of obesity and overweight on left ventricular diastolic function: a community-based study in an elderly cohort. J Am Coll Cardiol 57:1368–1374
13. Lavie CJ, De Schutter A, Parto P et al (2016) Obesity and prevalence of cardiovascular diseases and prognosis-the obesity paradox updated. Prog Cardiovasc Dis 58:537–547
14. Vecchie A, Dallegri F, Carbone F et al (2018) Obesity phenotypes and their paradoxical association with cardiovascular diseases. Eur J Intern Med 48:6–17
15. Kortelainen ML (2002) Myocardial infarction and coronary pathology in severely obese people examined at autopsy. Int J Obes Relat Metab Disord J the Int Assoc Study Obes 26:73–79
16. Obesity: preventing and managing the global epidemic. Report of a WHO Consultation, World Health Organization technical report series 894:i–xii, 1–253 (2000)
17. the use and interpretation of anthropometry (1995) Physical status. Report of a WHO Expert Committee, World Health Organization technical report series 854:1–452
18. Abate N, Garg A, Peshock RM et al (1995) Relationships of generalized and regional adiposity to insulin sensitivity in men. J Clin Investig 96:88–98
19. Deschenes D, Couture P, Dupont P, Tchernof A (2003) Subdivision of the subcutaneous adipose tissue compartment and lipid-lipoprotein levels in women. Obes Res 11:469–476
20. Liberale L, Bonaventura A, Vecchie A et al (2017) The role of adipocytokines in coronary atherosclerosis. Curr Atheroscler Rep 19:10
21. Makki K, Froguel P, Wolowczuk I (2013) Adipose tissue in obesity-related inflammation and insulin resistance: cells, cytokines, and chemokines. ISRN Inflamm 2013:139239
22. Lee YH, Mottillo EP, Granneman JG (1842) Adipose tissue plasticity from WAT to BAT and in between. Biochem Biophys Acta 358–369:2014

23. Granneman JG, Li P, Zhu Z, Lu Y (2005) Metabolic and cellular plasticity in white adipose tissue I: effects of beta3-adrenergic receptor activation. Am J Physiol Endocrinol Metab 289:E608–616
24. Himms-Hagen J, Cui J, Danforth E Jr et al (1994) Effect of CL-316,243, a thermogenic beta 3-agonist, on energy balance and brown and white adipose tissues in rats. Am J Physiol 266:R1371–1382
25. Cousin B, Cinti S, Morroni M et al (1992) Occurrence of brown adipocytes in rat white adipose tissue: molecular and morphological characterization. J Cell Sci 103(Pt 4):931–942
26. Esteve Rafols M (2014) Adipose tissue: cell heterogeneity and functional diversity. Endocrinologia y nutricion: organo de la Sociedad Espanola de Endocrinologia y Nutricion 61:100–112
27. Turer AT, Hill JA, Elmquist JK, Scherer PE (2012) Adipose tissue biology and cardiomyopathy: translational implications. Circ Res 111:1565–1577
28. Carbone F, Mach F, Montecucco F (2015) The role of adipocytokines in atherogenesis and atheroprogression. Curr Drug Targets 16:295–320
29. Patel VB, Basu R, Oudit GY (2016) ACE2/Ang 1-7 axis: a critical regulator of epicardial adipose tissue inflammation and cardiac dysfunction in obesity. Adipocyte 5:306–311
30. Patel VB, Shah S, Verma S, Oudit GY (2017) Epicardial adipose tissue as a metabolic transducer: role in heart failure and coronary artery disease. Heart Fail Rev 22:889–902
31. Lopaschuk GD, Ussher JR, Folmes CD et al (2010) Myocardial fatty acid metabolism in health and disease. Physiol Rev 90:207–258
32. Ansaldo AM, Montecucco F, Sahebkar A et al (2018) Epicardial adipose tissue and cardiovascular diseases. Int J Cardiol
33. Schindler TH (2018) Epicardial adipose tissue: a new cardiovascular risk marker? Int J Cardiol
34. Iacobellis G (2009) Epicardial and pericardial fat: close, but very different. Obesity (Silver Spring) 17:625; author reply 626–627
35. Zwick RK, Guerrero-Juarez CF, Horsley V, Plikus MV (2018) Anatomical, physiological, and functional diversity of adipose tissue. Cell Metab 27:68–83
36. Talman AH, Psaltis PJ, Cameron JD et al (2014) Epicardial adipose tissue: far more than a fat depot. Cardiovasc Diagn Therapy 4:416–429
37. Rabkin SW (2007) Epicardial fat: properties, function and relationship to obesity. Obes Rev Off J Int Assoc Study Obes 8:253–261
38. Iacobellis G (2015) Local and systemic effects of the multifaceted epicardial adipose tissue depot. Nat Rev Endocrinol 11:363–371
39. Izgi C (2015) Epicardial adipose tissue: just a predictor or a local player for coronary atherosclerosis? Anatol J Cardiol 15:360–362
40. Schlett CL, Ferencik M, Kriegel MF et al (2012) Association of pericardial fat and coronary high-risk lesions as determined by cardiac CT. Atherosclerosis 222:129–134
41. Marchington JM, Mattacks CA, Pond CM (1989) Adipose tissue in the mammalian heart and pericardium: structure, foetal development and biochemical properties. Comp Biochem Physiol 94B:225–232
42. Underwood MJ, El Khoury G, Deronck D et al (2000) The aortic root: structure, function, and surgical reconstruction. Heart 83:376
43. Prati F, Arbustini E, Labellarte A et al (2003) Eccentric atherosclerotic plaques with positive remodeling have a pericardial distribution: a permissive role of epicardial fat—a three-dimensional intravascular ultrasound study of left anterior descending artery lesions. Eur Heart J 24:329–336
44. Fillmore N, Mori J, Lopaschuk GD (2014) Mitochondrial fatty acid oxidation alterations in heart failure, ischaemic heart disease and diabetic cardiomyopathy. Br J Pharmacol 171:2080–2090
45. Mazurek T, Zhang L, Zalewski A et al (2003) Human epicardial adipose tissue is a source of inflammatory mediators. Circulation 108:2460–2466
46. Bambace C, Telesca M, Zoico E et al (2011) Adiponectin gene expression and adipocyte diameter: a comparison between epicardial and subcutaneous adipose tissue in men. Cardiovasc Pathol: Off J Soc Cardiovasc Pathol 20:e153–156

47. Stanley W, Recchia F, Lopaschuk G (2005) Myocardial substrate metabolism in the normal and failing heart. Physiol Rev 85:1093–1129
48. Kim TT, Dyck JR (1861) The role of CD36 in the regulation of myocardial lipid metabolism. Biochem Biophys Acta 1450–1460:2016
49. Burgeiro A, Fuhrmann A, Cherian S et al (2016) Glucose uptake and lipid metabolism are impaired in epicardial adipose tissue from heart failure patients with or without diabetes. Am J Physiol-Endocrinol Metab 310:E550–E564
50. Ruderman NB, Carling D, Prentki M, Cacicedo JM (2013) AMPK, insulin resistance, and the metabolic syndrome. J Clin Investig 123:2764–2772
51. Vural B, Atalar F, Ciftci C et al (2008) Presence of fatty-acid-binding protein 4 expression in human epicardial adipose tissue in metabolic syndrome. Cardiovascr Pathol Off J Soc Cardiovasc Pathol 17:392–398
52. Wozniak SE, Gee LL, Wachtel MS, Frezza EE (2009) Adipose tissue: the new endocrine organ? A review article. Dig Dis Sci 54:1847–1856
53. Yudkin JS, Eringa E, Stehouwer CDA (2005) "Vasocrine" signalling from perivascular fat: a mechanism linking insulin resistance to vascular disease. Lancet 365:1817–1820
54. Patel VB, Mori J, McLean BA et al (2016) ACE2 deficiency worsens epicardial adipose tissue inflammation and cardiac dysfunction in response to diet-induced obesity. Diabetes 65:85–95
55. Wang W, McKinnie SM, Patel VB et al (2013) Loss of apelin exacerbates myocardial infarction adverse remodeling and ischemia-reperfusion injury: therapeutic potential of synthetic apelin analogues. J Am Heart Assoc 2:e000249
56. McKinnie SMK, Wang W, Fischer C et al (2017) Synthetic modification within the "RPRL" region of apelin peptides: impact on cardiovascular activity and stability to neprilysin and plasma degradation. J Med Chem 60:6408–6427
57. Zhong JC, Zhang ZZ, Wang W et al (1863) Targeting the apelin pathway as a novel therapeutic approach for cardiovascular diseases. Biochimica et biophysica acta Mol Basis Dis 1942–1950:2017
58. Deshmane SL, Kremlev S, Amini S, Sawaya BE (2009) Monocyte chemoattractant protein-1 (MCP-1): an overview. J Interferon Cytokine Res Off J Int Soc Interferon Cytokine Res 29:313–326
59. Hansson GK (2005) Inflammation, atherosclerosis, and coronary artery disease. N Engl J Med 1685–1695
60. Borlaug BA, Paulus WJ (2011) Heart failure with preserved ejection fraction: pathophysiology, diagnosis, and treatment. Eur Heart J 32:670–679
61. Solomon SD, Biering-Sorensen T (2017) LA strain when ejection fraction is preserved: a new measure of diastolic function? JACC Cardiovasc Imaging 10:744–746
62. Ho JE, Enserro D, Brouwers FP et al (2016) Predicting heart failure with preserved and reduced ejection fraction: the international collaboration on heart failure subtypes. Circul Heart Fail 9
63. Mori J, Patel V, Abo Alrob O, Basu R, Altamimi T, DesAulniers J, Wagg C, Kassiri Z, Lopaschuk G, Oudit G (2014) Angiotensin 1–7 ameliorates diabetic cardiomyopathy and diastolic dysfunction in db/db mice by reducing lipotoxicity and inflammation. Circul: Heart Fail 7:327–339
64. Sanderson JE (2014) HFNEF, HFpEF, HF-PEF, or DHF: what is in an acronym? JACC Heart Fail 2:93–94
65. Boal F, Timotin A, Roumegoux J et al (2016) Apelin-13 administration protects against ischaemia/reperfusion-mediated apoptosis through the FoxO1 pathway in high-fat diet-induced obesity. Br J Pharmacol 173:1850–1863
66. Martinez FO, Gordon S (2014) The M1 and M2 paradigm of macrophage activation: time for reassessment. F1000prime Rep 6:13
67. Wentworth JM, Naselli G, Brown WA et al (2010) Pro-inflammatory CD11c+CD206+ adipose tissue macrophages are associated with insulin resistance in human obesity. Diabetes 59:1648–1656

68. Nawaz A, Aminuddin A, Kado T et al (2017) CD206(+) M2-like macrophages regulate systemic glucose metabolism by inhibiting proliferation of adipocyte progenitors. Nat Commun 8:286
69. Ho D, Zhao X, Gao S et al (2011) Heart rate and electrocardiography monitoring in mice. Curr protoc Mouse Biol 1:123–139
70. Fox K, Borer JS, Camm AJ et al (2007) Resting heart rate in cardiovascular disease. J Am Coll Cardiol 50:823
71. Leonard WR (2015) Physiological adaptations to environmental stressors, pp 251–272
72. Valente A, Jamurtas AZ, Koutedakis Y, Flouris AD (2015) Molecular pathways linking non-shivering thermogenesis and obesity: focusing on brown adipose tissue development. Biol Rev Camb Philos Soc 90:77–88
73. Suzuki T, Katz R, Jenny NS et al (2008) Metabolic syndrome, inflammation, and incident heart failure in the elderly: the cardiovascular health study. Circul Heart Fail 1:242–248
74. Wang W, Bodiga S, Das SK et al (2012) Role of ACE2 in diastolic and systolic heart failure. Heart Fail Rev 17:683–691
75. Patel VB, Zhong JC, Grant MB, Oudit GY (2016) Role of the ACE2/angiotensin 1–7 axis of the renin-angiotensin system in heart failure. Circ Res 118:1313–1326
76. Bussard RL, Busse LW (2018) Angiotensin II: a new therapeutic option for vasodilatory shock. Ther Clin Risk Manag 14:1287–1298
77. Patel VB, Lezutekong JN, Chen X, Oudit GY (2017) Recombinant human ACE2 and the angiotensin 1–7 axis as potential new therapies for heart failure. Can J Cardiol 33:943–946
78. Patel VB, Takawale A, Ramprasath T et al (2015) Antagonism of angiotensin 1-7 prevents the therapeutic effects of recombinant human ACE2. J Mol Med (Berl) 93:1003–1013
79. Castan-Laurell I, Masri B, Valet P (2018) The apelin/APJ system as a therapeutic target in metabolic diseases. Expert Opin Ther Targets 1–11
80. O'Harte FPM, Parthsarathy V, Hogg C, Flatt PR (2018) Long-term treatment with acylated analogues of apelin-13 amide ameliorates diabetes and improves lipid profile of high-fat fed mice. PLoS ONE 13:e0202350
81. Yang H, Zhao L, Zhang J et al (2015) Effect of treadmill running on apelin and APJ expression in adipose tissue and skeletal muscle in rats fed a high-fat diet. Int J Sports Med 36:535–541
82. Kuba K, Zhang L, Imai Y et al (2007) Impaired heart contractility in apelin gene-deficient mice associated with aging and pressure overload. Circ Res 101:e32–42
83. Zhang ZZ, Wang W, Jin HY et al (2017) Apelin is a negative regulator of angiotensin II-mediated adverse myocardial remodeling and dysfunction. Hypertension (Dallas Tex.: 1979) 70:1165–1175
84. Farkasfalvi K, Stagg MA, Coppen SR et al (2007) Direct effects of apelin on cardiomyocyte contractility and electrophysiology. Biochem Biophys Res Commun 357:889–895
85. Oudit GY, Liu GC, Zhong J et al (2010) Human recombinant ACE2 reduces the progression of diabetic nephropathy. Diabetes 59:529–538
86. Iacobellis G, Ribaudo MC, Zappaterreno A et al (2004) Relation between epicardial adipose tissue and left ventricular mass. Am J Cardiol 94:1084–1087
87. Fracanzani AL, Pisano G, Consonni D et al (2016) Epicardial adipose tissue (EAT) thickness is associated with cardiovascular and liver damage in nonalcoholic fatty liver disease. PLoS ONE 11:e0162473
88. Corradi D, Maestri R, Callegari S et al (2004) The ventricular epicardial fat is related to the myocardial mass in normal, ischemic and hypertrophic hearts. Cardiovasc Pathol: Off J Socr Cardiovasc Pathol 13:313–316
89. Paulus WJ, Tschope C (2013) A novel paradigm for heart failure with preserved ejection fraction: comorbidities drive myocardial dysfunction and remodeling through coronary microvascular endothelial inflammation. J Am Coll Cardiol 62:263–271
90. Kremen J, Dolinkova M, Krajickova J et al (2006) Increased subcutaneous and epicardial adipose tissue production of proinflammatory cytokines in cardiac surgery patients: possible role in postoperative insulin resistance. J Clin Endocrinol Metab 91:4620–4627

91. Borlaug BA, Redfield MM (2011) Diastolic and systolic heart failure are distinct phenotypes within the heart failure spectrum. Circulation 123:2006–2013; discussion 2014
92. Carbone S, Lavie CJ, Arena R (2017) Obesity and heart failure: focus on the obesity paradox. Mayo Clin Proc 92:266–279
93. Kitzman DW, Shah SJ (2016) The HFpEF obesity phenotype: the elephant in the room. J Am Coll Cardiol 68:200–203
94. Dalos D, Mascherbauer J, Zotter-Tufaro C et al (2016) Functional status, pulmonary artery pressure, and clinical outcomes in heart failure with preserved ejection fraction. J Am Coll Cardiol 68:189–199
95. Hsu WC, Araneta MR, Kanaya AM et al (2015) BMI cut points to identify at-risk Asian Americans for type 2 diabetes screening. Diabetes Care 38:150–158
96. Eaton CB, Pettinger M, Rossouw J et al (2016) Risk factors for incident hospitalized heart failure with preserved versus reduced ejection fraction in a multiracial cohort of postmenopausal women. Circ Heart Fail 9
97. James WPT (2009) WHO recognition of the global obesity epidemic. Int J Obes 32:S120
98. Brouwers FP, de Boer RA, van der Harst P et al (2013) Incidence and epidemiology of new onset heart failure with preserved vs. reduced ejection fraction in a community-based cohort: 11-year follow-up of PREVEND. Eur Heart J 34:1424–1431
99. Ingelsson E, Sundström J, Ärnlöv J et al (2005) Insulin resistance and risk of congestive heart failure. JAMA 294:334–341
100. Tsujimoto T, Kajio H (2017) Abdominal obesity is associated with an increased risk of all-cause mortality in patients with HFpEF. J Am Coll Cardiol 70:2739–2749
101. Vardeny O, Gupta DK, Claggett B et al (2013) Insulin resistance and incident heart failure the ARIC study (Atherosclerosis Risk in Communities). JACC Heart Fail 1:531–536

Chapter 11
Epicardial Adipose Tissue in the Progression and Calcification of the Coronary Artery Disease

María Luna-Luna, Alejandro Zentella-Dehesa, and Óscar Pérez-Méndez

Abstract The relationship between obesity and coronary artery disease (CAD) may be mediated by epicardial adipose tissue (EAT). EAT volume correlates with abdominal visceral adipose tissue and as a consequence EAT is increased in patients with obesity. The presence of EAT adjacent to the coronary atherosclerotic lesions suggests a paracrine participation of this tissue in the progression and calcification of the atheroma. EAT expresses cardioprotective adipocytokines and anti-calcifying factors, such as adiponectin and osteoprotegerin among others, whose expression declines in the setting of a hypertrophy of the EAT and CAD. In contrast, pro-inflammatory and pro-calcifying molecules such as TNF-alpha, and osteopontin, as well as some microRNAs, are expressed in a higher amount in patients with CAD than in control subjects. Therefore, the quantification of the EAT emerges as a potential and useful determination for evaluating the CAD risk. However, the understanding of the complexity of the secretory pattern of EAT is still under investigation; the knowledge derived from future studies in this field will provide new potential pharmacological targets to prevent and treat the CAD.

Keywords Coronary artery disease · Coronary artery calcium score · Abdominal visceral adipose tissue · Epicardial adipose tissue · Osteopontin · Osteoprotegerin · TNF-α · Bone morphogenetic proteins · microRNAs · Mesenchymal stem cells

M. Luna-Luna · Ó. Pérez-Méndez (✉)
Department of Molecular Biology, Instituto Nacional de Cardiología, Mexico City, Mexico
e-mail: opmendez@yahoo.com

A. Zentella-Dehesa
Biochemistry Unit, Instituto Nacional de Ciencias Médicas y Nutrición Salvador Zubirán, Mexico City, Mexico

Department of Genomic Medicine and Environmental Toxicology, Instituto de Investigaciones Biomédicas, Universidad Nacional Auntónoma de México, Mexico City, Mexico

Ó. Pérez-Méndez
School of Engineering and Sciences, Tecnológico de Monterrey, Campus Mexico City, Mexico

P. S. Tappia et al. (eds.), *Biochemistry of Cardiovascular Dysfunction in Obesity*, Advances in Biochemistry in Health and Disease 20,
https://doi.org/10.1007/978-3-030-47336-5_11

Introduction

The adipose tissue is a type of lax connective tissue composed of cells such as adipocytes, mesenchymal, fibroblasts, macrophages, endothelial cells and nerves. In addition, the extracellular matrix of adipose tissue contains collagen I, III, V and VI, fibronectin, hyluronan and thrombospondin-1 [1–5].

Nowadays, adipose tissue has been classified in three functional categories: brown, white and beige. The brown adipose tissue is specialized in thermogenesis; white adipose tissue is devoted to the storage of triglycerides, while beige adipose tissue has both functions. The white adipose tissue can be classified in subcutaneous and visceral; the former is found under skin (dermis) and the latter is surrounding organs.

In this way, obesity is mainly related with the increase of the abdominal visceral adipose tissue (aVAT) and represents a coronary artery disease (CAD) risk [6]. Obesity is characterized by a systemic inflammation associated to an altered secretion of pro-inflammatory adipocytokines and the decrease of anti-inflammatory and anti-atherogenic adipocytokines by the aVAT. For many years, the hypertrophy of aVAT has been considered one of the main risk factors of atherosclerosis, the main etiology of CAD; however, there are eutrophic patients with high risk of CAD, whereas not all the patients with obesity develop CAD. These apparent paradox may be explained by the epicardial adipose tissue (EAT); even if aVAT volume positively correlates with EAT, there are subjects with normal or even low aVAT but elevated EAT volume who are at high risk of CAD. In fact, several studies have associated the EAT volume with CAD severity and coronary arterial calcium (CAC) score in both, obese and euthrophic subjects. Considering that the topology of CAD coincides with EAT, this particular adipose tissue could play a more important role in the etiology of CAD than the aVAT because of its contiguity to coronary arteries.

Here, we provide the information supporting the idea that EAT may participate as a paracrine organ, and as a source of proteins, mesenchymal cells and probably microRNAs, which as a whole, promote the calcification of atherosclerotic plaque. We first briefly describe the types of adipose tissue known and the relation between the aVAT and EAT volumes. We also mention some studies about the association of EAT with atherosclerosis and coronary artery calcification. Finally, we further focus on proteins, mesenchymal stem cell and microRNA that could be released from EAT and regulate the calcification of the atherosclerotic plaque.

Types of Adipose Tissue

As above mentioned, the adipose tissue has been classified as brown, beige and white. Brown adipose tissue (BAT) is abundant in perinatal period and a small quantity is maintained in adults mainly in the neck, paravertebral and supraclavicular regions (Fig. 11.1) [7]. This tissue has multilocular adipocytes containing a large amount of mitochondria in their cytoplasm. BAT is regulated by sympathetic nerves and its

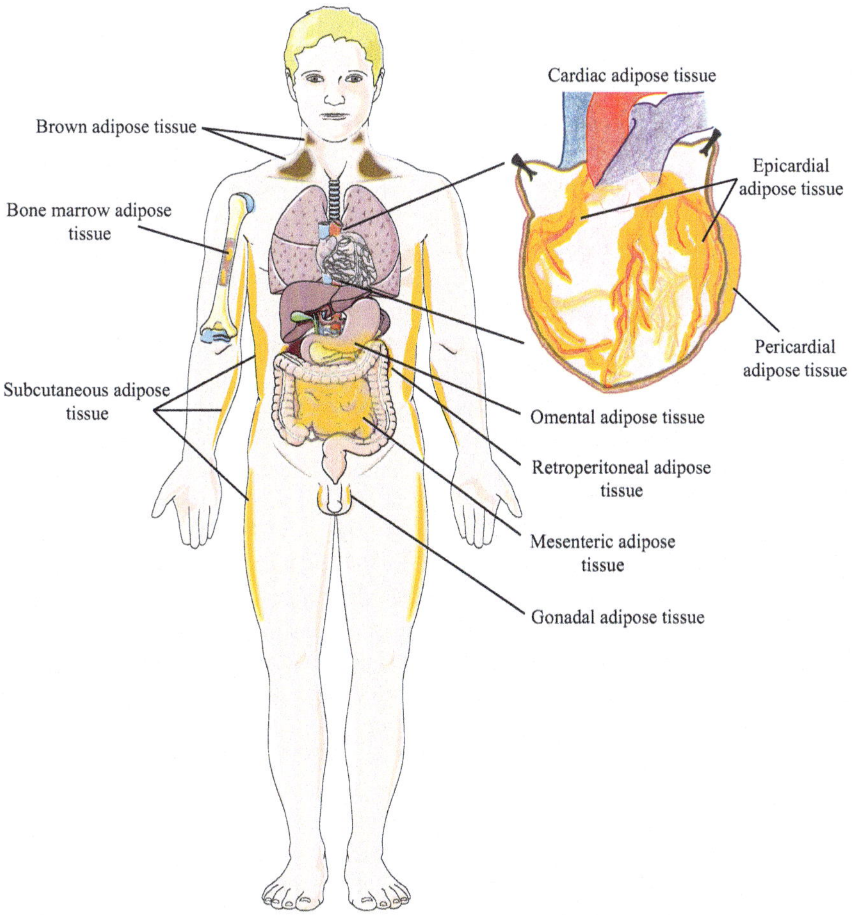

Fig. 11.1 Deposits of adipose tissue. The pericardial and epicaridal adipose tissues are two different fat deposits in the heart. The former is situated within the pericardium, whereas the latter is localized adjacent to the myocardium. In terms of obesity, the abdominal visceral adipose tissue has been widely associated with coronary artery disease (CAD), however, there are eutrophic patients with high risk of CAD. In this way, due to contiguity with coronary arteries, the epicardial adipose tissue could be a source of adipocytokines more important than abdominal visceral adipose tissue

main function is thermogenesis. This characteristic of BAT is mediated by several molecules which induce the expression and activity of UCP-1 (uncoupling protein-1) [8]. Briefly, UCP-1 is localized in the inner mitochondria membrane and uncouples the proton gradient of respiratory chain producing heat. The substrates for this process are the endogenous and exogenous fatty acids released in response to the sympathetic nervous system signaling via adrenergic β3 receptors.

In the past decades, it was believed that just the brown adipose tissue had this thermogenic capacity; however, the "browning" of the white adipose tissue has been of special interest for many research groups. In fact, beige adipose tissue is characterized for adipocytes morphologically similar to white adipocytes but, under certain stimulus, acquires the phenotype of brown adipocytes (multilocular, high mitochondrial content of UCP-1 and thermogenic activity) [9, 10]. The most common stimuli that induce browning are cold exposure, β3-adrenergic activators, thiazolidinediones, the presence of some peptides and hormones. Two theories have been proposed to explain the origin of beige adipose tissue; the first, it is de novo differentiation of precursor cells resident within of white adipose tissue. The second theory is the transdifferentiation of mature white adipocytes into brown-like cells (beige adipocytes) and vice versa, in response to the above-mentioned stimuli [11–13]. These adipocytes have been found mainly within subcutaneous adipose tissue [12] and their impact in the metabolism is being extensively studied.

Finally, the white adipose tissue is composed of unilocular adipocytes [14, 15] and it is subclassified in two types: subcutaneous and visceral. The subcutaneous adipose tissue (SAT) is mainly localized in femorogluteal regions, back and anterior abdominal wall and it represents over 80% of total body fat and its main function is storage of triglycerides [16, 17]. In contrast to SAT, visceral adipose tissue (VAT) has a ubiquitous location and it is divided in 4 subcategories; (1) intraperitoneal, also known as abdominal; comprises the omental and mesenteric adipose deposits, surrounding the stomach, and intestines, respectively; (2) retroperitoneal that surrounds the kidneys; (3) gonadal, covering uterus/ovaries or epididymis/testis; and (4) cardiac. Two different deposits constitute the cardiac adipose tissue: the pericardial situated within the pericardium, and epicardial that is localized adjacent to the myocardium (Fig. 11.1) [14, 18, 19].

The embryological origin of all types of adipose tissue is still controversial but it has been accepted that the BAT derives from paraxial mesoderm, VAT from lateral plate mesoderm and the SAT from lateral plate mesoderm as well as neural crest stem cells. (Fig. 11.2) [20–22]. In addition, it has been suggested that different VAT deposits could have, on their turn, different embryological origins. Therefore, each deposit of VAT could be considered as a mini-organ that fulfill independent functions [20].

In general, SAT is primarily involved in the storage of triglycerides, it is less metabolically active, and it contains smaller adipocytes than aVAT [16]. Jové et al. demonstrated that the expression of genes related with adipogenesis and formation of lipid droplets in SAT is higher than in aVAT [23]. These evidences are in agreement with the triglycerides storage capacity of the SAT.

When SAT capacity of triglycerides storage is saturated, the excess of fatty acids begins to be accumulated in ectopic deposits such as VAT, among others, inducing metabolic alterations [14]. In this way, several studies have associated the expansion of aVAT with an increased incidence of obesity, insulin resistance, type 2 diabetes mellitus and CAD [24–28].

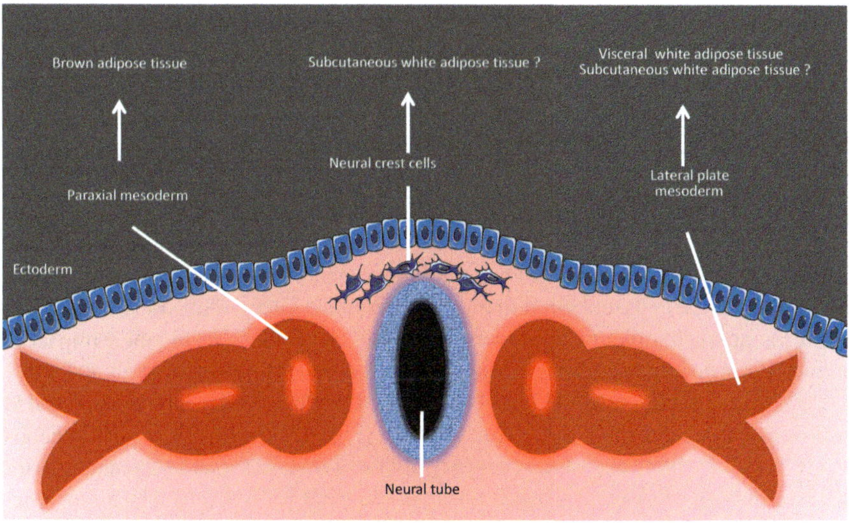

Fig. 11.2 Embryological origin of adipose tissues. Brown adipose tissue derives from paraxial mesoderm, whilst visceral adipose tissue from lateral plate of the mesoderm. The embryological origin of subcutaneous adipose tissue is still under debate, but it has been accepted that this adipose tissue derives from lateral mesoderm as well as from the stem cells belonging to the neural crest

Obesity and Epicardial Adipose Tissue

Epidemiological studies have demonstrated that obesity is a main cardiovascular risk factor [29]. Nowadays, it is widely accepted that the contribution of aVAT to CAD is more important than that of the subcutaneous adipose tissue. Obesity is traditionally defined by a BMI >30 kg/m^2, commonly associated with an excess of adiposity and characterized by structural and metabolic abnormalities of aVAT. On the setting of obesity, visceral adiposopathy is accompanied by an increase in the cellular size (hypertrophy), the production of pro-inflammatory adipocitokines such as TNF-α, MCP-1, IL-6, IL-8, PAI-1, chemerin, visfatin, resistin, lipocalin-2 and leptin. Concurrently, hypertrophic adipocytes secrete low levels of anti-atherogenic and anti-inflammamatory adipokines such as adiponectin, adrenomedulin and omentin-1 [14]. The chronic systemic inflammation, characteristic of obesity, has been one of the most important mechanisms that links obesity to CAD [14]. However, chronic inflammation in obesity does not explain why only certain arteries, not all, develop atherosclerosis, or why only some atherosclerotic plaques become calcified. The topology of atherosclerosis suggests a local participant that modifies the microenvironment of coronary arteries and promotes the progression and calcification of the atherosclerotic plaque. In this context, coronary atherosclerotic lesions are usually adjacent to epicardial adipose tissue (EAT) independently of the presence or absence of obesity [14, 30]. Therefore, EAT, but not aVAT, may be one of the most important local factors that promotes the progression and calcification of atherosclerotic plaque.

General Characteristics of Epicardial Adipose Tissue

Epicardial adipose tissue (EAT) has the same embryologic origin that omental and mesenteric adipose tissue; all of them derive from the lateral mesoderm, particularly from the splanchnopleuric mesoderm [18, 31–33]. EAT is localized between the myocardium and visceral pericardium [14, 18, 19] and it is mainly distributed around the coronary arteries, over the left ventricle apex, right ventricle lateral wall, and the atrioventricular and interventricular grooves (Fig. 11.1) [14, 18, 32–36]. EAT is vascularized by branches of the coronary arteries [18, 37, 38] and it is considered as the true deposit of visceral adipose tissue of the heart [35].

EAT is composed of adipocytes, smaller than subcutaneous and than other visceral adipose tissues [14, 39]. The small mean size of adipocytes in EAT has been attributed to a greater number of pre-adipocytes than that of other adipose tissues [32, 40]. EAT possesses a high capacity for storing and releasing FFA; therefore, this tissue might function as both, a buffer protecting the heart against the lipotoxicity of high levels of free fatty acids (FFA), and as a source of FFA for energy production, [14, 40, 41]. This dual behavior is important because the main source of energy in the heart is the β-oxidation of FFA [33, 40, 41]. For this reason, EAT is rich in saturated fatty acids such as myristic (14:0), palmitic (16:0) and stearic (18:0) acids whereas unsaturated fatty acids such as palmitoleic (16:1, Δ^{9}), oleic (18:1, Δ^{9}), linoleic (18:2, $\Delta^{9,12}$) and α-linolenic acids (18:3, $\Delta^{9,12,15}$) are scarce in EAT [14, 40].

Recently, the thermogenic potential of EAT has received increased attention; the expression of thermogenic genes such as PR domain containing-16 (PRDM-16), PPAR-γ coactivator 1-α (PGC1α), and UCP-1 are expressed at higher levels in EAT than in other deposits of lipids from patients with and without severe CAD [34, 36, 42]. These results strongly suggest that EAT could act similarly to brown adipose tissue, protecting the myocardium and coronary arteries against hypothermia.

EAT is a source of adipocytokines and its secretor profile may be modified in metabolic diseases (such as obesity and diabetes) becoming a tissue potentially harmful. Under such conditions, EAT favors inflammatory processes, insulin resistance, atherosclerosis progression and probably calcification of atheromatous plaques.

EAT and Coronary Artery Disease

Several studies have associated the EAT with coronary artery disease (CAD) [43–46]. Silagui et al. demonstrated that the extent of EAT was higher in patients with CAD than in controls, and that the extent of the tissue correlated with the coronary stenosis [43]. These results were further supported by a meta-analysis [47] showing that both, EAT thickness and volume, were higher in CAD patients compared with non-CAD subjects. In addition, patients with coronary plaque had an increased EAT volume compared with patients without plaque [47]. On the other hand, McKenney et al. demonstrated in a pig model of CAD that resection of EAT located in direct contiguity with coronary arteries decreased the evolution of atherosclerosis, thus supporting the role of EAT in the progression of the disease [48].

Inflammation is the major player in the onset and progression of the atherosclerotic plaque [14]. In this context, the analysis of transcriptome and secretome of EAT obtained from patients with CAD showed an adiposopathy characterized for an increased pro-inflammatory profile; several studies have demonstrated an increased expression of pro-inflammatory cytokines such as TNF-α, IL-6, MCP-1, IL-1β, PAI-1 in EAT from CAD patients [49–51].

It has been suggested that the lipopolysaccharides (LPS) may be recognized by TLR's on EAT inducing the expression of inflammatory cytokines by NF-κB pathway [52]. Thus, Baker et al. established that the systemic levels of LPS and the protein content of NF-κB, IKKβ, IKKγ and JNK1/2 in EAT from CAD patients is higher than that from non-CAD patients [53]. These data suggest that the innate immunity could have an important participation by stimulating EAT to a pro-inflammatory profile thus enhancing the local development of atherosclerosis in coronary arteries.

In addition to inflammatory mediators, the levels of expression of the cardioprotective adipocytokine, adiponectin, are lower, whereas two pro-atherogenic adipokines, leptin and visfatin are higher in EAT from patients with CAD compared with non-CAD subjects [50, 54, 55]. Furthermore, secretory type II phospholipase A2 (also known as sPLA2-II) was increased in EAT from CAD patients [56]; the presence of this enzyme suggests a contribution of EAT to the progression of CAD by retention of low-density lipoproteins (LDL) in the subendothelial space promoting the accumulation of lipids within atherosclerotic plaques.

In vitro studies have demonstrated that omentin-1 is able to promote an M2 macrophage phenotype and thus, this adipokine may limit the formation of foam cells [57]. In agreement with the potential pro-atherogenic role of EAT, omentin-1 is expressed at lower levels in EAT from CAD compared with non-CAD subjects and its expression is inversely associated with the presence of the disease [58].

The gene expression of PGC-1α and UCP-1 in EAT from patients with CAD and type 2 diabetes mellitus (DM2) is lower compared with both, CAD non-DM2 and non-CAD patients [59]. Moreover, the expression level of PGC-1α decreases with the number of injured coronary arteries. Also, a positive correlation between PGC-1α and UCP-1 in EAT suggests that PGC-1α could be a protective factor [59] for its role in the browning program [60].

All the previous evidences, support the active role of EAT in the etiology and progression of atherosclerosis, and open the possibility that EAT also contributes to other aspects of the disease, such as the plaque calcification.

EAT and Atherosclerotic Plaque Calcification

Coronary artery calcium (CAC) score is a well-known surrogate of atherosclerosis [61]. The calcification of atherosclerotic plaques might start in the second decade of life, concurrently with the formation of fatty streak, and represents up to 20% of plaque volume [62, 63]. Moreover, the deposits of calcium are frequent in older patients and in advanced lesions [64]. CAC scores may be useful in improving the algorithms used to evaluate the risk of CAD [62, 63]. However, the magnitude of CAC score is independent of the classical risk factors of atherosclerosis, such as dyslipidemia (excluding hypoalphalipoproteinemia), hypertension and obesity [64]. These observations strongly suggest that calcification of the atherosclerotic plaque and the etiology of the lipid laden of atheroma, are two concurrent aspects of atherosclerosis, but driven by different, and probably independent factors.

There is important evidence indicating that EAT plays an active role in the calcium deposits within the atheroma. Several studies have demonstrated that EAT volume measured by 64-multidetector computed tomography is higher in patients with coronary artery calcification [65–68]. Djaberi et al. found that EAT volume tends to increase with CAC score ($r = 0.33$, $p = 0.002$) and the average of EAT volume is higher in patients with a CAC score > 10 than in patients with CAC score ≤ 10 [66]. Data from the CAESAR study indicate that the prevalence of CAC > 0 increases with the EAT thickness and volume measured by echocardiography and multidetector computed tomography, respectively [67]. Recently, Iwasaki et al. demonstrated that the presence of CAC increases in patients with high EAT volume. This study included three groups of CAD patients classified according to tertiles of EAT volume; low-tertile (36–123 mL), mid-tertile (124–165 mL) and high-tertile (166–489 mL). The prevalence of CAC > 0 and CAC > 100 was higher in the high-tertile (83.9% and 59.2%, respectively), and CAC > 400 gradually increased as EAT volume augmented [68]. The CAC score increased 3.7% per each additional 10 mL of EAT volume [69] whereas in diabetic patients for same 10 mL of volume increase, the CAC score augmented by 13% and the probability of CAC progression went up to 12% [69].

The relationship between EAT and CAC is evident, however, the mechanisms responsible of calcium deposits in coronary arteries are still unknown. The actual evidence suggests that calcification is a process similar to the bone formation, which implicates the presence of several proteins, such as osteopontin, TNF-α, osteoprotegerin and BMPs that are produced by EAT, as well as microRNAs and different cell types; such pro-calcifying factors are detailed below.

Osteopontin

Several studies have demonstrated an increment of osteopontin (OPN) expression in arteries affected by atherosclerotic plaques, mainly associated with macrophages and foam cells [70, 71]. Studies in mice that over-expressed OPN and fed with atherogenic diet have demonstrated that this protein increases the atherosclerotic lesion size [72]. These results are consistent with other experiment using *OPN-/- APOE-/-* double knockout mice, in which the inflammation was attenuated and the extension of the atherosclerotic lesion was also lower compared with *APOE-/-* single knockout mice [73]. Besides its pro-inflammatory role, OPN has been associated with calcification of the atherosclerotic plaque; Hirota et al. demonstrated that the level of expression of *OPN* mRNA increases with the progression of atherosclerosis and it is higher in calcified atherosclerotic plaques. These observations suggest a pro-calcifying role of OPN [71]. This property of OPN seems to be dependent of its degree of phosphorylation: previous studies have demonstrated that recombinant phosphorylated OPN induced a lower degree of mineralization of human smooth muscle cells than the non-phosphorylated protein [74, 75].

It is important to emphasize that OPN may undergo different posttranscriptional and posttranslational modifications that may affect the functionality of this protein. Such modifications include alternative splicings, different protease cleavages, phosphorylation and glicosylation [76] (Fig. 11.3). In fact, OPN is released for many cellular types such as osteoblasts, macrophages, lymphocytes T and B and omental adipocytes [77–80]. However, the predominant isoform of OPN released by each tissue is unknown. We recently demonstrated that EAT from CAD patients is able to express levels of OPN mRNA higher than EAT from controls [81]. However, there was not a significant correlation between the expression of mRNA of *OPN* and the CAC score (unpublished data). The lack of correlation may be explained by two

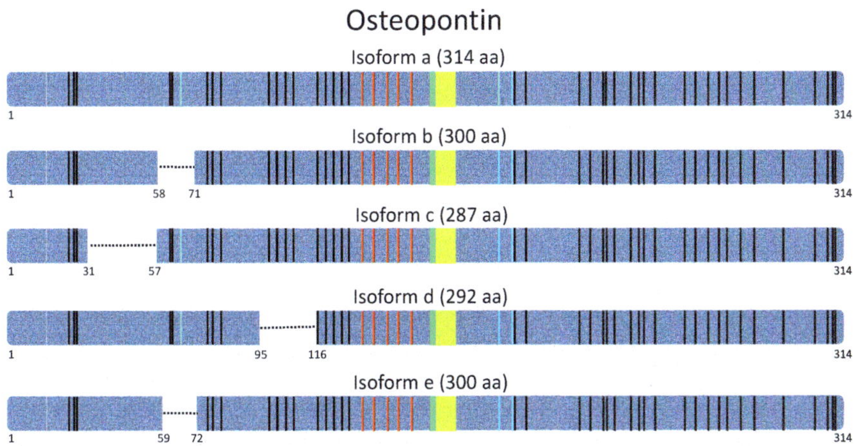

Fig. 11.3 Schematic representation of the different isoforms of osteopontin (OPN). OPN has several sites of phosphorylation (phosphoserine represented in black and phosphothreonine in blue) and glycosylation (indicated in red). In addition, OPN contains an RGD domain (green), which interacts with different integrins. The SVVYGLR domain (yellow) is a target for thrombin cleavage. The existence of the isoform "e" has not been experimentally confirmed

factors; the isoform of the secreted OPN, and the time of evolution of the calcification process that is widely variable among the patients. Nevertheless, to control the period of calcification, long-term follow-up studies are needed.

TNF-α

Besides the direct role of OPN on the artery calcification process, this protein is able to stimulate the secretion of TNF-α [82]. In terms of calcification, TNF-α promotes the artery mineralization by inducing the expression of pro-calcifying genes such as *ALP (alkaline phosphatase)* and *BMP-2* (*bone morphogenetic protein-2*) in vascular cells. BMP-2 released from endothelial cells could act on adjacent vascular smooth muscle cells (VSMC) inducing their transdifferentiation towards an osteoblastic genotype [83–85]. In addition, TNF-α is able to stimulate the mineralization of VSMC by inducing the expression of the transcriptional factor Runx2 [83]. In the same context, Lee et al. demonstrated that TNF-α in VSMC promoted the expression of Msx2, a transcription factor involved in the proliferation and differentiation of osteoblasts. The Msx2 effects are mediated by the NF-κB pathway, stimulating the expression of ALP [86].

TNF-α also inhibits the expression of the anti-calcifying gene, *MGP* (*matrix Gla protein*) [87]. In agreement with these observations, TNF-α induces the formation of calcified nodules in cultured myofibroblasts [88] as well as in human VSMC [84]. Interestingly, an apoptotic process was associated to the calcified nodules in human VSMC [89]. The inhibition of apoptosis in this model showed a reduction of the calcification process [89]. Therefore, VSCM-derived apoptotic bodies also accumulate calcium [89, 90] supporting the idea that the apoptosis precedes the calcification.

Osteoprotegerin

In terms of apoptosis, TNF-related apoptosis inducing ligand (TRAIL) secreted by T lymphocytes [91–93] may interact with DR5 (death receptor 5) on VSMC [93] promoting apoptosis and the consequent mineralization of the apoptotic bodies. In this context, osteoprotegerin (OPG) may play an anti-calcifying role by interacting with TRAIL and avoiding the previously mentioned apoptotic process [14].

Bennett et al. demonstrated that the size of the atheroma as well as the area of calcification of the lesions were higher in *APOE-/- OPG-/-* double knockout mice than in *APOE-/- OPG+/+* mice [94]. Moreover, in *LDLR-/-* mice treated with atherogenic diet, the administration of recombinant OPG reduced the calcification size of atherosclerotic plaque [95] supporting the anti-calcifying role of this protein.

OPG is expressed in healthy blood vessels and its expression decreases with the presence and progression of atherosclerosis whilst the expression of receptor activator of NF-κB ligand (RANKL) increases [96]. OPG is a decoy receptor that binds to RANKL thus preventing the interaction between RANK and RANKL [97]. The interaction RANKL-RANK activates NF-κB pathway and the transcription of pro-calcifying genes in VSMC, such as *BMP-4 (bone morphogenetic protein- 4)* and *ALP* [98, 99] which induce osteogenic differentiation of VSMC. In addition, RANKL is a pro-inflammatory molecule that stimulates the recruitment of monocytes to the subendothelial space; monocytes further differentiate into macrophages and secrete several pro-inflammatory cytokines that could act as pro-calcifying factors such as TNF-α and IL-6. These cytokines exacerbate the atherogenic process and promote the calcification [83] by upregulating pro-calcifying genes such as *Pit1*, *ALP*, *BMP-2* and *Runx2*, and by inhibiting anti-calcifying genes as *MGP* [83, 100]. Therefore, a deficiency of OPG promotes the calcification in atherosclerotic plaques.

BMPs

BMP-2 has been identified in calcified atherosclerotic plaques and it is considered as a potent osteogenic factor [101, 102]. BMP-2 is able to induce the osteogenic conversion of VSMC in vitro and in vivo [101–104]. The administration of LDN-193189, an inhibitor of BMP-2, prevented vascular calcification in *LDLR-/-* KO mice fed with high fat diet [105]. In accordance, *BMP-2* transgenic/*APOE-/-* KO mice showed an accelerated calcification compared with *APOE-/* KO mice [106]. BMP-2 up-regulates the expression of type III sodium-dependent Pi cotransporter-1 (Pit-1) increasing the uptake of phosphate and promoting the vascular calcification in VSMC [107]. In addition, BMP-2 inhibits the expression of the anti-calcifying protein, MGP [108]. MGP antagonizes BMP's (BMP-2 and BMP-4) by direct protein–protein interaction; it has been demonstrated that MGP over-expression reduced not only the BMPs activity but also the size, inflammation and calcification of atherosclerotic plaque [109]. Thus, the imbalance between these two proteins could promote an active calcifying process on the atherosclerotic plaque [108].

Concerning BMP-4, this protein promotes the progression of atherosclerosis by inhibiting the expression of ABCA1 and ABCG1 transporters [110]. BMP-4 expression is increased in calcified atherosclerotic plaques [111, 112]. In vitro, BMP-4 induces the calcification of VSMC [113] and the transdifferentiation of VSMC into osteoblastic lineage [114]. Data from our laboratory demonstrate that EAT from CAD patients expresses more *BMP-2* and *BMP-4* mRNA compared with control subjects (unpublished data). However, the role of BMP-4 is still unknown and warrants future studies to elucidate its mechanism of action.

Mesenchymal Stem Cells

Mesenchymal stem cells (MSC) may differentiate in several cellular types as chondrocytes, osteoblasts and adipocytes. The main sources of MSC are bone marrow, umbilical cordon and adipose tissue. It has been demonstrated that MSC from bone marrow contribute to calcification of atherosclerotic plaque [115]. These cells migrate from bone marrow to the lesion area in response to TGF-β1 released from the injured arteries [116]. Once the MSC have reached the lesion area, they differentiate to osteogenic linage, contributing to vascular calcification [115]. The role of MSC from other sources such as adipose tissue and its relationship with calcification of atherosclerotic plaque is still unknown. However, Chau et al. demonstrated that stromal vascular fraction (composed of several cell types, including MSC) from EAT is able to differentiate to osteoblasts [20], suggesting a possible role in the vascular calcification. Then, in the setting of CAD, MSC from EAT could be mobilized to coronary arteries; once in the arteries, the local microenvironment probably promotes its osteogenic differentiation. More studies are needed in this field.

microRNAs

The microRNAs regulate the activity of about 50% of the genes and participate in several processes such as cellular proliferation, differentiation, migration and apoptosis. MicroRNAs may regulate the arterial remodeling, angiogenesis and participate in the progression and calcification of the atherosclerotic plaque [117]. Below, we briefly mention some microRNAs and their potential role on atheroma calcification.

Xia et al. demonstrated that miR-3960 and miR-2861 induced the transdifferentiation of VSMC to osteogenic lineage by decreasing the protein level of its targets, Hoxa2 and HDAC5, respectively, and by increasing the expression of Runx2 [118]. Sudo et al. determined that miR-29 down-regulates the expression of elastin and stimulated the Ca^{2+} deposits in inorganic phosphate-treated VSMC [119]. In addition, the suppression of elastin promotes the osteoblastic transdifferentiation of VSMC, likely by activation of mTOR pathway [119]. Other pro-calcifying microRNA is miR-223; the over-expression of this microRNA induces the proliferation and migration of VSMC. Also, miR-223 reduces the production of Mef2c (myocyte enhancer factor 2c) and RhoB, both proteins involved in differentiation and the contractility of VSMC, promoting a secretory phenotype. Moreover, miRNA-223 stimulates the degradation of nuclear factor IA (NFIA), an inhibitor of calcification [120].

In terms of anti-calcifying microRNAs, the over-expression of miR-29b-3p in VSMC inhibits the expression of matrix metalloproteinase 2 (MMP2). MMP2 is characterized for stimulating calcification of VSMC in vitro and in vivo in different experimental models [121]. Accordingly, miR-29b-3p is down-regulated in vascular calcification [121]. Qiao et al. demonstrated that miR-205 inhibits the osteogenic differentiation in VSMC when interacts with its gene targets, Runx 2 and Smad

1. Moreover, this microRNA also regulates Runx2 in mesenchymal cells [122]. A decreased expression of miR-297a promotes the vascular calcification by fibroblast growth factor-23 (FGF-23) [123]. Osteogenic transdifferentiation of VSMC also was evaluated by Liao et al. and demonstrated that miR-133a inhibits this process for interacting with Runx2 [124].

miR-125b is another microRNA that inhibits the osteogenic transdifferentiation of VSMC and decreases the activity of ALP by interact with osteoblast transcription factor SP7 (osterix) [125]. Therefore, this miRNA may also contribute to the onset and progression of vascular calcification.

Additionally, miR-221 and miR-222 seem to participate in the intracellular balance of inorganic phosphate (Pi) and inorganic pyrophosphate (PPi) in VSMC under calcifying conditions [126]. Recent studies demonstrated that the transfection of VSMC with both microRNAs increases the *Enpp1* mRNA expression whilst decreases the expression of Pit-1 [126]. The resulting imbalance between Pi and PPi could promote the calcification of VSMC.

miR-30b and miR-30c both bind to Runx2 and trigger its down-regulation. Therefore, the activity of ALP is inhibited. In addition BMP-2 decreases the expression of miR-30b and miR-30c increasing the levels of Runx2 in VSMC and promoting osteoblastic transdifferentiation [127, 128]. Finally, the diminution of the expression of miR-204 increases the levels of Runx2 and ALP in VSMC and promotes the calcification of these cells [129].

All these studies demonstrate the importance of the miRNAs in the regulation of the osteoblastic transdifferentiation and calcification of VSMC. Recently, adipose tissue has been identified as a source of microRNAs via exosomal release [130]. In this way, it is possible that EAT may contribute also to the release of exosomes loaded with these microRNA.

Final Remarks

The relationship between obesity and cardiovascular risk is likely mediated, at least in part, by EAT. There are important evidences that point out to this adipose tissue as gland that secretes different pro-inflammatory molecules as well as pro- and anti-calcifying factors that contributes all together in a paracrine manner to the progression and calcification of the coronary atheroma.

Therefore, the quantification of the EAT emerges as a potential, more personalized and useful determination for evaluating the CAD risk. Unraveling the complexity of the secretory pattern of EAT in the setting of obesity is a task to be accomplished in the next years; these data will provide new potential, more specific pharmacological targets to fight against the atherosclerosis burden.

Acknowledgements This manuscript was supported by a CONACYT grant No. 233493. María de Jesús Luna-Luna is a doctoral student from "Programa de Doctorado en Ciencias Biomédicas de la Universidad Nacional Autónoma de México" and received fellowships from CONACYT, No. 408097 and EXP-AYTE-17345.

References

1. Esteve RM (2014) Adipose tissue: cell heterogeneity and functional diversity. Endocrinol Nutr 61:100–112
2. van Dam AD, Boon MR, Berbée JFP et al (2017) Targeting white, brown and perivascular adipose tissue in atherosclerosis development. Eur J Pharmacol 816:82–92
3. Alexopoulos N, Katritsis D, Raggi P (2014) Visceral adipose tissue as a source of inflammation and promoter of atherosclerosis. Atherosclerosis 233:104–112
4. Lin, Chun TH, Kang L (2016) Adipose extracellular matrix remodelling in obesity and insulin resistance. Biochem Pharmacol 119:8–16
5. Luo T, Nocon A, Fry J et al (2016) AMPK activation by metformin suppresses abnormal extracellular matrix remodeling in adipose tissue and ameliorates insulin resistance in obesity. Diabetes 65:2295–2310
6. Fox CS, Massaro JM, Hoffmann U et al (2007) Abdominal visceral and subcutaneous adipose tissue compartments: association with metabolic risk factors in the Framingham Heart Study. Circulation 116:39–48
7. Boon MR, van Marken Lichtenbelt WD (2016) Brown adipose tissue: a human perspective. Handb Exp Pharmacol 233:301–319
8. Song NJ, Chang SH, Li DY et al (2017) Induction of thermogenic adipocytes: molecular targets and thermogenic small molecules. Exp Mol Med 49:e353
9. Saito M (2014) Human brown adipose tissue: regulation and anti- obesity potential. Endocr J 61:409–416
10. Lidel ME, Betz MJ, Enerbäck S (2014) Brown adipose tissue and its therapeutic potential. J Intern Med 276:364–377
11. Cui XB, Chen SY (2016) White adipose tissue browning and obesity. J Biomed Res 31:1–2
12. Kiefer FW (2016) Browning and thermogenic programing of adipose tissue. Best Pract Res Clin Endocrinol Metab 30:479–485
13. Jeanson Y, Carrière A, Casteilla L (2015) A new role for browning as a redox and stress adaptive mechanism? Front Endocrinol (Lausanne) 6:158
14. Luna-Luna M, Medina-Urrutia A, Vargas-Alarcón G et al (2015) Adipose tissue in metabolic syndrome: onset and progression of atherosclerosis. Arch Med Res 46:392–407
15. Trayhurn P (2013) Hypoxia and adipose tissue function and dysfunction in obesity. Physiol Rev 93:1–21
16. Yamamoto A, Kikuchi Y, Kusakabe T et al (2020) Imaging spectrum of abnormal subcutaneous and visceral fat distribution. Insights Imaging 11:24
17. Passaro A, Miselli MA, Sanz JM et al (2017) Gene expression regional differences in human subcutaneous adipose tissue. BMC Genomics 18:202
18. Iacobellis G (2009) Epicardial and pericardial fat: close, but very different. Obesity (Silver Spring) 17:625
19. Chhabra L, Gurukripa KN (2015) Cardiac adipose tissue: distinction between epicardial and pericardial fat remains important! Int J Cardiol 201:274–275
20. Chau YY, Bandiera R, Serrels A et al (2014) Visceral and subcutaneous fat have different origins and evidence supports a mesothelial source. Nat Cell Biol 16:367–375
21. Gupta OT, Gupta RK (2015) Visceral adipose tissue mesothelial cells: living on the edge or just taking up space? Trends Endocrinol Metab 26:515–523
22. Sebo ZL, Jeffery E, Holtrup B, Rodeheffer MS (2018) A mesodermal fate map for adipose tissue. Development 145:dev166801

23. Jové M, Moreno-Navarrete JM, Pamplona R et al (2014) Human omental and subcutaneous adipose tissue exhibit specific lipidomic signatures. FASEB J 28:1071–1081
24. Fuster JJ, Ouchi N, Gokce N, Walsh K (2016) Obesity-induced changes in adipose tissue microenvironment and their impact on cardiovascular disease. Circ Res 118:1786–1807
25. Zhang C, Rexrode KM, van Dam RM et al (2008) Abdominal obesity and the risk of all-cause, cardiovascular, and cancer mortality: sixteen years of follow-up in US women. Circulation 117:1658–1667
26. Pinho CPS, Diniz ADS, Arruda IKG et al (2018) Waist circumference measurement sites and their association with visceral and subcutaneous fat and cardiometabolic abnormalities. Arch Endocrinol Metab 62:416–423
27. Sun Q, Townsend MK, Okereke OI et al (2009) Adiposity and weight change in mid-life in relation to healthy survival after age 70 in women: prospective cohort study. BMJ 339:b3796
28. Rallidis LS, Baroutsi K, Zolindaki M et al (2014) Visceral adipose tissue is a better predictor of subclinical carotid atherosclerosis compared with waist circumference. Ultrasound Med Biol 40:1083–1088
29. Prospective Studies Collaboration, Whitlock G, Lewington S et al (2009) Body-mass index and cause-specific mortality in 900 000 adults: collaborative analyses of 57 prospective studies. Lancet 373:1083–1096
30. Wu FZ, Wu CC, Kuo PL, Wu MT (2016) Differential impacts of cardiac and abdominal ectopic fat deposits on cardiometabolic risk stratification. BMC Cardiovasc Disord 16:20
31. Chistiakov DA, Grechko AV, Myasoedova VA et al (2017) Impact of the cardiovascular system-associated adipose tissue on atherosclerotic pathology. Atherosclerosis 263:361–368
32. Patel VB, Shah S, Verma S, Oudit GY (2017) Epicardial adipose tissue as a metabolic transducer: role in heart failure and coronary disease. Heart Fail Rev 22:889–902
33. Iacobellis G (2015) Local and system effects of the multifaceted epicardial adipose tissue depot. Nat Rev Endocrinol 11:363–371
34. González N, Moreno-Villegas Z, González-Bris A et al (2017) Regulation of visceral and epicardial adipose tissue for preventing cardiovascular injuries associated to obesity and diabetes. Cardiovasc Diabetol 16:44
35. Gaborit B, Sengenes C, Ancel P et al (2017) Role of epicardial adipose tissue in health and disease: a matter of fat? Compr Physiol 7:1051–1082
36. Sacks HS, Fain JN (2011) Human epicardial fat: what is new and what is missing? Clin Exp Pharmacol Physiol 38:879–887
37. Iacobellis G, Assael F, Ribaudo MC et al (2003) Epicardial fat from echocardiography a new method for visceral adipose tissue prediction. Obes Res 11:304–310
38. Sacks HS, Fain JN (2007) Human epicardial adipose tissue: a review. Am Heart J 153:907–917
39. Bambace C, Telesca M, Zoico E et al (2011) Adiponectin gene expression and adipocyte diameter: a comparison between epicardial and subcutaneous adipose tissue in men. Cardiovasc Pathol 20:e153–e156
40. Iacobellis G, Bianco AC (2011) Epicardial adipose tissue: emerging physiological, pathophysiological and clinical features. Trends Endocrinol Metab 22:450–457
41. Cherian S, Lopaschuk GD, Carvalho E (2012) Cellular cross-talk between epicardial adipose tissue and myocardium in relation to the pathogenesis of cardiovascular disease. Am J Physiol Endocrinol Metab 303:E937–E949
42. Sacks HS, Fain JN, Holman B et al (2009) Uncoupling protein-1 and related messenger ribonucleic acids in human epicardial and other adipose tissues: epicardial fat functioning as brown fat. J Clin Endocrinol Metab 94:3611–3615
43. Silaghi A, Piercecchi-Marti MD, Grino M et al (2008) Epicardial adipose tissue extent: relationship with age, body fat distribution, and coronaropathy. Obesity (Silver Spring) 16:2424–2430
44. de Feyter PJ (2011) Epicardial adipose tissue: an emerging role for the development of coronary atherosclerosis. Clin Cardiol 34:143–144
45. de Vos AM, Prokop M, Roos CJ et al (2008) Peri-coronary epicardial adipose tissue is related to cardiovascular risk factors and coronary artery calcification in post-menopausal women. Eur Heart J 29:777–783

46. Ueno K, Anzai T, Jinzaki M et al (2009) Increased epicardial fat volume quantified by 64-multidetector computed tomography is associated with coronary atherosclerosis and totally occlusive lesions. Circ J 73:1927–1933
47. Xu Y, Cheng X, Hong K et al (2012) How to interpret epicardial adipose tissue as a cause of coronary artery disease: a meta-analysis. Coron Artery Dis 23:227–233
48. McKenney ML, Schultz KA, Boyd JH, Byrd JP et al (2014) Epicardial adipose excision slows the progression of porcine coronary atherosclerosis. J Cardiothorac Surg 9:2
49. Iacobellis G, Barbaro G (2008) The double role of epicardial adipose tissue as pro- and anti-inflammatory organ. Horm Metab Res 40:442–445
50. Cheng KH, Chu CS, Lee KT et al (2008) Adipocytokines and proinflammatory mediators from abdominal and epicardial adipose tissue in patients with coronary artery disease. Int J Obes 32:268–274
51. Hirata Y, Kurobe H, Akaike M et al (2011) Enhanced inflammation in epicardial fat in patients with coronary artery disease. Int Heart J 52:139–142
52. Creely SJ, McTernan PG, Kusminski CM et al (2007) Lipopolysaccharide activates an innate immune system response in human adipose tissue in obesity and type 2 diabetes. Am J Physiol Endocrinol Metab. 292:E740–E747
53. Baker AR, Harte AL, Howell N et al (2009) Epicardial adipose tissue as a source of nuclear factor kappa B and c-Jun N-terminal kinase mediated inflammation in patients with coronary artery disease. J Clin Endocrinol Metab 94:261–267
54. Cesari M, Pessina AC, Zanchetta M et al (2006) Low plasma adiponectin is associated with coronary artery disease but not with hypertension in high-risk nondiabetic patients. J Intern Med 260:474–483
55. Iacobellis G, Pistilli D, Gucciardo M et al (2005) Adiponectin expression in human epicardial adipose tissue in vivo is lower in patients with CAD. Cytokine 29:251–255
56. Dutour A, Achard V, Sell H et al (2010) Secretory type II phospholipase A2 is produced and secreted by epicardial adipose tissue and over expressed in patients with CAD. J Clin Endocrinol Metab 95:963–967
57. Watanabe K, Watanabe R, Konii H et al (2016) Counteractive effects of omentin-1 against atherogenesis†. Cardiovasc Res 110:118–128
58. Du Y, Ji Q, Cai L et al (2016) Association between omentin-1 expression in human epicardial adipose tissue and coronary atherosclerosis. Cardiovasc Diabetol 15:90
59. Moreno-Santos I, Pérez-Belmonte LM, Macías-González M et al (2016) Type 2 diabetes is associated with decreased PGC1α expression in epicardial adipose tissue of patients with coronary artery disease. J Transl Med 14:243
60. Uldry M, Yang W, St-Pierre J et al (2006) Complementary action of the PGC1 coactivators in mitochondrial biogenesis and brown fat differentiation. Cell Metab 3:333–341
61. Hisamatsu T, Fujiyoshi A, Miura K (2019) Coronary artery calcium: its clinical utility in primary prevention. Clin Calcium 29:215–223
62. Erbel R, Möhlenkamp S, Moebus S et al (2010) Coronary risk stratification, discrimination, and reclassification improvement based on quantification of subclinical coronary atherosclerosis: the Heinz Nixdorf Recall study. J Am Coll Cardiol 56:1397–1406
63. Erbel R, Lehmann N, Churzidse S et al (2013) Gender-specific association of coronary artery calcium and lipoprotein parameters: the Heinz Nixdorf Recall Study. Atherosclerosis 229:531–540
64. García-Sánchez C, Posadas-Romero C, Posadas-Sánchez R et al (2015) Low concentrations of phospholipids and plasma HDL cholesterol subclasses in asymptomatic subjects with high coronary calcium scores. Atherosclerosis 238:250–255
65. Iwasaki K, Matsumoto T, Aono H et al (2011) Relationship between epicardial fat measured by 64-multidetector computed tomography and coronary artery disease. Clin Cardiol 34:166–171
66. Djaberi R, Schuijf JD, van Werkhoven JM et al (2008) Relation of epicardial adipose tissue to coronary atherosclerosis. Am J Cardiol 102:1602–1607
67. Kim BJ, Kang JG, Lee SH et al (2017) Relationship of echocardiographic epicardial fat thickness and epicardial fat volume by computed tomography with coronary artery calcification: data from the CAESAR Study. Arch Med Res. 48:352–359

68. Iwasaki K, Urabe N, Kitagawa A, Nagao T (2018) The association of epicardial fat volume with coronary characteristics and clinical outcome. Int J Cardiovasc Imaging 34:301–309
69. Yerramasu A, Dey D, Venuraju S et al (2012) Increased volume of epicardial fat is an independent risk factor for accelerated progression of sub-clinical coronary atherosclerosis. Atherosclerosis 220:223–230
70. Isoda K, Nishikawa K, Kamezawa Y et al (2002) Osteopontin plays an important role in the development of medial thickening and neointimal formation. Circ Res 91:77–782
71. Hirota S, Imakita M, Kohri K et al (1993) Expression of osteopontin messenger RNA by macrophages in atherosclerotic plaques. A possible association with calcification. Am J Pathol 143:1003–1008
72. Chiba S, Okamoto H, Kon S et al (2002) Development of atherosclerosis in osteopontin transgenic mice. Heart Vessels 16:111–117
73. Matsui Y, Rittling SR, Okamoto H et al (2003) Osteopontin deficiency attenuates atherosclerosis in female apolipoprotein E-deficient mice. Arterioscler Thromb Vasc Biol 23:1029–1034
74. Jono S, Peinado C, Giachelli CM (2000) Phosphorylation of osteopontin is required for inhibition of vascular smooth muscle cell calcification. J Biol Chem 275:20197–20203
75. Jiménez- Corona AE, Pérez-Torres A, Mas-Oliva J, Moreno A (2008) Effect of osteopontin, chondroitin sulfates (B, C), and human serum albumin in the crystalization behavior of hydroxiapatite in agarose and silica hydrogels. Cryst Growth Des 8:1335–1339
76. Wolak T (2014) Osteopontin—a multi-modal marker and mediator in atherosclerotic vascular disease. Atherosclerosis 236:327–337
77. Liaw L, Lindner V, Schwartz SM et al (1995) Osteopontin and beta 3 integrin are coordinately expressed in regenerating endothelium in vivo and stimulate Arg-Gly-Asp-dependent endothelial migration in vitro. Circ Res 77:665–672
78. Miyazaki Y, Setoguchi M, Yoshida S et al (1990) The mouse osteopontin gene. Expression in monocytic lineages and complete nucleotide sequence. J Biol Chem 265:14432–14438
79. Giachelli CM, Bae N, Almeida M et al (1993) Osteopontin is elevated during neointima formation in rat arteries and is a novel component of human atherosclerotic plaques. J Clin Invest 92:1686–1696
80. O'Brien ER, Garvin MR, Stewart DK et al (1994) Osteopontin is synthesized by macrophage, smooth muscle, and endothelial cells in primary and restenotic human coronary atherosclerotic plaques. Arterioscler Thromb 14:1648–1656
81. Luna-Luna M, Cruz-Robles D, Ávila-Vanzzini N et al (2017) Differential expression of osteopontin, and osteoprotegerin mRNA in epicardial adipose tissue between patients with severe coronary artery disease and aortic valvular stenosis: association with HDL subclasses. Lipids Health Dis 16:156
82. Zeyda M, Gollinger K, Todoric J et al (2011) Osteopontin is an activator of human adipose tissue macrophages and directly affects adipocyte function. Endocrinology 152:2219–2227
83. Deuell KA, Callegari A, Giachelli CM et al (2012) RANKL enhances macrophage paracrine pro-calcific activity in high phosphate-treated smooth muscle cells: dependence on IL-6 and TNF-alpha. J Vasc Res 49:510–521
84. Tintut Y, Patel J, Parhami F, Demer LL (2000) Tumor necrosis factor-alpha promotes in vitro calcification of vascular cells via the cAMP pathway. Circulation 102:2636–2642
85. Csiszar A, Smith KE, Koller A et al (2005) Regulation of bone morphogenetic protein-2 expression in endothelial cells: role of nuclear factor-kappaB activation by tumor necrosis factor-alpha, H2O2, and high intravascular pressure. Circulation 111:2364–2372
86. Lee HL, Woo KM, Ryoo HM, Baek JH (2010) Tumor necrosis factor-alpha increases alkaline phosphatase expression in vascular smooth muscle cells via MSX2 induction. Biochem Biophys Res Commun 391:1087–1092
87. Csiszar A, Ahmad M, Smith KE et al (2006) Bone morphogenetic protein-2 induces proinflammatory endothelial phenotype. Am J Pathol 168:629–638
88. Kaden JJ, Kiliç R, Sarikoç A et al (2005) Tumor necrosis factor alpha promotes an osteoblast-like phenotype in human aortic valve myofibroblasts: a potential regulatory mechanism of valvular calcification. Int J Mol Med 16:869–872

89. Proudfoot D, Skepper JN, Hegyi L et al (2000) Apoptosis regulates human vascular calcification in vitro: evidence for initiation of vascular calcification by apoptotic bodies. Circ Res 87:1055–1062
90. Proudfoot D, Skepper JN, Hegyi L et al (2001) The role of apoptosis in the initiation of vascular calcification. Z Kardiol 90(Suppl 3):43–46
91. Emery JG, McDonnell P, Burke MB et al (1998) Osteoprotegerin is a receptor for the cytotoxic ligand TRAIL. J Biol Chem 273:14363–14367
92. Gochuico BR, Zhang J, Ma BY et al (2000) TRAIL expression in vascular smooth muscle. Am J Physiol Lung Cell Mol Physiol 278:L1045–L1050
93. Sato K, Niessner A, Kopecky SL et al (2006) TRAIL-expressing T cells induce apoptosis of vascular smooth muscle cells in the atherosclerotic plaque. J Exp Med 203:239–250
94. Bennett BJ, Scatena M, Kirk EA et al (2006) Osteoprotegerin inactivation accelerates advanced atherosclerotic lesion progression and calcification in older ApoE-/- mice. Arterioscler Thromb Vasc Biol 26:2117–2124
95. Morony S, Tintut Y, Zhang Z, Cattley RC, Van G, Dwyer D et al (2008) Osteoprotegerin inhibits vascular calcification without affecting atherosclerosis in ldlr(-/-) mice. Circulation 117(3):411–420
96. Min H, Morony S, Sarosi I et al (2000) Osteoprotegerin reverses osteoporosis by inhibiting endosteal osteoclasts and prevents vascular calcification by blocking a process resembling osteoclastogenesis. J Exp Med 192:463–474
97. Schoppet M, Preissner KT, Hofbauer LC (2002) RANK ligand and osteoprotegerin: paracrine regulators of bone metabolism and vascular function. Arterioscler Thromb Vasc Biol 22:549–553
98. Kaden JJ, Bickelhaupt S, Grobholz R et al (2004) Receptor activator of nuclear factor kappaB ligand and osteoprotegerin regulate aortic valve calcification. J Mol Cell Cardiol 36:57–66
99. Kanegae Y, Tavares AT, Izpisúa Belmonte JC, Verma IM (1998) Role of Rel/NF-kappaB transcription factors during the outgrowth of the vertebrate limb. Nature 392:611–614
100. Sun M, Chang Q, Xin M et al (2017) Endegenous bone morphogenetic protein 2 plays a role in vascular smooth musccle cell calcification induced by interleukin 6 in vitro. Int J Immunopathol Pharmacol 30:227–237
101. Hruska KA, Mathew S, Saab G (2005) Bone morphogenetic proteins in vascular calcification. Circ Res 97:105–114
102. Shimizu T, Tanaka T, Iso T et al (2011) Notch signaling pathway enhances bone morphogenetic protein 2 (BMP2) responsiveness of Msx2 gene to induce osteogenic differentiation and mineralization of vascular smooth muscle cells. J Biol Chem 286:19138–19148
103. Cheng SL, Shao JS, Charlton-Kachigian N et al (2003) MSX2 promotes osteogenesis and suppresses adipogenic differentiation of multipotent mesenchymal progenitors. J Biol Chem 278:45969–45977
104. Shao JS, Cai J, Towler DA (2006) Molecular mechanisms of vascular calcification: lessons learned from the aorta. Arterioscler Thromb Vasc Biol 26:1423–1430
105. Derwall M, Malhotra R, Lai CS et al (2012) Inhibition of bone morphogenetic protein signaling reduces vascular calcification and atherosclerosis. Arterioscler Thromb Vasc Biol 32:613–622
106. Nakagawa Y, Ikeda K, Akakabe Y et al (2010) Paracrine osteogenic signals via bone morphogenetic protein-2 accelerate the atherosclerotic intimal calcification in vivo. Arterioscler Thromb Vasc Biol 30:1908–1915
107. Li X, Yang HY, Giachelli CM (2008) BMP-2 promotes phosphate uptake, phenotypic modulation, and calcification of human vascular smooth muscle cells. Atherosclerosis 199:271–277
108. Chiyoya M, Seya K, Yu Z et al (2018) Matrix Gla protein negatively regulates calcification of human aortic valve interstitial cells isolated from calcified aortic valves. J Pharmacol Sci 136:257–265
109. Yao Y, Bennett BJ, Wang X et al (2010) Inhibition of bone morphogenetic proteins protects against atherosclerosis and vascular calcification. Circ Res 107:485–494

110. Feng J, Gao J, Li Y et al (2014) BMP4 enhances foam cell formation by BMPR-2/Smad1/5/8 signaling. Int J Mol Sci 15:5536–5552
111. Dhore CR, Cleutjens JP, Lutgens E et al (2001) Differential expression of bone matrix regulatory proteins in human atherosclerotic plaques. Arterioscler Thromb Vasc Biol 21:1998–2003
112. Panizo S, Cardus A, Encinas M et al (2009) RANKL increases vascular smooth muscle cell calcification through a RANK-BMP4-dependent pathway. Circ Res 104:1041–1048
113. Mikhaylova L, Malmquist J, Nurminskaya M (2007) Regulation of in vitro vascular calcification by BMP4, VEGF and Wnt3a. Calcif Tissue Int 81:372–381
114. Hayashi K, Nakamura S, Nishida W, Sobue K (2006) Bone morphogenetic protein-induced MSX1 and MSX2 inhibit myocardin-dependent smooth muscle gene transcription. Mol Cell Biol 26:9456–9470
115. Wang W, Li C, Pang L et al (2014) Mesenchymal stem cells recruited by active TGFβ contribute to osteogenic vascular calcification. Stem Cells Dev 23:1392–1404
116. Wan M, Li C, Zhen G et al (2012) Injury-activated transforming growth factor β controls mobilization of mesenchymal stem cells for tissue remodeling. Stem Cells 30:2498–2511
117. Jamaluddin MS, Weakley SM, Zhang L et al (2011) miRNAs: roles and clinical applications in vascular disease. Expert Rev Mol Diagn 11:79–89
118. Xia ZY, Hu Y, Xie PL et al (2015) Runx2/miR-3960/miR-2861 positive feedback loop is responsible for osteogenic transdifferentiation of vascular smooth muscle cells. Biomed Res Int 2015:624037
119. Sudo R, Sato F, Azechi T, Wachi H (2015) MiR-29-mediated elastin down-regulation contributes to inorganic phosphorus-induced osteoblastic differentiation in vascular smooth muscle cells. Genes Cells 20:1077–1087
120. Rangrez AY, M'Baya-Moutoula E, Metzinger-Le Meuth V et al (2012) Inorganic phosphate accelerates the migration of vascular smooth muscle cells: evidence for the involvement of miR-223. PLoS ONE 7:e47807
121. Jiang W, Zhang Z, Yang H et al (2017) The involvement of miR-29b-3p in arterial calcification by targeting matrix metalloproteinase-2. Biomed Res Int 2017:6713606
122. Qiao W, Chen L, Zhang M (2014) MicroRNA-205 regulates the calcification and osteoblastic differentiation of vascular smooth muscle cells. Cell Physiol Biochem 33:1945–1953
123. Zheng S, Zhang S, Song Y et al (2016) MicroRNA-297a regulates vascular calcification by targeting fibroblast growth factor 23. Iran J Basic Med Sci 19:1331–1336
124. Liao XB, Zhang ZY, Yuan K et al (2013) MiR-133a modulates osteogenic differentiation of vascular smooth muscle cells. Endocrinology 154:3344–3352
125. Goettsch C, Rauner M, Pacyna N et al (2011) miR-125b regulates calcification of vascular smooth muscle cells. Am J Pathol 179:1594–1600
126. Mackenzie NC, Staines KA, Zhu D et al (2014) miRNA-221 and miRNA-222 synergistically function to promote vascular calcification. Cell Biochem Funct 32:209–216
127. Wu T, Zhou H, Hong Y et al (2012) miR-30 family members negatively regulate osteoblast differentiation. J Biol Chem 287:7503–7511
128. Balderman JA, Lee HY, Mahoney CE et al (2012) Bone morphogenetic protein-2 decreases microRNA-30b and microRNA-30c to promote vascular smooth muscle cell calcification. J Am Heart Assoc 1:e003905
129. Cui RR, Li SJ, Liu LJ et al (2012) MicroRNA-204 regulates vascular smooth muscle cell calcification in vitro and in vivo. Cardiovasc Res 96:320–329
130. Thomou T, Mori MA, Dreyfuss JM et al (2017) Adipose-derived circulating miRNAs regulate gene expression in other tissues. Nature 542:450–455

Part II
Modification of Cardiovascular Complications in Obesity

Chapter 12
Obesity and Coronary Artery Disease: Evaluation and Treatment

Marie-Eve Piché, Marie-Philippe Morin, and Paul Poirier

Abstract Individuals who are obese are more likely to develop cardiovascular disease (CVD) and manifestations of CVD, particularly coronary artery disease (CAD), angina, myocardial infarction, heart failure and sudden cardiac death. Susceptibility to obesity-related cardiovascular complications is not only mediated by overall body fatness, but is largely dependent upon individual differences in regional body fat distribution. Presence of CVD assessment can be challenging in obese patient. Baseline electrocardiogram may be influenced by obesity (false positive for inferior myocardial infarction, microvoltage, nonspecific ST-T changes) and obese patients may have impaired maximal exercise testing capacity (dyspnea, mechanical limitations, left ventricular diastolic dysfunction). Thus, other modalities may be of interest in the evaluation of clinically significant CAD in this population like nuclear medicine approaches, stress echocardiography using either physiological (treadmill exercise) or pharmacological stress (dobutamine) and stress cardiac magnetic resonance imaging. Coronary artery calcium screening can be use as well as computed tomography coronary angiography. At the end, coronary angiography is still considered the gold standard test for identifying the presence and extent of atherosclerotic CAD. The appropriate choice of test to assess CAD depends on local expertise, relative strengths and weaknesses of each modality as well as individual patient characteristics, pretest likelihood of CAD and finally the risk/benefit ratio of using a given modality.

M.-E. Piché · P. Poirier (✉)
Cardiology Department, Institut Universitaire de Cardiologie et de Pneumologie de Québec, Québec, Canada
e-mail: paul.poirier@criucpq.ulaval.ca

M.-E. Piché
e-mail: marie-eve.piche@criucpq.ulaval.ca

M.-P. Morin
Internal Medicine Department, Institut Universitaire de Cardiologie et de Pneumologie de Québec, Québec, Canada
e-mail: marie-philippe.morin.1@ulaval.ca

P. Poirier
Faculty of Pharmacy, Université Laval, Québec, Canada

P. S. Tappia et al. (eds.), *Biochemistry of Cardiovascular Dysfunction in Obesity*, Advances in Biochemistry in Health and Disease 20,
https://doi.org/10.1007/978-3-030-47336-5_12

Keywords Obesity · Coronary artery disease · Nuclear medicine · Calcium score · Angiography · Magnetic resonance imaging

Contribution of Obesity to Coronary Artery Disease

Large prospective studies have documented obesity as being an independent predictor of coronary artery disease (CAD) [1, 2]. This relationship was stronger in younger individuals. Post-mortem examination of arteries from young individuals who died from accidental injuries documented that obesity accelerates the progression of atherosclerosis decades before the appearance of clinical manifestations. Individuals with obesity are more likely to develop cardiovascular disease (CVD) and manifestations of CVD, such as CAD, angina, myocardial infarction (MI), heart failure and sudden cardiac death [1–4]. Obesity is also associated with a more rapid coronary artery calcification (CAC) process, a risk marker of coronary atherosclerosis [5]. Susceptibility to obesity co-existing cardiovascular complications is not only mediated by global adiposity, but is largely dependent upon individual differences in regional body fat distribution [6–9]. Studies using imaging techniques have identified excess abdominal visceral adipose tissue as a key driver of the cardiovascular risk and a predictor of CVD development over time [10–13].

Non-invasive Evaluation of Coronary Artery Disease

Electrocardiogram

Obesity is associated with a wide variety of ECG abnormalities; some being clinically relevent (Table 12.1). Most of these reflects structural changes related to obesity including: (1) displacement of the heart by the diaphragm elevation, (2) increased cardiac workload with associated cardiac hypertrophy and, (3) global fat accumulation as well as cardiac fat infiltration [4]. The position of the R wave may change, various arrhythmias may develop, or the QT interval may be prolonged. Nonspecific T wave flattening in the inferior and lateral leads (attributed to the horizontal displacement of the heart) are frequent [14, 15]. An increase incidence of false-positive criteria for inferior myocardial infarction in obese individuals due to the elevation of the diaphragm has been reported [16]. In obesity, the ECG signals and criteria of left ventricular hypertrophy are less informative and accurate due to the accumulation of epicardial and subcutaneous adipose tissue [14].

Since obesity may influence surface ECG (false positive for inferior myocardial infarction, microvoltage, nonspecific ST-T changes) and in light that obese patients may have impaired maximal exercise testing capacity, other modalities to assess the presence of significant atherosclerosis may be of interest in the evaluation of CAD in this population.

Table 12.1 ECG changes that may occur in obese individuals

Clinically significant
↑Heart rate
↑QRS interval
↑QTc interval
False-positive criteria for inferior myocardial infarction
Less clinically significant
↑PR interval
↑QRS voltage
↑QT dispersion
↑SAECG (late potentials)
↑ST-T abnormalities
↑ST depression
Left axis deviation
Flattening of the T wave (inferolateral leads)
Left atrial abnormalities

Invasive and Non-invasive Coronary Artery Disease Assessment

Treadmill Stress Test

Electrocardiographic exercise capacity testing is complex and limited in patients with obesity as from baseline, resting ECG abnormalities might limit interpretation. Obese patients also have a reduced ability to exercise as defined by low functional aerobic capacity during exercise treadmill testing. Patients with obesity are not able to achieve the minimal heart rate threshold (80–85% of the age-predicted heart rate) for a valid test result [17, 18] as the chronotropic response to exercise is altered, with peak heart rate, heart rate recovery and chronotropic index being lower in patients with obesity [17]. Nevertheless, both the standard Bruce and the ramped Bruce treadmill protocols achieve valid test results in most obese patients [19]. Excessive exercise blood pressure response is common and predicts long-term development of sustained hypertension and adverse cardiovascular outcomes [20, 21].

Nuclear Medicine

Single Photon Emission Tomography (SPECT)

Single-photon emission computed tomography (SPECT) myocardial perfusion imaging is commonly used in obese patients. Single-photon emission computed tomography can be used with exercise, vasodilator drugs (dipyridamole), and dobutamine stress [22, 23]. In obesity, SPECT diagnostic accuracy is limited by suboptimal image quality resulting from attenuation artefacts by the diaphragm or breast, and a decreased signal to noise ratio. The use of hybrid SPECT/computed tomography (CT) equipment, adoption of novel detectors and cameras, and image processing including CT-based attenuation correction algorithms allow a reduction of attenuation artefacts. Two-day protocols with larger tracer doses are recommended in obese patients who are 250–350 lb (113–160 kg). Accuracy and image quality increase with SPECT, when using 99 m-technetium-labelled radiotracers [24–26]. Disadvantages of SPECT include the reduced ability to detect triple-vessel or left-main stem coronary artery disease, and residual uncorrected attenuation.

Positron Emission Tomography (PET)—Rubidium

Positron emission tomography (PET) computed tomography rubidium is faster than SPECT, produces less radiation exposure, superior imaging quality and accuracy, and a reduced need for more invasive examinations [27]. Positron emission tomography myocardial imaging has demonstrated higher accuracy than SPECT for the detection of CAD, and its superiority was maintained in obese patients. In a meta-analysis (1442 patients), the mean sensitivity and specificity of PET myocardial imaging for CAD detection was 92% and 85% respectively [28]. It is important to emphasize that a normal PET myocardial perfusion imaging indicate a low risk annual cardiovascular events (<1% cardiovascular events, cardiovascular death, and non-fatal MI) in all categories of obese patients [29]. Additionally, PET myocardial imaging allow to quantify absolute coronary blood flow, adding to the diagnostic and prognostic capabilities beyond solely relative perfusion imaging, especially in the detection of triple-vessel and left-main CAD. Consequenetly, PET myocardial imaging represent the nuclear imaging technique of choice for patients with obesity.

Stress Echocardiography

Stress echocardiography using either exercise or pharmacological (usually dobutamine) stress is widely available, relatively inexpensive, well tolerated, has no weight limits and does not involve use of radiation. Overall, stress echocardiography has comparable accuracy to SPECT imaging for detection of CAD [30] but is highly operator-dependent and can be limited in the presence of poor acoustic window related to obesity. The sensitivity of stress echocardiography might be further enhanced by the addition of myocardial contrast agents [31]. In a prospective study of overweight and obese patients who underwent pharmacological stress echocardiography and coronary angiography, contrast imaging improved sensitivity and specificity of the test (82 vs. 70% and 78 vs. 67%, respectively with and without contrast) [32]. If severe technical limitations exist, transesophageal echocardiography with dobutamine might be useful [33, 34].

Stress Cardiac Magnetic Resonance

Stress cardiac magnetic resonance (CMR) allows assessment of ventricular function, regional wall motion abnormalities, rest and stress perfusion, and viability within a single examination. Ischemia on stress testing using CMR perfusion and dobutamine stress CMR is an independent predictor of cardiac events [35, 36] and inducible ischemia in a population with obesity (average body mass index [BMI] of 34 kg/m^2) has been reported as an independent predictor of major adverse cardiac events (MACE), and predict adverse cardiovascular events at 5 years follow-up, with over 89% of patients achieving good diagnostic image quality [37]. Absence of inducible ischemia was associated with a low annual rate of 0.6% for major adverse cardiac events in patients with obesity as it is the case in nonobese patients [35, 37]. Table weight limit, bore diameter, and claustrophobia may limit the feasibility of CMR in patients with obesity [38]. These limitations have been mostly overcome by the recent development of large-bore systems and open magnetic resonance imaging systems.

Computed Tomography Calcium Score-Coronary Artery Calcium Score

Coronary artery calcium (CAC) screening can enhance risk prediction in asymptomatic individuals. Obesity is associated with elevated CAC, a marker of coronary atherosclerosis that is predictive of cardiovascular events [39–42] and obesity appears to be a risk factor for more rapid progression of CAC over time [5]. The presence of extensive coronary calcification (CAC score) offers an inexpensive and reproducible technique to determine the presence and extent of calcified coronary artery plaque. Despite advances in CT scanners, obesity and high levels of coronary calcifications continue to limit the diagnostic accuracy and value of cardiac CT [43]. Nowadays, CT equipment have table weight limits of 350–450 lb (160–204 kg) in order to avoid equipment damage and examinations are additionally limited by gantry/bore diameter [44]. Coronary calcification score is highly sensitive for diagnosis of obstructive CAD as the negative predictive value is high. A systematic review of 28 studies and 3674 patients showed a sensitivity of 98% and specificity of 82% for obstructive CAD [45]. This led to the widely accepted conclusion that a normal CAC score reliably rules-out significant CAD and further invasive workup can be delayed.

Cardiac Computed Tomography Coronary Angiography

Computed tomography coronary angiography is emerging as an alternative approach for the quantification of both coronary calcified and non-calcified plaque [46–48]. This approach may be particularly useful in specific subsets of patients with obesity with unknown CAD and equivocal or uninterpretable stress tests or in case of discrepancies between clinical presentation and stress test results. Coronary artery calcium score allows risk stratification and plaque burden assessment whilst CT coronary angiography allows evaluation of luminal stenosis, plaque characterisation/quantification. Computed tomography coronary angiography requires the administration of intravenous contrast to visualize the non-calcified plaque and estimate the severity of luminal stenosis. Reported registry showed that symptomatic obese patients were more likely than non-obese patients to have CAD at CT coronary angiography [49]. Imai et al. studied 553 patients who underwent serial CT coronary angiography and observed that the risk of non-calcified plaques increased as abdominal visceral adiposity increased, with the highest quartile conferring the greatest risk, regardless of underlying CAD risk factors [50]. Technical challenges with CT coronary angiography plaque analysis included reproducibility, time-consuming analysis, and lack of biologic correlation. Despite proper preparation, artifacts and noisy scans are frequent, mainly caused by cardiac and respiratory motion and reduced signal to noise ratio. In addition, low vessel opacification may occur when injecting contrast in patients with obesity, due to differences in distribution of blood volume in peripheral venous and central circulation [51]. The rate of non-evaluable segments is also higher in patients with obesity. In order to improve image quality, medications such as beta-blockers for heart rate control and nitroglycerin for coronary dilatation may be used during the procedure. Others limitations include decreased accuracy in patients with extensive calcified plaque and exposure to radiation and contrast. In case where there is significant calcification, functional imaging is suggested. Nevertheless, sensitivity and negative predictive values are invariably high even in patients with obesity. New techniques of dual source CT and iterative image reconstruction are being developed that will improve image quality in obese patients.

Invasive Evaluation of Coronary Artery Disease

Coronary Angiography

Coronary angiography is the gold standard test for establishing the coronary anatomy and identifying the presence and extent of CAD. Individuals with obesity may suffered from several limitations and complications when comes the times to be evaluated in the catheterization laboratory. As compared to overweight patients (BMI 25–30 mg/m^2), patients with severe obesity from the Michigan Cardiovascular Consortium registry who undergone coronary angiography suffered significantly higher rates of vascular complications, contrast-induced nephropathy, nephropathy requiring dialysis and death, despite being younger and having a lower incidence of acute MI, cardiogenic shock and a lesser need for emergent intervention [52]. Coronary angiography difficulties included suboptimal radiographic coronary artery visualization that may prevent detection of significant angiographic lesions and stenosis, and potentially result in a greater risk of complications during percutaneous coronary interventions. Moreover, vascular access for the procedure may be difficult. Radial access is preferred in the obese population, because of fewer vascular complications, especially bleeding and hematomas [53–56]. If the femoral approach is used, vascular access closure devices should be used to accelerate ambulation [57]. Patients with obesity may require more radiation during coronary angiography to achieve adequate image quality, resulting in higher radiation exposure to both patient and staff [58]. The poor fluoroscopic image visualization associated with obesity often warrants multiple contrast agent injections, with increased risk of renal complications such as contrast-induced nephropathy. In addition to problems regarding vascular access and radiographic imaging, the engineering parameters and physical limitations of the angiographic table may limit obese patients' ability to undergo this procedure and to achieve the normal variety of angiographic views.

Intravascular Ultrasound (IVUS)

Intravascular ultrasound (IVUS) provides complementary diagnostic information about the coronary artery wall, which cannot be obtained by coronary angiography alone. Several intravascular imaging techniques such as intravascular ultrasound (IVUS), virtual histology IVUS (VH-IVUS), and optical coherence tomography (OCT) allow assessment of plaque burden, plaque morphology and response to therapy. Intravascular ultrasonography in interventional cardiology is an adjunctive procedure to coronary angiogram; as such, any contraindication to coronary angiography applies to IVUS as well. In general, the risks and complications associated with these procedures include those associated with all catheterization procedures.

The diagnosis of CAD in the obese population presents specific challenges; non-invasive testing is evolving rapidly to accommodate patients with obesity. Many non-invasive tests can detect anatomic coronary changes, impaired myocardial perfusion, or consequences of impaired perfusion such as abnormal myocardial contractile function. Non-invasive tests provide prognostic information that can improve cardiovascular risk stratification, further guiding subsequent testing and interventions. The choice of the optimal imaging modality depends on local expertise and availability, the relative strengths and weaknesses of each modality, as well as pretest probability of CAD. The dramatic rise in the proportion of young patients with obesity invokes the need for more agressive primary prevention and more upstream interventions, as well as better treatment of obesity.

Clinical Management and Treatment of Coronary Artery Disease in Obesity

Percutaneous Revascularization

Short-Term Outcomes After Percutaneous Coronary Interventions

Data from the CathPCI Registry showed that severe obesity was independently associated with a greater in-hospital mortality rate (OR, 1.14) and a lower in-hospital bleeding rate (OR, 0.80) after percutaneous coronary intervention [59]. In a large study (227,042 patients) including 37.2% obese and 7.4% severely obese patients, severe obesity increased the risk of contrast-induced nephropathy, nephropathy-requiring dialysis, and vascular complications compared to overweight patients [52]. The British Cardiovascular Intervention Society Registry reported adverse in hospital-outcomes and mortality of 345,192 patients undergoing percutaneous coronary interventions [60]. At 30 days post-percutaneous coronary interventions, lower mortality was seen with BMI 25–30 kg/m^2 (OR 0.86) and with BMI > 30 kg/m^2 (OR 0.90). At 1-year post-percutaneous coronary interventions, and up to 5 years, BMI > 25 kg/m^2 was an independent predictor of greater survival compared to normal weight patients; OR 0.70 at 1 year and 0.78 at 5 years independently of the clinical presentation (unstable angina, NSTEMI or STEMI) [60]. The APPROCH registry reported mortality about 30,258 patients who had percutaneous coronary interventions and shows that the 6-month mortality was lower in patients who were in the overweight or obese category compared to normal BMI patients [61].

Long-Term Outcomes After Percutaneous Coronary Interventions

A recent meta-analysis of 865,774 patients undergoing percutaneous coronary interventions shows a U-shape association across all BMI category for all causes mortality and risk of major cardiovascular events after percutaneous revascularization [62].

This obesity paradox seems to disappear when severe obesity is taken into consideration [61, 63]. In the APPROCH study, the 5 and 10 years mortality rates following percutaneous coronary interventions in patients with severe obesity with high risk coronary anatomy was increased compared to normal BMI patients [61].

Surgical Revascularization

Perioperative Mortality

Obesity has been inconsistently associated with an increased in-hospital mortality following coronary artery bypass graft (CABG) surgery. An analysis from the Society of Thoracic Surgeons' database (559,004 patients who underwent isolated CABG between 1997 and 2000) [64] showed an increased risk of in-hospital mortality in patients with moderate obesity (BMI = 35–39.9 kg/m^2) as well as with severe obesity (BMI > 40 kg/m^2). These results contrasted with others studies that found comparable postoperative mortality in patients with obesity following CABG [65, 66]. In a meta-analysis, in-hospital mortality after CABG was reported even less in the obese population [67]. In a retrospective multicentre study [68], the 30-day postoperative mortality was highest in extreme BMI groups (BMI < 20 kg/m^2; 4.0%, BMI > 40 kg/m^2; 3.8%) and lowest near a BMI of 30 kg/m^2 (3.1%), suggesting a "U-shaped" relationship [69]. Evidence is still conflicting regarding long-term mortality [67]. A meta-analysis found a reduction in long-term mortality (5 years) in the overweight and obese populations [67] which was confirmed by the latest meta-analysis [62]. On the contrary, a retrospective study showed that obesity was associated with a higher long-term mortality after CABG [70].

Several studies have reported greater postoperative complications after CABG with obesity, such as renal failure [71], respiratory failure, arrhythmias and greater intraoperative transfusion rate [4, 72, 73]. Postoperative cerebrovascular events, MI, and postoperative bleeding do not appear to be increased in patients with obesity [72, 73]. A greater incidence of postoperative atrial fibrillation was seen in obese vs non-obese patients [74]. In a large cohort study of patients who underwent isolated CABG, higher waist circumference was associated with an increased risk of postoperative atrial fibrillation, prolonged mechanical ventilation and reintubation, renal failure, sternal wound infections, longer intensive care unit and hospital stays,independently of BMI [75]. The large, poorly vascularized panniculus associated with dysglycemia in the obese patients may predispose to postoperative wound infections [61, 76]. Obesity has also been identified as a risk factor for superficial wound infection and saphenous vein harvest site infection [65].

Antiplatelet Therapy

Individuals with obesity display increased levels of coagulation factors, impaired fibrinolysis, increased platelet activity and higher platelet activation markers [77, 78]. While not fully understood, adipose tissue produces multiple bioactive substances and hormones such as leptin, adiponectin, TNF-a, interleukin-6, all of which may directly or indirectly influence coagulation and platelet function [78, 79]. Aspirin is the cornerstone of antiplatelet therapy for CAD but obesity is a risk factor for a reduced aspirin pharmacodynamic response [80]. Obesity-related inflammatory state, endothelial dysfunction and metabolic endotoxemia enhance a number of mechanisms that increase platelet reactivity and platelet turnover and decrease aspirin bioavailability, all contributing to a poor aspirin response [81, 82]. Studies investigating platelet reactivity in patients after acute coronary syndrome treated with thienopyridines, a class of selective irreversible ADP receptor/P2Y12 inhibitors, identified obesity as an important modulator of response to both clopidogrel [83, 84] and prasugrel [85, 86]. Obese patients without metabolic syndrome had a better response to thienopyridines (i.e. clopidogrel and prasugrel) compared with obese patients with metabolic syndrome and similar response to non-obese patients, suggesting metabolic conditions being better correlate of platelet inhibition than BMI [78, 87]. Conversely to thienopyridines, no correlation was reported between BMI and platelet activity with ticagrelor; obese patients do not express significantly higher level of platelet reactivity while ticagrelor seems to induce significantly higher platelet inhibition than prasugrel in obese patients [78, 88]. While studies suggest that obesity may promote platelet activation and blunt effects of anti-platelet medications, clinical observations have pointed to an "obesity paradox", namely that obese patients may have better post-acute coronary syndrome outcomes and may have a lower risk of re-infarction or death. However, data involving platelet assays is often conflicting and involved sample sizes too small to draw definitive conclusions about clinical outcomes and to make recommendations about dosing adjustment of antiplatelets therapy in obesity [78].

Benefits of Weight Loss on Coronary Artery Disease

The general goals of weight loss and weight management are at a minimum, to prevent further weight gain and preferentially to reduce body weight while maintaining a lower long term body weight. Patients should have their BMI and waist circumference measured not only for the initial assessment of the degree of overweight and obesity but also as a guide to evaluate the efficacy of weight loss treatment [88]. An elevated waist circumference ≥102 cm in Caucasian men and ≥88 cm in Caucasian women bring a higher risk of cardiovascular disease [89]. This target is lower for others ethnic group like the Asian population [90]. The 2013 ACC/AHA guidelines recommend a sustained weight loss of 3–5% to result in a clinically meaningful reduction in triglycerides, blood glucose, HBAIC and the risk of developing type 2 diabetes. Greater amount of weight loss is recommended to improve HDL, LDL, blood pressure and to decrease oral hypoglycemic medications, insulin and antihypertensive medications [91]. Obesity management and treatment include lifestyle modification as diet, exercise and counselling, pharmacotherapy and bariatric surgery. In order to achieve weight reduction, ACC/AHA recommend a comprehensive high-intensity lifestyle modification. This include behavioral interventions (self-monitoring of food intake and physical activity), regular counselling (≥14 sessions over a 6-month period by a trained interventionist), physical activity (≥150 min/week of aerobic exercise) and a calorie-restricted diet (1200–1500 kcal/day for women and 1500–1800 kcal/day for men or a 500–750 kcal deficit/day) [91]. No diet is proved to be superior to another to achieve long term weight reduction [91, 92] as adherence to the diet seems to be the main factor associated with success [93]. Mediterranean diet decreases major cardiovascular events (MACE) in patients with high cardiovascular risk and is an interesting option for this population [94]. For weight loss maintenance, long-term behavioral interventions, monthly face to face or phone contact with the interventionist, continuation of the restricted-calorie diet and a higher level of physical activity (≥200–300 min/week) are recommended [91]. The weight maintenance phase could be particularly challenging for patients because of the long term change of hunger hormones (increase of ghrelin and decrease of GLP-1) [95] and decreased basal metabolic rate [96].

No weight loss studies have shown reduction of cardiovascular disease or mortality through lifestyle modification. LOOK AHEAD was one of the biggest trial and failed to show any reduction of major adverse cardiac events or cardiovascular mortality after 9.6 years [97]. Participants in the intervention group lost 8.6 versus 0.7% in the control group in the first year but experiment weight regain thereafter and at 9.6 years, there was only a 2.5% of weight loss difference between groups. On the other hand, the intervention group had a significant improvement of some cardiovascular risk factor (HBAIC, lipid profile, etc.). Pharmacotherapy is recommended as adjunct to lifestyle modification in patients with BMI $\geq$ 27 kg/m^2 with comorbidities or BMI $\geq$ 30 kg/m^2. Three medications are available in Canada for weight loss: Liraglutide, Naltrexone-Bupropion and Orlistat. In United States, few other medications are also available: phentermine and phentermine-topiramate [91]. These medications can achieve an average of 5–10% weight loss, but interindividual variability exists and some people are low whereas others are high responders [91]. Liraglutide has been shown to reduce the major adverse cardiac events and cardiovascular death in the LEADER trial but this was in a type 2 diabetes population who were using the 1.8 mg dosing [98]. Lorcaserin appears to be safe regarding cardiovascular disease but no benefits regarding cardiovascular mortality or cardiovascular disease was demonstrated [99]. It has been withdrawn from the market due to noncardiovasular safety concern. The cardiovascular safety of the others medications has not been established yet. Some interim analysis of the LIGHT trial shows that Naltrexone-Bupropion has a cardiovascular safety but no solid conclusion can be draw of this trial that was terminated early due to public release of the interim data by the sponsor [100].

Bariatric surgery is also part of the ACC/AHA recommendations for patients with BMI $\geq$ 35 kg/m^2 with comorbidities or BMI $\geq$ 40 kg/m^2 [91]. Weight loss, remission of comorbidities and weight loss maintenance varies regarding the type of the procedures (gastric banding < sleeve gastrectomy < Roux-y-gastric bypass < biliopancreatic diversion with duodenal switch) [101]. Bariatric surgery is superior to intensive medical treatment to decrease HBAIC or to resolve type 2 diabetes [102]. A retrospective study of 20,235 surgical and non surgical patient [103] documents that bariatric surgery was associated to a lower incidence of macrovascualar disease (defined as first occurrence of CAD or cerebrovascular events) mainly driven by a lower incidence of CAD (defined by acute MI, unstable angina, percutaneous coronary intervention, or CABG [103]. The SOS study which is a non-randomized prospective controlled study demonstrated a reduction of cardiovascular death in the bariatric surgery group compared to the control group [104]. There is no randomized control trial regarding the effects of bariatric surgery on major adverse cardiac event incidence.

Conclusions

Obesity is a serious medical condition both at the individual and population level and promotes numbers of health-related medical conditions like systemic hypertension, dyslipidemia, sleep apnea, type 2 diabetes and other CVD. Overweight, obesity and severe obesity are all an independent risk factor of CVD including CAD. However, assessment of CAD may be challenging as well as the medical management of these Patients in order to reduce the progression/impact of CAD.

References

1. Rabkin SW, Mathewson FA, Hsu PH (1977) Relation of body weight to development of ischemic heart disease in a cohort of young North American men after a 26 year observation period: the Manitoba Study. Am J Cardiol 39:452–458
2. Hubert HB, Feinleib M, McNamara PM, Castelli WP (1983) Obesity as an independent risk factor for cardiovascular disease: a 26-year follow-up of participants in the Framingham Heart Study. Circulation 67:968–977
3. Plourde B, Sarrazin JF, Nault I, Poirier P (2014) Sudden cardiac death and obesity. Expert Rev Cardiovasc Ther 12:1099–1110
4. Poirier P, Giles TD, Bray GA et al (2006) Obesity and cardiovascular disease: pathophysiology, evaluation, and effect of weight loss: an update of the 1997 American Heart Association Scientific Statement on Obesity and Heart Disease from the Obesity Committee of the Council on Nutrition, Physical Activity, and Metabolism. Circulation 113:898–918
5. Kronmal RA, McClelland RL, Detrano R et al (2007) Risk factors for the progression of coronary artery calcification in asymptomatic subjects: results from the Multi-Ethnic Study of Atherosclerosis (MESA). Circulation 115:2722–2730

6. Sam S, Haffner S, Davidson MH et al (2008) Relationship of abdominal visceral and subcutaneous adipose tissue with lipoprotein particle number and size in type 2 diabetes. Diabetes 57:2022–2027
7. Kragelund C, Hassager C, Hildebrandt P, Torp-Pedersen C, Kober L, Group Ts (2005) Impact of obesity on long-term prognosis following acute myocardial infarction. Int J Cardiol 98:123–131
8. Yusuf S, Hawken S, Ounpuu S et al (2004) Effect of potentially modifiable risk factors associated with myocardial infarction in 52 countries (the INTERHEART study): case-control study. Lancet 364:937–952
9. Neeland IJ, Poirier P, Despres JP (2018) Cardiovascular and metabolic heterogeneity of obesity: clinical challenges and implications for management. Circulation 137:1391–1406
10. Abraham TM, Pedley A, Massaro JM, Hoffmann U, Fox CS (2015) Association between visceral and subcutaneous adipose depots and incident cardiovascular disease risk factors. Circulation 132:1639–1647
11. Liu J, Fox CS, Hickson DA et al (2010) Impact of abdominal visceral and subcutaneous adipose tissue on cardiometabolic risk factors: the Jackson Heart Study. J Clin Endocrinol Metab 95:5419–5426
12. Fox CS, Massaro JM, Hoffmann U et al (2007) Abdominal visceral and subcutaneous adipose tissue compartments: association with metabolic risk factors in the Framingham Heart Study. Circulation 116:39–48
13. Preis SR, Massaro JM, Robins SJ et al (2010) Abdominal subcutaneous and visceral adipose tissue and insulin resistance in the Framingham heart study. Obesity (Silver Spring) 18:2191–2198
14. Alpert MA, Terry BE, Cohen MV, Fan TM, Painter JA, Massey CV (2000) The electrocardiogram in morbid obesity. Am J Cardiol 85(908–10):A10
15. Eisenstein I, Edelstein J, Sarma R, Sanmarco M, Selvester RH (1982) The electrocardiogram in obesity. J Electrocardiol 15:115–118
16. Starr JW, Wagner GS, Behar VS, Walston A 2nd, Greenfield JC Jr (1974) Vectorcardiographic criteria for the diagnosis of inferior myocardial infarction. Circulation 49:829–836
17. Gondoni LA, Titon AM, Nibbio F, Augello G, Caetani G, Liuzzi A (2009) Heart rate behavior during an exercise stress test in obese patients. Nutr Metab Cardiovasc Dis 19:170–176
18. Lear SA, Brozic A, Myers JN, Ignaszewski A (1999) Exercise stress testing. An overview of current guidelines. Sports Med 27:285–312
19. Bires AM, Lawson D, Wasser TE, Raber-Baer D (2013) Comparison of Bruce treadmill exercise test protocols: is ramped Bruce equal or superior to standard Bruce in producing clinically valid studies for patients presenting for evaluation of cardiac ischemia or arrhythmia with body mass index equal to or greater than 30? J Nucl Med Technol 41:274–278
20. Chrysohoou C, Skoumas J, Georgiopoulos G et al (2017) Exercise capacity and haemodynamic response among 12,327 individuals with cardio-metabolic risk factors undergoing treadmill exercise. Eur J Prev Cardiol 24:1627–1636
21. Schultz MG, Otahal P, Cleland VJ, Blizzard L, Marwick TH, Sharman JE (2013) Exercise-induced hypertension, cardiovascular events, and mortality in patients undergoing exercise stress testing: a systematic review and meta-analysis. Am J Hypertens 26:357–366
22. Wackers FJ (1991) Myocardial perfusion imaging in ischemic heart disease anno 1990. Curr Opin Cardiol 6:590–601
23. Iskandrian AS, Heo J, Askenase A, Segal BL, Helfant RH (1987) Thallium imaging with single photon emission computed tomography. Am Heart J 114:852–865
24. Korbee RS, Boiten HJ, Ottenhof M, Valkema R, van Domburg RT, Schinkel AF (2013) What is the value of stress (99m)Tc-tetrofosmin myocardial perfusion imaging for the assessment of very long-term outcome in obese patients? J Nucl Cardiol 20:227–233
25. Ferraro S, Perrone-Filardi P, Desiderio A et al (1996) Left ventricular systolic and diastolic function in severe obesity: a radionuclide study. Cardiology 87:347–353
26. Freedman N, Schechter D, Klein M, Marciano R, Rozenman Y, Chisin R (2000) SPECT attenuation artifacts in normal and overweight persons: insights from a retrospective comparison

of Rb-82 positron emission tomography and TI-201 SPECT myocardial perfusion imaging. Clin Nucl Med 25:1019–1023
27. Yoshinaga K, Chow BJ, Williams K et al (2006) What is the prognostic value of myocardial perfusion imaging using rubidium-82 positron emission tomography? J Am Coll Cardiol 48:1029–1039
28. Nandalur KR, Dwamena BA, Choudhri AF, Nandalur SR, Reddy P, Carlos RC (2008) Diagnostic performance of positron emission tomography in the detection of coronary artery disease: a meta-analysis. Acad Radiol 15:444–451
29. Chow BJ, Dorbala S, Di Carli MF et al (2014) Prognostic value of PET myocardial perfusion imaging in obese patients. JACC Cardiovasc Imaging 7:278–287
30. Fleischmann KE, Hunink MG, Kuntz KM, Douglas PS (1998) Exercise echocardiography or exercise SPECT imaging? A meta-analysis of diagnostic test performance. JAMA 280:913–920
31. Shah BN, Senior R (2016) Stress echocardiography in patients with morbid obesity. Echo Res Pract 3:R13–R18
32. Hu SJ, Liu SX, Katus HA, Luedde M (2007) The value of contrast dobutamine stress echocardiography on detecting coronary artery disease in overweight and obese patients. Can J Cardiol 23:885–889
33. Madu EC (2000) Transesophageal dobutamine stress echocardiography in the evaluation of myocardial ischemia in morbidly obese subjects. Chest 117:657–661
34. Legault S, Senechal M, Bergeron S et al (2010) Usefulness of an accelerated transoesophageal stress echocardiography in the preoperative evaluation of high risk severely obese subjects awaiting bariatric surgery. Cardiovasc Ultrasound 8:30
35. Jahnke C, Nagel E, Gebker R et al (2007) Prognostic value of cardiac magnetic resonance stress tests: adenosine stress perfusion and dobutamine stress wall motion imaging. Circulation 115:1769–1776
36. Lipinski MJ, McVey CM, Berger JS, Kramer CM, Salerno M (2013) Prognostic value of stress cardiac magnetic resonance imaging in patients with known or suspected coronary artery disease: a systematic review and meta-analysis. J Am Coll Cardiol 62:826–838
37. Shah RV, Heydari B, Coelho-Filho O et al (2014) Vasodilator stress perfusion CMR imaging is feasible and prognostic in obese patients. JACC Cardiovasc Imaging 7:462–472
38. Lim SP, Arasaratnam P, Chow BJ, Beanlands RS, Hessian RC (2015) Obesity and the challenges of noninvasive imaging for the detection of coronary artery disease. Can J Cardiol 31:223–226
39. Greenland P, Bonow RO, Brundage BH et al (2007) ACCF/AHA 2007 clinical expert consensus document on coronary artery calcium scoring by computed tomography in global cardiovascular risk assessment and in evaluation of patients with chest pain: a report of the American College of Cardiology Foundation Clinical Expert Consensus Task Force (ACCF/AHA Writing Committee to Update the 2000 Expert Consensus Document on Electron Beam Computed Tomography) developed in collaboration with the Society of Atherosclerosis Imaging and Prevention and the Society of Cardiovascular Computed Tomography. J Am Coll Cardiol 49:378–402
40. Blaha MJ, Rivera JJ, Budoff MJ et al (2011) Association between obesity, high-sensitivity C-reactive protein >/=2 mg/L, and subclinical atherosclerosis: implications of JUPITER from the Multi-Ethnic Study of Atherosclerosis. Arterioscler Thromb Vasc Biol 31:1430–1438
41. Burke GL, Bertoni AG, Shea S et al (2008) The impact of obesity on cardiovascular disease risk factors and subclinical vascular disease: the Multi-Ethnic Study of Atherosclerosis. Arch Intern Med 168:928–935
42. Chang Y, Kim BK, Yun KE et al (2014) Metabolically-healthy obesity and coronary artery calcification. J Am Coll Cardiol 63:2679–2686
43. Westwood M, Al M, Burgers L et al (2013) A systematic review and economic evaluation of new-generation computed tomography scanners for imaging in coronary artery disease and congenital heart disease: Somatom Definition Flash, Aquilion ONE, Brilliance iCT and Discovery CT750 HD. Health Technol Assess 17:1–243

44. Uppot RN (2007) Impact of obesity on radiology. Radiol Clin North Am 45:231–246
45. Paech DC, Weston AR (2011) A systematic review of the clinical effectiveness of 64-slice or higher computed tomography angiography as an alternative to invasive coronary angiography in the investigation of suspected coronary artery disease. BMC Cardiovasc Disord 11:32
46. Sun J, Zhang Z, Lu B et al (2008) Identification and quantification of coronary atherosclerotic plaques: a comparison of 64-MDCT and intravascular ultrasound. AJR Am J Roentgenol 190:748–754
47. Papadopoulou SL, Neefjes LA, Schaap M et al (2011) Detection and quantification of coronary atherosclerotic plaque by 64-slice multidetector CT: a systematic head-to-head comparison with intravascular ultrasound. Atherosclerosis 219:163–170
48. Voros S, Rinehart S, Qian Z et al (2011) Coronary atherosclerosis imaging by coronary CT angiography: current status, correlation with intravascular interrogation and meta-analysis. JACC Cardiovasc Imaging 4:537–548
49. Labounty TM, Gomez MJ, Achenbach S et al (2013) Body mass index and the prevalence, severity, and risk of coronary artery disease: an international multicentre study of 13,874 patients. Eur Heart J Cardiovasc Imaging 14:456–463
50. Imai A, Komatsu S, Ohara T et al (2012) Visceral abdominal fat accumulation predicts the progression of noncalcified coronary plaque. Atherosclerosis 222:524–529
51. Husmann L, Leschka S, Boehm T et al (2006) Influence of body mass index on coronary artery opacification in 64-slice CT angiography. Rofo 178:1007–1013
52. Buschur ME, Smith D, Share D et al (2013) The burgeoning epidemic of morbid obesity in patients undergoing percutaneous coronary intervention: insight from the Blue Cross Blue Shield of Michigan Cardiovascular Consortium. J Am Coll Cardiol 62:685–691
53. Benamer H, Louvard Y, Sanmartin M et al (2007) A multicentre comparison of transradial and transfemoral approaches for coronary angiography and PTCA in obese patients: the TROP registry. EuroIntervention 3:327–332
54. Cox N, Resnic FS, Popma JJ, Simon DI, Eisenhauer AC, Rogers C (2004) Comparison of the risk of vascular complications associated with femoral and radial access coronary catheterization procedures in obese versus nonobese patients. Am J Cardiol 94:1174–1177
55. McNulty PH, Ettinger SM, Field JM et al (2002) Cardiac catheterization in morbidly obese patients. Catheter Cardiovasc Interv 56:174–177
56. Hibbert B, Simard T, Wilson KR et al (2012) Transradial versus transfemoral artery approach for coronary angiography and percutaneous coronary intervention in the extremely obese. JACC Cardiovasc Interv 5:819–826
57. Wong P, Harding S, Walters D, Hull ML, Jang IK (2001) Vascular complications after hemostatic puncture closure device (Angio-Seal) are not higher in overweight patients. J Invasive Cardiol 13:623–625
58. Plourde G, Pancholy SB, Nolan J et al (2015) Radiation exposure in relation to the arterial access site used for diagnostic coronary angiography and percutaneous coronary intervention: a systematic review and meta-analysis. Lancet 386:2192–2203
59. Payvar S, Kim S, Rao SV et al (2013) In-hospital outcomes of percutaneous coronary interventions in extremely obese and normal-weight patients: findings from the NCDR (National Cardiovascular Data Registry). J Am Coll Cardiol 62:692–696
60. Holroyd EW, Sirker A, Kwok CS et al (2017) The relationship of body mass index to percutaneous coronary intervention outcomes: does the obesity paradox exist in contemporary percutaneous coronary intervention cohorts? Insights from the British Cardiovascular Intervention Society Registry. JACC Cardiovasc Interv 10:1283–1292
61. Terada T, Forhan M, Norris CM et al (2017) Differences in short- and long-term mortality associated With BMI following coronary revascularization. J Am Heart Assoc 6
62. Ma WQ, Sun XJ, Wang Y, Han XQ, Zhu Y, Liu NF (2018) Does body mass index truly affect mortality and cardiovascular outcomes in patients after coronary revascularization with percutaneous coronary intervention or coronary artery bypass graft? A systematic review and network meta-analysis. Obes Rev 19:1236–1247

63. Li YH, Lin GM, Lin CL, Wang JH, Han CL (2013) Relation of body mass index to mortality among patients with percutaneous coronary intervention longer than 5 years follow-up: a meta-analysis. Int J Cardiol 168:4315–4318
64. Prabhakar G, Haan CK, Peterson ED, Coombs LP, Cruzzavala JL, Murray GF (2002) The risks of moderate and extreme obesity for coronary artery bypass grafting outcomes: a study from the Society of Thoracic Surgeons' database. Ann Thorac Surg 74:1125–1130; discussion 1130–1
65. Moulton MJ, Creswell LL, Mackey ME, Cox JL, Rosenbloom M (1996) Obesity is not a risk factor for significant adverse outcomes after cardiac surgery. Circulation 94:II87–92
66. Birkmeyer NJ, Charlesworth DC, Hernandez F et al (1998) Obesity and risk of adverse outcomes associated with coronary artery bypass surgery. Northern New England Cardiovascular Disease Study Group. Circulation 97:1689–1694
67. Oreopoulos A, Padwal R, Norris CM, Mullen JC, Pretorius V, Kalantar-Zadeh K (2008) Effect of obesity on short- and long-term mortality postcoronary revascularization: a meta-analysis. Obesity (Silver Spring) 16:442–450
68. Prapas SN, Panagiotopoulos IA, Salama Ayyad MA et al (2010) Impact of obesity on outcome of patients undergoing off-pump coronary artery bypass grafting using aorta no-touch technique. Interact Cardiovasc Thorac Surg 11:234–237
69. Wagner BD, Grunwald GK, Rumsfeld JS et al (2007) Relationship of body mass index with outcomes after coronary artery bypass graft surgery. Ann Thorac Surg 84:10–16
70. Benedetto U, Danese C, Codispoti M (2014) Obesity paradox in coronary artery bypass grafting: myth or reality? J Thorac Cardiovasc Surg 147:1517–1523
71. Virani SS, Nambi V, Lee VV et al (2009) Obesity: an independent predictor of in-hospital postoperative renal insufficiency among patients undergoing cardiac surgery? Tex Heart Inst J 36:540–545
72. Nolan HR, Davenport DL, Ramaiah C (2013) BMI is an independent preoperative predictor of intraoperative transfusion and postoperative chest-tube output. Int J Angiol 22:31–36
73. Totaro P (2008) Obesity and coronary surgery: new concepts for an old problem. Expert Rev Cardiovasc Ther 6:897–903
74. Hernandez AV, Kaw R, Pasupuleti V et al (2013) Association between obesity and postoperative atrial fibrillation in patients undergoing cardiac operations: a systematic review and meta-analysis. Ann Thorac Surg 96:1104–1116
75. Chasse M, Mathieu P, Voisine P et al (2016) The underestimated belly factor: waist circumference is linked to significant morbidity following isolated coronary artery bypass grafting. Can J Cardiol 32:327–335
76. Parisian Mediastinitis Study G (1996) Risk factors for deep sternal wound infection after sternotomy: a prospective, multicenter study. J Thorac Cardiovasc Surg 111:1200–1207
77. Unek IT, Bayraktar F, Solmaz D et al (2010) The levels of soluble CD40 ligand and C-reactive protein in normal weight, overweight and obese people. Clin Med Res 8:89–95
78. Beavers CJ, Heron P, Smyth SS, Bain JA, Macaulay TE (2015) Obesity and antiplatelets-does one size fit all? Thromb Res 136:712–716
79. Farb MG, Bigornia S, Mott M et al (2011) Reduced adipose tissue inflammation represents an intermediate cardiometabolic phenotype in obesity. J Am Coll Cardiol 58:232–237
80. Tamminen M, Lassila R, Westerbacka J, Vehkavaara S, Yki-Jarvinen H (2003) Obesity is associated with impaired platelet-inhibitory effect of acetylsalicylic acid in nondiabetic subjects. Int J Obes Relat Metab Disord 27:907–911
81. Stohlawetz P, Folman CC, von dem Borne AE et al (1999) Effects of endotoxemia on thrombopoiesis in men. Thromb Haemost 81:613–617
82. Guthikonda S, Alviar CL, Vaduganathan M et al (2008) Role of reticulated platelets and platelet size heterogeneity on platelet activity after dual antiplatelet therapy with aspirin and clopidogrel in patients with stable coronary artery disease. J Am Coll Cardiol 52:743–749
83. Sibbing D, von Beckerath O, Schomig A, Kastrati A, von Beckerath N (2007) Impact of body mass index on platelet aggregation after administration of a high loading dose of 600 mg of clopidogrel before percutaneous coronary intervention. Am J Cardiol 100:203–205

84. Cuisset T, Frere C, Quilici J et al (2009) Relationship between aspirin and clopidogrel responses in acute coronary syndrome and clinical predictors of non response. Thromb Res 123:597–603
85. Cuisset T, Grosdidier C, Loundou AD et al (2013) Clinical implications of very low on-treatment platelet reactivity in patients treated with thienopyridine: the POBA study (predictor of bleedings with antiplatelet drugs). JACC Cardiovasc Interv 6:854–863
86. Cayla G, Cuisset T, Silvain J et al (2013) Prasugrel monitoring and bleeding in real world patients. Am J Cardiol 111:38–44
87. Pankert M, Quilici J, Loundou AD et al (2014) Impact of obesity and the metabolic syndrome on response to clopidogrel or prasugrel and bleeding risk in patients treated after coronary stenting. Am J Cardiol 113:54–59
88. Deharo P, Pankert M, Bonnet G et al (2014) Body mass index has no impact on platelet inhibition induced by ticagrelor after acute coronary syndrome, conversely to prasugrel. Int J Cardiol 176:1200–1202
89. Emerging Risk Factors Collaboration, Wormser D, Kaptoge S et al (2011) Separate and combined associations of body-mass index and abdominal adiposity with cardiovascular disease: collaborative analysis of 58 prospective studies. Lancet 377:1085–1095
90. Gray LJ, Yates T, Davies MJ et al (2011) Defining obesity cut-off points for migrant South Asians. PLoS ONE 6:e26464
91. Jensen MD, Ryan DH, Apovian CM et al (2014) 2013 AHA/ACC/TOS guideline for the management of overweight and obesity in adults: a report of the American College of Cardiology/American Heart Association Task Force on Practice Guidelines and The Obesity Society. Circulation 129:S102–S138
92. Sacks FM, Bray GA, Carey VJ et al (2009) Comparison of weight-loss diets with different compositions of fat, protein, and carbohydrates. N Engl J Med 360:859–873
93. Dansinger ML, Gleason JA, Griffith JL, Selker HP, Schaefer EJ (2005) Comparison of the Atkins, Ornish, Weight Watchers, and Zone diets for weight loss and heart disease risk reduction: a randomized trial. JAMA 293:43–53
94. Estruch R, Ros E, Salas-Salvado J et al (2018) Retraction and republication: primary prevention of cardiovascular disease with a mediterranean diet. N Engl J Med 368:1279–1290. N Engl J Med 378:2441–2442
95. Sumithran P, Prendergast LA, Delbridge E et al (2011) Long-term persistence of hormonal adaptations to weight loss. N Engl J Med 365:1597–1604
96. Fothergill E, Guo J, Howard L et al (2016) Persistent metabolic adaptation 6 years after "The Biggest Loser" competition. Obesity (Silver Spring) 24:1612–1619
97. Look Ahead Research Group, Wing RR, Bolin P et al (2013) Cardiovascular effects of intensive lifestyle intervention in type 2 diabetes. N Engl J Med 369:145–154
98. Marso SP, Daniels GH, Brown-Frandsen K et al (2016) Liraglutide and cardiovascular outcomes in type 2 diabetes. N Engl J Med 375:311–322
99. Bohula EA, Wiviott SD, Scirica BM (2019) Lorcaserin safety in overweight or obese patients. N Engl J Med 380:100
100. Nissen SE, Wolski KE, Prcela L et al (2016) Effect of naltrexone-bupropion on major adverse cardiovascular events in overweight and obese patients with cardiovascular risk factors: a randomized clinical trial. JAMA 315:990–1004
101. Piche ME, Auclair A, Harvey J, Marceau S, Poirier P (2015) How to choose and use bariatric surgery in 2015. Can J Cardiol 31:153–166
102. Schauer PR, Bhatt DL, Kirwan JP et al (2017) Bariatric surgery versus intensive medical therapy for diabetes—5-year outcomes. N Engl J Med 376:641–651
103. Fisher DP, Johnson E, Haneuse S et al (2018) Association between bariatric surgery and macrovascular disease outcomes in patients with type 2 diabetes and severe obesity. JAMA 320:1570–1582
104. Sjostrom L, Peltonen M, Jacobson P et al (2012) Bariatric surgery and long-term cardiovascular events. JAMA 307:56–65

Chapter 13
Inflammation and Epicardial Adipose Tissue in the Pathobiology of Atherogenesis and Neointimal Hyperplasia Following Coronary Intervention

Finosh G. Thankam, Mohamed M. Radwan, and Devendra K. Agrawal

Abstract The global incidence of coronary heart diseases (CHDs) has been increasing at an alarmin rate that demands an increasing attention to develop more effective therapeutic interventions and preventive strategies. The basic and applied research have significantly advanced the understanding on the molecular pathology of CHDs and opened multiple translational avenues in the management. Despite the significant enhancement of knowledge in the underlying pathophysiology, atherosclerosis remains a leading cause of global death and disability which is mainly attributed to alterations in LDL phenotype, endothelial dysfunction, inflammation and neointimal hyperplasia. Also, the integration of multidisciplinary elements of medical sciences, immunobiology, nutritional science, intervention biology, molecular signaling, vascular cell biology, animal models and translational medicine are warranted in designing improved management strategies. The critical discussion in this article insights into the underlying mechanisms associated with the degenerative changes and inflammatory events leading to atheroma and subsequent CHDs. In addition, the current understanding on the influence of high calorie diets is highlighted in relation to the molecular pathology of CHDs. Also, the prospectus and novel opportunities are discussed regarding next generation management strategies to address the pathological challenges associated with CHDs.

Keywords Atherosclerosis · Coronary heart diseases · Endothelial dysfunction · Neointimal hyperplasia · Oxidized LDL

F. G. Thankam · M. M. Radwan · D. K. Agrawal (✉)
Department of Translational Research, Western University of Health Sciences, 309 E. Second Street, Pomona, CA 91766, USA
e-mail: DAgrawal@WesternU.edu

P. S. Tappia et al. (eds.), *Biochemistry of Cardiovascular Dysfunction in Obesity*, Advances in Biochemistry in Health and Disease 20,
https://doi.org/10.1007/978-3-030-47336-5_13

Introduction

A recent report by American Heart Association (AHA) points that once in every 40 seconds an American experience myocardial infarction (MI). Annually, nearly 0.6 million newer cases of MI and around 200,000 recurrent heart attacks have been reported. Also, approximately 18.2 million people over the age of 20 suffer from coronary heart diseases (CHD) which accounts for 6.7% of the total US population [15]. These reports are alarming and reveal the increasing demand for more effective therapeutic interventions and preventive strategies. The past few decades witnessed significant evolution in the understanding of CHD pathology which resulted in the advancement of clinical approaches for its management. Atherosclerosis is the primary cause for CHDs and the current concepts identify atherosclerosis to be an inflammatory disorder rather than mere cholesterol storage disease, which is an outdated concept [89]. Despite the significant enhancement of knowledge in the underlying pathophysiology, atherosclerosis remains a leading cause of global death and disability.

Early atheroma formation begins during childhood, but seems to aggravate with age. The word atheroma is derived from two Greek words '*athera*' which means porridge and '*oma*' which means lump. CHDs are mostly due to atheromatous narrowing and subsequent occlusion of the lumen of coronary arteries or its branches [52]. Atherosclerosis is a degenerative disorder in which aging and chronic inflammatory components aggravate the pathology by altering the expression status of various regulatory and functional genes across different vascular beds. Even though most studies have focused on the inner (intimal) and the middle (medial) vascular layers as the target sites for atheroma formation, recent evidence revealed the implications of the outer (adventitial) vascular layers in the pathogenesis of CHDs [2, 159].

The association of nutritional factors in the pathogenesis of CHDs are relevant, as caloric restriction reveals promising results by lowering CHD risks. The higher dietary cholesterol intake promotes atherosclerosis in rabbit models and is one of the earliest clues regarding the influence of nutrition in CHDs [31]. The identification and isolation of cholesterol from atheromatous patients marks another breakthrough in the history of CHDs [3]. Another interesting observation supporting the nutritional influence of CHDs revealed a sharp drop in the mortality rate caused by CHDs in Northern Europe during the second World War owing to the shortage of food. However, the mortality rate increased drastically after the war [157]. These observations suggest that the consumption of a high calorie diet is an alarming risk for the development of CHDs.

In this chapter, we critically discussed the underlying mechanisms associated with the degenerative changes and inflammatory events leading to atheroma and subsequent CHDs. In addition, the current understanding on the influence of high calorie diets is highlighted in relation to the molecular pathology of CHDs.

Biology of Coronary Arteries

Coronary arteries that provide blood supply to the myocardium originate from the root of the aorta and form two primary coronary arteries, the right coronary artery (RCA) and the left main coronary artery (LMCA). The RCA originates from the anterior ascending aorta and supplies blood to the right atrium, the right ventricle, and the sinoatrial (SA) node and the atrioventricular (AV) nodes. The RCA branches into right posterior descending artery (PDA) and marginal artery (MA). Also, the RCA supplies to the septum of the heart in association with the left anterior descending artery (LAD). On the other hand, the LMCA branches off to the left anterior descending (LAD) and left circumflex (LCX) coronary arteries. The LAD supplies blood to both the front and the left side of the heart while the LCX is responsible for the blood supply to the left atrium and the posterior-lateral aspect of left ventricle. In addition, the coronary arteries give rise to small branches such as obtuse marginal artery (OMA), diagonal arteries (DA) and septal perforator (SP) [73, 92, 112]. Figure 13.1 depicts the main branches of the coronary artery.

The wall of the coronary artery is composed of various cell types and extracellular matrix (ECM) components including collagenous and elastic fibers. The continuous and rapid movement of blood within the coronary artery provides very limited opportunity for the blood to sufficiently nourish the vascular tissue. In addition, the increased thickness of the arterial wall hinders the easy diffusion/trafficking of metabolites and metabolic exhausts. The small blood vessels in the arterial wall tissue, called vasa vasorum, are responsible for nourishing the coronary artery. The vasa vasorum are located on the outer layer to withstand the pressure due to the

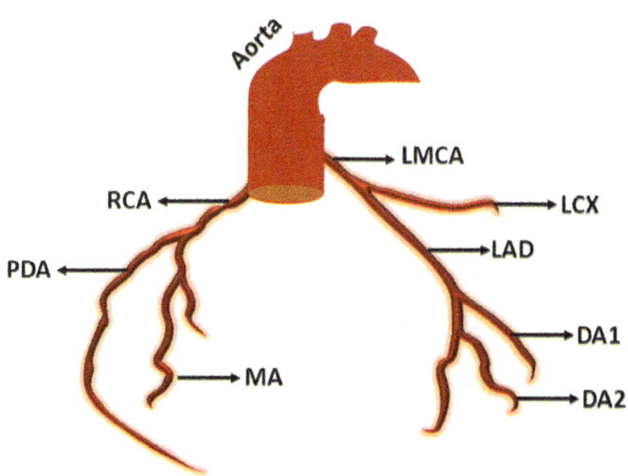

Fig. 13.1 Vascular branching of coronary arteries in the heart: the coronary arteries arise from the base of aorta and composed of two main arteries—the left main coronary artery (LMCA) and right coronary artery (RCA). LMCA branches into left circumflex artery (LCX) and left anterior descending (LAD). LAD gives rise to diagonal arteries (DA1 and DA2) and several minor branches. RCA branches mainly into right posterior descending artery (PDA) and marginal artery (MA)

blood flow. The remote location of vasa vasorum is thought to be a reason for the increased susceptibility of arteriosclerosis. In addition, there are minute nerve fibers called nervi vasorum which regulate the action of smooth muscle cells (SMCs) of the medial layer. The coronary artery is made up of three layers; the outer layer is tunica externa (adventitia), the middle layer is tunica media and the internal layer is tunica intima, as shown in Fig. 13.2.

The tunica adventitia is made up of a connective tissue that is mainly composed of collagen and elastic fibers, in which fibroblasts form the major cell type. The elastic nature of these fibers allows them to withstand the pressure exerted due to the blood flow. The vasa vasorum and nervi vasorum are found in the adventitial layer that aid to control the lumen size by regulating the inward and the outward remodeling responses [153]. In addition, the adventitia acts as the site for the recruitment of immune cells and subsequent inflammatory responses, especially in the case of vascular damage [182]. Moreover, the adventitia communicates with the surrounding tissue to facilitate the exchange of signals to and from the coronary circulation and forms a potential niche for the stem/progenitor cells which plays significant role in vascular growth, repair and regeneration. The plethora of cell population including

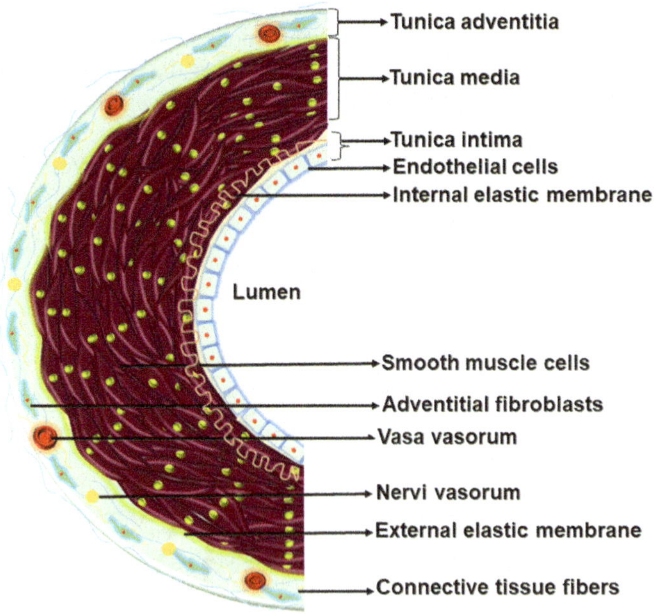

Fig. 13.2 Histological organization and structure of coronary artery: The coronary arteries composed of three major layers—tunica adventitia, tunica media and tunica intima. Tunica adventitia the outer layer constituted mainly by connective tissues and are characterized by the presence of small blood vessels and nerves. The medial layer is mainly constituted of smooth muscle cells (SMCs) whereas the intimal layer is constituted of endothelial cells. The internal elastic membrane separates intima from media and the external elastic membrane separates adventitia from media

fibroblasts, macrophages, T- and B- lymphocytes, mast cells, dendritic cells, pericytes, adipocytes and progenitor cells represents diverse cellular interactions and the homeostasis of these cellular functions is the prerequisite to maintain the vascular integrity [95].

The tunica media of coronary artery is mainly composed of multiple layers of SMCs (typically up to 40 layers of circumferentially or helically arranged cells) and connective tissue composed of collagen, proteoglycans and elastic fibers [144, 175]. The average thickness of medial layer of coronary artery is ~200 μm. The medial layer of coronary artery is equipped with lesser amount of elastic tissue and a greater number of SMCs when compared with other elastic arteries. The medial layer is separated from adventitia by external elastic membrane which is composed of interrupted layers of elastin. On the other hand, the internal elastic membrane separates media from the intima. Unmyelinated nerve axons are closely adherent to the outer border of the elastic membrane and the depolarization of SMCs are propagated throughout the media along the low-resistance gap junctions/nexus [175]. The mature SMCs express a unique repertoire of contractile proteins including smooth muscle myosin heavy chain (SM-MHC), alpha-smooth muscle actin (α-SMA), ion channel components, and other mediators required for maintaining the contractile function [117].

The tunica intima is composed of an elastic membranous lining embedded with smooth endothelial cells, a sub-endothelial layer of connective tissue and SMCs. The endothelium is responsible for the smooth luminal lining that prevents the adhesion of platelets and leukocytes and act as a selectively permeable diffusion barrier between the blood and the wall of coronary artery. The endothelial cells are aligned in a longitudinal fashion relative to the artery and communicate each other by zonula occulodens (tight junctions) and gap junctions [175]. Apart from the diffusion function, the endothelial cells exhibit metabolic and endocrine functions. The endothelium secretes anti-thrombotic agent prostacyclin (PGI_2), prothrombotic agent (von Willebrand factor), fibrinolytic agents, inflammatory mediators, growth factors, nitric oxide (NO), angiotensin II, and endothelin-1 and is the common site in the pathology of CHDs [176]. Also, these cells bear the receptors for low-density lipoproteins (LDL), thrombin and factor X. The endothelial cells (ECs) maintain homeostasis for the secretion of pro/anti-thrombotic substances and for the proliferation of SMCs by maintaining a collagenous basement membrane [145]. The histological architecture of coronary artery is displayed in Fig. 13.2.

Coronary Interventions

The current available treatment options of coronary heart diseases (CHD) are either: (i) Medical treatments which may include, coronary dilators such as nitrates, antiplatelets such as aspirin and clopidogrel, beta-blockers, calcium-channel blockers, angiotensin converting enzyme (ACE) inhibitors, ranolazine as metabolic modulator and statins, (ii) Percutaneous coronary intervention (PCI), or (iii) Coronary artery bypass graft (CABG) surgery [20, 123]. The selection of the treatment depends on the extent of the disease (e.g. the number of the affected vessels) the age of the patient and the presence of associated comorbidities such as diabetes mellitus [147]. The goal of the treatment, either the surgical (CABG) or interventional (PCI), is to restore the blood flow to the ischemic myocardium supplied by the affected segment, which subsequently, resolves the cardiac ischemic pain (known as angina pectoris), avoid the loss of the cardiomyocytes due to ischemic necrosis (known as myocardial infarction) and prevents the occurrence of the serious ischemic complications such as the cardiac dysrhythmias and heart failure. The outcome of these procedures, either surgical or interventional, is limited by the occurrence of vascular remodeling following intervention.

The PCI treatment consists of using a small balloon to dilate the narrowing in the coronary artery produced by the atherosclerotic lesion (atheroma) [53]. The recoil of the coronary artery after the balloon angioplasty has been a major challenge of this revolutionary procedure that lead to the invention of the coronary stents. The coronary stent is an expandable tube deployed into the affected segment of the coronary artery, following the balloon angioplasty, to prevent its recoil and keep it patent [33]. The stent may be bare metal, drug eluting or biodegradable [146]. The inevitable vessel trauma associated with the dilatation of the coronary artery, by inflating the balloon, and implantation of the stent, leads to denudation of the vascular endothelium and exposure of the rough subendothelial tissues to the blood stream [55]. The exposed subendothelial fibrous tissue leads to activation of blood platelets, release of the adhesion molecules and activation of the leukocytes which set the stage to a local inflammatory reaction [55]. The resultant neointimal hyperplasia, as mentioned above, impairs the effectiveness of the CHD treatment. It also progresses with time, resulting in recurrence of patient's symptoms which call for reintervention [190]. The surgical treatment (CABG) is also afflicted by the recurrence of the disease, either in the native coronary arteries (due to atherosclerosis) or in the surgically implanted venous graft(s) due to the development of NIH [18, 160]. This may require another interventional or surgical procedure. The CAD is commonly the disease of old age, and the physiological fitness of the patient to sustain an additional surgical intervention as CABG is of major concern. On the other hand, the percutaneous coronary intervention (PCI), although minimally invasive, is associated with several complications [154]. Based on these facts, the need for innovative strategies to mitigate the recurrence of the atherosclerotic coronary heart disease and the development of neointimal hyperplasia (NIH), or more optimally to prevent its occurrence all together is obvious.

Mechanism of Atherosclerosis

Atherosclerosis occurs due to the deposition of fat and/or fibrous tissue at the intimal layer of arteries. The subsequent accumulation of more fibrous components and calcium lead to the formation of atherosclerotic plaques which encroach the arterial lumen impeding/blocking the blood flow and lead to tissue ischemia. In addition, the atheroma triggers the thrombus formation which occludes the artery and offers an alternative route for tissue ischemia. Elevated serum cholesterol, LDL, genetic factors and sedentary lifestyle have been the major risk factors for the initiation and progression of atherosclerosis. The atheroma formation of coronary artery results in the ischemic death of myocardial tissue leading to MI. The clinical atherosclerosis has been classified into different subtypes based on the pathological features suggesting that significant progress has been attained in understanding the pathology of atherosclerosis as seen in Table 13.1 [184]. However, the translational advancements in the management of CHDs have not been achieved. This section focuses on the molecular pathology of atherosclerosis regarding the biochemical and immunological alterations associated with the structure of coronary artery.

History

Being an inflammatory disorder, the atherosclerosis has a very long history of 5,000 years in which the Egyptian papyri referred that heat and redness to be the concomitants of diseases which clearly signified inflammation. However, a classical definition of inflammation was given by the Roman encyclopedist, Anulus Cornelius Celsus in the first century as (Latin) *rubor* (redness), *calor* (heat), *dolor* (pain), and *tumor* (swelling) [87]. Centuries later, Jean Lobstein in 1829 coined the term arteriosclerosis [99]. Decades later, the German physician/pathologist Rudolf Virchow postulated the cellular pathology of atherosclerosis [99]. During the same time period, Carl von Rokitansky hypothesized the involvement of mechanical injury to vessel wall, endothelial dysfunction and inflammation in the pathogenesis of atherosclerosis [101]. This was proven by Mayerl et al.; after two centuries using the same human specimen collected von Rokitansky by identifying the accumulation of T cell population in the atherosclerotic lesion [99]. In 1910, the German chemist Windaus characterized and identified calcified connective tissue and cholesterol the atherosclerotic plaques [181]. The successful development of a rabbit atherosclerosis model by feeding a cholesterol rich diet conducted by Anitschkow and Chaltow in 1913, is a turning point in the history of atherosclerosis research, as these results paved the way for the identification of classical risk factors for atherosclerosis and CHDs [99]. The '*response to injury hypothesis*' proposed by Ross in the late twentieth century revealed the association of endothelial and intimal layer dysfunction and the immunological alterations in the pathogenesis of atherosclerosis [136]. The '*altered lipoprotein hypothesis*' suggested the priming role of oxidized-LDL (oxLDL) in the formation of foam cells in the intima [156]. Later, it was proven that the native LDL is transported to intima via endothelium where it undergoes oxidation and acts as chemoattractant for monocytes and SMCs. These cells uptake oxLDL leading to the formation of foam cells and this concept is called '*retention of modified LDL hypothesis*' [111]. A large body of literature have emphasized on the association

Table 13.1 Classification and features of atherosclerotic lesions

Classification	Subtype	Characteristics
Non-atherosclerotic intimal lesions	Intimal thickening	Presence of SMCs and absence of lipids, foam cells and thrombosis
	Intimal xanthoma	Minimal foam cells, absence of necrotic core, fibrous cap and thrombosis
Progressive atherosclerotic lesions	Pathological intimal thickening	Increased SMCs and GAGs, lipid accumulation and absence of thrombosis
	Fibroatheroma	Presence of necrosis, foam cells, increased lipid pool, increased cellular debris, ECM disorganization, with/without calcification and absence of thrombosis
	Plaque fissure	Increase in the size of necrotic core, hemorrhage, angiogenesis, plaque tear and absence of thrombosis
	Thin cap fibroatheroma	Thin fibrous cap, infiltration of macrophages and T cells, minimal SMCs, increased hemorrhage and absence of thrombosis
Thrombus-lesions	Plaque rupture	Disruption of fibrous cap, and thrombosis
	Plaque erosion	Intense intimal thickening, and thrombosis
	Calcified nodule	Eruptive calcification, fibro-calcification, necrosis and thrombosis
Healed lesions	Healed rupture/erosion/nodule	Increased SMCs, proteoglycans, collagen type III and necrosis, large sized calcification, inflammation, minimal necrotic core, luminal stenosis and absence of thrombosis

of chronic inflammatory components with endothelial function, lipid metabolism, and damage-associated molecular patterns (DAMPs)-mediated sterile inflammation in the pathogenesis of atherosclerosis [180]. However, the exact molecular events underlying the initiation of plaque formation are largely unknown and the history is awaiting to add groundbreaking findings in the field of atherosclerosis research.

Molecular Pathogenesis

LDL Modification and Initiation of Atherosclerosis

Lipoproteins function in the transport of cholesterol in the blood and the atherosclerosis results due to the accumulation of lipoproteins in the arterial intima. The lipoproteins <70 nm diameter (HDL, IDL and LDL and remnants of VLDL and chylomicrons) are permeable to the endothelial barrier and enter the intimal layer from the circulation [185]. In addition, the apoB-containing (non-HDL) lipoproteins interact with the proteoglycans (sulphated glycosaminoglycans, sGAGs) of intimal matrix resulting in the halting and subsequent accumulation of these lipoproteins around the subendothelial layer of intima [12]. The accumulated lipoproteins are highly susceptible to oxidation due to the action of oxidizing agents, proteases and lipases at the intimal layer [113]. The oxidized phospholipids (oxPL), especially oxLDL, and their derivatives are pro-inflammatory and pro-atherogenic as they trigger the recruitment and the activation of leukocytes [189]. The oxPLs moieties from oxLDL activate the endothelial cells and impair the permeability barrier. The CD14 receptors of dendritic cells act as ligands for oxPL leading to inflammasome-dependent hyperactivation of phagocytes [191].

The oxidation of LDL takes place in two phases. Initial phase represents the oxidation of lipid components without or little alteration in apoB100 and such partially oxidized LDL is referred as minimally-oxLDL (mini-oxLDL). The mini-oxLDL is negatively charged, possess affinity to LDL-R, activates anti-apoptotic pathways and upregulates pro-inflammatory cytokines [188]. The cytokine-mediated activation of inflammatory cells continues the oxidation of mini-oxLDL by promoting the oxidation of the remaining lipid moieties and protein components resulting in the loss of affinity to LDL-R and leads to complete oxidation of oxLDL. The oxLDL is recognized by scavenging receptors (oxLDL-R) and accelerates foam cell formation [155, 188]. In general, the circulatory LDL is resistant to oxidation as the serum lipoproteins are protected by the antioxidant defense and is comprised of the antioxidant vitamin, alpha tocopherol, as a transport vehicle [68]. However, the LDL encounters several cell/tissue derived pro-oxidants in the subendothelial space including transition metal ions (iron and copper), hemin and oxidizing enzymes such as myeloperoxidase, and lipoxygenase [121]. In addition, nitric oxide (NO) in presence of superoxide, released during oxidative stress, triggers the oxidation of LDL [188]. Moreover, the hyperglycemia results in the glucose-mediated peroxidation of LDL by superoxide and advanced glycation end products (AGEs) mediated glycation of apo-B100 by Maillard reaction [69]. The common mediators and the mechanism of LDL oxidation are displayed in Table 13.2.

The increased accumulation of oxLDL in the intima leads to its aggregation which in turn nucleates the atherosclerotic plaque formation. The uptake of oxidized, proteolyzed and/or lipolyzed LDL or cholesterol crystals by macrophages, DCs and SMCs results in cytoplasmic lipid droplets leading to foam cell formation; the typical hallmark of atherosclerosis [130]. Intracellular cholesterol crystals have also been

Table 13.2 Mediators and mechanism of LDL oxidation

Mediators	Mechanism	Reference
Copper ion	Free radical generation	[121]
Ferric ion	Free radical generation	[121]
Ceruloplasmin	Providing catalytically active copper ions	[81]
Lipoxygenase	Direct oxidation of lipid components and production of hydroperoxides	[188]
Myeloperoxidase	Generation of secondary radicals	[188]
Peroxynitrate	Hypochlorous acid mediated oxidation	[118]
Thiols	Oxidation of -SH groups	[120]
Xanthine oxidase	Generation of superoxide	[121]
Reactive oxygen species	Peroxidation of lipids	[121]

reported in foam cells [150]. Similar to oxidation, the proteolysis of LDL by the serene protease cathepsin G (CatG, also called neutrophil protease) triggers its interaction with the intimal matrix. In addition to neutrophils, the CatG is also expressed in mast cells, mononuclear leukocytes and B lymphocytes (non-plasma) [23]. Interestingly, CatG stimulates the degradation of LDL in blood plasma and prevents the risk of atherosclerosis in LDL-R (LDL-receptor)-deficient mouse model [177]. Contrastingly, the information regarding the mechanism for the pro-atherogenic role of CatG in the intima by promoting the LDL trap is unavailable and warrants further research.

The phagocytosis of matrix-bound oxLDL by macrophages, the pinocytosis of native LDL cholesterol by SMCs and the upregulation of the lectin-like oxLDL-R (LOX-1) are tightly associated with the scavenging of ox-LDL from the intima [24]. Also, the cholesterol clearance from the intima occurs via the HDL-mediated reverse cholesterol transport. Moreover, under normal physiology the LDL-C (LDL-cholesterol) move to the target tissue without being trapped in the intima [62]. Hence, lowering the cholesterol burden in the intima using the strategies which prevent LDL trap and immune system activation could help the safer exit of LDL-C prior to the uptake by macrophages and SMCs and prevent the risk of foam cell formation.

Inflammation

The accumulation of oxLDL alters the cellular function and elicits inflammatory responses in arterial wall which drives atherosclerosis. The early innate immune responses are characterized by the increased population of macrophages followed by the adaptive responses mediated by T and B cells. Macrophage colony-stimulating factor (M-CSF) signaling activates the macrophages to internalize the LDL leading to foam cell formation. OxLDL at the sub-endothelial space acts as DAMPs to activateTLR4 receptor and triggers the downstream inflammasome signaling in macrophages [142, 168]. The activation of inflammation by oxLDL upregulates the chemokine monocyte chemoattractant factor protein-1 (MCP-1) resulting in the recruitment of more monocytes to the atherosclerotic lesion [28]. The classical monocytes ($CD14^+CD16^-$) are the major contributors for the plaque macrophages and play a key role in atherogenesis [135]. The recruited monocytes undergo local proliferation and differentiation to macrophages and activates the scavenging of oxLDL via scavenging receptors (SR) including SR-A1, SR-B2 (CD36), and LOX-1 (E1) [26]. These SRs are regulated by NF-κB which is the master switch for inflammatory cytokines [58].

The cholesterol uptake via oxLDL is counterbalanced by ATP-binding cassette (ABC) transporter A1 and/or G1-mediated cholesterol efflux by the macrophages [186]. ABC transporters facilitate the incorporation of cholesterol to apolipoprotein A1 and HDLs containing apoA1 or apoE for reverse cholesterol transport and subsequent metabolism in liver. The imbalance between cholesterol intake and uptake results in foam cell formation [104]. In addition, the precipitation of cholesterol within the cells forms crystals that trigger inflammasome activation, apoptosis and necrosis which in turn lead to the development of atherosclerotic necrotic core. The necrotic core characterized with cellular debris and lipids and is highly thrombogenic in nature which is separated from the bloodstream by a fibrous cap. The irregularity or the rupture of the fibrous cap initiates the intraluminal thrombus formation and pave the ways to the development of cerebrovascular stroke and MI [104]. Interestingly, the lipid accumulation has been identified to begin in circulating monocytes to develop a foamy phenotype which subsequently migrates to the atherosclerotic lesion and aggravates the pathology [183].

The retention of local inflammatory signals following the LDL trap triggers the influx of more monocytes/macrophages and lymphocytes to the intimal layer [88]. The recruited $CD4^+$ T cells, especially the T helper type 1 cells (T_H1), recognize the oxLDL as an antigen and secretes the proinflammatory mediators such as interferon-γ (IFN-γ) and tumor necrosis factor-α (TNF-α) [44]. In addition, the IL-6 released from the cells of arterial wall triggers acute phase response by secreting C-reactive protein (CRP) by the liver to the systemic circulation [56]. Apart from T_H1 cells, T_H2 cells, natural killer cells and $CD8^+$ T cells have been identified from the atherosclerotic lesions [56]. The recruitment of lymphocytes and the sustenance of cytokines persist the localized inflammation resulting in the inefficient efferocytosis (the phagocytic clearance of apoptotic or necrotic cells) which eventually initiates the formation of atherosclerotic plaques with a central necrotic core of lipids and a fibrous cap [171]. Usually, the population of T_H1 cells exceeds the number of T_H2 cells in the atherosclerotic lesions. Moreover, IL-12 and IL-18 secretion by the activated macrophages trigger the activation of T-bet and T_H1-skewing in the progenitor cells leading to their differentiation into T_H1 cells and subsequent IFN-γ release [38]. This suggests that the IFN-γ-IL-12-IL18 axis of pro-inflammatory cytokine signaling accelerates the formation of plaques. The exact role of T_H2 cells in atherosclerosis is debatable, however is considered to be athero-protective as these cells prevent the differentiation of IFN-γ secreting T_H1 cells via IL-4 signaling. Contrastingly, the IL-4 is responsible for the upregulation of class A scavenging receptors, vascular cell adhesion molecule-1 (VCAM-1), MMP1 and MCP1 in macrophages which are pro-atherogenic mediators [171]. IL-5, another cytokine released from T_H2 cells, elicits its athero-protective role by enhancing IgM secretion by B-1 cells to neutralize the oxLDL [17]. However, Davenport et al.; reported that the T_H2 cells elicit atherogenic responses at the chronic stage of plaque formation [32]. Information regarding the potential role of T_H2 cells in the pathology of atherosclerosis is vague and warrants further investigation.

The upregulation of IL-6 and TGF-β in the vascular lesion activates T_H17 cells to secrete IL-17 family of cytokines, especially IL-17A and IL-17F, and IL-22. IL-17 signaling activates NF-κB, ERK1/2, CCAAT/enhancer binding protein β and C/EBPδ which results in the production of pro-inflammatory cytokines including TNF, IL-1β, and IFN-γ [171]. In addition, the immune responses triggered by oxLDL and collagen V promote T_H17 activation via IL-6 signaling. Moreover, the increased ROS and subsequent activation of cAMP response element binding protein (CREBP) associated with atherosclerosis induce IL-17 suggesting its role in vascular tissue inflammation [165]. Neutralization of IL-17A, using antibody, prevented the expansion of fibrous cap in the atherosclerotic lesion suggesting its role in plaque stability [50]. Contrastingly, the upregulation of IL-17 was shown to inhibit the effects of IFN-γ along with the upregulation of athero-protective mediators such as IL-5 and IL-10 [163]. Also, the VCAM-1 inhibitory effects of IL-17 result in reduced T cell accumulation within the lesion suggesting the athero-protective function of T_H17 cells [163]. Collectively, these evidences suggest that T_H17 cells exhibit contrasting roles as atherogenic and athero-protective agent depending on the inflammatory status of the vascular tissue [164].

T_{reg} cells are immuno-suppressors and minimize the immune responses without antigen exposure thereby playing an athero-protective role. The sub-population of T_{reg} cells named natural T_{reg} cells (nT_{reg} cells) are characterized with the expression of CD4, CD25 (IL-2R) and FoxP3 and secrete the anti-inflammatory cytokines IL-10 and TGF-β [171]. Specifically, the IL-10 is released by the type 1 regulatory T cells (T_R1 cells) and TGF-β by T_H3 subset of T_{reg} cells [132]. T_R1 cells significantly decreased T_H1 response and IFN-γ secretion, increased IL-10 production, inhibited foam cell formation, activated the polarization of M2 macrophages and subsequently prevented the plaque formation [8, 90, 96, 116]. The T_{reg} cells exhibit a protective role in atherosclerosis which were proven in animal models, however human data is limited on this aspect which impedes their translational potential.

NK cells are the subset of T cells which express NK1.1, Ly49, CD4 and TCR surface markers and secrete anti-inflammatory cytokines such as IL-4, IL-10 and IL-13 [77]. NK cells were reported to augment the plaque formation and have been detected in human carotid artery plaques and the atherosclerotic tissue from abdominal aortic aneurysms [171]. MCP-1 and fractalkine (CX3CL1) are the key signals for the recruitment of NK cells to the lesion site. In human patients the circulating NK cells express CD160 and NKG2C (NK cell Group 2 isoform C) which trigger the cytotoxicity and cytokine secretion suggesting its contribution to atherosclerosis. In addition, NK cells were reported to promote atherosclerosis via $CD4^+$ T cell dependent mechanism [79]. The overall cellular and cytokine transit at the lesion site is

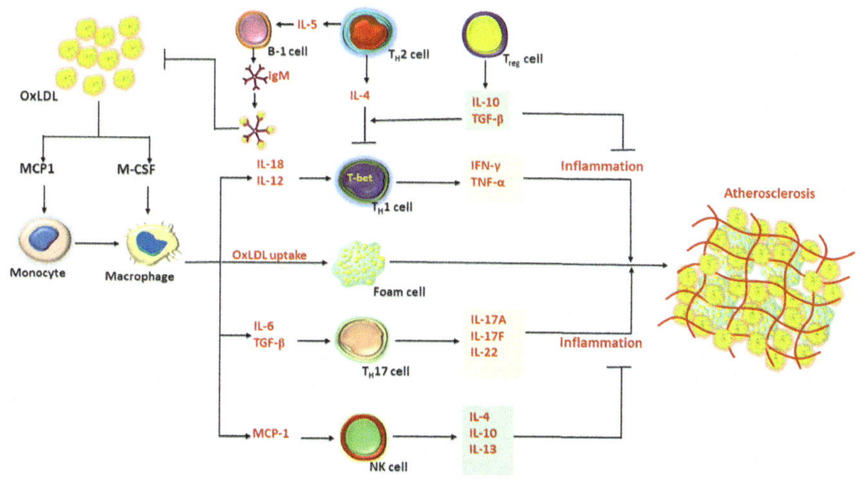

Fig. 13.3 The immune cells and cytokine transit at the site of atherosclerotic lesion: the ox-LDL at the intima of coronary artery recruits and activates monocytes and macrophages via MCP1 and M-CSF signaling resulting in the increased uptake of ox-LDL to form foam cells. The secretion of macrophages especially IL-18 and IL-12 activates T_H1 cells and upregulates the secretion of pro-inflammatory cytokines IFN-γ and TNF-α leading to inflammation. Similarly, IL-6 and TGF-β trigger the release of another school of pro-inflammatory cytokines including IL-17A, IL-17-F and IL-22 from T_H17 cells. The persistence of inflammatory signals accelerates atherosclerosis. On the other hand, the MCP1 signaling activates NK cells to secrete anti-inflammatory cytokines IL-4, IL-10 and IL-13. In addition, the activation of T_H2 cells and T_{reg} cells prevents the progression of inflammation via the inhibition of T_H1 signaling. IL-5 released from T_H2 cells stimulates B cells to secrete IgM against ox-LDL

depicted in Fig. 13.3. Even though, most reports reveal the association of NK cells with the pathology of atherosclerosis, the mechanistic studies to reveal the precise role of NK cells are limited, which warrants further investigations.

Progression of Atherosclerosis

The formation of atheromatous lesions imparts hypoxia and subsequent hypoxic responses including neo-angiogenesis [122]. The hypervascularity within the plaque due to angiogenesis results in plaque rupture and leads to acute complications. In addition, the intra-plaque hemorrhage leads to the formation of unstable plaques, which are at the higher risk of rupture. Furthermore, the RBCs within the plaque act as reservoirs for cholesterol and phospholipids which facilitate the expansion of the necrotic core and foster inflammation [75]. The cells of innate immune system secrete the proangiogenic factors including vascular endothelial growth factor (VEGF) and basic fibroblast growth factor (bFGF) and MMPs for facilitating ECM degradation and activating the growth factors [25, 104, 124]. Apart from the lipid laden macrophages, metaplastic SMCs also contribute to foam cell formation. These SMCs are believed to be migrated from the media to the growing atherosclerotic plaque in the intima and proliferate over years leading to the deposition of ECM components to the plaque [16].

Human pre-atherosclerotic lesions (diffused intimal thickening, DIT) are mainly composed of SMCs, proteoglycans and elastin, however lack macrophages and thrombus. These SMCs (synthetic phenotype) in the intimal layer are heterogeneous, characterized by the increased content of synthetic organelles such as ER, ribosomes and mitochondria (decreased expression of contractile proteins and increased ECM secretion) [4, 9, 105, 161]. In addition, the synthetic SMCs express minimal level of cholesterol esterase and ABC transporter A1 resulting in foam cell formation [71]. The ECM deposited by these SMCs plays a central role in the initiation and progression of atherosclerosis owing to the electrostatic interactions between the negatively charged proteoglycans especially chondroitin sulfate and heparan sulfate and the positively charged apolipoproteins especially ApoB leading to the lipid burden in the intima [14]. In addition, the SMCs in the intima express minimal level of α-SMA when compared with that of medial layer [149].

The continuous deposition of ECM and lipid trap leads to fibroatheroma which are characterized by fibrous cap, necrotic core and impaired efferocytosis of SMCs and macrophages that subsequently leads to secondary necrosis, release of DAMPs and subsequent sterile inflammation. Eventually, the proteoglycans are replaced with collagens mainly type 1 and 3 which are secreted by SMCs [14]. The proteomics of lipid loaded secretome of SMCs challenged with TGF-β, PDGF, IL-1β, Ang II, cholesterol and mechanical stretch revealed increased collagen synthesis confirming the fibrotic potential of SMCs in the intimal layer [7]. The phenotype switch of SMCs and their migration to intima are identified to be the key events associated with the progression of atherosclerosis. Little is known regarding the underlying molecular mechanism; however, offers immense translational potential.

Calcification

Vascular calcification is associated with CVDs including hypertension, congestive heart failure, cardiac hypertrophy, and ischemia [5]. The evidence of atherosclerotic calcification has a history of more than 5000 years which was documented in an autopsy of the mummy of an Egyptian woman and also in the radiographs of naturally mummified man discovered in the Tyrolean Alps [85, 107]. Generally, the calcification of arteries occurs in the intimal layer, however calcium deposits were also detected in the adventitial and medial layers [82, 86]. In the early stages of fibroatheroma, the calcification occurs as granules in the necrotic core and adjacent ECM which involves macrophage- and SMC-derived calcifying microvesicles along with the activity of osteochondrogenic cells [63, 131]. The osteochondrogenic differentiation of SMCs are confirmed by the upregulation of pro-osteogenic transcription factors Runt-related transcription factor 2 (RUNX2) and core binding factor a1 (Cbfa1), and the osteochondrogenic markers including osteocalcin, alkaline phosphatase and type II collagen which in turn are upregulated by oxLDL and inflammation [14, 126]. The microcalcification coalesces to form larger speckles which form sheets in the matrix of fibroatheroma. Fibrin encapsulation on the fragments of these sheets leading to the formation of calcium nodules which project to the lumen and accelerate thrombosis [174].

The calcification of the intima occurs as a result of several mechanisms including the nucleation of apatite crystals by the cellular debris followed by apoptosis/necrosis, circulating nucleational components released by remodeling bone or locally by the secretome of macrophages or SMCs, downregulation of the inhibitors of mineralization and osteogenic trans-differentiation of SMCs [47]. As a protective mechanism, the endothelial cells secrete matrix GLA protein which mediates the vitamin-K dependent inhibition of vascular calcification [27]. Even though the inflammation triggers calcification, the advanced calcified plaques are resistant/tolerant to inflammation leading to the propagation of the lesion and impaired healing [148]. This leads to plaque rupture due to the unfavorable mechanical integrity caused by the microscale calcium deposition and increased local stress resulting from the thinning of fibrous cap owing to the interfacial debonding [173]. Also, the persistence of adaptive immune responses promotes fibrosis and stabilizes the plaque. In addition, the transdifferentiated SMCs acquire osteoblasts like phenotypes and orchestrate regulated calcification which in turn acts a barrier to prevent the inflammation [148].

Usually, the calcification is limited to the subintimal layer and begins in the second decade of life following the formation of fatty streaks which increases with the age [6]. In addition, the microcalcification alters endothelial function and leads to plaque stabilization [6]. The unstable inflammatory microcalcification (0.5–2 μm) proceeds to more stable fibrotic macro-calcified (>2 to >5 μm) plaques by subsiding inflammation due to the phenotype switch of T_H1 to T_H2 lymphocytes, M1 to M2 macrophages and SMCs to osteoblast like phenotypes [57, 115]. The T_H2 and M2a macrophages trigger TGF-β signaling to enhance fibrosis via the cytokines IL-4 and IL-13 [108]. For accelerating the wound healing, the M2a macrophages undergo further phenotype shift to form M_{reg} or M2c sub-type by the activity of T_{reg} cells [36, 84]. The proteins secreted by M2c macrophages include IL-10, resistin-like molecule-α (RELMα), MMPs (MMP-9, MMP-13, MMP-2) and arginase-1, which suppress the inflammation, inactivate the myofibroblasts and promote fibrolysis [126]. The mechanism of pathogenesis of atherosclerosis is displayed in Fig. 13.4.

Endothelial Dysfunction

Endothelial lining of the whole human body is roughly equivalent to the weight of liver and occupies the total area of nearly 6 tennis courts [61]. Endothelium is the major regulator of vascular wall homeostasis which physiologically maintains the relaxed vascular tone, minimal oxidative stress regulated by releasing nitric oxide (NO), PGI_2, and endothelin-1, regulation of vascular permeability, platelet and WBC

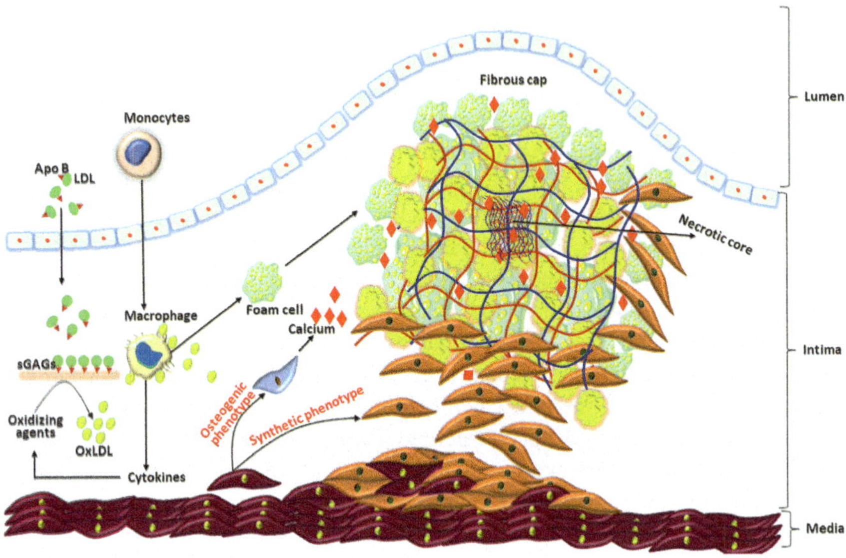

Fig. 13.4 The general mechanism of the development of atherosclerosis: the circulatory LDLs diffused to the intimal layer are trapped by the GAGs of the intimal ECM making it available for the oxidizing machineries leading to ox-LDL burden. The subsequent activation of macrophages and inflammatory episodes leads to the formation of foam cells followed by the creation of necrotic core which grows and matures to atherosclerotic plaques. The phenotypic switch of SMCs to osteogenic and synthetic lineages leads to the calcification and infiltration of SMCs to intima, respectively. The continuation of these cellular and biochemical events results in the increased thickness of intima and ultimately leads to occlusion of the lumen

adhesion and control of AngII activity [152]. The atherogenic risk factors disturb the endothelial homeostasis and affect the production of endogenous vasodilators such as NO by the endothelial cells, upregulate the adhesion molecules such as vascular cell adhesion molecule-1 (VCAM-1) on endothelial cells (ECs) and binds to circulating WBCs and chemoattractants that facilitate the entry of adhered WBCs to the intima [30, 65]. In addition, alterations in the local hemodynamics, are sensed by flow-dependent ion channels or surface molecules including integrins, affecting the endothelial function. Such flow patterns disturb the atheroprotective functions of the endothelium such as vasodilation, anti-thrombotic, and anti-inflammatory properties [49].

Under normal conditions the ECs remain in quiescent phase, however, the pathological stimuli including disturbances in blood flow and increased unidirectional laminar shear stress result in their activation. ECs respond to the pre-atherogenic stimuli by activating the expression of a battery of atheroprotective genes by upregulating the transcription factor Kruppel-like factor 2 (KLF2) resulting in their anti-inflammatory and anti-thrombotic phenotype [169]. However, the persistence of atherogenic signal and subsequent activation of NF-κB-HIF-1α axis increase the level of pro-inflammatory cytokines and stimulate the mechano-transducers, Yes-associated protein (YAP) and transcriptional coactivator with PDZ-binding motif (TAZ), resulting in the shift of ECs to pro-inflammatory phenotype [41, 178]. The increased hypoxia and the accumulation of hemoglobin:haptoglobin complexes at the lesion site activate the secretion of VEGF by macrophages which in turn stimulate intra plaque angiogenesis and disturbs the endothelial integrity leading to vascular permeability and unstable plaques [54]. In addition, the ECs at the lesion undergoes endothelial-mesenchymal transition leading to the increased pool of mesenchymal cells which in turn contribute to plaque instability by upregulating MMPs [40].

On encountering the DAMPs released as a result of pre-atherogenic stress, the ECs stimulate the production of effector proteins via the activation of NF-κB which include inducible endothelial-leukocyte adhesion molecules (ELAM) especially E-selectin and VCAM-1, procoagulants such as tissue factor, and the chemokines including IL-8, IL-18, and MCP-1 leading to the pro-inflammatory endothelial phenotype [49]. VCAM-1 exhibits selective adhesion potential to mononuclear WBCs and lymphocytes via VLA-4 receptor and aggravates the inflammation [39]. OxLDL and oxidized lysophosphatidylcholine trigger the expression of VCAM-1 paving the way for the atherogenesis and VCAM-1 is implicated as a biomarker for endothelial dysfunction, [78]. Moreover, the continuous activation of NF-κB results in the formation of super-enhanced (multiple enhancers bound by an array of transcription factors to drive the expression of diverse genes associated with cellular events such as inflammation) complex in EC genome signifying an epigenetic level of regulation of proinflammatory endothelial cells in atherogenesis [22].

The early lesions in human and animal models are characterized by the formation of distinct geometry associated with arterial branch points and the regions of altered hemodynamics [29, 48]. These conditions suggest the longer dwell time allowing the LDL to permeate the endothelial lining as the disturbed flow physically damages the endothelial integrity. The shear stress response elements (SSREs) in the promoter region of VCAM-1 and other atheroprone genes including PDGF and eNOS regulate the gene expression responding to the alterations in hemodynamics [49]. The KLF2 expression antagonizes NF-κB-dependent pro-inflammatory pathways and elicits the atheroprotective effects by restoring the barrier function, metabolism, and the release of reparative miRNAs via exosomes [10, 59, 91]. Furthermore, the KLF2-mediated production of autocoids (paracrine molecules which act similar to local hormones) such as NO and natriuretic-peptide C (CNP) elicits atheroprotection [11].

The vascular wall is highly susceptible to oxidative stress due to the presence of active pro-oxidant systems including xanthine oxidase, mitochondrial respiratory chain, lipoxygenases, uncoupled eNOS and NADPH oxidases. The antioxidant defense counter balances the oxidant system and the disturbances in the oxidant-antioxidant homeostasis lead to oxidative stress [97]. Increased oxidative stress and impaired antioxidant defense induce structural modification to LDL leading to the formation and exposure of new antigenic epitopes to the macrophages which give rise to clonal expansion of LDL specific T cells [106]. In addition, the oxidative stress alters the structure and function of β2-glycoprotein which in turn triggers T_H1

cell response. Also, the hyperlipidemia impairs L-arginine-NO pathway inducing AngII-type-1 receptor and subsequent vasoconstriction which alters the hemodynamics [106]. The increased administration of antioxidants and regular exercise are beneficial to improve the endothelial healing following the oxidative stress [61].

Neointimal Hyperplasia (NIH)

Coronary interventions including balloon angioplasty (with or without stenting), endarterectomy and/or CABG have benefitted millions of CHD sufferers across the world. However, these therapeutic interventions fail from restenosis due to neointimal hyperplasia (NIH) [190]. It has been demonstrated that the development of NIH and atherosclerosis share similar pathological mechanisms there by referring NIH as in-stent neo-atherosclerosis. Since NIH occurs in an accelerated rate than atherosclerosis, it has been referred as accelerated atherosclerosis [70]. In addition, NIH exhibits a prevalence of 30% within 1 year following the coronary interventions and still remain as the leading cause of thrombosis associated with stent/graft failure [114, 184]. NIH refers to post-intervention remodeling of coronary artery due to the unregulated proliferation and subsequent migration of SMCs from the medial layer to intimal layer resulting in wall thickening and gradual occlusion of the lumen [18]. The SMCs associated with NIH switch contractile phenotype to secretory/synthetic phenotype to release the mediators, growth factors, receptors, ECM components, MMPs and cytokine, accelerating the progression of NIH. Since the coronary interventions and other therapeutic manipulations exert the risk of vascular injury, the subsequent inflammatory cascade plays a significant role in the development of NIH [190]. This section throws light to the current understanding of molecular pathogenesis, the association of epicardial adipose tissue (EAT), influence of high calorie diet and translational avenues in NIH.

The accumulation of SMCs and fibroblasts with subsequent increase in the deposition of ECM in the intimal layer leading to the narrowing of lumen is the major histological hallmark of NIH. In general, the thickness of medial layer remains unaltered while the intimal layer expands [110]. The growth factors including PDGF, EGF and FGF and the cytokines such as IL-6 and IL-8 are responsible for the dedifferentiation of SMCs to form the synthetic phenotype, however, IGF-1 reverses the phenotypic switch [103]. In addition, the adventitial fibroblasts transform to myofibroblasts and migrate through media to acquire SMC-like phenotype contributing to NIH [137]. Furthermore, the circulating bone marrow-derived progenitor cells, platelets and mononuclear WBCs are involved in the formation of neointima [103, 141]. However, the exact molecular mechanisms underlying the phenotype switch of SMCs and NIH formation are unknown.

The cellular proliferation signaling including Ras–MAPK and PI3K–Akt are associated with NIH development. Also, MAPK-mediated activation of the transcription factors such as Elk-1 and Sap1 are believed to stimulate the proliferation, migration and dedifferentiation of SMCs. In addition, the survival of secretory SMCs are facilitated by insulin-dependent Akt pathway in response to the nutritional status via mTOR signaling [109]. The cellular events associated with SMC migration such as actin polymerization, cell–cell and cell-ECM adhesion, microtubule remodeling, and myosin force generation are promoted by growth mediators such as VEGF, PDGF, bFGF and TGF-β [46]. The growth factors and AngII enhance the ECM components secreted by SMCs and facilitate the ECM deposition at the intima [109]. The inhibition of SMC migration has been achieved by controlling the activity of MMPs using TIMP-1, -2 and -3 which in turn prevents NIH formation and enhanced reendothelialization [138].

In addition to the cells of vascular wall, the tissues adjacent to the coronary artery also influences the pathogenesis of atherosclerosis and NIH. For example, the epicardial adipose tissue (EAT) has proven to be associated with the formation of atherosclerosis and subsequent cardiac complications [1, 192]. Generally, the EAT is directly associated with the adventitial layer of coronary artery (Pericoronary epicardial adipose tissue) and pericardium (myocardial epicardial adipose tissue) which elicit both autocrine and paracrine signaling to secrete several pro-inflammatory mediators. Also, the EAT is composed of adipocytes, ganglia and network of nerves, stromovascular and immune cells [100]. However, the metabolism and the interactions among these cells contributing to NIH are largely unknown. However, it has been reported that the EAT-derived mediators contribute to endothelial dysfunction by aggravating inflammation and oxidative stress which ultimately result in the migration of SMCs [67].

Interestingly, EAT is characterized with immunopositivity of several inflammatory biomarkers such as CD3 (lymphocytes), CD68 (macrophages), and that for mast cells suggesting the inflammatory phenotype of EAT. Furthermore, the inflammatory cells in EAT reflect the vascular pathology and plaque instability or NIH following coronary intervention [100]. The vasocrine secretion of EAT includes the adipokines such as TNF-α, MCP-1, IL-6, IL-1β, resistin and others which contribute to the inflammatory milieu and promotes NIH [128, 143]. In addition, the EAT triggers innate immune responses via TLRs leading to the activation and nuclear translocation of NF-κB resulting in the expression of a battery of proinflammatory genes. Lipopolysaccharide (LPS), the classical ligand for TLR, has been found to be increased in the systemic circulation of CHD patients, however, the origin of LPS in these patients are largely unknown [13]. Moreover, the concomitant increase in the expression of genes associated with oxidative stress in EAT of CHD subjects suggest the possible interaction between the vascular tissue and EAT [139]. Similar to LDL, the epicardial fat also affects the endothelial function as the secretory type II phospholipase A2 (sPLA2-IIA), the key enzyme involved in the retention of LDL in the subendothelial space, was found to be upregulated in the EAT of CHD patients [37].

Apart from the secretion of adipokines, the EAT exerts mechanical effects to contribute to atherosclerosis and NIH. Under physiological condition EAT attenuates coronary artery torsion whereas under pathological circumstances EAT leads to vessel expansion owing to its compressibility [125]. EAT-derived angiopoietin-like protein (Angptl) 2, a pro-inflammatory mediator, augments NIH following the vascular interventions by upregulating TNF-α, IL-1β and MMP2 [170]. In addition, EAT-derived mediators including TGF-β, leptin, and viafastin facilitate the migration and proliferation of SMCs leading to NIH [102, 127]. Accumulating evidence from literature suggests that EAT operates diverse pathways which are linked to the pathogenesis of NIH following the vascular injury [64]. However, the currently available experimental models and research strategies are incomplete which warrant further investigations for the translation to clinical arena.

High Fructose Diet and NIH/Atherosclerosis

Chronic overnutrition is a major risk factor for cardiovascular disease and is a prevalent complication following reparative surgery and/or coronary interventions [98]. In addition, the high sugar intake and subsequent hyperglycemia promote the formation of NIH following the vascular intervention [21]. The global consumption of sugar-sweetened beverages, and processed foods have drastically increased in all age groups which remain as the major source of dietary fructose [187]. Increased consumption of fructose has been correlated with the increased body weight, hyperlipidemia, upregulation of circulating and tissue level pro-inflammatory cytokines, hypertension and endothelial dysfunction [162, 187]. Methylglyoxal (MG), an intermediate of fructose metabolism, has been tightly associated with atherosclerosis and diabetes [93]. The impact of high fructose intake has raised serious concerns in CVD research and this section outlines the association of high fructose diet and NIH.

It has been found that the ingestion of high fructose diet has negative effects on the human health, particularly cardiovascular health [94]. High fructose diet is an inflammatory diet that induces oxidative stress through uninhibited hepatic fructose metabolism [140]. Fructose metabolism does not involve the glycolysis rate limiting enzyme phosphofructokinase [140]. Unlimited fructose metabolism continues to add calories and overloads the mitochondria with its metabolic intermediates, resulting in mitochondrial stress and increased reactive oxygen species [94, 167]. Moreover, it fuels the hepatic lipogenesis [94]. High fructose intake (e.g. in corn syrup) induces the production of the hepatic glucose, probably through promoting the transcription of glucose 6 phosphatase gene via activating the Carbohydrate Response Element Binding Protein (ChREBP) [72]. High fructose diet was reported to promote hepatic insulin resistance in a differential manner, it impairs the insulin's inhibition of hepatic gluconeogenesis (glucose production) while maintaining the hepatic lipogenic effect (hepatic do novo lipogenesis). This leads to hyperlipidemia, hepatic steatosis (fatty infiltration of the liver), and hyperglycemia [72]. These metabolic derangements result in obesity and metabolic syndrome.

Metabolic syndrome and obesity are characterized by chronic persistent inflammation and high levels of the systemic proinflammatory cytokines [179]. Free fatty acids (FFAs) which are released from the dead hypertrophied fat cells of the obese person activate the immune cells and induce innate immunological reaction [129]. These FFAs act as ligands for TLR receptors, producing and perpetuating a chronic systemic inflammatory process [158]. High fructose intake alters the gut microbiome and bypasses the liver tolerance producing a state of systemic endotoxemia [34, 35, 129]. This endotoxemia further induces generalized inflammatory response. High fat diet results in insulin resistance via desensitizing the insulin receptors, by serine phosphorylation of the insulin receptor substrates (IRSs) via activation of protein kinase C (PKC) isoenzymes, producing type 2 diabetes mellitus (T2DM) [45, 80]. Obesity, T2DM, hyperlipidemia and inflammation induce neointimal hyperplasia (NIH) and coronary artery diseases (CAD) [66, 74, 134].

As mentioned in the above section, the EAT changes its phenotype in obese, T2DM patients and becomes more inflammatory with M1 proinflammatory macrophage predominance [166]. The inflammation and the oxidative stress produced by the high fructose contribute to the NIH through promotion of medial smooth muscle cell proliferation, migration and phenotype switching [119]. Taken together, high fructose diet plays an important role in the development of insulin resistance, oxidative stress, inflammation, hyperlipidemia and hypertension [42, 60], which result in intimal dysfunction, NIH and atherosclerosis [19, 133].

The inflammatory milieu caused by obesity and metabolic syndrome, as mentioned above plays a pivotal role in atherosclerosis and NIH. Inflammatory cytokines diffuse from the inflammatory EAT as well as from the vascular lumen and induce phenotypic changes of the medial vascular smooth muscle cells (VSMCs). Subsequently these cells change from quiescent and contractile cells to migratory, proliferative and secretory cells that migrate to the subintimal space, proliferate and lay down the extracellular matrix component of the neointimal hyperplasia. The NIH tissue impinges on the vascular lumen, impairing the blood flow and reducing the tissue perfusion by the affected arterial segment of the coronary vasculature. The representative images of coronary artery OCT (optical coherence tomography) and coronary angiography from CHD pig model are shown in Fig. 13.5. This results in tissue ischemia and necrosis (myocardial infarction) if the blood supply is not enough to sustain the vital functions of the cardiomyocytes [151]. The lack of understanding of the underlying molecular mechanisms and the influence of contributing factors limit the development of translational strategies to prevent the NIH. Medical science is looking forward for novel and outstanding discoveries in the management of CHDs and their treatment-related complications such as NIH.

Translational Avenues and Future Directions

Recent advancements in the medical science and technology have unveiled several pathological mechanisms underlying the atherosclerosis and NIH. The similarities in pathological mechanism of both atherosclerosis and NIH converge to the endothelial dysfunction and sterile inflammation suggesting the possibilities of common targets of intervention for both. However, the lack of proper knowledge regarding the exact underlying molecular pathogenesis limits the development of effective translational avenues in the management of atherosclerosis and NIH. The molecular signaling

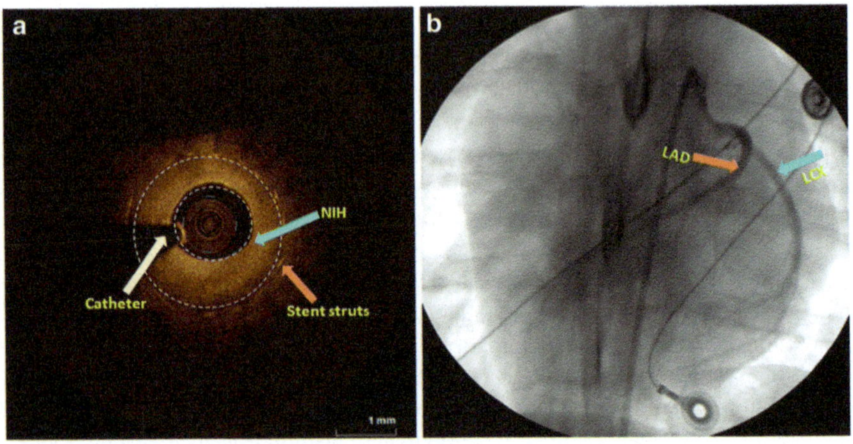

Fig. 13.5 A representative imaging of coronary artery by optical coherence tomography (**a**) and by angiography (**b**). Formation of neointimal hyperplasia beyond the stent struts leading to in-stent restenosis 5-month after implanting a bare metal stent is clearly evident in Yucatan microswine, and (**b**) Imaging of the coronary arteries by angiography during percutaneous coronary intervention showing LAD and LCX

leading to the phenotypic switch among immune and non-immune cell types is still an enigma. Several dietary and metabolic components have been associated with the pathogenesis of atherosclerosis and NIH, however the integration of diverse molecular signals triggering the upregulation of pro-atherogenic genes needs to be unveiled. Also, the modified surgical techniques for coronary interventions promising to retain the endothelial wall integrity is need of the hour. The knowledge regarding the alteration of pathological mediators at genetic and epigenetic levels exhibits immense translational potential, however, warrants further investigations. Stem cell and gene therapies have opened promising translational potential; however, the clinical data on these aspects are scarce. The regenerative strategies for coronary artery repair remain unexplored field in translational medicine. The therapeutic exosomes in cardiology is an emerging trend, unfortunately, the available literature mainly focuses on myocardial regeneration [83, 172]. The advent of cardiovascular tissue engineering provides abundant translational potential, even though this field of medical research is still in infancy [43, 51, 76]. Tissue engineering-based stimuli/environmental sensitive smart/intelligent matrices have been emerging which would form a strong basis for next generation personalized medicine in the management of atherosclerosis and NIH. The medical world is looking forward to outstanding discoveries to alleviate the sufferings of millions of CVD patients across the globe and expecting an improved quality of life.

Acknowledgements This work was supported by the research funds of Western University of Health Sciences to FGT, and research Grants R01 HL144125 and R01HL147662 to DKA from the National Institutes of Health, USA. The contents of this chapter are solely the responsibility of the authors and do not necessarily represent the official views of the National Institutes of Health.

Conflict of Interests All authors have read the Springer's policy on disclosure of potential conflicts of interest. Author C (DKA) has received grants from the National Institutes of Health. Author A (FGT) received start-up funds from Western University of Health Sciences. Author B (MMR) has no relevant affiliations or financial or non-financial involvement with any organization or entity with financial or non-financial interest or conflict with the subject matter or materials discussed in the manuscript apart from those disclosed. Author A (FGT), Author B (MMR), and Author C (DKA) declare that they have no conflict of interest. No writing assistance was utilized in the production of this manuscript.

References

1. Agrawal D, Swier V, Gunasekar P, et al (2018) Vitamin D deficiency induces pro-inflammatory phenotype of epicardial adipose tissue accelerating neointimal hyperplasia following coronary intervention. Atherosclerosis 275:e48. https://doi.org/10.1016/j.atherosclerosis.2018.06.126
2. Agrawal DK, Siddique A (2019) Commentary: rejuvenation of "broken heart" with bioengineered gel. J Thoracic Cardiovasc Surg 157:1491–1493. https://doi.org/10.1016/j.jtcvs.2018.08.076
3. Ahrens EH (1957) Dietary control of serum lipids in relation to atherosclerosis. JAMA 164:1905. https://doi.org/10.1001/jama.1957.62980170017007d
4. Aikawa M, Sivam PN, Kuro-o M et al (1993) Human smooth muscle myosin heavy chain isoforms as molecular markers for vascular development and atherosclerosis. Circ Res 73:1000–1012. https://doi.org/10.1161/01.res.73.6.1000
5. Albanese I, Khan K, Barratt B et al (2018) Atherosclerotic calcification: Wnt is the hint. J Am Heart Assoc 7:e007356. https://doi.org/10.1161/JAHA.117.007356
6. Alexopoulos N, Raggi P (2009) Calcification in atherosclerosis. Nat Rev Cardiol 6:681–688. https://doi.org/10.1038/nrcardio.2009.165
7. Amento EP, Ehsani N, Palmer H, Libby P (1991) Cytokines and growth factors positively and negatively regulate interstitial collagen gene expression in human vascular smooth muscle cells. Arterioscler Thromb J Vasc Biol 11:1223–1230. https://doi.org/10.1161/01.ATV.11.5.1223
8. Andersson J, Libby P, Hansson GK (2010) Adaptive immunity and atherosclerosis. Clin Immunol 134:33–46. https://doi.org/10.1016/j.clim.2009.07.002
9. Andreeva ER, Pugach IM, Orekhov AN (1997) Collagen-synthesizing cells in initial and advanced atherosclerotic lesions of human aorta. Atherosclerosis 130:133–142. https://doi.org/10.1016/s0021-9150(96)06056-x
10. Atkins GB, Simon DI (2013) Interplay between NF-κB and Kruppel-like factors in vascular inflammation and atherosclerosis: location, location, location. J Am Heart Assoc 2:e000290. https://doi.org/10.1161/JAHA.113.000290
11. Atkins GB, Wang Y, Mahabeleshwar GH et al (2008) Hemizygous deficiency of Krüppel-like factor 2 augments experimental atherosclerosis. Circ Res 103:690–693. https://doi.org/10.1161/CIRCRESAHA.108.184663
12. Bäck M, Yurdagul A, Tabas I et al (2019) Inflammation and its resolution in atherosclerosis: mediators and therapeutic opportunities. Nat Rev Cardiol 16:389–406. https://doi.org/10.1038/s41569-019-0169-2
13. Baker AR, Harte AL, Howell N et al (2009) Epicardial adipose tissue as a source of nuclear factor-kappaB and c-Jun N-terminal kinase mediated inflammation in patients with coronary artery disease. J Clin Endocrinol Metab 94:261–267. https://doi.org/10.1210/jc.2007-2579
14. Basatemur GL, Jørgensen HF, Clarke MCH et al (2019) Vascular smooth muscle cells in atherosclerosis. Nat Rev Cardiol. https://doi.org/10.1038/s41569-019-0227-9
15. Benjamin EJ, Muntner P, Alonso A et al (2019) Heart disease and stroke statistics—2019 update: a report from the American heart association. Circulation. https://doi.org/10.1161/CIR.0000000000000659

16. Bennett MR, Sinha S, Owens GK (2016) Vascular smooth muscle cells in atherosclerosis. Circ Res 118:692–702. https://doi.org/10.1161/CIRCRESAHA.115.306361
17. Binder CJ, Chou M-Y, Fogelstrand L et al (2008) Natural antibodies in murine atherosclerosis. Curr Drug Targets 9:190–195
18. Bonatti J, Oberhuber A, Schachner T et al (2004) Neointimal hyperplasia in coronary vein grafts: pathophysiology and prevention of a significant clinical problem. Heart Surg Forum 7:72–87
19. Bonora E, Tessari R, Micciolo R et al (1997) Intimal-medial thickness of the carotid artery in nondiabetic and NIDDM patients: relationship with insulin resistance. Diab Care 20:627–631. https://doi.org/10.2337/diacare.20.4.627
20. Booth J, Clayton T, Pepper J et al (2008) Randomized, controlled trial of coronary artery bypass surgery versus percutaneous coronary intervention in patients with multivessel coronary artery disease: six-year follow-up from the stent or surgery trial (SoS). Circulation 118:381–388. https://doi.org/10.1161/CIRCULATIONAHA.107.739144
21. Breen DM, Dhaliwall JK, Chan KK et al (2010) Insulin inhibits and oral sucrose increases neointimal growth after arterial injury in rats. JVR 47:412–422. https://doi.org/10.1159/000281581
22. Brown JD, Lin CY, Duan Q et al (2014) NF-κB directs dynamic super enhancer formation in inflammation and atherogenesis. Mol Cell 56:219–231. https://doi.org/10.1016/j.molcel.2014.08.024
23. Burster T, Macmillan H, Hou T et al (2010) Cathepsin G: roles in antigen presentation and beyond. Mol Immunol 47:658–665. https://doi.org/10.1016/j.molimm.2009.10.003
24. Chellan B, Reardon CA, Getz GS, Hofmann Bowman MA (2016) Enzymatically modified low-density lipoprotein promotes foam cell formation in smooth muscle cells via macropinocytosis and enhances receptor-mediated uptake of oxidized low-density lipoprotein. Arterioscler Thromb Vasc Biol 36:1101–1113. https://doi.org/10.1161/ATVBAHA.116.307306
25. Chenicheri S, Usha R, Ramachandran R, et al (2017) Insight into oral biofilm: primary, secondary and residual caries and phyto-challenged solutions. Open Dent J 11:312–333. https://doi.org/10.2174/1874210601711010312
26. Chistiakov DA, Bobryshev YV, Orekhov AN (2016) Macrophage-mediated cholesterol handling in atherosclerosis. J Cell Mol Med 20:17–28. https://doi.org/10.1111/jcmm.12689
27. Cola C, Almeida M, Li D et al (2004) Regulatory role of endothelium in the expression of genes affecting arterial calcification. Biochem Biophys Res Commun 320:424–427. https://doi.org/10.1016/j.bbrc.2004.05.181
28. Combadière C, Potteaux S, Rodero M et al (2008) Combined inhibition of CCL2, CX3CR1, and CCR5 Abrogates Ly6C ^{hi} and Ly6C ^{lo} monocytosis and almost abolishes atherosclerosis in hypercholesterolemic mice. Circulation 117:1649–1657. https://doi.org/10.1161/CIRCULATIONAHA.107.745091
29. Cornhill JF, Roach MR (1976) A quantitative study of the localization of atherosclerotic lesions in the rabbit aorta. Atherosclerosis 23:489–501. https://doi.org/10.1016/0021-9150(76)90009-5
30. Cybulsky MI, Gimbrone MA (1991) Endothelial expression of a mononuclear leukocyte adhesion molecule during atherogenesis. Science 251:788–791. https://doi.org/10.1126/science.1990440
31. Dalen JE, Devries S (2014) Diets to prevent coronary heart disease 1957–2013: what have we learned? Am J Med 127:364–369. https://doi.org/10.1016/j.amjmed.2013.12.014
32. Davenport P, Tipping PG (2003) The role of interleukin-4 and interleukin-12 in the progression of atherosclerosis in apolipoprotein E-deficient mice. Am J Pathol 163:1117–1125. https://doi.org/10.1016/S0002-9440(10)63471-2
33. de Feyter PJ, de Jaegere PPT, Serruys PW (1994) Incidence, predictors, and management of acute coronary occlusion after coronary angioplasty. Am Heart J 127:643–651. https://doi.org/10.1016/0002-8703(94)90675-0

34. Do M, Lee E, Oh M-J et al (2018) High-glucose or -fructose diet cause changes of the gut microbiota and metabolic disorders in mice without body weight change. Nutrients 10:761. https://doi.org/10.3390/nu10060761
35. Doherty DG (2016) Immunity, tolerance and autoimmunity in the liver: a comprehensive review. J Autoimmun 66:60–75. https://doi.org/10.1016/j.jaut.2015.08.020
36. Duffield JS, Lupher M, Thannickal VJ, Wynn TA (2013) Host responses in tissue repair and fibrosis. Annu Rev Pathol Mech Dis 8:241–276. https://doi.org/10.1146/annurev-pathol-020712-163930
37. Dutour A, Achard V, Sell H et al (2010) Secretory type II phospholipase A2 is produced and secreted by epicardial adipose tissue and overexpressed in patients with coronary artery disease. J Clin Endocrinol Metab 95:963–967. https://doi.org/10.1210/jc.2009-1222
38. Elhage R (2003) Reduced atherosclerosis in interleukin-18 deficient apolipoprotein E-knockout mice. Cardiovasc Res 59:234–240. https://doi.org/10.1016/S0008-6363(03)00343-2
39. Elices MJ, Osborn L, Takada Y et al (1990) VCAM-1 on activated endothelium interacts with the leukocyte integrin VLA-4 at a site distinct from the VLA-4/fibronectin binding site. Cell 60:577–584. https://doi.org/10.1016/0092-8674(90)90661-w
40. Evrard SM, Lecce L, Michelis KC et al (2016) Endothelial to mesenchymal transition is common in atherosclerotic lesions and is associated with plaque instability. Nat Commun 7:11853. https://doi.org/10.1038/ncomms11853
41. Feng S, Bowden N, Fragiadaki M et al (2017) Mechanical activation of hypoxia-inducible factor 1α drives endothelial dysfunction at atheroprone sites. Arterioscler Thromb Vasc Biol 37:2087–2101. https://doi.org/10.1161/ATVBAHA.117.309249
42. Ferder L, Ferder MD, Inserra F (2010) The role of high-fructose corn syrup in metabolic syndrome and hypertension. Curr Hypertens Rep 12:105–112. https://doi.org/10.1007/s11906-010-0097-3
43. Finosh GT, Jayabalan M (2012) Regenerative therapy and tissue engineering for the treatment of end-stage cardiac failure. Biomatter 2:1–14. https://doi.org/10.4161/biom.19429
44. Frostegård J, Ulfgren A-K, Nyberg P et al (1999) Cytokine expression in advanced human atherosclerotic plaques: dominance of pro-inflammatory (Th1) and macrophage-stimulating cytokines. Atherosclerosis 145:33–43. https://doi.org/10.1016/S0021-9150(99)00011-8
45. Gao Z, Zhang X, Zuberi A et al (2004) Inhibition of insulin sensitivity by free fatty acids requires activation of multiple serine kinases in 3T3-L1 adipocytes. Mol Endocrinol 18:2024–2034. https://doi.org/10.1210/me.2003-0383
46. Gerthoffer WT (2007) Mechanisms of vascular smooth muscle cell migration. Circ Res 100:607–621. https://doi.org/10.1161/01.RES.0000258492.96097.47
47. Giachelli CM (2004) Vascular Calcification Mechanisms. JASN 15:2959–2964. https://doi.org/10.1097/01.ASN.0000145894.57533.C4
48. Giddens DP, Zarins CK, Glagov S (1993) The role of fluid mechanics in the localization and detection of atherosclerosis. J Biomech Eng 115:588–594. https://doi.org/10.1115/1.2895545
49. Gimbrone MA, García-Cardeña G (2016) Endothelial cell dysfunction and the pathobiology of atherosclerosis. Circ Res 118:620–636. https://doi.org/10.1161/CIRCRESAHA.115.306301
50. Gisterå A, Robertson A-KL, Andersson J, et al (2013) Transforming growth factor–β signaling in T cells promotes stabilization of atherosclerotic plaques through an interleukin-17–dependent pathway. Sci Transl Med 5:196ra100–196ra100. https://doi.org/10.1126/scitranslmed.3006133
51. Gnanaprakasam Thankam F, Muthu J (2013) Influence of plasma protein–hydrogel interaction moderated by absorption of water on long-term cell viability in amphiphilic biosynthetic hydrogels. RSC Adv 3:24509. https://doi.org/10.1039/c3ra43710h
52. Grech ED (2003) ABC of interventional cardiology: pathophysiology and investigation of coronary artery disease. BMJ 326:1027–1030. https://doi.org/10.1136/bmj.326.7397.1027
53. Grüntzig AR, Senning Å, Siegenthaler WE (1979) Nonoperative dilatation of coronary-artery stenosis: percutaneous transluminal coronary angioplasty. N Engl J Med 301:61–68. https://doi.org/10.1056/NEJM197907123010201

54. Guo L, Akahori H, Harari E, et al (2018) CD163+ macrophages promote angiogenesis and vascular permeability accompanied by inflammation in atherosclerosis. J Clin Invest 128:1106–1124. https://doi.org/10.1172/JCI93025
55. Hanke H, Strohschneider T, Oberhoff M, et al (1990) Time course of smooth muscle cell proliferation in the intima and media of arteries following experimental angioplasty. Circ Res 67:651–659. https://doi.org/10.1161/01.RES.67.3.651
56. Hansson GK (2005) Inflammation, atherosclerosis, and coronary artery disease. N Engl J Med 352:1685–1695. https://doi.org/10.1056/NEJMra043430
57. Hansson GK, Robertson A-KL, Söderberg-Nauclér C (2006) Inflammation and atherosclerosis. Annu Rev Pathol Mech Dis 1:297–329. https://doi.org/10.1146/annurev.pathol.1.110304.100100
58. Hashizume M, Mihara M (2012) Blockade of IL-6 and TNF-α inhibited oxLDL-induced production of MCP-1 via scavenger receptor induction. Eur J Pharmacol 689:249–254. https://doi.org/10.1016/j.ejphar.2012.05.035
59. Hergenreider E, Heydt S, Tréguer K et al (2012) Atheroprotective communication between endothelial cells and smooth muscle cells through miRNAs. Nat Cell Biol 14:249–256. https://doi.org/10.1038/ncb2441
60. Herman MA, Samuel VT (2016) The sweet path to metabolic demise: fructose and lipid synthesis. Trends Endocrinol Metab 27:719–730. https://doi.org/10.1016/j.tem.2016.06.005
61. Higashi Y, Noma K, Yoshizumi M, Kihara Y (2009) Endothelial function and oxidative stress in cardiovascular diseases. Circ J 73:411–418. https://doi.org/10.1253/circj.CJ-08-1102
62. Houde M, Van Eck M (2018) Escaping the atherogenic trap: preventing LDL fusion and binding in the intima. Atherosclerosis 275:376–378. https://doi.org/10.1016/j.atherosclerosis.2018.05.032
63. Hutcheson JD, Goettsch C, Bertazzo S et al (2016) Genesis and growth of extracellular-vesicle-derived microcalcification in atherosclerotic plaques. Nat Mater 15:335–343. https://doi.org/10.1038/nmat4519
64. Iacobellis G, Bianco AC (2011) Epicardial adipose tissue: emerging physiological, pathophysiological and clinical features. Trends Endocrinol Metab 22:450–457. https://doi.org/10.1016/j.tem.2011.07.003
65. Ignarro LJ, Napoli C (2005) Novel features of nitric oxide, endothelial nitric oxide synthase, and atherosclerosis. Curr Diab Rep 5:17–23
66. Ioannou GN, Weiss NS, Boyko EJ et al (2006) Elevated serum alanine aminotransferase activity and calculated risk of coronary heart disease in the United States. Hepatology 43:1145–1151. https://doi.org/10.1002/hep.21171
67. Iozzo P (2011) Myocardial, perivascular, and epicardial fat. Diab Care 34:S371–S379. https://doi.org/10.2337/dc11-s250
68. Itabe H (2009) Oxidative modification of LDL: its pathological role in atherosclerosis. Clinic Rev Allerg Immunol 37:4–11. https://doi.org/10.1007/s12016-008-8095-9
69. Kawamura M, Heinecke JW, Chait A (1994) Pathophysiological concentrations of glucose promote oxidative modification of low density lipoprotein by a superoxide-dependent pathway. J Clin Invest 94:771–778. https://doi.org/10.1172/JCI117396
70. Kijani S, Vázquez AM, Levin M, et al (2017) Intimal hyperplasia induced by vascular intervention causes lipoprotein retention and accelerated atherosclerosis. Physiol Rep 5. https://doi.org/10.14814/phy2.13334
71. Kim DN, Imai H, Schmee J et al (1985) Intimal cell mass-derived atherosclerotic lesions in the abdominal aorta of hyperlipidemic swine. Part 1. Cell of origin, cell divisions and cell losses in first 90 days on diet. Atherosclerosis 56:169–188. https://doi.org/10.1016/0021-9150(85)90017-6
72. Kim M-S, Krawczyk SA, Doridot L et al (2016) ChREBP regulates fructose-induced glucose production independently of insulin signaling. J Clin Invest 126:4372–4386. https://doi.org/10.1172/JCI81993
73. Kini S, Bis KG, Weaver L (2007) Normal and variant coronary arterial and venous anatomy on high-resolution CT angiography. Am J Roentgenol 188:1665–1674. https://doi.org/10.2214/AJR.06.1295

74. Klepchick PR, Sun H, Ozanich BA et al (2005) The contribution of systemic factors to intimal hyperplasia in type 2 diabetes mellitus. J Am Coll Surg 201:S105. https://doi.org/10.1016/j.jamcollsurg.2005.06.249
75. Kolodgie FD, Gold HK, Burke AP et al (2003) Intraplaque hemorrhage and progression of coronary atheroma. N Engl J Med 349:2316–2325. https://doi.org/10.1056/NEJMoa035655
76. Komeri R, Thankam FG, Muthu J (2015) Influence of matrix and bulk behaviour of an injectable hydrogel on the survival of encapsulated cardiac cells. RSC Adv 5:31439–31449. https://doi.org/10.1039/C4RA16254D
77. Kronenberg M, Gapin L (2002) The unconventional lifestyle of NKT cells. Nat Rev Immunol 2:557–568. https://doi.org/10.1038/nri854
78. Kume N, Cybulsky MI, Gimbrone MA (1992) Lysophosphatidylcholine, a component of atherogenic lipoproteins, induces mononuclear leukocyte adhesion molecules in cultured human and rabbit arterial endothelial cells. J Clin Invest 90:1138–1144. https://doi.org/10.1172/JCI115932
79. Kyaw T, Peter K, Li Y, et al (2017) Cytotoxic lymphocytes and atherosclerosis: significance, mechanisms and therapeutic challenges. Br J Pharmacol. https://doi.org/10.1111/bph.13845. https://bpspubs.onlinelibrary.wiley.com. Accessed 3 Sept 2019
80. Lam TKT, Yoshii H, Haber CA et al (2002) Free fatty acid-induced hepatic insulin resistance: a potential role for protein kinase C-δ. Am J Physiol Endocrinol Metab 283:E682–E691. https://doi.org/10.1152/ajpendo.00038.2002
81. Lamb DJ, Leake DS (1994) Acidic pH enables caeruloplasmin to catalyse the modification of low-density lipoprotein. FEBS Lett 338:122–126. https://doi.org/10.1016/0014-5793(94)80348-X
82. Lanzer P, Boehm M, Sorribas V et al (2014) Medial vascular calcification revisited: review and perspectives. Eur Heart J 35:1515–1525. https://doi.org/10.1093/eurheartj/ehu163
83. Lazar E, Benedek T, Korodi S et al (2018) Stem cell-derived exosomes: an emerging tool for myocardial regeneration. World J Stem Cells 10:106–115. https://doi.org/10.4252/wjsc.v10.i8.106
84. Lech M, Anders H-J (2013) Macrophages and fibrosis: How resident and infiltrating mononuclear phagocytes orchestrate all phases of tissue injury and repair. Biochimica et Biophysica Acta (BBA) Mol Basis Dis 1832:989–997. https://doi.org/10.1016/j.bbadis.2012.12.001
85. Leibowitz JO (1970) The history of coronary heart disease. University of California Press
86. Li N, Cheng W, Huang T et al (2015) Vascular adventitia calcification and its underlying mechanism. PLoS ONE. https://doi.org/10.1371/journal.pone.0132506
87. Libby P (2012) Inflammation in atherosclerosis. Arterioscler Thromb Vasc Biol 32:2045–2051. https://doi.org/10.1161/ATVBAHA.108.179705
88. Libby P, Ridker PM, Hansson GK (2011) Progress and challenges in translating the biology of atherosclerosis. Nature 473:317–325. https://doi.org/10.1038/nature10146
89. Libby P, Theroux P (2005) Pathophysiology of coronary artery disease. Circulation 111:3481–3488. https://doi.org/10.1161/CIRCULATIONAHA.105.537878
90. Lin J, Li M, Wang Z, et al (2010) The role of CD4+CD25+ regulatory T cells in macrophage-derived foam-cell formation. J Lipid Res 51:1208–1217. https://doi.org/10.1194/jlr.D000497
91. Lin Z, Natesan V, Shi H, et al (2010) Kruppel-like factor 2 regulates endothelial barrier function. Arterioscler Thromb Vasc Biol 30:1952–1959. https://doi.org/10.1161/ATVBAHA.110.211474
92. Loukas M, Sharma A, Blaak C, et al (2013) The clinical anatomy of the coronary arteries. J Cardiovasc Trans Res 6:197–207. https://doi.org/10.1007/s12265-013-9452-5
93. Lu J, Ji J, Meng H, et al (2013) The protective effect and underlying mechanism of metformin on neointima formation in fructose-induced insulin resistant rats. Cardiovasc Diabetol 12:58. https://doi.org/10.1186/1475-2840-12-58
94. Lustig RH (2013) Fructose: it's "Alcohol Without the Buzz." Adv Nutr 4:226–235. https://doi.org/10.3945/an.112.002998
95. Majesky MW, Dong XR, Hoglund V, et al (2011) The adventitia: a dynamic interface containing resident progenitor cells. Arterioscler Thromb Vasc Biol 31:1530–1539. https://doi.org/10.1161/ATVBAHA.110.221549

96. Mallat Z, Gojova A, Brun V, et al (2003) Induction of a regulatory T cell type 1 response reduces the development of atherosclerosis in apolipoprotein E-knockout mice. Circulation 108:1232–1237. https://doi.org/10.1161/01.CIR.0000089083.61317.A1
97. Marchio P, Guerra-Ojeda S, Vila JM, et al (2019) Targeting early atherosclerosis: a focus on oxidative stress and inflammation. Oxidative Med Cell Longevity. https://www.hindawi.com/journals/omcl/2019/8563845/. Accessed 16 Sept 2019
98. Mauro CR, Tao M, Yu P, et al (2016) Pre-operative dietary restriction reduces intimal hyperplasia and protects from ischemia reperfusion injury. J Vasc Surg 63:500-509.e1. https://doi.org/10.1016/j.jvs.2014.07.004
99. Mayerl C, Lukasser M, Sedivy R, et al (2006) Atherosclerosis research from past to present—on the track of two pathologists with opposing views, Carl von Rokitansky and Rudolf Virchow. Virchows Arch 449:96–103. https://doi.org/10.1007/s00428-006-0176-7
100. Mazurek T, Zhang L, Zalewski A, et al (2003) Human epicardial adipose tissue is a source of inflammatory mediators. Circulation 108:2460–2466. https://doi.org/10.1161/01.CIR.0000099542.57313.C5
101. Methe H, Weis M (2007) Atherogenesis and inflammation–was Virchow right? Nephrol Dial Transplant 22:1823–1827. https://doi.org/10.1093/ndt/gfm112
102. Miao C-Y, Li Z-Y (2012) The role of perivascular adipose tissue in vascular smooth muscle cell growth. Br J Pharmacol 165:643–658. https://doi.org/10.1111/j.1476-5381.2011.01404.x
103. Moreno K, Murray-Wijelath J, Yagi M, et al (2011) Circulating inflammatory cells are associated with vein graft stenosis. J Vasc Surg 54:1124–1130. https://doi.org/10.1016/j.jvs.2011.04.039
104. Moroni F, Ammirati E, Norata GD, et al (2019) The role of monocytes and macrophages in human atherosclerosis, plaque neoangiogenesis, and atherothrombosis. Mediators Inflamm 2019:1–11. https://doi.org/10.1155/2019/7434376
105. Mosse PR, Campbell GR, Wang ZL, Campbell JH (1985) Smooth muscle phenotypic expression in human carotid arteries. I. Comparison of cells from diffuse intimal thickenings adjacent to atheromatous plaques with those of the media. Lab Invest 53:556–562
106. Murdaca G, Spanò F, Cagnati P, Puppo F (2013) Free radicals and endothelial dysfunction: Potential positive effects of TNF-α inhibitors. Redox Rep 18:95–99. https://doi.org/10.1179/1351000213Y.0000000046
107. Murphy WA, Nedden zur D, Gostner P, et al (2003) The iceman: discovery and imaging. Radiology 226:614–629. https://doi.org/10.1148/radiol.2263020338
108. Murray PJ, Wynn TA (2011) Protective and pathogenic functions of macrophage subsets. Nat Rev Immunol 11:723–737. https://doi.org/10.1038/nri3073
109. Muto A, Fitzgerald TN, Pimiento JM, et al (2007) Smooth muscle cell signal transduction: implications of vascular biology for vascular surgeons. J Vasc Surg 45 Suppl A:A15–A24. https://doi.org/10.1016/j.jvs.2007.02.061
110. Muto A, Model L, Ziegler K et al (2010) Mechanisms of vein graft adaptation to the arterial circulation: Insights into the neointimal algorithm and management strategies. Circ J 74:1501–1512
111. Navab M, Berliner JA, Watson AD et al (1996) The Yin and Yang of oxidation in the development of the fatty streak: a review based on the 1994 George Lyman Duff memorial lecture. Arterioscler Thromb Vasc Biol 16:831–842. https://doi.org/10.1161/01.ATV.16.7.831
112. Ogobuiro I, Tuma F (2019) Anatomy, thorax, heart coronary arteries. StatPearls. StatPearls Publishing, Treasure Island (FL)
113. Oörni K, Pentikäinen MO, Ala-Korpela M, Kovanen PT (2000) Aggregation, fusion, and vesicle formation of modified low density lipoprotein particles: molecular mechanisms and effects on matrix interactions. J Lipid Res 41:1703–1714
114. Otsuka F, Byrne RA, Yahagi K et al (2015) Neoatherosclerosis: overview of histopathologic findings and implications for intravascular imaging assessment. Eur Heart J 36:2147–2159. https://doi.org/10.1093/eurheartj/ehv205
115. Otsuka F, Sakakura K, Yahagi K et al (2014) Has our understanding of calcification in human coronary atherosclerosis progressed? Arterioscler Thromb Vasc Biol 34:724–736. https://doi.org/10.1161/ATVBAHA.113.302642

116. Ou H, Guo B, Liu Q et al (2018) Regulatory T cells as a new therapeutic target for atherosclerosis. Acta Pharmacol Sin 39:1249–1258. https://doi.org/10.1038/aps.2017.140
117. Owens GK (1995) Regulation of differentiation of vascular smooth muscle cells. Physiol Rev 75:487–517. https://doi.org/10.1152/physrev.1995.75.3.487
118. Panasenko OM, Briviba K, Klotz L-O, Sies H (1997) Oxidative modification and nitration of human low-density lipoproteins by the reaction of hypochlorous acid with nitrite. Arch Biochem Biophys 343:254–259. https://doi.org/10.1006/abbi.1997.0171
119. Papaharalambus CA, Griendling KK (2007) Basic mechanisms of oxidative stress and reactive oxygen species in cardiovascular injury. Trends Cardiovasc Med 17:48–54. https://doi.org/10.1016/j.tcm.2006.11.005
120. Parthasarathy S (1987) Oxidation of low-density lipoprotein by thiol compounds leads to its recognition by the acetyl LDL receptor. Biochimica et Biophysica Acta (BBA) Lipids Lipid Metab 917:337–340. https://doi.org/10.1016/0005-2760(87)90139-1
121. Parthasarathy S, Raghavamenon A, Garelnabi MO, Santanam N (2010) Oxidized low-density lipoprotein. In: Uppu RM, Murthy SN, Pryor WA, Parinandi NL (eds) Free radicals and antioxidant protocols. Humana Press, Totowa, NJ, pp 403–417
122. Pelham CJ, Drews EM, Agrawal DK (2016) Vitamin D controls resistance artery function through regulation of perivascular adipose tissue hypoxia and inflammation. J Mol Cell Cardiol 98:1–10. https://doi.org/10.1016/j.yjmcc.2016.06.067
123. Plitt GD, Spring JT, Moulton MJ, Agrawal DK (2018) Mechanisms, diagnosis, and treatment of heart failure with preserved ejection fraction and diastolic dysfunction. Expert Rev Cardiovasc Ther 16:579–589. https://doi.org/10.1080/14779072.2018.1497485
124. Pourcet B, Staels B (2018) Alternative macrophages in atherosclerosis: not always protective! Journal of Clinical Investigation 128:910–912. https://doi.org/10.1172/JCI120123
125. Prati F, Arbustini E, Labellarte A et al (2003) Eccentric atherosclerotic plaques with positive remodelling have a pericardial distribution: a permissive role of epicardial fat? A three-dimensional intravascular ultrasound study of left anterior descending artery lesions. Eur Heart J 24:329–336. https://doi.org/10.1016/s0195-668x(02)00426-8
126. Pugliese G, Iacobini C, Fantauzzi CB, Menini S (2015) The dark and bright side of atherosclerotic calcification. Atherosclerosis 238:220–230. https://doi.org/10.1016/j.atherosclerosis.2014.12.011
127. Qi X-Y, Qu S-L, Xiong W-H et al (2018) Perivascular adipose tissue (PVAT) in atherosclerosis: a double-edged sword. Cardiovasc Diabetol 17:134. https://doi.org/10.1186/s12933-018-0777-x
128. Rabkin SW (2007) Epicardial fat: properties, function and relationship to obesity. Obes Rev 8:253–261. https://doi.org/10.1111/j.1467-789X.2006.00293.x
129. Radwan MM, Radwan BM, Nandipati KC et al (2013) Immunological and molecular basis of nonalcoholic steatohepatitis and nonalcoholic fatty liver disease. Exp Rev Clin Immunol 9:727–738. https://doi.org/10.1586/1744666X.2013.816484
130. Rajamäki K, Lappalainen J, Öörni K et al (2010) Cholesterol crystals activate the NLRP3 inflammasome in human macrophages: a novel link between cholesterol metabolism and inflammation. PLoS ONE 5:e11765. https://doi.org/10.1371/journal.pone.0011765
131. Rattazzi M, Bennett BJ, Bea F et al (2005) Calcification of advanced atherosclerotic lesions in the innominate arteries of ApoE-deficient mice: potential role of chondrocyte-like cells. Arterioscler Thromb Vasc Biol 25:1420–1425. https://doi.org/10.1161/01.ATV.0000166600.58468.1b
132. Ray A, Khare A, Krishnamoorthy N et al (2010) Regulatory T cells in many flavors control asthma. Mucosal Immunol 3:216–229. https://doi.org/10.1038/mi.2010.4
133. Rectenwald JE, Moldawer LL, Huber TS et al (2000) Direct evidence for cytokine involvement in neointimal hyperplasia. Circulation 102:1697–1702. https://doi.org/10.1161/01.CIR.102.14.1697
134. Rekhi U, Piche JE, Immaraj L, Febbraio M (2019) Neointimal hyperplasia: are fatty acid transport proteins a new therapeutic target? Curr Opin Lipidol 30:377–382. https://doi.org/10.1097/MOL.0000000000000627

135. Rogacev KS, Cremers B, Zawada AM et al (2012) CD14++ CD16+ monocytes independently predict cardiovascular events. J Am Coll Cardiol 60:1512–1520. https://doi.org/10.1016/j.jacc.2012.07.019
136. Ross R (1993) The pathogenesis of atherosclerosis: a perspective for the 1990s. Nature 362:801–809. https://doi.org/10.1038/362801a0
137. Roy-Chaudhury P, Wang Y, Krishnamoorthy M et al (2009) Cellular phenotypes in human stenotic lesions from haemodialysis vascular access. Nephrol Dial Transplant 24:2786–2791. https://doi.org/10.1093/ndt/gfn708
138. Rutanen J, Puhakka H, Ylä-Herttuala S (2002) Post-intervention vessel remodeling. Gene Ther 9:1487–1491. https://doi.org/10.1038/sj.gt.3301866
139. Salgado-Somoza A, Teijeira-Fernández E, Fernández AL et al (2010) Proteomic analysis of epicardial and subcutaneous adipose tissue reveals differences in proteins involved in oxidative stress. Am J Physiol Heart Circ Physiol 299:H202-209. https://doi.org/10.1152/ajpheart.00120.2010
140. Samuel VT (2011) Fructose induced lipogenesis: from sugar to fat to insulin resistance. Trends Endocrinol Metab 22:60–65. https://doi.org/10.1016/j.tem.2010.10.003
141. Sata M, Saiura A, Kunisato A et al (2002) Hematopoietic stem cells differentiate into vascular cells that participate in the pathogenesis of atherosclerosis. Nat Med 8:403–409. https://doi.org/10.1038/nm0402-403
142. Satish M, Agrawal DK (2019) Atherothrombosis and the NLRP3 inflammasome—endogenous mechanisms of inhibition. Transl Res. https://doi.org/10.1016/j.trsl.2019.08.003
143. Satish M, Saxena SK, Agrawal DK (2019) Adipokine dysregulation and insulin resistance with atherosclerotic vascular disease: metabolic syndrome or independent sequelae? J Cardiovasc Trans Res. https://doi.org/10.1007/s12265-019-09879-0
144. Saxton A, Manna B (2019) Anatomy, thorax, heart right coronary arteries. StatPearls. StatPearls Publishing, Treasure Island (FL)
145. Schwartz BG, Economides C, Mayeda GS et al (2009) The endothelial cell in health and disease: its function, dysfunction, measurement and therapy. Int J Impot Res 22:77
146. Serruys PW, Kutryk MJB, Ong ATL (2006) Coronary-artery stents. N Engl J Med 354:483–495. https://doi.org/10.1056/NEJMra051091
147. Serruys PW, Morice M-C, Kappetein AP et al (2009) Percutaneous coronary intervention versus coronary-artery bypass grafting for severe coronary artery disease. N Engl J Med 360:961–972. https://doi.org/10.1056/NEJMoa0804626
148. Shanahan CM (2007) Inflammation ushers in calcification: a cycle of damage and protection? Circulation 116:2782–2785. https://doi.org/10.1161/CIRCULATIONAHA.107.749655
149. Shankman LS, Gomez D, Cherepanova OA et al (2015) KLF4-dependent phenotypic modulation of smooth muscle cells has a key role in atherosclerotic plaque pathogenesis. Nat Med 21:628–637. https://doi.org/10.1038/nm.3866
150. Sheedy FJ, Grebe A, Rayner KJ et al (2013) CD36 coordinates NLRP3 inflammasome activation by facilitating intracellular nucleation of soluble ligands into particulate ligands in sterile inflammation. Nat Immunol 14:812–820. https://doi.org/10.1038/ni.2639
151. Shoji M, Koba S, Kobayashi Y (2014) Roles of bone-marrow-derived cells and inflammatory cytokines in neointimal hyperplasia after vascular injury. Biomed Res Int 2014:1–8. https://doi.org/10.1155/2014/945127
152. Sitia S, Tomasoni L, Atzeni F et al (2010) From endothelial dysfunction to atherosclerosis. Autoimmun Rev 9:830–834. https://doi.org/10.1016/j.autrev.2010.07.016
153. Smith JD, Bryant SR, Couper LL et al (1999) Soluble transforming growth factor-β type II receptor inhibits negative remodeling, fibroblast transdifferentiation, and intimal lesion formation but not endothelial growth. Circ Res 84:1212–1222. https://doi.org/10.1161/01.RES.84.10.1212
154. Smith SC, Feldman TE, Hirshfeld JW et al (2006) ACC/AHA/SCAI 2005 guideline update for percutaneous coronary intervention. J Am Coll Cardiol 47:e1–e121. https://doi.org/10.1016/j.jacc.2005.12.001

155. Steinberg D (2009) The LDL modification hypothesis of atherogenesis: an update. J Lipid Res 50:S376–S381. https://doi.org/10.1194/jlr.R800087-JLR200
156. Steinberg D, Witztum JL (1990) Lipoproteins and atherogenesis. Current concepts. JAMA 264:3047–3052
157. Strøm A, Jensen RA, Oslo MD, Oslo MD (1951) Mortality from circulatory diseases in Norway 1940–1945. Lancet 257:126–129. https://doi.org/10.1016/S0140-6736(51)91210-X
158. Suganami T, Tanimoto-Koyama K, Nishida J et al (2007) Role of the toll-like receptor 4/NF-κB pathway in saturated fatty acid-induced inflammatory changes in the interaction between adipocytes and macrophages. ATVB 27:84–91. https://doi.org/10.1161/01.ATV.0000251608.09329.9a
159. Sun Z (2014) Atherosclerosis and atheroma plaque rupture: normal anatomy of vasa vasorum and their role associated with atherosclerosis. Sci World J 2014:285058. https://doi.org/10.1155/2014/285058
160. Sur S, Swier VJ, Radwan MM, Agrawal DK (2016) Increased expression of phosphorylated polo-like kinase 1 and histone in bypass vein graft and coronary arteries following angioplasty. PLoS ONE 11:e0147937. https://doi.org/10.1371/journal.pone.0147937
161. Swier VJ, Tang L, Krueger KD et al (2015) Coronary injury score correlates with proliferating cells and alpha-smooth muscle actin expression in stented porcine coronary arteries. PLoS ONE 10:e0138539. https://doi.org/10.1371/journal.pone.0138539
162. Swier VJ, Tang L, Radwan MM, et al (2016) The role of high cholesterol-high fructose diet on coronary arteriosclerosis. Histol Histopathol 31:167–176. https://doi.org/10.14670/HH-11-652
163. Taleb S, Romain M, Ramkhelawon B et al (2009) Loss of SOCS3 expression in T cells reveals a regulatory role for interleukin-17 in atherosclerosis. J Exp Med 206:2067–2077. https://doi.org/10.1084/jem.20090545
164. Taleb S, Tedgui A (2018) IL-17 in atherosclerosis: the good and the bad. Cardiovasc Res 114:7–9. https://doi.org/10.1093/cvr/cvx225
165. Taleb S, Tedgui A, Mallat Z (2015) IL-17 and Th17 cells in atherosclerosis: subtle and contextual roles. Arterioscler Thromb Vasc Biol 35:258–264. https://doi.org/10.1161/ATVBAHA.114.303567
166. Tateya S, Kim F, Tamori Y (2013) Recent advances in obesity-induced inflammation and insulin resistance. Front Endocrinol. https://doi.org/10.3389/fendo.2013.00093
167. Thankam FG, Chandra IS, Kovilam AN et al (2018) Amplification of mitochondrial activity in the healing response following rotator cuff tendon injury. Sci Rep 8:1–14. https://doi.org/10.1038/s41598-018-35391-7
168. Thankam FG, Roesch ZK, Dilisio MF et al (2018) Association of inflammatory responses and ECM disorganization with HMGB1 upregulation and NLRP3 inflammasome activation in the injured rotator cuff tendon. Sci Rep 8:1–14. https://doi.org/10.1038/s41598-018-27250-2
169. Theodorou K, Boon RA (2018) Endothelial cell metabolism in atherosclerosis. Front Cell Dev Biol. https://doi.org/10.3389/fcell.2018.00082
170. Tian Z, Miyata K, Tazume H et al (2013) Perivascular adipose tissue-secreted angiopoietin-like protein 2 (Angptl2) accelerates neointimal hyperplasia after endovascular injury. J Mol Cell Cardiol 57:1–12. https://doi.org/10.1016/j.yjmcc.2013.01.004
171. Tse K, Tse H, Sidney J et al (2013) T cells in atherosclerosis. Int Immunol 25:615–622. https://doi.org/10.1093/intimm/dxt043
172. Vandergriff A, Huang K, Shen D et al (2018) Targeting regenerative exosomes to myocardial infarction using cardiac homing peptide. Theranostics 8:1869–1878. https://doi.org/10.7150/thno.20524
173. Vengrenyuk Y, Carlier S, Xanthos S et al (2006) A hypothesis for vulnerable plaque rupture due to stress-induced debonding around cellular microcalcifications in thin fibrous caps. PNAS 103:14678–14683. https://doi.org/10.1073/pnas.0606310103
174. Virmani R, Kolodgie FD, Burke AP et al (2000) Lessons from sudden coronary death: a comprehensive morphological classification scheme for atherosclerotic lesions. Arterioscler Thromb Vasc Biol 20:1262–1275. https://doi.org/10.1161/01.atv.20.5.1262

175. Waller BF, Orr CM, Slack JD et al (1992) Anatomy, histology, and pathology of coronary arteries: A review relevant to new interventional and imaging techniques-Part I. Clin Cardiol 15:451–457. https://doi.org/10.1002/clc.4960150613
176. Wang D, Wang Z, Zhang L, Wang Y (2017) Roles of cells from the arterial vessel wall in atherosclerosis. Mediators Inflamm 2017:1–9. https://doi.org/10.1155/2017/8135934
177. Wang J, Sjöberg S, Tang T-T et al (2014) Cathepsin G activity lowers plasma LDL and reduces atherosclerosis. Biochimica et Biophysica Acta (BBA) Mol Basis Dis 1842:2174–2183. https://doi.org/10.1016/j.bbadis.2014.07.026
178. Wang L, Luo J-Y, Li B et al (2016) Integrin-YAP/TAZ-JNK cascade mediates atheroprotective effect of unidirectional shear flow. Nature 540:579–582. https://doi.org/10.1038/nature20602
179. Welsh P, Polisecki E, Robertson M et al (2010) Unraveling the directional link between adiposity and inflammation: a bidirectional mendelian randomization approach. J Clin Endocrinol Metab 95:93–99. https://doi.org/10.1210/jc.2009-1064
180. Wick G, Knoflach M, Xu Q (2004) Autoimmune and inflammatory mechanisms in atherosclerosis. Annu Rev Immunol 22:361–403. https://doi.org/10.1146/annurev.immunol.22.012703.104644
181. Windaus A (1910) Über den Gehalt normaler und atheromatöser Aorten an Cholesterin und Cholesterinestern. Hoppe-Seyler´s Zeitschrift für physiologische Chemie 67:174–176. https://doi.org/10.1515/bchm2.1910.67.2.174
182. Witter K, Tonar Z, Schöpper H (2017) How many layers has the adventitia? Structure of the arterial tunica externa revisited. Anat Histol Embryol 46:110–120. https://doi.org/10.1111/ahe.12239
183. Xu L, Dai Perrard X, Perrard JL et al (2015) Foamy monocytes form early and contribute to nascent atherosclerosis in mice with hypercholesterolemia. Arterioscler Thromb Vasc Biol 35:1787–1797. https://doi.org/10.1161/ATVBAHA.115.305609
184. Yahagi K, Kolodgie FD, Otsuka F et al (2016) Pathophysiology of native coronary, vein graft, and in-stent atherosclerosis. Nat Rev Cardiol 13:79–98. https://doi.org/10.1038/nrcardio.2015.164
185. Yin K, Agrawal DK (2014) High-density lipoprotein: a novel target for antirestenosis therapy: HDL and antirestenosis therapy. Clin Transl Sci 7:500–511. https://doi.org/10.1111/cts.12186
186. Yin K, You Y, Swier V et al (2015) Vitamin D protects against atherosclerosis via regulation of cholesterol efflux and macrophage polarization in hypercholesterolemic swine. Arterioscler Thromb Vasc Biol 35:2432–2442. https://doi.org/10.1161/ATVBAHA.115.306132
187. Yoo S, Ahn H, Park YK (2016) High dietary fructose intake on cardiovascular disease related parameters in growing rats. Nutrients. https://doi.org/10.3390/nu9010011
188. Yoshida H, Kisugi R (2010) Mechanisms of LDL oxidation. Clin Chim Acta 411:1875–1882. https://doi.org/10.1016/j.cca.2010.08.038
189. Yurdagul A, Finney AC, Woolard MD, Orr AW (2016) The arterial microenvironment: the where and why of atherosclerosis. Biochem J 473:1281–1295. https://doi.org/10.1042/BJ20150844
190. Zain MA, Jamil RT, Siddiqui WJ (2019) Neointimal hyperplasia. StatPearls. StatPearls Publishing, Treasure Island (FL)
191. Zanoni I, Tan Y, Di Gioia M et al (2017) By capturing inflammatory lipids released from dying cells, the receptor CD14 induces inflammasome-dependent phagocyte hyperactivation. Immunity 47:697-709.e3. https://doi.org/10.1016/j.immuni.2017.09.010
192. Zhou Y, Zhang H-W, Tian F, et al (2016) Influence of increased epicardial adipose tissue volume on 1 year in-stent restenosis in patients who received coronary stent implantation. J Geriatr Cardiol 13:768–775. https://doi.org/10.11909/j.issn.1671-5411.2016.09.012

Chapter 14
Therapeutic Interventions of Endocannabinoid Signaling in Obesity-Related Cardiovascular Dysfunction

Vivek S. Kumawat, Siddhi Bagwe-Parab, Meena Chintamaneni, and Ginpreet Kaur

Abstract Obesity is a major risk factor involved in the progression of cardiovascular diseases. The aggravation of cardiovascular diseases is primarily linked to the release of pro-inflammatory cytokines. The endocannabinoid system (ECS) plays an important role in various diseases such as obesity, inflammation, Type II diabetes mellitus (T2DM) and cardiovascular dysfunction (CVD). ECS comprises of endocannabinoid receptors; cannabinoid 1 (CB_1), cannabinoid 2 (CB_2) and cannabinoid enzymes which are responsible for regulation of signaling pathways involved in obesity and CVD. The potential therapeutic interventions by drugs acting on cannabinoid receptors and their usage in the mitigation of obesity and CVD has been demonstrated in various pre-clinical and clinical studies. The expression of CB_1 inhibition and CB_2 activation receptors has been observed in the CVDs and obesity. Several reports suggest that; inhibition of centrally acting CB_1 receptor antagonist show a significant reduction of obesity, atherosclerosis and modulate blood pressure. These therapeutic agents have also been reported to be associated with unwanted side effects. Rimonabant (SR141716/Acomplia/Zimulti), a selective CB_1 inverse agonist had shown promising anti-obesity and anti-CVD effects, but later was withdrawn due to severe neuropsychiatric side-effects. The researchers are now exploring alternative approaches to block CB_1 receptors by avoiding psychotropic effects. Drugs like TXX-522 have been developed to exhibit good binding capacity to CB_1 receptors and have lesser brain penetration. Numerous studies have reported that, CB_2 receptor agonists prevent the onset of cardio metabolic diseases by scavenging free radicals and attenuating inflammation. Fatty acid amide hydrolase (FAAH) and monoacylglycerol lipase (MAGL) enzymes inhibition lead to reduction in the reactive oxygen species and regulation of apoptotic pathway, thus ameliorating the elevated biomarkers responsible for obesity and CVD. The central idea of this review is to examine the mechanism of action of cannabinoid receptors and to explore the therapeutic interventions

V. S. Kumawat · S. Bagwe-Parab · M. Chintamaneni · G. Kaur (✉)
Shobhaben Pratapbhai Patel School of Pharmacy & Technology Management, SVKM's NMIMS, V. L. Mehta Road, Vile Parle (W), Mumbai 400056, India
e-mail: ginpreet.kaur@nmims.edu

P. S. Tappia et al. (eds.), *Biochemistry of Cardiovascular Dysfunction in Obesity*, Advances in Biochemistry in Health and Disease 20,
https://doi.org/10.1007/978-3-030-47336-5_14

at cellular and molecular levels with respect to their anti-obesity and cardiovascular effects.

Keywords Obesity · Phytochemicals · Endocannabinoids · Signal transduction · Cardiovascular dysfunction · Therapeutic interventions

Abbreviations

2-AG	2-Arachidonoylglycerol
AEA	Anandamide
BP	Blood Pressure
CB_1	Cannabinoid receptor 1
CB_2	Cannabinoid receptor 2
CCK	Cholecystokinin
CVD	Cardiovascular dysfunctions
ECS	Endocannabinoid system
FAAH	Fatty acid amide hydrolase
GLP-1	Glucagon-like peptide-1
GPCR	G Protein-coupled receptor
HR	Heart rate
IL-1β	Interleukin-1 beta
IL-6	Interleukin-6
LDL	Low-density lipoprotein
MAGL	Monoacylglycerol lipase
MCP-1	Monocyte Chemoattractant Protein-1
PKC	Protein kinase C
ROS	Reactive oxygen species
T2DM	Type 2 diabetes mellitus
TNF-α	Tumor necrosis factor-alpha
WHO	World Health Organization

Introduction

Obesity is defined as an excess body weight for a given height due to abnormal and excessive fat accumulation, thus impairing health. It is a global epidemic consequence that potentially increases the risk of morbidity and mortality [1]. According to World Health Organization (WHO) global estimates in 2016, more than 1.9 billion adults (39%) were overweight and 650 million (13%) people suffered with obesity [2]. The increasing incidences of obesity is alarming for increase in the financial burden related to healthcare utilization. Numerous studies have linked obesity to increase in the risk of CVDs [3], metabolic disorders such as type 2 diabetes mellitus (T2DM) and dyslipidemia [4]. Association of obesity, hypertension, T2DM and dyslipidemias have also shown to increase the incidences of CVDs.

The pathogenesis of obesity is a complex mechanism involving many intervening factors such as diet, environment, physiological differences, medical interventions, behavioral factors, genetic and epigenetic composition [5]. Regulation of body weight and appetite is mainly carried out by the; nervous system, endocannabinoid system, adipose tissues and gastrointestinal hormones. Essential hormones involved in the regulation of appetite signals include; leptin, insulin, ghrelin, cholecystokinin (CCK) and glucagon-like peptide-1 (GLP-1) [6, 7]. These hormones transmit energy status information to brain cells and hypothalamus as a signal for hunger. Obesity is the major risk associated with CVD in both adults and children. Insulin resistance, increased blood pressure, dyslipidemia, atherosclerosis are the primary causes of obesity-associated CVDs. Inflammation of adipose tissue and decreased levels of adiponectin are considered the major factors associated with obesity and CVDs [8, 9]. In the past few decades, detailed knowledge of the pathophysiological processes of obesity and CVDs have been revealed. Strategies like lifestyle management, pharmacotherapies, bariatric surgery etc. are used in the treatment of obesity and obesity-induced CVDs. In pharmacotherapy intervention; Orlistat (inhibits gastric and pancreatic lipases), Lorcaserin (serotonin 2C receptor agonist) and Liraglutide (GLP-1 receptor agonist) are used for the treatment/management of obesity [10]. Novel therapeutic interventions that were indicated for the treatment of obesity were rapidly withdrawn from the market, because of unacceptable side-effects such as; gastrointestinal disturbances, neuropsychotropic effects and cardiovascular myopathies. ECS plays an important role in the control of food intake, energy balance, inflammation, oxidative stress [11], and lipid/glucose metabolism [12]. These factors are involved in the progression of CVDs. The potentials of ECS in appetite regulation and obesity makes it an important physiological strategy to be expended for the treatment of obesity and obesity-induced CVDs. The present review discusses the pathophysiological role of ECS receptors and their role in obesity-induced CVDs. Also, ECS receptors involvement in the disease and therapeutic approaches targeting ECS are elaborated for the treatment of obesity-induced CVD.

Obesity and Cardiovascular Risk

Many studies link obesity as the major cause of CVDs with causes including; insulin resistance, dyslipidemia, hypertension, increased cardiac load and atherosclerosis. The hemodynamic changes and metabolic changes in the body are the major risk factors of heart failure induced by obesity. Many details of the pathophysiological relationship between obesity and atherosclerosis have been revealed from last few decades [13]. These diseases have been displayed as lipid storage disorders with triglyceride accumulation in the tissues and cholesterol esters in atherosclerotic plaques. Adipokine imbalance is suggested as a strong link between obesity and atherosclerosis, in which there occurs an imbalance between levels of pro-inflammatory and anti-inflammatory adipokines (adiponectin). These factors may lead to the development of insulin resistance and endothelial dysfunction. Adiponectin is an anti-inflammatory and vasculo-protective adipokine released by adipose tissues [14]. Additionally, adiponectin suppresses the superoxide generation and enhances the oxidation of low-density lipoprotein (LDL). Also, adiponectin limits the initiation of atherosclerotic plaque formation in atherosclerosis.

Inflammation is the common bridge between obesity and atherosclerosis [9]. Obesity-induced inflammation is resulted by the activation of adipose tissue macrophages, and T cells/B cells within the adipose tissues. Various inflammatory cytokines such as interleukin-1β (IL-1β), interleukin-6 (IL-6), tumor necrosis factor-alpha (TNF-α), Monocyte Chemoattractant Protein-1 (MCP-1) and leptin majorly play an important role in atherosclerosis [14]. Atherosclerosis causes the narrowing and hardening of arteries that carry blood to the heart muscle which leads to coronary artery disease. In obesity, 70% of men and 60% of women are associated with hypertension due to excess of adiposity [10]. This may lead to insulin resistance and hyperinsulinemia, thus followed by the activation of the sympathetic nervous system. This activation finally increases hypertension by increase in vasoconstriction and cardiac output [15]. Hypertension and myocardial fibrosis eventually leads to diastolic and systolic dysfunction in the heart. Overall, these factors contribute to the heart failure induced by obesity. Obesity-induced cardiovascular disease is estimated to result in 11% of heart failure cases in males and up to 14% in women. The rise in prevalence of obesity may also increase the risk of heart failure [16, 17]. Figure 14.1 shows the linkage between obesity and cardiovascular diseases.

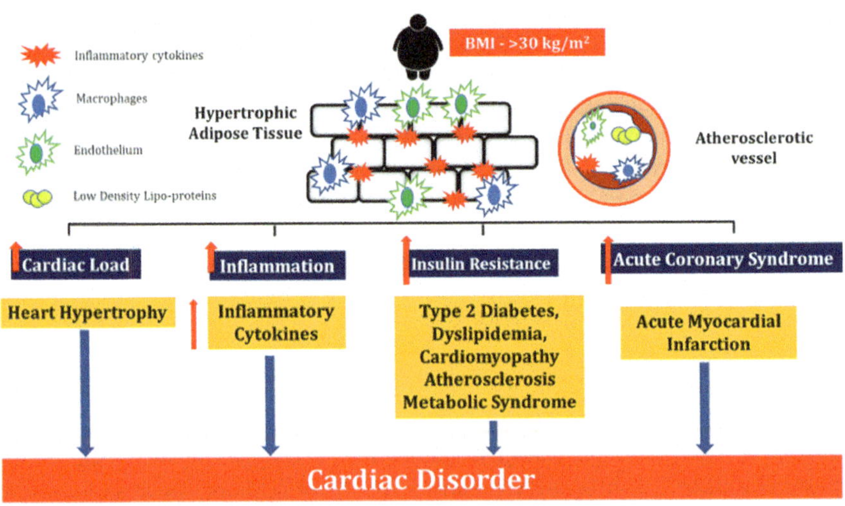

Fig. 14.1 Pathogenesis of Obesity-induced cardiac disorder

Endocannabinoid System

Since 1990's, discovery of the ECS [18], its receptors, and the potential of pharmacotherapy in the treatment of many clinical conditions has gained more attention. ECS consists of three major constituents such as endocannabinoid molecules, cannabinoid receptors, and enzymes involved in regulation of ECS. The endogenous endocannabinoid molecules are Anandamide (AEA) and 2-arachidonoylglycerol (2-AG) [19], which are responsible for maintaining signaling pathway with cannabinoid receptors. ECS consists of two G Protein-coupled receptor (GPCR) [20] as cannabinoid receptor 1 (CB_1) and cannabinoid receptor 2 (CB_2). Fatty acid amide hydrolase (FAAH) and monoacylglycerol lipase (MAGL) enzymes are mainly involved in enzymatic degradation of AEA and 2-AG respectively [7, 21]. The ECS acts on metabolism and physiology of multiple systems such as the nervous system, immune system and endocrine system.

The main function of the ECS is the regulation of protein synthesis, glycogen synthesis, and fat deposition. Activation of the ECS receptors stimulates appetite, glucose homeostasis and insulin secretion [22]. Evidences from the pre-clinical and clinical studies indicate that, targeting ECS receptors play an important role in the management of obesity, hyperglycemia and atherosclerosis (Fig. 14.2). The combined action of drugs on receptors and enzymes of the ECS display a major role in

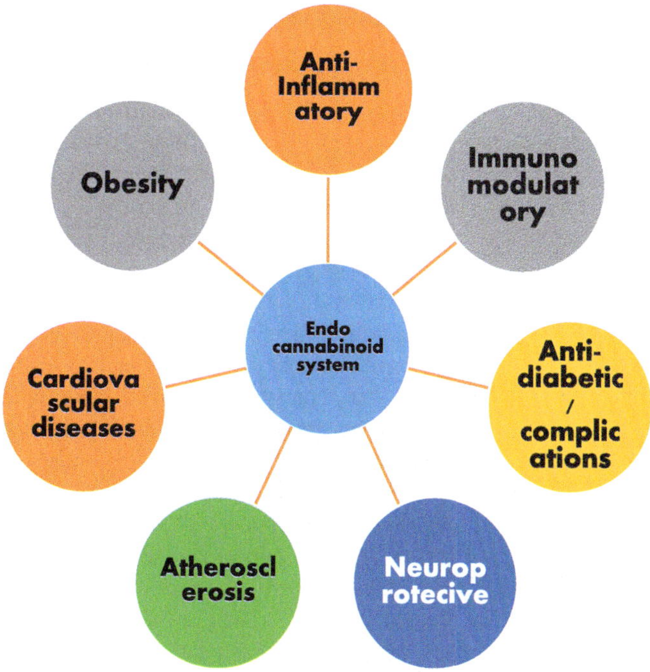

Fig. 14.2 Pathophysiological role of the Endocannabinoid system in human body

the management of appetite in obesity. The CB_1 receptors are primarily distributed in the areas of the brain related to motor control and sites associated with; pain processing, cognition, emotional responses, motivated behavior, and homeostasis. CB_2 receptors are located primarily on the immune cells and are also expressed in all the hematopoietic cells such as lymphocytes, natural killer cells, macrophages, and neutrophils [23]. Both CB_1 and CB_2 receptors are G-protein coupled receptors that possess seven transmembrane domains. The expression of these receptors vary considerably across tissues and cell types [24].

Pathophysiological Functions of Endocannabinoid System

Cannabinoid receptors and endocannabinoids are distributed throughout the body. The role of ECS has been observed in multiple physiological and pathological functions. *Cannabis Sativa* L. was the first plant to stimulate the interest in the research of ECS [25]. Tetrahydrocannabinol (THC) and cannabidiol are the major components in *C. Sativa* L. THC activates the CB_1 receptors and inhibit the release of the neurotransmitters in the brain. CB_1/CB_2 receptors and cannabinoid ligands play an important role in the various pathophysiological conditions such as; obesity, pain, inflammation, T2DM, cancer and atherosclerosis. However, it is noted that cannabinoids are highly prone to produce undesirable neuropsychotropic side-effects.

CB_1 receptors are centrally located receptors which play an important role in appetite regulation.

Administration of 2-AG in the limbic forebrain have been reported to be linked to control; motivation and appetite [26]. Also, a comprehensive study by Kirkham et al. showed that CB_1 receptor antagonist Rimonabant (SR141716) significantly decreased the feed intake in rats. CB_2 receptors also show anti-obesity activity by peripheral restriction of cannabinoidal action by AM6545 [27]. Another study presents FAAH-deficient mice which exhibits increased appetite due to elevated AEA levels in brain, liver, and small intestine [28]. CB_2 receptors majorly play an important role in inflammation and oxidative stress. Several in vitro and in vivo studies report that, CB_2 receptor agonists reduce neuroinflammation, diabetic complications and neuropathic pain by inhibiting the inflammatory cytokines such as Tumor necrosis factor α (TNF-α), Interleukin-1β (IL-1β), Transforming growth factor-β (TGF-β) and Nuclear factor -$\kappa\beta$ (Nf-$\kappa\beta$) [29]. Many pre-clinical studies suggests the modulation of cannabinoid receptors which may alleviate atherosclerosis [30]. Anti-atherosclerosis effect of ECS is mainly due to anti-inflammatory action and inhibition of accumulating macrophages in the artery [31]. Endocannabinoids and cannabinoid receptors show a promising role in receptor signaling during cardiac hypertrophy and heart failure. AEA suppresses the cardiac hypertrophy and cardiomyopathy, mediated by the action of CB_1 and CB_2 receptors. Many studies reveal that, activation of CB_2 receptors protect the heart from ischemic reperfusion injury [32]. Similarly, endocannabinoids such as AEA, 2-AG [33] and CB_2 receptor agonist such as JWH-015

[34] show a protective effect in ischemic reperfusion injury. Many possible mechanisms are involved in the protection of heart by ECS such as inhibition of inflammatory cytokines, scavenging of reactive oxygen species (ROS), and modulation of; PI3K/Akt, p38/ERK1/2, and Protein kinase C (PKC) activation [35]. Therapeutic interventions to be expended for endocannabinoid receptor signaling which targets obesity and cardiac disorders should be explored.

Components of the ECS (CB_1/CB_2 receptors and enzymes) which are involved in the endocannabinoid signaling and their role in pathological mechanisms of cardiovascular diseases; including atherosclerosis, myocardial infarction, and cardiac hypertrophy is explained further. Here, we discuss the potential roles of the ECS in the cardiovascular diseases.

Role of Cannabinoid Receptor 1 (CB_1) in Obesity-induced Cardiovascular Diseases

Smoking and ingestion of cannabis have detrimental effects on heart by raising the heart rate. This is due to the presence of THC and Phyto-cannabinoids which modulate the autonomic nervous system via CB_1 receptors [36]. Since the identification of cannabinoid receptors in the heart, extensive research has been carried out to study the modulation and activity of these receptors. Expression of CB_1 receptors were also identified in the human coronary artery, smooth muscle cells and myocardium [37].

CB_1 receptors are over-expressed in obesity, having increased concentrations of endocannabinoid enzymes and receptors in the adipose tissues. Many in vivo studies have confirmed that, the blockade of the CB_1 receptor by Rimonabant, a selective CB_1 receptor inverse agonist; decreased the food intake, body weight, adiposity and calorie intake [38, 39]. CB_1 receptor blocking by taranabant (CB_1 receptor inverse agonist) also showed significant improvement in insulin sensitivity [40]. An important finding from the research study of Schaich et al. demonstrates that, oral administration of Rimonabant in transgenic rats resulted in increased insulin resistance and hypertension. Acute systemic blockade of CB_1 receptor significantly reduced the blood pressure, weight gain and increased the secretion of insulin. The overall study concluded that CB_1 receptor blockade may be an effective therapeutic strategy for the treatment of angiotensin II dependent hypertension and associated metabolic syndrome [41]. The clinical study conducted by Ruilope et al. demonstrated that, CB_1 receptor blockade showed minimal effects on the blood pressure of normotensive subjects but significantly decreased the blood pressure in obese and T2DM patients [42]. Therefore, these studies suggest the role of CB_1 receptor inhibition in the regulation of hypertension. Several studies have reported positive correlation of ECS and CD.

Rimonabant, a potent CB_1 receptor inverse agonist has been withdrawn from the market, due to extensive neuropsychiatric effects [43]. Novel molecules should be developed for the blockade of CB_1 receptor which will benefit the patients suffering from obesity and related metabolic disorders. Research should be carried out to find out alternative ways to block the CB_1 receptors. After Rimonabant withdrawal from the market, non-central CB_1 blocking agents are reaching the research pipeline. The peripheral blockade of CB_1 receptors restrict Blood–brain barrier crossing, which prevent neuronal toxicity [44]. Chen et al. demonstrated that, novel peripheral CB_1 receptor antagonist TXX-522 showed molecular docking similarities with Rimonabant but has minimum penetration in the brain. TXX-522 has showed a significant decrease in the rat body weight with no neuropsychiatric effects [45]. CB_1 receptors have shown to inhibit macrophage accumulation and anti-inflammatory action in coronary artery diseases [46]. CB_1 receptor antagonist with peripheral blocking and combinatorial therapy in low doses are devoid of the side effects and are more effective for the treatment of obesity and cardiovascular disease (Table 14.1).

Role of Cannabinoid Receptor 2 (CB_2) in Obesity-induced Cardiovascular Disease

The CB_2 receptor was first cloned by Munro et al. from the human leukemia cell lines (HL-60) [53]. The expression of CB_2 receptors is mainly in the immune cells, spleen cells, peripheral tissues and organs like heart, liver and pancreas [54]. CB_2 receptors are also expressed in the cardiovascular system. Many research findings have demonstrated the expression of CB_2 receptors in rat cardiomyocytes, myocardium, smooth muscles and endothelial cells [55, 56]. Elevation in CB_2 receptor expression is seen in pathophysiological conditions such as inflammation and injury. This directs the protective effect of CB_2 receptors in inflammation and injury [57]. CB_2 receptor agonists display an important role in the inhibition of macrophages and leucocyte accumulation in the coronary artery [58]. Figure 14.3 summarizes the involvement and modulation of ECS in the management of obesity and its induced CVDs.

A study conducted by Zhao et al. has demonstrated that, administration of a CB_2 receptor agonist WIN55212-2 (prototypic aminoalkylindole) in apolipoprotein E-knockout (ApoE(−/−)) mouse with a high-fat diet, reduces the size of atherosclerotic lesions and also decreases the expression of Nf-κβ [59]. Verty et al. has demonstrated the anti-obesity activity of CB_2 receptor agonist JWH-015 in C57BL/6 mice. The results from this study suggests that, administration of JWH-015 reduced the feed intake, fat mass and adipocyte cells by modulating energy homeostasis and obesity-associated metabolic pathologies in the absence of any adverse neurological effects [60]. A CB_2 receptor agonist, Beta-caryophyllene has been reported to show anti-diabetic activity, by decreasing plasma glucose levels and inflammatory cytokines (TNF-α and IL-6) [61, 62]. Collectively, activation of CB_2 receptors exhibit a beneficial role in the treatment of inflammation, lipid metabolism, CVDs, T2DM and its related complications (Table 14.2) [63].

Table 14.1 Role of CB_1 antagonist in the obesity and CVD

Sr. No	Study	Drug	Methods	Effects	Reference
1	Obesity	Rimonabant (SR141716)	ob/ob mice	Improved carbohydrate metabolism, decreased feed intake	Jourdan et al. [47]
2	Obesity	Rimonabant (SR141716)	Diet-Induced obesity in Rats	Decreased food intake and body weight	Nogueiras et al. [48]
3	Obesity	AM251	Diet-induced obesity in rats	Decreased expression of GPI, LDH, body weight	Arrabal et al. [49]
4	Atherosclerosis	Rimonabant	LDL Receptor–Deficient Mice	Decreased Pro-inflammatory cytokines, reduced atherosclerosis development in the aortic sinus	Frédérique Dol-Gleizes et al. [50]
5	Hypertension	Rimonabant	Hypertensive (mRen2) 27 rats	Decreased weight gain, Serum leptin, insulin, SBP	Schaich et al. [41]
6	Cardio metabolic abnormalities by obesity	AM6545 (Peripheral Antagonist)	Diet-induced obesity in mice	Decreased body weight, adiposity, insulin resistance, hyperleptinemia, and food intake	Tam et al. [51]
7	Diabetes Mellitus	Ibipinabant	Male Zucker diabetic fatty rats	Decreased glucose, glucose excursion area under the curve, HbA1c, Increased Islets insulin	Rohrbach et al. [52]

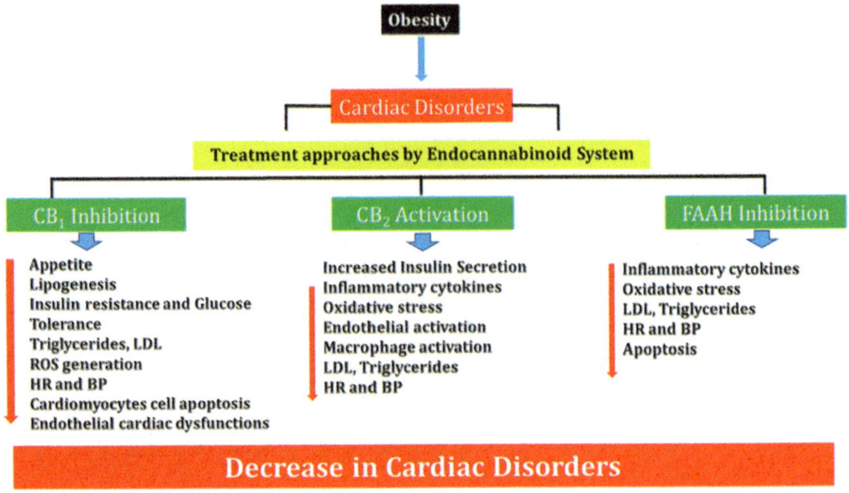

Fig. 14.3 Treatment approaches by targeting the ECS in obesity-induced cardiac disorders

Conclusion

Dysregulation in the ECS leads to the development of obesity and obesity-induced cardiovascular disorders. CB_1 receptor activation aggravates the uncontrolled metabolic pathway, thus leading to fat accumulation and in turn obesity. Inhibition of CB_1 receptors has displayed a potential role in controlling obesity and fat accumulation. Several studies have reported that, inhibition of CB_1 receptors led to decrease in food intake, body weight reduction, decrease in insulin sensitivity, reduction in the blood pressure, macrophage de-aggregation in coronary artery diseases etc. Drugs like Rimonabant (SR141716/Acomplia/Zimulti), having anorectic anti-obesity property had shown promising effects in the management of obesity and linked CVDs. Rimonabant was withdrawn globally within two years of its approval, for its severe neuropsychiatric side effects. A novel compound TXX-522 has been recently developed based on the parent compound rimonabant. TXX-522 has good binding capacity to CB_1 receptors, and has also exhibited lesser brain peneteration. Therefore, phytochemical compounds with similar structures can be screened for anti-obesity and linked CVDs. Also, CB_1 receptor activation in the heart causes depressed contractility of the heart and hypotension; inhibition of the same can ameliorate the cardiovascular diseases such as atherosclerosis, ischemia heart reperfusion injury and heart failure. CB_2 receptor activation has shown promising effects in the amelioration of inflammation and oxidative stress related injury. Therefore, CB_2 agonists can be expended for regulation of insulin secretion, inflammation and atherosclerosis. FAAH inhibition decreases inflammatory cytokines, regulates heart rate/blood pressure, reduces reactive oxygen species and downregulates apoptotic cell death. Therefore, this review hypothesized that ECS is a better treatment strategy for obesity and its linked cardiovascular diseases. Thus, the novel approach to treat

Table 14.2 Role of CB_2 agonist in the obesity and CVD

Sr. No	Study	Drug	Methods	Effects	Reference
1	Obesity	JWH-015	Diet-induced Obesity in Mice	Decreased body weight, reduction in food intake, and inflammation	Verty et al. [64]
2	Obesity	JWH-133	Obese Italian children	Decreased fat accumulation, IL-6, increased Adiponectin, IL-4	Rossi et al. [65]
3	Obesity	HU308	Diet-induced obesity in transgenic mice	Decreased inflammation, body weight No psychotropic affects	Schmitz et al. [66]
4	Cardiotoxicity	Beta-Caryophyllene	Doxorubicin-induced cardiotoxicity	Decreased inflammatory cytokines, iNOS, COX, decreased in heart oxidative stress	Meeran et al. [67]
5	Atherosclerosis	delta-9-tetrahydrocannabinol	Atherosclerosis in ApoE2/2 C57Bl/6 mice	Decreased IFN-ɣ, TNF-α, Decreased atherosclerosis plaque	Steffenes et al. [68]
6	Type 2 diabetes	Beta-caryophyllene	Streptozotocin and Nicotinamide induced T2DM	Decreased Glucose levels, triglycerides, Cholesterol	Kaur et al. [69]
7	Infarct and Ischemia-Reperfusion Heart Injury	HU308	Infarct and IR heart injury patient	Decreased the infarct size and the levels of reactive oxygen species and TNF-α	Wang et al. [32]

obesity-induced cardiovascular diseases by ECS will be useful in clinical research and will be a better therapeutic option for the treatment of the same.

References

1. Cercato C, Fonseca FA (2019) Cardiovascular risk and obesity. Diabetol Metab Syndr 11(1):1–15
2. Levesque RJR (2011) Obesity and overweight. Encyclopedia of Adolescence. [Online]. Available: https://www.who.int/en/news-room/fact-sheets/detail/obesity-and-overweight. Accessed 03 Feb 2020
3. Poirier P et al (2006) Obesity and cardiovascular disease: pathophysiology, evaluation, and effect of weight loss: an update of the 1997 American Heart Association Scientific Statement on obesity and heart disease from the Obesity Committee of the Council on Nutrition, Physical Activity, and Metabolism. Circulation 113(6):898–918
4. Al-Goblan AS, Al-Alfi MA, Khan MZ (2014) Mechanism linking diabetes mellitus and obesity. Diabetes Metab Syndr Obes Targets Ther 7:587–591
5. Wong C, Marwick TH (2007) Obesity cardiomyopathy: diagnosis and therapeutic implications. Nat Clin Pract Cardiovasc Med 4(9):480–490
6. Suzuki K, Jayasena CN, Bloom SR (2011) The Gut hormones in appetite regulation. J Obes
7. Ye L, Cao Z, Wang W, Zhou N (2019) New insights in cannabinoid receptor structure and signaling. Curr Mol Pharmacol 12(3):239–248
8. Ho WSV, Kelly MEM (2017) Cannabinoids in the cardiovascular system, 1st edn, vol. 80. Elsevier Inc
9. Ellulu MS, Patimah I, Khaza'ai H, Rahmat A, Abed Y (2017) Obesity & inflammation: the linking mechanism & the complications. Arch Med Sci 13(4):851–863
10. Heymsfield SB, Wadden TA (2017) Mechanisms, pathophysiology, and management of obesity. N Engl J Med 376(3):254–266
11. Han D et al (2017) Activation of cannabinoid receptor type II by AM1241 protects adipose-derived mesenchymal stem cells from oxidative damage and enhances their therapeutic efficacy in myocardial infarction mice via Stat3 activation. Oncotarget 8(39):64853–64866
12. Engeli S (2008) Dysregulation of the endocannabinoid system in obesity. J Neuroendocrinol 20(SUPPL 1):110–115
13. Csige I et al (2018) The impact of obesity on the cardiovascular system. J Diab Res
14. Lovren F, Teoh H, Verma S (2015) Obesity and atherosclerosis: mechanistic insights. Can J Cardiol 31(2):177–183
15. Santos C, Marques da Silva P (2018) Hemodynamic patterns in obesity associated hypertension. BMC Obes 5(1)
16. Re RN (2009) Obesity-related hypertension. Ochsner J 9(3):133–136
17. Ebong IA, Goff DC, Rodriguez CJ, Chen H, Bertoni AG (2014) Mechanisms of heart failure in obesity. Obesity Res Clin Pract 8(6):e540–e548
18. Di Marzo V, Fontana A (1995) Anandamide, an endogenous cannabinomimetic eicosanoid: 'Killing two birds with one stone'. Prostaglandins, Leukot Essent Fatty Acids 53(1):1–11
19. Demuth DG, Molleman A (2006) Cannabinoid signalling. Life Sci 78(6):549–563
20. Dalton GD, Bass CE, Van Horn CG, Howlett AC (2009) Signal transduction via cannabinoid receptors. CNS Neurol Disord Drug Targets 8(6):422–431
21. Howlett AC, Abood ME (2017) CB1 and CB2 receptor pharmacology. Adv Pharmacol 80:169–206
22. de Fonseca FR, del Arco I, Bermudez-Silva FJ, Bilbao A, Cippitelli A, Navarro M (2005) The endocannabinoid system: physiology and pharmacology. Alcohol Alcohol 40(1):2–14

23. Svíženská I, Dubový P, Šulcová A (2008) Cannabinoid receptors 1 and 2 (CB1 and CB2), their distribution, ligands and functional involvement in nervous system structures: a short review. Pharmacol Biochem Behav 90(4):501–511
24. Howlett AC, Abood ME (2017) CB1 and CB2 receptor pharmacology. Adv Pharmacol 169–206
25. Pellati F, Borgonetti V, Brighenti V, Biagi M, Benvenuti S, Corsi L (2018) *Cannabis sativa* L. and nonpsychoactive cannabinoids: their chemistry and role against oxidative stress, inflammation, and cancer. BioMed Res Int 1–10
26. Kirkham TC, Williams CM, Fezza F, Di Marzo V (2002) Endocannabinoid levels in rat limbic forebrain and hypothalamus in relation to fasting, feeding and satiation: Stimulation of eating by 2-arachidonoyl glycerol. Br J Pharmacol 136(4):550–557
27. Cluny NL et al (2010) A novel peripherally restricted cannabinoid receptor antagonist, AM6545, reduces food intake and body weight, but does not cause malaise, in rodents. Br J Pharmacol 161(3):629–642
28. Tourĩo C, Oveisi F, Lockney J, Piomelli D, Maldonado R (2010) FAAH deficiency promotes energy storage and enhances the motivation for food. Int J Obes 34(3):557–568
29. Turcotte C, Blanchet MR, Laviolette M, Flamand N (2016) The CB2 receptor and its role as a regulator of inflammation. Cell Mol Life Sci 73(23):4449–4470
30. Steffens S, Mach F (2006) Cannabinoid receptors in atherosclerosis. Curr Opin Lipidol 17(5):519–526
31. Durst R, Lotan C (2011) The potential for clinical use of cannabinoids in treatment of cardiovascular diseases. Cardiovasc Ther 29(1):17–22
32. Wang PF et al (2012) Cannabinoid-2 receptor activation protects against infarct and ischemia-reperfusion heart injury. J Cardiovasc Pharmacol 59(4):301–307
33. Pacher P, Haskó G (2008) Endocannabinoids and cannabinoid receptors in ischaemia-reperfusion injury and preconditioning. Br J Pharmacol 153(2):252–262
34. Montecucco F, Burger F, Mach F, Steffens S (2008) CB2 cannabinoid receptor agonist JWH-015 modulates human monocyte migration through defined intracellular signaling pathways. Am J Physiol Hear Circ Physiol 294(3)
35. Sun HJ et al (2017) Activation of endocannabinoid receptor 2 as a mechanism of propofol pretreatment-induced cardioprotection against ischemia-reperfusion injury in rats. Oxid Med Cell Longev 2:1–10
36. Jones RT (2002) Cardiovascular system effects of marijuana. J Clin Pharmacol 42(S1):58S-63S
37. Kaschina E (2016) Cannabinoid CB1/CB2 receptors in the heart: expression, regulation, and function. Cannabinoids Health Dis 169–186
38. Pi-Sunyer FX, Aronne LJ, Heshmati HM, Devin J, Rosenstock J (2006) Effect of rimonabant, a cannabinoid-1 receptor blocker, on weight and cardiometabolic risk factors in overweight or obese patients: RIO-North America: a randomized controlled trial. J Am Med Assoc 295(7):761–775
39. Guillamat-Prats R, Rami M, Herzig S, Steffens S (2019) Endocannabinoid signalling in atherosclerosis and related metabolic complications. Thromb Haemost 119(4):567–575
40. Aronne LJ et al (2010) A clinical trial assessing the safety and efficacy of taranabant, a CB1R inverse agonist, in obese and overweight patients: a high-dose study. Int J Obes 34(5):919–935
41. Schaich CL, Shaltout HA, Brosnihan KB, Howlett AC, Diz DI (2014) Acute and chronic systemic CB1 cannabinoid receptor blockade improves blood pressure regulation and metabolic profile in hypertensive (mRen2)27 rats. Physiol Rep 2(8):e12108
42. Ruilope LM et al (2008) Effect of rimonabant on blood pressure in overweight/obese patients with/without co-morbidities: analysis of pooled RIO study results. J Hypertens 26(2):357–367
43. Sam AH, Salem V, Ghatei MA (2011) Rimonabant: from RIO to ban. J Obes 1–4
44. Serrano A et al (2012) Obesity and the endocannabinoid system: is there still a future for CB1 antagonists in obesity? Curr Obes Rep 1(4):216–228
45. Chen W et al (2017) Novel peripherally restricted cannabinoid 1 receptor selective antagonist TXX-522 with prominent weight-loss efficacy in diet induced obese mice. Front Pharmacol 8:1–11

46. Sugamura K et al (2009) Activated endocannabinoid system in coronary artery disease and antiinflammatory effects of cannabinoid 1 receptor blockade on macrophages. Circulation 119(1):28–36
47. Jourdan T et al (2012) Antagonism of peripheral hepatic cannabinoid receptor-1 improves liver lipid metabolism in mice: evidence from cultured explants. Hepatology 55(3):790–799
48. Nogueiras R et al (2008) Peripheral, but not central, CB1 antagonism provides food intake-independent metabolic benefits in diet-induced obese rats. Diabetes 57(11):2977–2991
49. Arrabal S et al (2015) Pharmacological blockade of cannabinoid CB1 receptors in diet-induced obesity regulates mitochondrial dihydrolipoamide dehydrogenase in muscle. PLoS One 10(12)
50. Dol-Gleizes F et al (2009) Rimonabant, a selective cannabinoid CB1 receptor antagonist, inhibits atherosclerosis in LDL receptor-deficient mice. Arterioscler Thromb Vasc Biol 29(1):12–18
51. Tam J et al (2010) Peripheral CB1 cannabinoid receptor blockade improves cardiometabolic risk in mouse models of obesity. J Clin Invest 120(8):2953–2966
52. Rohrbach K et al (2012) Ibipinabant attenuates β-cell loss in male Zucker diabetic fatty rats independently of its effects on body weight. Diabetes. Obes Metab 14(6):555–564
53. Munro S, Thomas KL, Abu-Shaar M (1993) Molecular characterization of a peripheral receptor for cannabinoids. Nature 365(6441):61–65
54. Galiègue S et al (1995) Expression of central and peripheral cannabinoid receptors in human immune tissues and leukocyte subpopulations. Eur J Biochem 232(1):54–61
55. Bouchard JF, Lépicier P, Lamontagne D (2003) Contribution of endocannabinoids in the endothelial protection afforded by ischemic preconditioning in the isolated rat heart. Life Sci 72(16):1859–1870
56. Weis F et al (2010) Substantially altered expression pattern of cannabinoid receptor 2 and activated endocannabinoid system in patients with severe heart failure. J Mol Cell Cardiol 48(6):1187–1193
57. Pacher P, Mechoulam R (2011) Is lipid signaling through cannabinoid 2 receptors part of a protective system? Progress Lipid Res 50(2):193–211
58. Steffens S, Mach F (2006) Towards a therapeutic use of selective CB2 cannabinoid receptor ligands for atherosclerosis. Future Cardiol 2(1):49–53
59. Zhao Y et al (2010) WIN55212-2 ameliorates atherosclerosis associated with suppression of pro-inflammatory responses in ApoE-knockout mice. Eur J Pharmacol 649(1–3):285–292
60. Verty ANA, Stefanidis A, McAinch AJ, Hryciw DH, Oldfield B (2015) Anti-obesity effect of the CB2 receptor agonist JWH-015 in diet-induced obese mice. PLoS One 10(11)
61. Basha RH, Sankaranarayanan C (2014) β-Caryophyllene, a natural sesquiterpene, modulates carbohydrate metabolism in streptozotocin-induced diabetic rats. Acta Histochem 116(8):1469–1479
62. Basha RH, Sankaranarayanan C (2016) β-Caryophyllene, a natural sesquiterpene lactone attenuates hyperglycemia mediated oxidative and inflammatory stress in experimental diabetic rats. Chem Biol Interact 245:50–58
63. Kumawat VS, Kaur G (2019) Therapeutic potential of cannabinoid receptor 2 in the treatment of diabetes mellitus and its complications. Eur J Pharmacol 862:1–5
64. Verty ANA, Stefanidis A, McAinch AJ, Hryciw DH, Oldfield B (2015) Anti-obesity effect of the CB2 receptor agonist JWH-015 in diet-induced obese mice. PLoS ONE 10(11):1–19
65. Rossi F et al (2016) Cannabinoid receptor 2 as antiobesity target: Inflammation, fat storage, and browning modulation. J Clin Endocrinol Metab 101(9):3469–3478
66. Schmitz K, Mangels N, Häussler A, Ferreirós N, Fleming I, Tegeder I (2016) Pro-inflammatory obesity in aged cannabinoid-2 receptor-deficient mice. Int J Obes 40(2):366–379
67. Meeran MFN, Al Taee H, Azimullah S, Tariq S, Adeghate E, Ojha S (2019) β-Caryophyllene, a natural bicyclic sesquiterpene attenuates doxorubicin-induced chronic cardiotoxicity via activation of myocardial cannabinoid type-2 (CB2) receptors in rats. Chem Biol Interact 304:158–167

68. Steffens S et al (2005) Low dose oral cannabinoid therapy reduces progression of atherosclerosis in mice. Nature 434(7034):782–786
69. Kumawat VS, Kaur G (2020) Insulinotropic and antidiabetic effects of β-caryophyllene with l-arginine in type 2 diabetic rats. J Food Biochem

Chapter 15
Obesity and Cardiovascular Disease: Impact of Resveratrol as a Therapeutic

Xavier Lieben Louis, Sampath Krishnan, Jeffrey T. Wigle, and Thomas Netticadan

Abstract Obesity is a global epidemic and obese populations are at a much higher risk of developing diseases such as hypertension, type 2 diabetes, stroke and congestive heart failure. Increased adiposity is the hallmark of this physiological alteration of the body in response to excess intake of energy rich food, and this condition has far reaching health consequences in humans. Adipose dysfunction develops over time leading to increased secretion of inflammatory cytokines that cause inflammation and oxidative stress, which are independent risk factors for cardiovascular disease. Diastolic dysfunction is characteristic of the cardiac pathology associated with obesity. Obesity is a manageable condition and in some cases completely reversible with lifestyle modifications such as increased physical activity and a calorically restricted diet. In other cases, obesity can be reversed with either medications or surgery. In this regard, food derived compounds have been reported to have therapeutic benefits. Resveratrol is one such compound; it belongs to a family of plant compounds called polyphenols. In this chapter, we will review the causes and consequences of

X. Lieben Louis · S. Krishnan · J. T. Wigle
Institute of Cardiovascular Sciences, St. Boniface Hospital Albrechtsen Research Centre, Winnipeg, MB, Canada

X. Lieben Louis · J. T. Wigle
Department of Biochemistry and Medical Genetics, University of Manitoba, Winnipeg, MB, Canada

T. Netticadan (✉)
Department of Physiology and Pathophysiology, University of Manitoba, Winnipeg, MB, Canada
e-mail: tnetticadan@sbrc.ca

S. Krishnan · T. Netticadan
Canadian Centre for Agri-Food Research in Health and Medicine, St. Boniface Hospital Albrechtsen, Research Centre, Winnipeg, MB, Canada

T. Netticadan
Agriculture and Agri-Food Canada, Winnipeg, MB, Canada

Heart Failure Research Laboratory, Canadian Centre for Agri-Food Research in Health and Medicine, 351 Tache Avenue, Winnipeg, MB, Canada

P. S. Tappia et al. (eds.), *Biochemistry of Cardiovascular Dysfunction in Obesity*, Advances in Biochemistry in Health and Disease 20,
https://doi.org/10.1007/978-3-030-47336-5_15

obesity, obesity associated cardiovascular disease and the potential of resveratrol in prevention/treatment of obesity and obesity associated cardiovascular disease.

Keywords Obesity · Cardiovascular disease · Resveratrol · Polyphenols · Lipotoxicity · Adipose dysfunction

Introduction

Cardiovascular disease (CVD) and subsequent heart failure claims at least half a billion lives every year, around the globe. According to estimates available through the World Health Organization, around 17.9 million people died of CVD in 2016 [1]. There are a number of risk factors for the development of CVD such as hypertension, diabetes, hyperlipidemia, coronary disease, valvular disease and certain genetic mutations [2–4]. In combination with neurohormonal and cellular changes, the heart is capable of acutely compensating for many of the stresses arising from these pathophysiological conditions through a combination of structural and functional remodeling [5]. However, prolonged stress leads to maladaptive remodeling and the permanent loss of function of the heart and culminates in heart failure [6]. There are a number of therapeutic strategies currently being used to prevent, abate or reverse the development of heart failure [7]. For mild cases of CVD, life style modifications is the first step of therapy and pharmaceutical agents are prescribed when the disease has already progressed beyond what is manageable with life style changes [8, 9]. In certain cases such as valvular disease, electrical abnormalities and structural disabilities of the heart at birth, surgeries or devices such as pacemakers are used as the first step in treatment of the disease [7]. However, despite all of the modern biomedical inventions and advanced therapeutic methodologies, heart failure still claims millions of lives every year around the globe. This scenario leads us to think of alternative strategies that may more effectively prevent the development or arrest the progression of CVD into overt heart failure and mortality.

The obesity epidemic is directly and indirectly associated with millions of deaths every year [10]. Obesity is a condition wherein excess fat gets deposited under the skin and in other major organs of the body. This increased adiposity further increases risk of development of diabetes, hypertension, vascular diseases and other independent risk factors of cardiovascular disease [11]. Heart failure is a major cause of death in obese patients and the millions of individuals becoming obese every year are at risk of developing some form of heart disease [12].

Family genetics, gene abnormalities, diet, level of physical activity and other environmental factors are considered to be the major risk factors of obesity [13–16]. Genetic predisposition, together with diet plays a significant role in the development of obesity [17]. The development and increased availability of energy dense foods has certainly boosted the incidence of obesity around the globe (Fig. 15.1) [18]. High caloric intake accompanied by low physical activity results in increasing lipids stored as fat. After a certain point, cells that store fat (adipocytes) cannot keep up

Fig. 15.1 Evolution of obesity. From: https://www.flickr.com/photos/christopherdombres/7350782488

with the demand and become dysfunctional [19]. Healthy, smaller adipocytes secrete cardioprotective beneficial adipokines such as adiponectin. In contrast, dysfunction, enlarged adipocytes secrete pro-inflammatory adipokines such as leptin [20]. This adipocyte dysfunction leads to high levels of circulating lipids which subsequently results in increased uptake of lipids by other organs. Increased lipid deposition contributes to cellular stresses resulting in dysfunction of the organ [21]. The heart is one of the target organs in hyperlipdemic situations and the resulting stress causes cardiac dysfunction that culminates in heart failure [22].

Earlier forms of therapies were all derived from natural products. Different types of plants were used to treat all human ailments [23, 24]. Many of these medications were either eaten as the food by itself or mixed in combination with other foods [25, 26]. Nutraceuticals is the modern term for compounds that are naturally derived that, together with nutrition, delivers medicinal effects [27]. By this definition, resveratrol (RES) can be considered a nutraceutical as it is a polyphenol that is mainly found in grapes and other berries. Over the years, research has shown that RES has strong medicinal properties against a variety of human ailments including cancer, cardiovascular diseases, diabetes and some types of infections [28]. Cardioprotective properties of RES have been well documented and has recently also been shown to be beneficial in human clinical trials [29, 30].

Obesity

Obesity is a condition wherein excess fat accumulates in the body. A way to assess obesity is to classify individuals using the body mass index (BMI) which is calculated by dividing body weight by the square of that individual's height. For adults the classification of obesity states is as follows (Table 15.1).

However, BMI is not a direct measure of fat stores and hence additional measurements are required to calculate the levels of body fat [31]. Waist circumference, skin fold test, waist to hip ratio calculation and modern techniques such as whole-body

Table 15.1 Classification based on BMI

BMI	Classification
≤24.9 kg/m^2	Normal
25–29.9 kg/m^2	Overweight
≥30 kg/m^2	Obese
≥39 kg/m^2	Severely obese

air displacement plethysmography (ADP), underwater weighing, dual energy X-ray absorptiometry (DXA) and even ultrasound techniques can be used to accurately determine body fat composition [32–36]. A small percentage of the obese population are completely healthy despite being obese and are often termed as being "healthy-obese" [37]. However, in most cases obesity is accompanied by several comorbidities and obese individuals experience poor quality of life and suffer psychologically from social stigma [38, 39]. Obesity is a manageable condition and in some cases completely reversible with life style changes including increasing physical activity, adopting a caloric restricted diet, taking medications and in some cases surgery [40–42]. Irrespective of the status of national development (underdeveloped, developing or developed), the growing incidence of obesity and increased dependence on health care systems has become a major worldwide concern [43–47].

World Statistics

At least 2.8 million deaths that occur every year are associated with being obese or overweight. The prevalence of obesity worldwide has almost tripled since 1975 and as of 2016, a total of more than 1.9 billion adults were considered obese or overweight. Shockingly, more people reside in countries where obesity causes more deaths than being underweight [48]. The prevalence of obesity was highest in the Americas (approximately 62% overweight and 26% obese) and lowest in South East Asia (approximately 14% overweight and 3% obese). Women are reported to be more prone to obesity when compared to men around the world [48, 49]. In some African and Mediterranean countries, the prevalence of obesity in women is almost double that of men. Income is also correlated with the increased prevalence of obesity [18].

Canadian Statistics

According to the latest data published by Statistics Canada, roughly 1 in 4 Canadian adults (27%) are obese [50]. Statistics Canada data from 2018 shows that 69.4% of men and 56.7% of women 18 years and older are overweight or obese [51]. These estimates are higher than from the self-reported data in 2009 which showed 59.2% of Canadian men and 43.9% of women in the overweight or obese category [52].

The number of obese children and youth is also increasing with 30% between the ages of 5 and 17 being obese or overweight [53–55]. The geographic distribution of obesity varies within the country. The lowest reported prevalence being 22% in British Columbia, while the highest in New Brunswick along with Newfoundland and Labrador where prevalence is as high as 38% [50]. Obesity is also seen more commonly in certain ethnic populations [56, 57].

Obesity in Children

Childhood obesity is another major area of concern as it possesses several health risks including cardiovascular disease and early mortality in their adulthood [54, 58, 59]. Based on data from Statistics Canada in 2018, 23.7% of children between the ages 12–17 years are overweight or obese [51]. Obesity in children and youth is measured using a different set of BMI cut-offs. According to the International Obesity Task Force (IOTF), a BMI greater than 21.22 and 26.02 kg/m^2 for 12 year old boys and girls respectively, is categorized as obese. There are other systems of BMI categories so the estimates of childhood obesity may vary accordingly. For example, in the 2004 Canadian Community Health Survey based on the IOTF system, obesity rates were reported to be 8.2% among children and youth (2–17 years). However, based on the Centers for Disease Control system, the estimate increased to 12.7%, while based on the WHO system it was estimated to be 12.5% [60].

Health Costs

Obesity is associated with a number of co-morbidities. The risk of type 2 diabetes, hypertension, cardiovascular diseases and cancers increases significantly with being obese. Due to social stigma around obese individuals, a number of psychological conditions are also prevalent among the obese [61, 62]. Premature mortality rates are also reported to increase alongside the severity of obesity [10]. All these factors contribute to an increased life time dependence on the health care system by these individuals when compared to the non-obese. This directly results in higher than normal health expenditure per obese individual and puts a burden on the health care system. Estimates based on 2008 data put health care costs related to obesity between $4.6 billion to $7.1 billion per year [63]. These cost figures show how important it is to ramp up the awareness and fight against obesity, especially among those who are in the higher risk categories [64].

Overall, the increasing prevalence of obesity is a major global health concern and needs immediate attention to protect future generations. Race, sex, income and other socioeconomic factors have been seen to be associated with increased risk for obesity [65]. Childhood obesity is also increasing which is particularly alarming given that the risk of obesity increases with age. Childhood obesity also results in early onset of comorbidities such as diabetes, hypertension and reduced life expectancy [66].

Pathology of Obesity

Adiposity/obesity and Adipose Dysfunction

Obesity is a condition that develops due to excess fat deposition in the body overtime. Generally, excess energy in the body is stored as fat. Accordingly, the amount of fat stored is the difference between food (energy) intake and energy expenditure (Fig. 15.2).

Human body stores excess circulating lipids in adipocytes [67, 68]. These stored fats are kept as a reserve energy bank that is used during a state of fasting [69]. Adipocytes are normally an integral part of the human body and are required for normal physiology. However, when circulating lipids are chronically higher than normal, the amount of fat deposits also increases. Overtime, the increased adiposity results in pathological changes of obesity [70]. There are many factors such as energy dense diet, physical inactivity, gene mutations and/or other pathophysiological conditions that contribute to increased circulating lipids [71, 72].

Adiposity could be increased in two different ways, either by increasing the number (hyperplasia) or the size (hypertrophy) of adipocytes [73, 74]. The first is considered to be physiological; while the second, wherein the adipocyte enlarges, is considered pathological [75]. Earlier, adipose tissues were considered as just a store of fat. However, later it was discovered that adipose tissues are also an endocrine organ and adipocytes secrete hormones (adiponectin, leptin and resistin), cytokines (TNF-α, IL-6) and proteins (cholesteryl ester treansfer protein, angiotensin II, plasminogen activator inhibitor 1) involved in the metabolism and functions of the liver, muscles, vasculature, brain and other organs of the body [76]. Cytokine secretions from adipocytes are generally known as adipokines [77]. Some adipokines are beneficial while others cause unhealthy effects on biological functions [20]. In normal or healthy conditions, adipocytes secrete more adipokines which are beneficial, while in pathophysiological conditions where adipocyte dysfunction occurs, the balance will be shifted to an increased release of detrimental adipokines [75, 77]. The origin of adipocyte dysfunction is mainly associated to the physiological demands of storing very high levels of fat. Resident macrophages are also present among the adipocytes and are involved in fat storage and secretion of cytokines. The

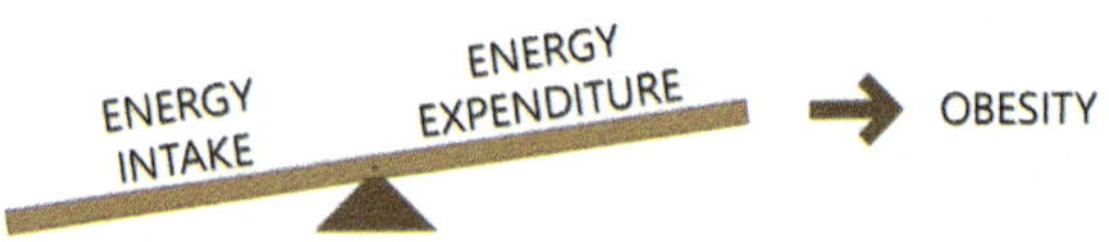

Fig. 15.2 Energy balance

adipocytes and macrophages are highly involved in the genesis of chronic inflammation in the adipose tissue and release of pro-inflammatory factors into the blood [78]. Consequently, there is also increased lipolysis and release of free fatty acids into circulation. Accordingly, adipose tissue is considered to be the major source of the pathophysiological effects in obesity [19, 79]. This also makes adipocytes a potential drug target to ameliorate the metabolic disarray in obese conditions. To some extent, targeting adipocyte dysfunction has shown promise in preventing or improving metabolic imbalances in obesity [80].

Obesity and Cardiovascular Disease

According to seminal Framingham Heart Study, the risk of developing heart failure in obese individuals was 2 times that of normal weight subjects [81]. It has been found that 32–49% of heart failure patients are considered obese; furthermore, obese patients develop heart failure up to 10 years earlier than their normal BMI counterparts [82]. Cardiovascular complications are one of the major contributors to poor health and lower life expectancy among obese populations [83, 84]. Obesity, especially abdominal obesity, is an independent risk factor for CVD [85, 86]. Higher BMI is directly associated with adipocyte dysfunction, increased release of adipokines, insulin resistance, hypertension, increased inflammation and oxidative stress that promotes the development of cardiovascular disease [87]. Although not unanimously accepted, the 'obese paradox' theory claims that obese individuals have a better prognosis to CVD when compared to normal weight individuals [82]. A possible explanation for this paradoxical theory is that the BMI measurements are insufficient to accurately assess the state of obesity and adipocyte dysfunction in an individual [88]. Additional measurements such as waist circumference and waist to hip ratio would better classify the subjects based on the levels of fat deposition. A study on the Monza population has shown that with every 1 kg/m^2 increase in BMI, the risk of developing left ventricular (LV) hypertrophy increases by 5.1%, and for every 1 cm increase in waist circumference the risk increase by 2.5% [89]. Visceral adiposity and subcutaneous fat deposits contribute to increased waist circumference which has been found to be an independent risk factor for developing heart disease [90]. For increased risk of CVD, the WHO's cut-off values for waist circumference are 102 cm in men and 88 cm in women [91].

Obesity exerts stress on the heart by increasing the blood volume and cardiac output simultaneously, placing a larger workload on the heart [10]. This results in adverse changes to hemodynamics, cardiovascular structure and function. Obesity increases total body area and volume by additional fat tissue and the changes in cardiovascular system are aimed at maintaining sufficient blood supply to the whole body [10]. Adipose tissue contains a large volume of fluid which is present in the interstitial spaces of the tissue. The interstitial space adds up to approximately 10% of the total adipose tissue weight. Obesity also increases lean body mass which independently elevates cardiac output [92]. A combination of increase in lean and fat mass could account for a large increase in stroke volume and cardiac output. The expansion in volume of blood increases the preload on the heart and shifts the Frank-Starling curve to the left. A significant change in vascular structure and function is also observed in obesity. Obesity causes arterial stiffness [93], increased intima-media thickness [94, 95] and increased calcification [96]. All these vascular changes are also independent predictors of CVD. Further, these vascular changes may also contribute to the development of hypertension in obese individuals [97].

These changes in hemodynamics increases wall tension and induces LV dilation and hypertrophy [82]. Prolonged exposure to these stressful conditions reduces LV wall compliance and then diastolic dysfunction ensues. Initial adaptations by the LV help preserve LV systolic function in the early stages of cardiac remodeling. Overtime, impairment in systolic function will develop and heart failure will be initiated [98]. It was also found that the fatty heart, as a result of increased fat deposits is more prone to cardiomyopathy [99]. Damage to heart muscles by fat accumulation happens in two ways, metaplasia and lipotoxicity [100]. In metaplasia, some cells (epithelial or mesenchymal) are replaced by fat cells, disrupting the cardiac electroconduction. In lipotoxicity, free fatty acid accumulation in cardiomyocytes induces cell death in the myocardium. In either case, damage to cells results in myocardial weakening, resulting in the development of cardiomyopathy [10]. The obesity associated increase in blood volume also induces left atrial enlargement, which increases the risk of developing atrial fibrillation. Based on the findings from Women Health Study, obesity was associated with increased risk for atrial fibrillation [101]. Other types of arrhythmias and sudden cardiac death are also found at higher rates in obese populations [82, 102]. Obesity and metabolic dysfunction increases the risk of coronary artery disease. The incidence of coronary atherosclerosis is very high in adult obesity and is a major risk factor for heart disease [103]. Obesity is also directly linked to increased incidence of stroke. The INTESTROKE study has found that waist to hip ratio was strongly associated with increased risk for stroke [104]. Obstructive sleep apnea is another risk factor for hypertension and CVD [105]. Obesity is one of the major risk factors for obstructive sleep apnea, and in many sleep apnea patients are undiagnosed which increases the risk of heart disease.

Adipose tissue is also an endocrine organ releasing a number of molecules into the blood stream. TNF-α, IL-6, leptin, angiotensinogen, resistin and plasminogen activator inhibitor-1 are released from adipose tissues and have direct or indirect effect on promoting development or progression of heart disease [106]. A significant proportion of the circulating concentrations of these molecules have originated from adipose tissue. Most of these are mediators of the inflammatory response and may be involved in progression of coronary artery diseases [107].

Indirect effect of obesity on cardiovascular pathology involves impairments of kidney structure and function. Glomerular hyperfiltration, increased albumin loss, glomerulosclerosis and progressive loss of kidney function are associated with obesity-induced kidney damages [108]. Population studies, PREVEND [109] and Framingham Heart Study [110] have found direct correlation between kidney damage and obesity.

Lipids and Heart

Fatty acids are the primary energy source of the heart. In normal physiological conditions, approximately 70% of energy is derived from the oxidation of fatty acids [111]. The remaining energy is derived from glucose, lactate and ketones. Generation of ATP from fatty acid oxidation is a comparatively more oxygen demanding process than generating ATP from glucose. The heart has the ability to switch to glucose as the major energy source during oxygen deficient conditions such as ischemia, hypertension and other pathological conditions [112, 113]. This allows the heart to adapt to difficult conditions, preserve available oxygen and minimize the damage to the tissue. Fetal hearts also depend more on glucose and lactate for energy, while adult hearts shift to fatty acid oxidation to meet their energy needs [112]. Diet, hepatic fatty acid synthesis and lipolysis in adipose tissue are the major sources of lipid for the heart. Heart tissues can use both non-esterified (free fatty acids) and esterified (bound to lipoproteins) fatty acids. Circulating triglycerides undergo lipolysis mediated by endothelium-bound lipoprotein lipase and are then internalized via membrane receptors, transporters or simply by diffusion [114]. Internalized free fatty acids are then converted to fatty acyl-CoAs and then either stored as acyl glycerides (mono, di or tri) or transported to mitochondria for ATP generation [115]. Triglycerides stored intracellularly are processed to free fatty acids by hormone sensitive lipase and adipose triglyceride lipase [116].

Lipotoxicity

The balancing act of energy homeostasis is much more complex than the earlier mentioned equation (Fig. 15.1). For example, there is a significant difference between white adipose tissue (WAT) and brown adipose tissue (BAT). While both are fat deposits, their physiological roles are different. WAT is associated with the genesis of metabolic syndrome while, BAT contributes to thermogenesis [117]. Diet is the major source of fatty acids (FA); it is also synthesized from other sources through de novo lipogenesis [118]. Depending on physiologic demands FAs are released into circulation from adipose tissue by lipolysis and will be used by other organs. FAs can also be transported into the cells by different protein transport mechanisms [115, 119]. These FAs are then used for a variety of cellular mechanisms involving synthesis of membrane, signaling molecules, post-translational protein modification, transcriptional regulation and more importantly for energy production through beta oxidation [120]. Normally, a balance is maintained between lipid uptake and oxidation thereby preventing lipid accumulation. Metabolic disturbances often results in increased lipid levels in the circulation. Higher circulating FAs in obesity and type 2 diabetes causes excess deposition of FAs in non-adipose tissues such as kidney, liver, skeletal muscles and heart [121]. The excess lipid accumulating inside the cell may cause cellular dysfunction through ER stress, mitochondrial dysfunction, oxidative stress and ultimately results in cell death [122]. This process of lipid induced cellular damage and death is known as lipotoxicity [123–125].

Lipotoxicity in the Heart

The heart is one of the major organs affected by lipid accumulation [21]. Lipid accumulation is also observed in cardiac pathologies wherein the myocardium reverts to glucose as the primary source of energy [112]. Lipotoxic effects leads to cardiomyocyte dysfunction, contractile abnormalities, cell death and pathogenesis of heart failure [111, 126]. Cardiomyopathies observed in in the setting of type 2 diabetes and obesity are often a result of lipotoxic damage to the myocardium [116]. Long chain fatty acids like palmitate have been found to induce lipotoxicity in the heart muscle cells when compared to short chain fatty acids [127, 128]. Some pathological cellular changes associated with lipid accumulation are ER stress, mitochondrial dysfunction and oxidative stress. Increased ceramide accumulation has been observed as a contributor towards cell death in the heart [129]. Insulin is involved in regulation of glucose metabolism, activation of survival pathways in ischemia and also in intracellular Ca^{2+} handling. Lipotoxicity induces insulin resistance and thereby causes cardiomyocyte dysfunction. Activation of protein kinase C (PKC), mitogen activated protein kinases and reduced peroxisome proliferator-activated receptors are also considered to be involved in the process of lipid accumulation and cellular responses in the cell [21]. Lipid accumulation also induces contractile abnormalities through the degeneration of myofibrils [130].

Resveratrol

RES is a phytoalexin compound produced by plants mainly in response to fungal infections, UV radiation and other environmental stresses such as cold temperatures [131]. RES is present in significant amounts in grapes, peanuts, soy beans, pomegranates, mulberry and bilberry [132] and to a lesser extend in pine, eucalyptus and spruce trees, and in a few flowering plants, such as *Veratrum grandiflorum* and *Veratrum formosanum* [133, 134].

RES was discovered in the roots of white hellebore plants [135]. Later it was also found in roots of *Polyganum cuspidatum*, a Japaneese knotweed which was also called Ko-jo-kon and is the richest known source of RES. It was used in the preparation of Japanese and Chinese herbal medicines against skin infections like warts, dermatitis and athletes foot [136]. This was followed by reports on presence of RES in eucalyptus and pine [137, 138]. In 1976 Langcake and Price reported the presence of RES in grape vines for the first time [139]. During this period RES was mainly investigated for its anti-fungal properties and used as a screening marker for disease resistant grape cultivars [140, 141]. The first report linking RES to potential cardiovascular benefits was from a Japanese group which showed that RES administration reduced triglyceride synthesis in mice [142]. Later in 1992, moderate red wine consumption was linked to the reduced incidence of cardiovascular disease among the French population; this theory is known as the 'French Paradox' [143]. At the same time, Siemann and Creasy reported that RES might be one of the bioactive ingredients in wine [144]. Further, Frankel et al. showed that the phenolic component of red wine inhibited LDL oxidation which is a risk factor for atherogenesis and thereby ischemic heart disease [145]. The association of RES to the French Paradox generated a greater interest in RES research wherein either purified RES or food containing significant amount of RES were tested on a wide range of research models of human disease [131]. The highlights of RES research outcomes are its beneficial effects against different types of cancers, cardiovascular diseases and also against metabolic diseases such as diabetes and obesity [146–148].

Resveratrol Chemistry

RES is a stilbene derivative, produced in plants by stilbene synthase. Stilbene synthase catalyze the synthesis of RES from one molecule of p-coumaroyl CoA and three molecules of malonyl CoA. RES exists in two structural isomeric forms, *cis*- and *trans*-RES (Fig. 15.3) (molecular weight: 228.24); both isomers are lipophilic in nature [149]. RES has a melting point around 260 °C. It is insoluble in water but soluble in ethanol and DMSO. The *trans*-RES isomer is relatively more stable as compared to the *cis*-RES isomer; however, the *trans* form can get converted to the *cis* for when exposed to heat or UV radiation [150]. The *trans*-RES in the powder form is stable in normal atmosphere at room temperature and undergoes negligible oxidation in these conditions. RES is susceptible to photolysis if exposed to direct sunlight. Due to its structural similarity to the synthetic estrogen diethylstilbestrol, RES is also considered to be a phytoestrogen [133].

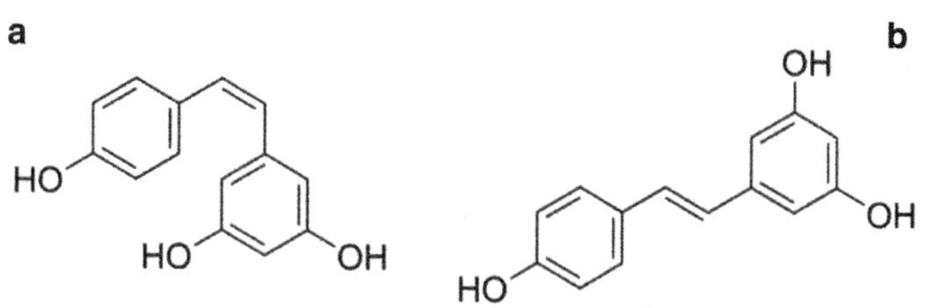

Fig. 15.3 *Cis*-resveratrol (**a**) and *trans*-resveratrol (**b**). From: https://en.wikipedia.org/wiki/File:Cis_and_trans_resveratrol_notext.svg

Cardioprotection with Resveratrol

RES is present at high amounts in grape skins and subsequently in red wines (Fig. 15.4). This led to the 'French Paradox' theory in which lower incidence of cardiovascular disease in the French population was associated with a higher consumption of red wine [150]. Subsequent research confirmed the cardioprotective properties of RES [151]. RES was reported to exhibit cardioprotective properties by reducing cardiac abnormalities in hypertension, ischemic heart disease, obesity, diabetes and cardiomyopathies; RES has been shown to improve heart and blood vessel structure and function in animal models of cardiovascular disease [30, 152, 153]. Major cellular mechanisms that mediate effects of RES range from improving cardiovascular risk factors such as hyperlipidemia and insulin resistance to reducing oxidative stress and inflammation. Among the many molecules identified as RES targets in the heart, AMPK, SIRT1 and nitric oxide (NO) are most frequently reported [152]. RES is found to enhance AMPK activity and thereby its downstream signaling pathway which indirectly results in increased NO production [154, 155]. AMPK activation could also be involved in RES-mediated decrease in fibrosis [156]. An increase in SIRT1 expression is associated with RES administration. SIRT1 could improve cardiac function by increasing SERCA2A expression and thereby improving Ca^{2+} handling. SIRT1 could also induce AMPK activation and improve mitochondrial function [157]. Anti-inflammatory and antioxidant activities were the first properties to be identified and associated to health benefits of RES. RES inhibits NFkB activation and translocation into the nucleus, preventing the transcription of a variety of genes detrimental to the cell [158]. RES also helps preserve major antioxidant enzyme activities such as superoxide dismutase, catalase, and glutathione peroxidase, while reducing NADPH oxidase activity. Nuclear factor erythroid 2-related factor 2 (Nrf2) is involved in maintaining an antioxidant environment inside the cells and RES is

Fig. 15.4 Major sources of resveratrol: grapes and red wine. From: https://www.publicdomainfiles.com/show_file.php?id=13534675212806 (wine glass). https://www.publicdomainpictures.net/en/view-image.php?image=299825&picture=grapesvintage-illustration (grapes)

found to promote Nrf2 activation [159]. Modulation of L-type calcium channel is also a potential mechanism by which RES could improve Ca^{2+} irregularities in cardiac cells [160].

Resveratrol in Obesity-Induced Heart Disease

There is sufficient evidence showing that RES improves metabolic abnormalities in animals [161]. However, there are only a few studies that have explored the cardioprotective property of RES in obesity. The first study showed that RES administration in rats for 2 weeks reduced both infarct size and cardiac apoptosis in ex-vivo ischemic-reperfused hearts [162]. In another study RES prevented an increase in blood pressure and preserved vascular function in an animal model of diet-induced obesity, the high fat fed rats [163]. Louis et al. reported that RES reversed diastolic heart dysfunction in high fat fed rats [164] and Qin et al. showed a significant decrease in cardiac hypertrophy and improvement in diastolic heart function in obese mice exhibiting characteristics of early stage type II diabetes [165]. Cardiomyopathy is the major form of heart disease affecting the obese and overweight population. Yingjie et al. showed that RES attenuated high fat diet-induced cardiomyopathy in a mouse model [166]. This effect was associated with upregulation of estrogen receptor alpha which has been proposed to mediate RES actions in vivo. As discussed earlier, obesity pathologically affects vascular function. NO is an endogenous vasodilator and an established target of RES [167]. Huang et al. reported that RES treatment mitigated vascular dysfunction in high fat fed mice through upregulating eNOS/NO mechanism [168].

To date, no study has examined the impact of resveratrol in preventing obesity induced deterioration of heart function in humans. However, there are a few clinical studies which have examined the potential of resveratrol in reducing obesity. A recent clinical trial reported significant reductions in weight loss in obese patients who were on a combination of resveratrol and orlistat (a standard anti-obesity medication) [169]. Similarly a combination of resveratrol and hesperetin (a bioactive compound) was reported to reduce glucose levels and improved vascular function in overweight and obese patients [170]. A combination of epigallocatechin-3-gallate (a bioactive compound) and resveratrol was also shown to increase mitochondrial capacity and stimulate fat oxidation in overweight and obese patients [171]. Earlier studies reported that resveratrol supplementation reduced glucose, triglycerides and markers of inflammation in obese men [172], improved cerebral blood flow in obese subjects [173], and reduced intestinal and hepatic lipoprotein production in obese individuals with mild hypertriglyceridemia [174].

Overall, there is some evidence that RES improves heart structure and function in animal models of obesity. However, there is no information on impact of RES on heart structure and function in humans with obesity, therefore future studies should examine this aspect. Given that diet and genetic predisposition are major contributors towards the development of obesity, future research is necessary to examine if RES can protect the human heart in the settings of diet induced obesity. It is also important to know if RES can improve diastolic heart dysfunction in obese animals and thereby prevent the progression to heart failure. Finally, more research exploring cellular mechanisms involved in the cardioprotective action of RES is needed. Given the success of RES in combination with standard medication or other bioactive compounds in improving metabolic parameters in overweight and obese subjects, it would be interesting to examine the potential of combination therapy in preventing heart dysfunction in these subjects.

References

1. WHO (2017) Cardiovascular diseases. WHO
2. Berry JD, Dyer A, Cai X, Garside DB, Ning H, Thomas A, Greenland P, Van Horn L, Tracy RP, Lloyd-Jones DM (2012) Lifetime risks of cardiovascular disease. N Engl J Med 366:321–329. https://doi.org/10.1056/NEJMoa1012848
3. Lloyd-Jones DM, Nam BH, D'Agostino RB Sr, Levy D, Murabito JM, Wang TJ, Wilson PW, O'Donnell CJ (2004) Parental cardiovascular disease as a risk factor for cardiovascular disease in middle-aged adults: a prospective study of parents and offspring. JAMA 291:2204–2211. https://doi.org/10.1001/jama.291.18.2204
4. Murabito JM, Nam BH, D'Agostino RB Sr, Lloyd-Jones DM, O'Donnell CJ, Wilson PW (2004) Accuracy of offspring reports of parental cardiovascular disease history: the Framingham Offspring Study. Ann Intern Med 140:434–440
5. Cohn JN, Ferrari R, Sharpe N (2000) Cardiac remodeling–concepts and clinical implications: a consensus paper from an international forum on cardiac remodeling. Behalf of an international forum on cardiac remodeling. J Am Coll Cardiol 35:569–582
6. Grossman W, Paulus WJ (2013) Myocardial stress and hypertrophy: a complex interface between biophysics and cardiac remodeling. J Clin Invest 123:3701–3703. https://doi.org/10.1172/JCI69830
7. Liu SS, Monti J, Kargbo HM, Athar MW, Parakh K (2013) Frontiers of therapy for patients with heart failure. Am J Med 126(6–12):e6. https://doi.org/10.1016/j.amjmed.2012.04.033
8. Wexler R, Pleister A, Raman SV, Borchers JR (2012) Therapeutic lifestyle changes for cardiovascular disease. Phys Sportsmed 40:109–115. https://doi.org/10.3810/psm.2012.02.1957
9. Guyatt GH, Devereaux PJ (2004) A review of heart failure treatment. Mt Sinai J Med 71:47–54
10. Bastien M, Poirier P, Lemieux I, Despres JP (2014) Overview of epidemiology and contribution of obesity to cardiovascular disease. Prog Cardiovasc Dis 56:369–381
11. Guh DP, Zhang W, Bansback N, Amarsi Z, Birmingham CL, Anis AH (2009) The incidence of co-morbidities related to obesity and overweight: a systematic review and meta-analysis. BMC Public Health 9:88
12. Artham SM, Lavie CJ, Patel HM, Ventura HO (2008) Impact of obesity on the risk of heart failure and its prognosis. J Cardiometab Syndr 3:155–161. https://doi.org/10.1111/j.1559-4572.2008.00001.x
13. Walley AJ, Blakemore AI and Froguel P (2006) Genetics of obesity and the prediction of risk for health. Hum Mol Genet 15 Spec No 2:R124–R130. doi: https://doi.org/10.1093/hmg/ddl215
14. Hebebrand J, Hinney A (2009) Environmental and genetic risk factors in obesity. Child Adolesc Psychiatr Clin N Am 18:83–94. https://doi.org/10.1016/j.chc.2008.07.006
15. Dasouki MJ, Youngs EL, Hovanes K (2011) Structural chromosome abnormalities associated with obesity: report of four new subjects and review of literature. Curr Genomics 12:190–203. https://doi.org/10.2174/138920211795677930

16. Youngson NA, Whitelaw E (2011) The effects of acquired paternal obesity on the next generation. Asian J Androl 13:195–196. https://doi.org/10.1038/aja.2010.163
17. Drewnowski A (2004) Obesity and the food environment: dietary energy density and diet costs. Am J Prev Med 27:154–162. https://doi.org/10.1016/j.amepre.2004.06.011
18. Drewnowski A, Darmon N (2005) The economics of obesity: dietary energy density and energy cost. Am J Clin Nutr 82:265S-273S
19. Bluher M (2013) Adipose tissue dysfunction contributes to obesity related metabolic diseases. Best Pract Res Clin Endocrinol Metab 27:163–177. https://doi.org/10.1016/j.beem.2013.02.005
20. Northcott JM, Yeganeh A, Taylor CG, Zahradka P, Wigle JT (2012) Adipokines and the cardiovascular system: mechanisms mediating health and disease. Can J Physiol Pharmacol 90:1029–1059. https://doi.org/10.1139/y2012-053
21. Drosatos K, Schulze PC (2013) Cardiac lipotoxicity: molecular pathways and therapeutic implications. Curr Heart Fail Rep 10:109–121. https://doi.org/10.1007/s11897-013-0133-0
22. Ussher JR (2014) The role of cardiac lipotoxicity in the pathogenesis of diabetic cardiomyopathy. Expert Rev Cardiovasc Ther 12:345–358. https://doi.org/10.1586/14779072.2014.891939
23. Fabricant DS, Farnsworth NR (2001) The value of plants used in traditional medicine for drug discovery. Environ Health Perspect 109(Suppl 1):69–75
24. Mosihuzzaman M (2012) Herbal medicine in healthcare–an overview. Nat Prod Commun 7:807–812
25. Santos AK, Costa JG, Menezes IR, Cansancao IF, Santos KK, Matias EF, Coutinho HD (2010) Antioxidant activity of five Brazilian plants used as traditional medicines and food in Brazil. Pharmacogn Mag 6:335–338. https://doi.org/10.4103/0973-1296.71789
26. Bhamarapravati S, Pendland SL, Mahady GB (2003) Extracts of spice and food plants from Thai traditional medicine inhibit the growth of the human carcinogen Helicobacter pylori. Vivo 17:541–544
27. Brower V (2005) A nutraceutical a day may keep the doctor away. Consumers are turning increasingly to food supplements to improve well-being when pharmaceuticals fail. EMBO Rep 6:708–711. https://doi.org/10.1038/sj.embor.7400498
28. Raederstorff D, Kunz I, Schwager J (2013) Resveratrol, from experimental data to nutritional evidence: the emergence of a new food ingredient. Ann N Y Acad Sci 1290:136–141. https://doi.org/10.1111/nyas.12147
29. Tome-Carneiro J, Larrosa M, Gonzalez-Sarrias A, Tomas-Barberan FA, Garcia-Conesa MT, Espin JC (2013) Resveratrol and clinical trials: the crossroad from in vitro studies to human evidence. Curr Pharm Des 19:6064–6093
30. Raj P, Louis XL, Thandapilly SJ, Movahed A, Zieroth S, Netticadan T (2014) Potential of resveratrol in the treatment of heart failure. Life Sci 95:63–71
31. Rothman KJ (2008) BMI-related errors in the measurement of obesity. Int J Obes (Lond) 32(Suppl 3):S56–S59. https://doi.org/10.1038/ijo.2008.87
32. Schneider HJ, Friedrich N, Klotsche J, Pieper L, Nauck M, John U, Dorr M, Felix S, Lehnert H, Pittrow D, Silber S, Volzke H, Stalla GK, Wallaschofski H, Wittchen HU (2010) The predictive value of different measures of obesity for incident cardiovascular events and mortality. J Clin Endocrinol Metab 95:1777–1785. https://doi.org/10.1210/jc.2009-1584
33. Carmienke S, Freitag MH, Pischon T, Schlattmann P, Fankhaenel T, Goebel H, Gensichen J (2013) General and abdominal obesity parameters and their combination in relation to mortality: a systematic review and meta-regression analysis. Eur J Clin Nutr 67:573–585. https://doi.org/10.1038/ejcn.2013.61
34. Das SK (2005) Body composition measurement in severe obesity. Curr Opin Clin Nutr Metab Care 8:602–606
35. Deurenberg P, Deurenberg-Yap M (2002) Validation of skinfold thickness and hand-held impedance measurements for estimation of body fat percentage among Singaporean Chinese, Malay and Indian subjects. Asia Pac J Clin Nutr 11:1–7

36. de Koning L, Merchant AT, Pogue J, Anand SS (2007) Waist circumference and waist-to-hip ratio as predictors of cardiovascular events: meta-regression analysis of prospective studies. Eur Heart J 28:850–856. https://doi.org/10.1093/eurheartj/ehm026
37. Hamer M, Stamatakis E (2012) Metabolically healthy obesity and risk of all-cause and cardiovascular disease mortality. J Clin Endocrinol Metab 97:2482–2488. https://doi.org/10.1210/jc.2011-3475
38. Wang J, Sereika SM, Styn MA, Burke LE (2013) Factors associated with health-related quality of life among overweight or obese adults. J Clin Nurs 22:2172–2182. https://doi.org/10.1111/jocn.12280
39. Gouveia MJ, Frontini R, Canavarro MC, Moreira H (2014) Quality of life and psychological functioning in pediatric obesity: the role of body image dissatisfaction between girls and boys of different ages. Qual Life Res. https://doi.org/10.1007/s11136-014-0711-y
40. Dalle Grave R, Calugi S, El Ghoch M (2013) Lifestyle modification in the management of obesity: achievements and challenges. Eat Weight Disord 18:339–349. https://doi.org/10.1007/s40519-013-0049-4
41. Lagerros YT, Rossner S (2013) Obesity management: what brings success? Therap Adv Gastroenterol 6:77–88. https://doi.org/10.1177/1756283X12459413
42. Picot J, Jones J, Colquitt JL, Gospodarevskaya E, Loveman E, Baxter L and Clegg AJ (2009) The clinical effectiveness and cost-effectiveness of bariatric (weight loss) surgery for obesity: a systematic review and economic evaluation. Health Technol Assess 13:1–190, 215–357, iii–iv. doi: https://doi.org/10.3310/hta13410
43. Pawloski LR, Ruchiwit M, Markham SM (2011) The growing burden of obesity in Thailand: a review of current trends and policies. Pediatr Nurs 37:256–261
44. Wang YC, McPherson K, Marsh T, Gortmaker SL, Brown M (2011) Health and economic burden of the projected obesity trends in the USA and the UK. Lancet 378:815–825. https://doi.org/10.1016/S0140-6736(11)60814-3
45. Tsai AG, Abbo ED, Ogden LG (2011) The time burden of overweight and obesity in primary care. BMC Health Serv Res 11:191. https://doi.org/10.1186/1472-6963-11-191
46. Ramachandran A, Snehalatha C (2010) Rising burden of obesity in Asia. J Obes. https://doi.org/10.1155/2010/868573
47. Withrow D, Alter DA (2011) The economic burden of obesity worldwide: a systematic review of the direct costs of obesity. Obes Rev 12:131–141. https://doi.org/10.1111/j.1467-789X.2009.00712.x
48. WHO (2018) Obesity and overweight.
49. Borders TF, Rohrer JE, Cardarelli KM (2006) Gender-specific disparities in obesity. J Community Health 31:57–68
50. Canada S (2018) Obesity in Canadian Adults, 2016 and 2017.
51. Canada S (2018) Health characteristics, annual estimates. June 25
52. Canada S (2009) Overweight and obese adults (self-reported)
53. Karen C, Roberts MS, de Groh M, Aziz A, Gilbert J-A (2012) Overweight and obesity in children and adolescents: results from the 2009 to 2011 Canadian Health Measures Survey. Health Reports
54. Roberts KC, Shields M, de Groh M, Aziz A, Gilbert JA (2012) Overweight and obesity in children and adolescents: results from the 2009 to 2011 Canadian Health Measures Survey. Health Rep 23:37–41
55. Canada PHAo (2017) Tackling obesity in Canada: childhood obesity and excess weight rates in Canada
56. Ng C, Corey PN, Young TK (2011) Socio-economic patterns of obesity among aboriginal and non-Aboriginal Canadians. Can J Public Health 102:264–268
57. Bruce SG, Riediger ND, Zacharias JM, Young TK (2011) Obesity and obesity-related comorbidities in a Canadian First Nation population. Prev Chronic Dis 8:A03
58. Must A (1996) Morbidity and mortality associated with elevated body weight in children and adolescents. Am J Clin Nutr 63:445S-447S

59. Franks PW, Hanson RL, Knowler WC, Sievers ML, Bennett PH, Looker HC (2010) Childhood obesity, other cardiovascular risk factors, and premature death. N Engl J Med 362:485–493. https://doi.org/10.1056/NEJMoa0904130
60. PHAC C (2013) Obesity in Canada
61. Dalrymple KL, Galione J, Hrabosky J, Chelminski I, Young D, O'Brien E, Zimmerman M (2011) Diagnosing social anxiety disorder in the presence of obesity: implications for a proposed change in DSM-5. Depress Anxiety 28:377–382. https://doi.org/10.1002/da.20794
62. Zimmerman M, Hrabosky JI, Francione C, Young D, Chelminski I, Dalrymple K and Galione JN (2011) Impact of obesity on the psychometric properties of the Diagnostic and Statistical Manual of Mental Disorders, Fourth Edition criteria for major depressive disorder. Compr Psychiatry 52:146–150. doi: https://doi.org/10.1016/j.comppsych.2010.05.001
63. Canada PHAo (2011) Obesity in Canada—Health and economic implications.
64. Finkelstein EA, Khavjou OA, Thompson H, Trogdon JG, Pan L, Sherry B, Dietz W (2012) Obesity and severe obesity forecasts through 2030. Am J Prev Med 42:563–570. https://doi.org/10.1016/j.amepre.2011.10.026
65. Rengma MS, Sen J, Mondal N (2015) Socio-economic, demographic and lifestyle determinants of overweight and obesity among adults of Northeast India. Ethiop J Health Sci 25:199–208. https://doi.org/10.4314/ejhs.v25i3.2
66. Lee EY, Yoon KH (2018) Epidemic obesity in children and adolescents: risk factors and prevention. Front Med 12:658–666. https://doi.org/10.1007/s11684-018-0640-1
67. Fischer-Posovszky P, Wabitsch M, Hochberg Z (2007) Endocrinology of adipose tissue: an update. Horm Metab Res 39:314–321. https://doi.org/10.1055/s-2007-976539
68. Lapid K, Graff JM (2017) Form(ul)ation of adipocytes by lipids. Adipocyte 6:176–186. https://doi.org/10.1080/21623945.2017.1299298
69. Viscarra JA, Ortiz RM (2013) Cellular mechanisms regulating fuel metabolism in mammals: role of adipose tissue and lipids during prolonged food deprivation. Metabolism 62:889–897. https://doi.org/10.1016/j.metabol.2012.12.014
70. Trayhurn P (2014) Hypoxia and adipocyte physiology: implications for adipose tissue dysfunction in obesity. Annu Rev Nutr. https://doi.org/10.1146/annurev-nutr-071812-161156
71. Lin JD, Wu TY, Lin LP, Hsu SW, Liu CT, Wu CL (2013) An exploratory study of health behaviors and the risks for triple H (hypertension, hyperlipidemia, and hyperglycemia) in young adults with disabilities between 20 and 39 years of age. Res Dev Disabil 34:3211–3217. https://doi.org/10.1016/j.ridd.2013.06.044
72. Nestruck AC, Davignon J (1986) Risks for hyperlipidemia. Cardiol Clin 4:47–56
73. Jo J, Gavrilova O, Pack S, Jou W, Mullen S, Sumner AE, Cushman SW, Periwal V (2009) Hypertrophy and/or hyperplasia: dynamics of adipose tissue growth. PLoS Comput Biol 5:e1000324. https://doi.org/10.1371/journal.pcbi.1000324
74. Longo M, Zatterale F, Naderi J, Parrillo L, Formisano P, Raciti GA, Beguinot F, Miele C (2019) Adipose tissue dysfunction as determinant of obesity-associated metabolic complications. Int J Mol Sci. https://doi.org/10.3390/ijms20092358
75. Skurk T, Alberti-Huber C, Herder C, Hauner H (2007) Relationship between adipocyte size and adipokine expression and secretion. J Clin Endocrinol Metab 92:1023–1033. https://doi.org/10.1210/jc.2006-1055
76. Hajer GR, van Haeften TW, Visseren FL (2008) Adipose tissue dysfunction in obesity, diabetes, and vascular diseases. Eur Heart J 29:2959–2971. https://doi.org/10.1093/eurheartj/ehn387
77. Coelho M, Oliveira T, Fernandes R (2013) Biochemistry of adipose tissue: an endocrine organ. Arch Med Sci 9:191–200. https://doi.org/10.5114/aoms.2013.33181
78. Suganami T, Nishida J, Ogawa Y (2005) A paracrine loop between adipocytes and macrophages aggravates inflammatory changes: role of free fatty acids and tumor necrosis factor alpha. Arterioscler Thromb Vasc Biol 25:2062–2068. https://doi.org/10.1161/01.ATV.0000183883.72263.13
79. Bluher M (2009) Adipose tissue dysfunction in obesity. Exp Clin Endocrinol Diabetes 117:241–250. https://doi.org/10.1055/s-0029-1192044

80. Wang GX, Cho KW, Uhm M, Hu CR, Li S, Cozacov Z, Xu AE, Cheng JX, Saltiel AR, Lumeng CN, Lin JD (2014) Otopetrin 1 protects mice from obesity-associated metabolic dysfunction through attenuating adipose tissue inflammation. Diabetes 63:1340–1352. https://doi.org/10.2337/db13-1139
81. Hubert HB, Feinleib M, McNamara PM, Castelli WP (1983) Obesity as an independent risk factor for cardiovascular disease: a 26-year follow-up of participants in the Framingham Heart Study. Circulation 67:968–977
82. Csige I, Ujvarosy D, Szabo Z, Lorincz I, Paragh G, Harangi M, Somodi S (2018) The impact of obesity on the cardiovascular system. J Diabetes Res 2018:3407306. https://doi.org/10.1155/2018/3407306
83. Jiang J, Ahn J, Huang WY, Hayes RB (2013) Association of obesity with cardiovascular disease mortality in the PLCO trial. Prev Med 57:60–64. https://doi.org/10.1016/j.ypmed.2013.04.014
84. Flegal KM, Graubard BI, Williamson DF, Gail MH (2007) Cause-specific excess deaths associated with underweight, overweight, and obesity. JAMA 298:2028–2037. https://doi.org/10.1001/jama.298.17.2028
85. Shields M, Tremblay MS, Connor Gorber S, Janssen I (2012) Abdominal obesity and cardiovascular disease risk factors within body mass index categories. Health Rep 23:7–15
86. Ortega-Loubon C, Fernandez-Molina M, Singh G, Correa R (2019) Obesity and its cardiovascular effects. Diabetes Metab Res Rev 35:e3135. https://doi.org/10.1002/dmrr.3135
87. Taube A, Schlich R, Sell H, Eckardt K, Eckel J (2012) Inflammation and metabolic dysfunction: links to cardiovascular diseases. American journal of physiology. Heart Circulatory Physiol 302:H2148–H2165
88. Chrysant SG, Chrysant GS (2013) New insights into the true nature of the obesity paradox and the lower cardiovascular risk. J Am Soc Hypertens 7:85–94. https://doi.org/10.1016/j.jash.2012.11.008
89. Bombelli M, Facchetti R, Sega R, Carugo S, Fodri D, Brambilla G, Giannattasio C, Grassi G, Mancia G (2011) Impact of body mass index and waist circumference on the long-term risk of diabetes mellitus, hypertension, and cardiac organ damage. Hypertension 58:1029–1035. https://doi.org/10.1161/HYPERTENSIONAHA.111.175125
90. Britton KA, Fox CS (2011) Ectopic fat depots and cardiovascular disease. Circulation 124:e837–e841. https://doi.org/10.1161/CIRCULATIONAHA.111.077602
91. Klein S, Allison DB, Heymsfield SB, Kelley DE, Leibel RL, Nonas C, Kahn R, Association for Weight M, Obesity P, Naaso TOS, American Society for N and American Diabetes A (2007) Waist circumference and cardiometabolic risk: a consensus statement from Shaping America's Health: Association for Weight Management and Obesity Prevention; NAASO, The Obesity Society; the American Society for Nutrition; and the American Diabetes Association. Am J Clin Nutr 85:1197–1202
92. Poirier P, Giles TD, Bray GA, Hong Y, Stern JS, Pi-Sunyer FX, Eckel RH, American Heart A, Obesity Committee of the Council on Nutrition PA and Metabolism (2006) Obesity and cardiovascular disease: pathophysiology, evaluation, and effect of weight loss: an update of the 1997 American Heart Association Scientific Statement on Obesity and Heart Disease from the Obesity Committee of the Council on Nutrition, Physical Activity, and Metabolism. Circulation 113:898–918. https://doi.org/10.1161/CIRCULATIONAHA.106.171016
93. Zebekakis PE, Nawrot T, Thijs L, Balkestein EJ, van der Heijden-Spek J, Van Bortel LM, Struijker-Boudier HA, Safar ME, Staessen JA (2005) Obesity is associated with increased arterial stiffness from adolescence until old age. J Hypertens 23:1839–1846
94. Iannuzzi A, Licenziati MR, Acampora C, Salvatore V, Auriemma L, Romano ML, Panico S, Rubba P, Trevisan M (2004) Increased carotid intima-media thickness and stiffness in obese children. Diab Care 27:2506–2508
95. Ozcetin M, Celikyay ZR, Celik A, Yilmaz R, Yerli Y, Erkorkmaz U (2012) The importance of carotid artery stiffness and increased intima-media thickness in obese children. S Afr Med J 102:295–299

96. Lee CD, Jacobs DR Jr, Schreiner PJ, Iribarren C, Hankinson A (2007) Abdominal obesity and coronary artery calcification in young adults: the Coronary Artery Risk Development in Young Adults (CARDIA) Study. Am J Clin Nutr 86:48–54
97. Kotsis V, Stabouli S, Papakatsika S, Rizos Z, Parati G (2010) Mechanisms of obesity-induced hypertension. Hypertens Res 33:386–393. https://doi.org/10.1038/hr.2010.9
98. Aurigemma GP, de Simone G, Fitzgibbons TP (2013) Cardiac remodeling in obesity. Circ Cardiovasc Imaging 6:142–152. https://doi.org/10.1161/CIRCIMAGING.111.964627
99. Goldberg IJ, Trent CM, Schulze PC (2012) Lipid metabolism and toxicity in the heart. Cell Metab 15:805–812. https://doi.org/10.1016/j.cmet.2012.04.006
100. Carpenter HM (1962) Myocardial fat infiltration. Am Heart J 63:491–496
101. Karasoy D, Bo Jensen T, Hansen ML, Schmiegelow M, Lamberts M, Gislason GH, Hansen J, Torp-Pedersen C, Olesen JB (2013) Obesity is a risk factor for atrial fibrillation among fertile young women: a nationwide cohort study. Europace 15:781–786. https://doi.org/10.1093/europace/eus422
102. Chrostowska M, Szyndler A, Hoffmann M, Narkiewicz K (2013) Impact of obesity on cardiovascular health. Best Pract Res Clin Endocrinol Metab 27:147–156. https://doi.org/10.1016/j.beem.2013.01.004
103. Steyn K, Sliwa K, Hawken S, Commerford P, Onen C, Damasceno A, Ounpuu S, Yusuf S, Africa IIi (2005) Risk factors associated with myocardial infarction in Africa: the INTERHEART Africa study. Circulation 112:3554–3561. https://doi.org/10.1161/CIRCULATIONAHA.105.563452
104. O'Donnell MJ, Xavier D, Liu L, Zhang H, Chin SL, Rao-Melacini P, Rangarajan S, Islam S, Pais P, McQueen MJ, Mondo C, Damasceno A, Lopez-Jaramillo P, Hankey GJ, Dans AL, Yusoff K, Truelsen T, Diener HC, Sacco RL, Ryglewicz D, Czlonkowska A, Weimar C, Wang X, Yusuf S, Investigators I (2010) Risk factors for ischaemic and intracerebral haemorrhagic stroke in 22 countries (the INTERSTROKE study): a case-control study. Lancet 376:112–123. https://doi.org/10.1016/S0140-6736(10)60834-3
105. Drager LF, Togeiro SM, Polotsky VY, Lorenzi-Filho G (2013) Obstructive sleep apnea: a cardiometabolic risk in obesity and the metabolic syndrome. J Am Coll Cardiol 62:569–576. https://doi.org/10.1016/j.jacc.2013.05.045
106. Van de Voorde J, Pauwels B, Boydens C, Decaluwe K (2013) Adipocytokines in relation to cardiovascular disease. Metabolism 62:1513–1521. https://doi.org/10.1016/j.metabol.2013.06.004
107. Luc G, Empana JP, Morange P, Juhan-Vague I, Arveiler D, Ferrieres J, Amouyel P, Evans A, Kee F, Bingham A, Machez E, Ducimetiere P (2010) Adipocytokines and the risk of coronary heart disease in healthy middle aged men: the PRIME Study. Int J Obes (Lond) 34:118–126. https://doi.org/10.1038/ijo.2009.204
108. Wickman C, Kramer H (2013) Obesity and kidney disease: potential mechanisms. Semin Nephrol 33:14–22. https://doi.org/10.1016/j.semnephrol.2012.12.006
109. de Jong PE, Verhave JC, Pinto-Sietsma SJ (2002) Obesity and target organ damage: the kidney. Int J Obes Relat Metab Disord 26(Suppl 4):S21–S24. https://doi.org/10.1038/sj.ijo.0802213
110. Foster MC, Hwang SJ, Massaro JM, Hoffmann U, DeBoer IH, Robins SJ, Vasan RS, Fox CS (2011) Association of subcutaneous and visceral adiposity with albuminuria: the Framingham Heart Study. Obesity (Silver Spring) 19:1284–1289. https://doi.org/10.1038/oby.2010.308
111. Schulze PC, Drosatos K, Goldberg IJ (2016) Lipid use and misuse by the heart. Circ Res 118:1736–1751. https://doi.org/10.1161/CIRCRESAHA.116.306842
112. Taegtmeyer H, Sen S, Vela D (2010) Return to the fetal gene program: a suggested metabolic link to gene expression in the heart. Ann N Y Acad Sci 1188:191–198. https://doi.org/10.1111/j.1749-6632.2009.05100.x
113. Pascual F, Coleman RA (2016) Fuel availability and fate in cardiac metabolism: a tale of two substrates. Biochim Biophys Acta 1861:1425–1433. https://doi.org/10.1016/j.bbalip.2016.03.014
114. Trent CM, Yu S, Hu Y, Skoller N, Huggins LA, Homma S, Goldberg IJ (2014) Lipoprotein lipase activity is required for cardiac lipid droplet production. J Lipid Res 55:645–658. https://doi.org/10.1194/jlr.M043471

115. van der Vusse GJ, van Bilsen M, Glatz JF (2000) Cardiac fatty acid uptake and transport in health and disease. Cardiovasc Res 45:279–293
116. Lopaschuk GD, Ussher JR, Folmes CD, Jaswal JS, Stanley WC (2010) Myocardial fatty acid metabolism in health and disease. Physiol Rev 90:207–258. https://doi.org/10.1152/physrev.00015.2009
117. Saely CH, Geiger K, Drexel H (2012) Brown versus white adipose tissue: a mini-review. Gerontology 58:15–23. https://doi.org/10.1159/000321319
118. Ameer F, Scandiuzzi L, Hasnain S, Kalbacher H, Zaidi N (2014) De novo lipogenesis in health and disease. Metabolism. https://doi.org/10.1016/j.metabol.2014.04.003
119. Schwenk RW, Holloway GP, Luiken JJ, Bonen A, Glatz JF (2010) Fatty acid transport across the cell membrane: regulation by fatty acid transporters. Prostaglandins Leukot Essent Fatty Acids 82:149–154. https://doi.org/10.1016/j.plefa.2010.02.029
120. Das UN (2006) Essential fatty acids: biochemistry, physiology and pathology. Biotechnol J 1:420–439. https://doi.org/10.1002/biot.200600012
121. Nishi H, Higashihara T, Inagi R (2019) Lipotoxicity in kidney, heart, and skeletal muscle dysfunction. Nutrients. https://doi.org/10.3390/nu11071664
122. Turkieh A, Caubere C, Barutaut M, Desmoulin F, Harmancey R, Galinier M, Berry M, Dambrin C, Polidori C, Casteilla L, Koukoui F, Rouet P, Smih F (2014) Apolipoprotein O is mitochondrial and promotes lipotoxicity in heart. J Clin Invest 124:2277–2286. https://doi.org/10.1172/JCI74668
123. Nolan CJ (2014) Lipotoxicity, beta cell dysfunction, and gestational diabetes. Cell Metab 19:553–554. https://doi.org/10.1016/j.cmet.2014.03.020
124. Eguchi K, Manabe I (2014) Toll-like receptor, lipotoxicity and chronic inflammation: the pathological link between obesity and cardiometabolic disease. J Atheroscler Thromb
125. Lake AD, Novak P, Hardwick RN, Flores-Keown B, Zhao F, Klimecki WT, Cherrington NJ (2014) The adaptive endoplasmic reticulum stress response to lipotoxicity in progressive human nonalcoholic fatty liver disease. Toxicol Sci 137:26–35. https://doi.org/10.1093/toxsci/kft230
126. Schulze PC (2009) Myocardial lipid accumulation and lipotoxicity in heart failure. J Lipid Res 50:2137–2138
127. de Vries JE, Vork MM, Roemen TH, de Jong YF, Cleutjens JP, van der Vusse GJ, van Bilsen M (1997) Saturated but not mono-unsaturated fatty acids induce apoptotic cell death in neonatal rat ventricular myocytes. J Lipid Res 38:1384–1394
128. Hickson-Bick DL, Buja LM, McMillin JB (2000) Palmitate-mediated alterations in the fatty acid metabolism of rat neonatal cardiac myocytes. J Mol Cell Cardiol 32:511–519. https://doi.org/10.1006/jmcc.1999.1098
129. Park TS, Hu Y, Noh HL, Drosatos K, Okajima K, Buchanan J, Tuinei J, Homma S, Jiang XC, Abel ED, Goldberg IJ (2008) Ceramide is a cardiotoxin in lipotoxic cardiomyopathy. J Lipid Res 49:2101–2112. https://doi.org/10.1194/jlr.M800147-JLR200
130. Dyntar D, Eppenberger-Eberhardt M, Maedler K, Pruschy M, Eppenberger HM, Spinas GA, Donath MY (2001) Glucose and palmitic acid induce degeneration of myofibrils and modulate apoptosis in rat adult cardiomyocytes. Diabetes 50:2105–2113
131. Juhasz B, Varga B, Gesztelyi R, Kemeny-Beke A, Zsuga J, Tosaki A (2010) Resveratrol: a multifunctional cytoprotective molecule. Curr Pharm Biotechnol 11:810–818
132. Giovinazzo G, Ingrosso I, Paradiso A, De Gara L, Santino A (2012) Resveratrol biosynthesis: plant metabolic engineering for nutritional improvement of food. Plant Foods for Human Nutrition (Dordrecht, Netherlands) 67:191–199
133. Pervaiz S (2003) Resveratrol: from grapevines to mammalian biology. FASEB J 17:1975–1985. https://doi.org/10.1096/fj.03-0168rev17/14/1975
134. de Lorgeril M, Salen P, Guiraud A, Boucher F, de Leiris J (2003) Resveratrol and non-ethanolic components of wine in experimental cardiology. Nutr Metab Cardiovasc Dis 13:100–103
135. Takaoka MJ (1940) Of the phenolic substances of white hellebore (Veratrum grandiflorum Loes. fil.). J Faculty Sci Hokkaido Imperial University 3:1–16

136. Tsai WH, Yang CC, Li PC, Chen WC, Chien CT (2013) Therapeutic potential of traditional chinese medicine on inflammatory diseases. J Tradit Complement Med 3:142–151. https://doi.org/10.4103/2225-4110.114898
137. Hillis WE, Isoi K (1965) Variation in the chemical composition of *Eucalyptus sideroxylon*. Phytochemistry 4:541–550
138. Tyukavkina NA, Gromova AS, Lutskii VI, Voronov VK (1974) Hydroxystilbenes from the bark of *Pinus sibirica*. Chem Nat Compd 8:570–572
139. Langcake P, Pryce RJ (1976) The production of resveratrol by *Vitis vinifera* and other members of the Vitaceae as a response to infection or injury. Physiol Plant Pathol 9:77–86
140. Pool RM, Creasy LL, Frackelton AS (1981) Resveratrol and the viniferins, their application to screening for disease resistance in grape breeding programs. Vitis 20:136–145
141. Kubo M, Shin H, Haneda T, Tani T, Namba K (1981) Studies on the antifungal substance of crude drug: 2. On the roots of *Polygonum cuspidatum* (Polygonaceae). Shoyakugaku Zasshi 35:58–61
142. Arichi H, Kimura Y, Okuda H, Baba K, Kozawa M, Arichi S (1982) Effects of stilbene components of the roots of *Polygonum cuspidatum* Sieb. et Zucc. on lipid metabolism. Chem Pharm Bull 30:1766–1770
143. Renaud S, de Lorgeril M (1992) Wine, alcohol, platelets, and the French paradox for coronary heart disease. Lancet 339:1523–1526
144. Creasy EHSaLL (1992) Concentration of the phytoalexin resveratrol in wine. Am J Enol Vitic 43:49–52
145. Frankel EN, Kanner J, German JB, Parks E, Kinsella JE (1993) Inhibition of oxidation of human low-density lipoprotein by phenolic substances in red wine. Lancet 341:454–457
146. Xu Q, Si LY (2012) Resveratrol role in cardiovascular and metabolic health and potential mechanisms of action. Nutrition research (New York, N.Y.) 32:648–658.
147. Tome-Carneiro J, Gonzalvez M, Larrosa M, Yanez-Gascon MJ, Garcia-Almagro FJ, Ruiz-Ros JA, Tomas-Barberan FA, Garcia-Conesa MT, Espin JC (2013) Resveratrol in primary and secondary prevention of cardiovascular disease: a dietary and clinical perspective. Ann N Y Acad Sci 1290:37–51
148. Ko JH, Sethi G, Um JY, Shanmugam MK, Arfuso F, Kumar AP, Bishayee A, Ahn KS (2017) The role of resveratrol in cancer therapy. Int J Mol Sci. https://doi.org/10.3390/ijms18122589
149. Wallerath T, Deckert G, Ternes T, Anderson H, Li H, Witte K, Forstermann U (2002) Resveratrol, a polyphenolic phytoalexin present in red wine, enhances expression and activity of endothelial nitric oxide synthase. Circulation 106:1652–1658
150. Opie LH, Lecour S (2007) The red wine hypothesis: from concepts to protective signalling molecules. Eur Heart J 28:1683–1693. https://doi.org/10.1093/eurheartj/ehm149
151. Li H, Forstermann U (2012) Red wine and cardiovascular health. Circ Res 111:959–961. https://doi.org/10.1161/CIRCRESAHA.112.278705
152. Yu W, Fu YC, Wang W (2012) Cellular and molecular effects of resveratrol in health and disease. J Cell Biochem 113:752–759. https://doi.org/10.1002/jcb.23431
153. Zheng X, Zhu S, Chang S, Cao Y, Dong J, Li J, Long R, Zhou Y (2013) Protective effects of chronic resveratrol treatment on vascular inflammatory injury in streptozotocin-induced type 2 diabetic rats: Role of NF-kappa B signaling. Eur J Pharmacol. https://doi.org/10.1016/j.ejphar.2013.10.034
154. Chan AY, Dolinsky VW, Soltys CL, Viollet B, Baksh S, Light PE, Dyck JR (2008) Resveratrol inhibits cardiac hypertrophy via AMP-activated protein kinase and Akt. J Biol Chem 283:24194–24201. https://doi.org/10.1074/jbc.M802869200
155. Thandapilly SJ, Louis XL, Yang T, Stringer DM, Yu L, Zhang S, Wigle J, Kardami E, Zahradka P, Taylor C, Anderson HD, Netticadan T (2011) Resveratrol prevents norepinephrine induced hypertrophy in adult rat cardiomyocytes, by activating NO-AMPK pathway. Eur J Pharmacol 668:217–224. https://doi.org/10.1016/j.ejphar.2011.06.042
156. Beauloye C, Bertrand L, Horman S, Hue L (2011) AMPK activation, a preventive therapeutic target in the transition from cardiac injury to heart failure. Cardiovasc Res 90:224–233. https://doi.org/10.1093/cvr/cvr034

157. Price NL, Gomes AP, Ling AJ, Duarte FV, Martin-Montalvo A, North BJ, Agarwal B, Ye L, Ramadori G, Teodoro JS, Hubbard BP, Varela AT, Davis JG, Varamini B, Hafner A, Moaddel R, Rolo AP, Coppari R, Palmeira CM, de Cabo R, Baur JA, Sinclair DA (2012) SIRT1 is required for AMPK activation and the beneficial effects of resveratrol on mitochondrial function. Cell Metab 15:675–690. https://doi.org/10.1016/j.cmet.2012.04.003
158. Penumathsa SV, Maulik N (2009) Resveratrol: a promising agent in promoting cardioprotection against coronary heart disease. Can J Physiol Pharmacol 87:275–286. https://doi.org/10.1139/Y09-013
159. Li H, Xia N, Forstermann U (2012) Cardiovascular effects and molecular targets of resveratrol. Nitric Oxide 26:102–110. https://doi.org/10.1016/j.niox.2011.12.006
160. Zhang LP, Yin JX, Liu Z, Zhang Y, Wang QS, Zhao J (2006) Effect of resveratrol on L-type calcium current in rat ventricular myocytes. Acta Pharmacol Sin 27:179–183. https://doi.org/10.1111/j.1745-7254.2006.00250.x
161. Poulsen MM, Jorgensen JO, Jessen N, Richelsen B, Pedersen SB (2013) Resveratrol in metabolic health: an overview of the current evidence and perspectives. Ann N Y Acad Sci 1290:74–82. https://doi.org/10.1111/nyas.12141
162. Lekli I, Szabo G, Juhasz B, Das S, Das M, Varga E, Szendrei L, Gesztelyi R, Varadi J, Bak I, Das DK, Tosaki A (2008) Protective mechanisms of resveratrol against ischemia-reperfusion-induced damage in hearts obtained from Zucker obese rats: the role of GLUT-4 and endothelin. Am J Physiol Heart Circ Physiol 294:H859–H866. https://doi.org/10.1152/ajpheart.01048.2007
163. Aubin MC, Lajoie C, Clement R, Gosselin H, Calderone A, Perrault LP (2008) Female rats fed a high-fat diet were associated with vascular dysfunction and cardiac fibrosis in the absence of overt obesity and hyperlipidemia: therapeutic potential of resveratrol. J Pharmacol Exp Ther 325:961–968. https://doi.org/10.1124/jpet.107.135061
164. Louis XL, Thandapilly SJ, MohanKumar SK, Yu L, Taylor CG, Zahradka P, Netticadan T (2012) Treatment with low-dose resveratrol reverses cardiac impairment in obese prone but not in obese resistant rats. J Nutr Biochem 23:1163–1169. https://doi.org/10.1016/j.jnutbio.2011.06.010
165. Qin F, Siwik DA, Luptak I, Hou X, Wang L, Higuchi A, Weisbrod RM, Ouchi N, Tu VH, Calamaras TD, Miller EJ, Verbeuren TJ, Walsh K, Cohen RA, Colucci WS (2012) The polyphenols resveratrol and S17834 prevent the structural and functional sequelae of diet-induced metabolic heart disease in mice. Circulation 125:1757–1764. https://doi.org/10.1161/circulationaha.111.067801
166. Lu Y, Lu X, Wang L, Yang W (2019) Resveratrol attenuates high fat diet-induced mouse cardiomyopathy through upregulation of estrogen related receptor-alpha. Eur J Pharmacol 843:88–95. https://doi.org/10.1016/j.ejphar.2018.10.018
167. Xia N, Forstermann U, Li H (2014) Resveratrol and endothelial nitric oxide. Molecules 19:16102–16121. https://doi.org/10.3390/molecules191016102
168. Huang JP, Hsu SC, Li DE, Chen KH, Kuo CY, Hung LM (2018) Resveratrol mitigates high-fat diet-induced vascular dysfunction by activating the Akt/eNOS/NO and Sirt1/ER pathway. J Cardiovasc Pharmacol 72:231–241. https://doi.org/10.1097/FJC.0000000000000621
169. Arzola-Paniagua MA, Garcia-Salgado Lopez ER, Calvo-Vargas CG, Guevara-Cruz M (2016) Efficacy of an orlistat-resveratrol combination for weight loss in subjects with obesity: a randomized controlled trial. Obesity (Silver Spring) 24:1454–1463. https://doi.org/10.1002/oby.21523
170. Xue M, Weickert MO, Qureshi S, Kandala NB, Anwar A, Waldron M, Shafie A, Messenger D, Fowler M, Jenkins G, Rabbani N, Thornalley PJ (2016) Improved glycemic control and vascular function in overweight and obese subjects by glyoxalase 1 inducer formulation. Diabetes 65:2282–2294. https://doi.org/10.2337/db16-0153
171. Most J, Timmers S, Warnke I, Jocken JW, van Boekschoten M, de Groot P, Bendik I, Schrauwen P, Goossens GH, Blaak EE (2016) Combined epigallocatechin-3-gallate and resveratrol supplementation for 12 wk increases mitochondrial capacity and fat oxidation, but not insulin sensitivity, in obese humans: a randomized controlled trial. Am J Clin Nutr 104:215–227. https://doi.org/10.3945/ajcn.115.122937

172. Timmers S, Konings E, Bilet L, Houtkooper RH, van de Weijer T, Goossens GH, Hoeks J, van der Krieken S, Ryu D, Kersten S, Moonen-Kornips E, Hesselink MK, Kunz I, Schrauwen-Hinderling VB, Blaak EE, Auwerx J, Schrauwen P (2011) Calorie restriction-like effects of 30 days of resveratrol supplementation on energy metabolism and metabolic profile in obese humans. Cell Metab 14:612–622. https://doi.org/10.1016/j.cmet.2011.10.002
173. Wong RH, Berry NM, Coates AM, Buckley JD, Bryan J, Kunz I, Howe PR (2013) Chronic resveratrol consumption improves brachial flow-mediated dilatation in healthy obese adults. J Hypertens 31:1819–1827. https://doi.org/10.1097/HJH.0b013e328362b9d6
174. Dash S, Xiao C, Morgantini C, Szeto L, Lewis GF (2013) High-dose resveratrol treatment for 2 weeks inhibits intestinal and hepatic lipoprotein production in overweight/obese men. Arterioscler Thromb Vasc Biol 33:2895–2901. https://doi.org/10.1161/ATVBAHA.113.302342

Chapter 16
Role of Flavonoids in Obesity Induced Cardiovascular Dysfunction

Zoha Ahmed, Antara Pal, Siddhi Bagwe-Parab, Ginpreet Kaur, Harpal Singh Buttar, and Hardeep Singh Tuli

Abstract Obesity is found to be a contributing risk factor for Cardiovascular Dysfunction (CVD), coronary heart disease (CHD) and heart failure (HF). Obesity is found to be linked to the release of several proinflammatory mediators. The fluctuation in the levels of these mediators and adipocytokines released by the adipocytes further leads to cardiovascular complications. The increase in the risk of CVD and mortality caused by obesity is due to increasing levels of atherosclerotic plaques in the arteries and blood vessels leading to the heart. The current available treatment strategies leads to several unwanted side effects for the patient which leads to a decreased quality of life. Flavonoids are polyphenolic compounds which occur naturally in nature. Several clinical and preclinical studies have demonstrated that flavonoids are beneficial in decreasing the cardiovascular risks presented by obesity. The mechanism of action primarily depends on the antioxidant and anti-inflammatory actions of flavonoids. In view of these observations, the intake of foods containing flavonoids have enormous potential in preventing obesity induced cardio-metabolic diseases. However, randomized and placebo-controlled clinical trials are needed to determine the long-term safety and efficacy of different flavonoids. The focus of this chapter is to highlight the cost-effective health benefits of flavonoid-rich-foods and dietary supplements containing flavonoids for the prevention and cure of obesity linked cardiovascular diseases (CVDs).

Keywords Cardiovascular diseases · Inflammation · Oxidative stress · Flavonoids · Obesity

Z. Ahmed · A. Pal · S. Bagwe-Parab · G. Kaur (✉)
Shobhaben Pratapbhai Patel School of Pharmacy and Technology Management, SVKM's NMIMS, Mumbai 400056, Maharashtra, India
e-mail: Ginpreet.kaur@nmims.edu

H. S. Buttar
Department of Pathology and Laboratory Medicine, Faculty of Medicine, University of Ottawa, Ottawa, ON, Canada

H. S. Tuli
Department of Biotechnology, Maharishi Markandeshwar (Deemed to be University), Mullana, Ambala, Haryana, India

P. S. Tappia et al. (eds.), *Biochemistry of Cardiovascular Dysfunction in Obesity*, Advances in Biochemistry in Health and Disease 20,
https://doi.org/10.1007/978-3-030-47336-5_16

Abbreviations

CVDs	Cardiovascular diseases
CHD	Congenital heart disease
LDL	Low density lipoproteins
HDL	High density lipoproteins
BMI	Body mass index
TNF-α	Tumor necrosis factor alpha
IL-1	Interleukin-1
IL-6	Interleukin-6
SAA	Serum amyloid A
MI	Myocardial infarction

Introduction

Association Between the Occurrence of Atherosclerosis and Cardiovascular Diseases

Cardiovascular diseases (CVDs) are an assemblage of different pathophysiological conditions related to the heart and its connecting vessels. These conditions cause defects in both the structures and functionality of the heart. The damage induced by these diseases may be temporary or permanent and, in some cases, may even lead to death. The heart may experience medical conditions such as arrythmias, myocardial infarction, flutter and fibrillation. These act as a major causative factor for hypoxia and ischemia in the body. The most prevalent cardiovascular disorder is hypertension or high blood pressure. Reports suggest that racial disparities are a major risk factor for hypertension. It has been elucidated that the prevalence of hypertension is higher in blacks (48.4%) and relatively lower in whites (28.1%) [1]. Prevalence studies of hypertension conducted around the world revealed that economically-developed countries have a prevalence rate ranging between 20 and 50%. Rural India has the smallest fraction of its population suffering from cardiovascular diseases with prevalence in women being 6.8% and 3.4% in men [2]. Studies carried out in India in 2014, found significant differences between the disease rates between rural and urban India. The overall cardiovascular diseases and co-morbidities in India were 29.8%, which ranged from 27.6% in rural parts to 33.8% in urban areas [3]. This difference can be attributed to the difference in lifestyles led by the people in these regions. The urban population of India has shifted towards a more sedentary lifestyle, spending long hours either at home or in the office. Harmful dietary practices have been adapted as an influence of rapid urbanization and the western culture. Increases in salt consumption through the use of packaged foods is also a significant contributing factor to the higher prevalence of hypertension and other CVDs in urban areas [4].

Cardiovascular diseases (CVDs) have posed immense health and economic burdens. Among the CVDs, congenital heart disease (CHD) is the leading cause of death in the US, causing deaths in 43.8% CVD patients. This is followed by increased mortality caused by stroke, high blood pressure and arterial diseases. CVDs were responsible for deaths of 17.9 million people in 2016, among who, 85% of deaths were caused by a heart attack or stroke [5].

The current therapeutic strategies for management of CVDs include several pharmacological and non-pharmacological approaches. The non-pharmacological strategies comprise of interventions in the daily lives of patients. This can be done through smoking cessation therapies, increase in physical activity, dietary changes with the attempt to reduce excess weight gain, and other lifestyle modifications. Although these approaches may not bring about marked effects, they are necessary in addition to the various pharmacological therapies. Government initiatives, including media campaigns, are launched to target youth and highlight the importance of smoking cessation as a measure to reduce the risk factors associated with CVDs. Incorporation of a low-calorie diet in obese patients suffering from hypertension is especially effective in decreasing their weight and blood pressure.

In addition to lifestyle modifications, physicians employ medications in order to increase the effectiveness and speed of therapy. These pharmacological approaches includes drugs of various classes which counteract CVDs through different mechanisms. Antiplatelet agents, including aspirin, remains the drug of choice for the immediate treatment following a heart attack. Antithrombotic agents such as hirudin and bivalirudin act as direct thrombin IIa inhibitors while rivaroxaban acts as an oral direct Xa inhibitor used for the treatment of thrombocytopenia and related disorders. Lipid lowering agents targeted at lowering low density lipoproteins (LDL) or increasing high density lipoproteins (HDL) levels are often employed as an adjuvant therapy [6]. In cases where medications are not effective in treating the CVD, surgical interventions or use of external devices is required. Common examples of these interventional techniques include bypass surgery or stents.

In response to this, research and clinical trials are being conducted to use flavonoids such as quercetin, naringenin, apigenin, eriodyctiol and myricetin in our diet to counter the manifestation of CVDs [7]. Several studies have shown that dietary flavonoids improve metabolic function, reduce oxidative stress, improve cardiorespiratory and mental health and may extend lifespan in humans. Preclinical studies on quercetin and its derivatives (rutin, hyperoside, quercitrin) has shown to cause the uncoupling of oxidative phosphorylation in rat heart mitochondria at minute concentrations. This flavonoid-induced partial uncoupling activity of oxidative phosphorylation in mitochondria, without affecting the respiration rate, could have a cardioprotective effect [8]. Results from epidemiological studies suggest that regular intake of fruit and vegetables containing flavonoids improve endothelial function and reduce stiffening of arteries caused by atherosclerosis, prevent blood platelet aggregation and consequently reduce the risk of coronary artery disease and stroke [9].

It is now believed that oxidative stress in the mitochondria plays a crucial role in the overproduction of reactive oxygen species (ROS) and nitrogen species (NOS) which are mainly formed by NADPH oxidase. The oxidative stress state can be reversed by both endogenous antioxidants (glutathione, L-cysteine, superoxide dismutase, catalase), and exogenous micronutrients (vitamin C and E, Zn and Cu), and flavonoids, polyphenolic compounds, and carotenoids.

The incidence of obesity is on the rise in most developed and developing nations of the world. Especially pertaining to increased stress, improper work life balance, sedentary lifestyle, increased consumption of packaged foods or food rich in saturated fats. Obesity is a major risk factor for CVDs and results in high rates of CVD-related mortality. The current available therapies, though effective, have several deleterious effects on health and decrease quality of life for the patient. Previously conducted studies have demonstrated an existing link between flavonoids and subsequent decrease in obesity. The dietary intervention of flavonoids in the reduction of obesity and decrease the overall risk of CVD is primarily attributed to the anti-inflammatory and antioxidant effects of flavonoids, both of which are major pathways for obesity-induced CVDs. This new area of study opens doors for further investigation and research in incorporating flavonoids as an economical, safe and effective option in the treatment strategies for obesity-induced CVDs. Several preclinical trials cited in the chapter account for the positive effects of flavonoids on obesity induced CVDs. However, further clinical trials need to be conducted to assess their efficacy and commercial use.

Link Between Obesity and CVDs

Obesity-associated disorders are a growing concern all around the world. Obesity is a primary cause of hyperlipidemia, dyslipidemia, insulin resistance and diabetes. Obesity is measured with the use of Body Mass Index (BMI). BMI values of 30.0 kg/m^2 and higher classify as obesity [10]. The African-American youth population has been found to have the highest rates of obesity [11]. The association of obesity with CVD related risk and mortality remains controversial. Not all individuals suffering from obesity experience cardiovascular complications. Similarly, it is not necessary that individuals with CVDs are obese. However, there appears to be a significant association between occurrence of obesity and increased risk of CVDs [12]. Studies have found that overweight and obese individuals showed a greater risk of developing heart failures when compared to normal weighted people. The risk showed an increase with a unit increase in BMI [13]. The central deposition of adipose tissues in obese individuals is associated with elevated levels of fibrinogen, C-reactive protein and several proinflammatory mediators [14]. These play a major role in the development of atherosclerotic plaques and deposits in the walls of the arteries and blood vessels leading to the heart. This leads to an increased risk of high blood pressure. Persistent hypertension decreases life expectancy while increasing the risk of stroke and myocardial infarction (MI).

Role of Inflammation and Atherosclerosis in Obesity-Induced CVDs

Obesity was found to be associated with hypertriglyceridemia-increased level of low-density lipoprotein (LDL) and low levels of high-density lipoproteins (HDL) in individuals [15]. Inflammation is a primary underlying mechanism that links obesity with disorders of the cardiovascular system. Adipose tissue is made up of adipocytes which promote metabolic homeostasis in normal individuals. Increased number and enlargement of adipocytes alters this homeostasis, resulting in abnormal levels of metabolites, such as lipids, fatty acids and cytokines (Fig. 16.1) [16]. The major contributors to obesity-induced inflammation are the adipocytokines produced by the adipocytes. The most abundant among these are adiponectin and leptin. Resistin and visafatin are also produced by the adipocytes in significant amounts [17]. This leads to activation of monocytes and macrophages which along with adipocytes produce several inflammatory factors such as leptin, tumor necrosis factor alpha (TNF-α), interleukin-6 (IL-6), interleukin-1 (IL-1), endothelin, serum amyloid A (SAA) and angiotensinogen. The accumulation of macrophages progresses to local inflammation. The amount of proinflammatory factors produced by the adipose tissue also increases as BMI increases [18].

Obesity is attributed to a chronic inflammatory response which is caused by the abnormal upregulation of cytokines. Acute phase proteins have also emerged as

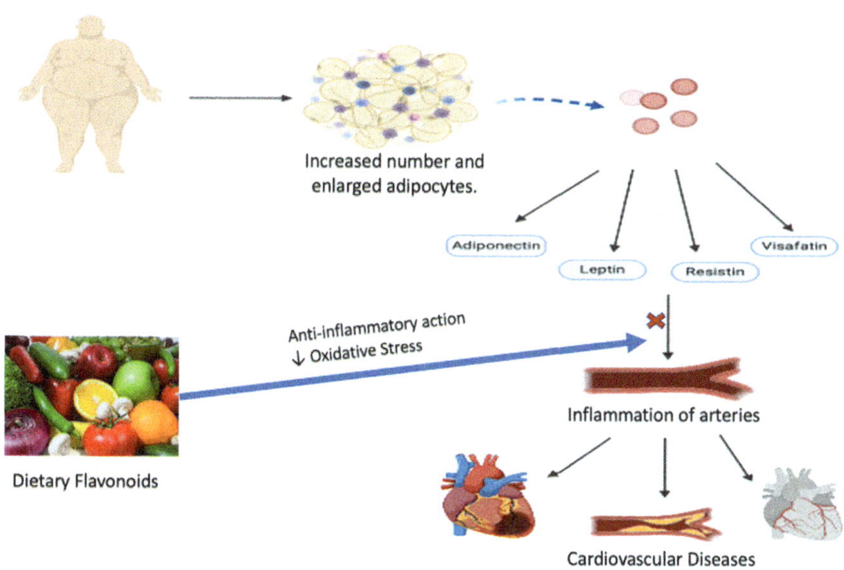

Fig. 16.1 Diagrammatic representation of obesity, adipocytes secreted bioactive substances, and influence of dietary flavonoids in the reduction of cardiovascular diseases

useful biomarkers for obesity induced CVD. These acute phase proteins give rise to myocardial fibrosis which further leads to both systolic and diastolic heart failure.

The pathological role of various adipocytokines in the occurrence of obesity-induced CVDs is described below.

Adiponectin

Adiponectin is an important mediator in the suppression of inflammation and regulation of insulin resistance. It demonstrates inhibitory effects on specific molecules including adhesion molecules which lead to atherogenic phenomenon. Individuals suffering from obesity have shown to express low levels of adiponectin. There is a marked decrease in blood serum levels of adiponectin in individuals with visceral obesity, insulin resistance and related disorders. An inverse relation has been observed between adiponectin levels and insulin resistance. Visceral obesity results in an increase in the levels of inflammatory mediators such as TNF-α and IL-1. A combination of these leads to vascular changes which adds to the stress on the heart contributing to metabolic syndrome [19]. Adiponectin is widely known to exert anti-inflammatory effects in atherosclerotic plaques as a virtue of suppression of proinflammatory cytokines such as IL-6, TNF-α and other inflammatory factors [20, 21]. The physiological levels of adiponectin possess the ability to inhibit the activity and expression of adhesion molecules including E-selectins, intracellular adhesion molecule-1, vascular cellular adhesion molecule. This phenomenon results in an overall decrease in the adherence of monocytes to the arterial endothelium [22]. In vitro studies have demonstrated that adiponectin exhibits a dose dependent suppression of the TNF-α–induced surface expression of E-selectin, Intercellular Adhesion Molecule 1 (ICAM-1), and vascular cell adhesion molecule 1

(VCAM-1) [23]. The modification of macrophages into foam cells is carried out by adiponectin. It also induces vasodilation by regulation of endothelial nitric oxide synthase which stimulates the production of nitric oxide [24]. Insulin resistance in individuals also expresses low levels of adiponectin. In vivo studies have demonstrated that adiponectin enhances tissue insulin sensitivity by alleviating triglyceride content attributable to the enhanced fatty acid oxidation and energy consumption in the liver and muscle [25].

Leptin

Leptin plays a major role in modulating the responses of the immune system and inflammation cascading. It acts on the hypothalamus and regulates food intake levels thereby inducing metabolic effects and regulating energy homeostasis [26]. The presence of infectious and inflammatory stimuli such as cytokines and lipopolysaccharides cause a marked increase in leptin levels in the body [27]. This observation suggests that leptin is an important mediator of inflammation and immune responses. However, Leptin is a product of the *ob* gene which is an adipose specific gene. Mice which are devoid of the leptin coding gene due to autosomal recessive mutation, are designated as *ob/ob* mice. They are commonly found to be obese and diabetic. When these mice are subjected to regular administration of leptin doses they exhibit reduced food intake, elevated rates of metabolism, and significant weight loss [28]. Leptin exerts its activity through the central nervous system.

Chemistry of Flavonoids

Flavonoids are polyphenols occurring in nature as secondary plant metabolites. They comprise benzo-γ-pyran rings with phenol and pyran groups making up the primary segments (Fig. 16.2).

Their synthesis arises from aromatic amino acid precursors such as phenylalanine, tyrosine, malonate through shikimic acid pathway [29, 30]. It bases the chemistry on the 15-carbon skeleton comprising $C_6{-}C_3{-}C_6$ units. The primary structure comprises two benzene rings, A and B, which are linked by a heterocyclic pyran ring-C [31]. This structure undergoes substitutions by diverse groups such as hydroxylation, methylation, and glycosylation to produce distinct subgroups [32]. Based on the specific substitutions and linkages, there are five major classes of flavonoids. Table 16.1 depicts specific illustrations of these major flavonoid classes along with their dietary sources and primary structures.

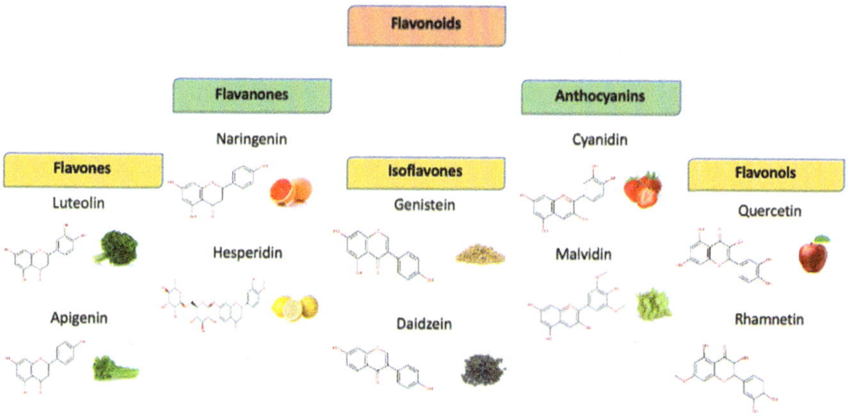

Fig. 16.2 Illustrates the structures and the dietary sources of the flavonoids

Table 16.1 Dietary sources and chemical structures of major flavonoid classes [33, 34]

Flavonoid Class	Examples	Food sources	Basic structure	Reference
Flavones	Luteolin Apigenin Tangeritin Chrysin	Parsley, thyme, celery, vegetable oils	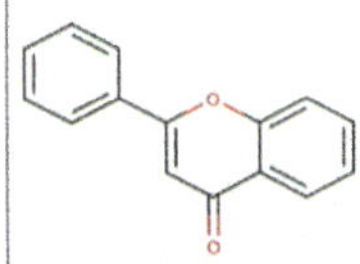	[35]
Flavanones	Naringenin Hesperidin Eriodictyol Butin Sylibin	Citrus fruits, grapefruit, tomato, orange, lemon	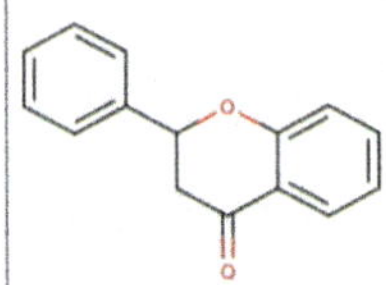	[36, 37]
Isoflavones	Daidzein Genistein Glycitin	Soybeans, fava beans, psoralea, legumes	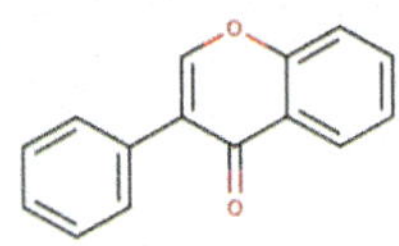	[38]
Anthocyanins	Cyanidin Delphinidin Pelargonidin Peonidin Malvidin Petunidin	Blackcurrant, strawberry, raspberry, blueberry, cherry, grapes, apple, peach, plum	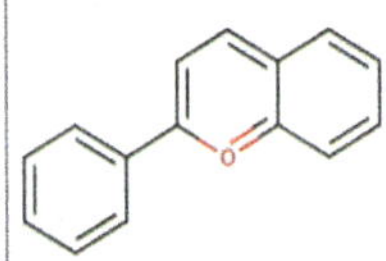	[32]
Flavonols	Quercetin Kaempferol Myricetin Rhamnetin Galangin	Apple, potato, celery, eggplant, red wine, walnuts, almonds raw spinach, cocoa	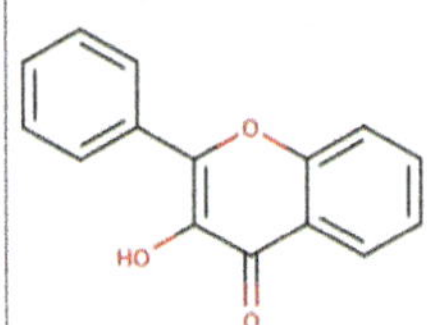	[39]

Flavones

These types of flavonoids include a 3-carbon unsaturated group and have a double bond between C2 and C3. The backbone has a 2-phenylchromen-2-one structure [35]. In plants, they conjugate as 7-*O*-glycosides [40]. Widely known flavones are apigenin, luteolin, etc. The most prevalent origins of flavones are parsley, thyme, celery. Polymethoxylated flavones are in the peels and tissues of citrus fruits [35]. In an investigation directed by W. Gong and J. Wu, flavones present in *Scutellaria baicalensis*, were found helpful in the management of nicotine-induced cell proliferation and lung cancer [41]. This brought to the revelation of alternative benefits of flavones, in the management of cardiovascular diseases.

Flavanones

Flavanones possess antioxidant and radical scavenging activities [42]. The major flavanone aglycones are hesperidin, naringenin, and eriodyctiol. Naringenin is often detected in grapefruit and grapefruit juice [37]. It has been learned for its role in the treatment of liver diseases because of the buildup of oxidative stress [43]. Citrus fruits are used for their antioxidant properties. This is because of the rich content of flavanones present in these fruits [36]. Certain studies established that the antioxidant potential of flavanones results from their planar character [44].

Isoflavones

Isoflavones are a class of phytoestrogens. These are plant-derived compounds that show estrogenic activity. This class of flavonoids has a phenyl group substituted on a position meta to the oxygen atom. Of all available phytoestrogens, soy isoflavones are used in the prevention and management of cardiovascular diseases. These show the visible effects of arterial vasodilation, inhibition of atherosclerosis, and lowering levels of serum cholesterol. These are effective in the treatment of atherosclerosis, which is by the character of their antioxidant effects. On smooth muscle cells, these compounds show antiproliferative and antimigratory effects along with effects on thrombus formation [45]. Major isoflavones available from soy protein are daidzein, genistein, and glycitin [38]. Certain investigations have confirmed the contribution of Isoflavone in the diminishing of blood pressure [46].

Anthocyanins

Anthocyanins form the principal part of many pigments found in flower petals, fruits, vegetables, grains, etc. They exist in plants in glycosidic forms called anthocyanins [32]. Structural differences, particularly, the difference in the number of hydroxyl groups present, methylation and glucosylation and the distribution of positive charge can induce color variations in anthocyanins [47]. Cyanidin, Delphinidin, Malvidin etc. are the most commonly used anthocyanins. Cell culture studies, animal models and human trials that have been conducted proved that anthocyanins possess anti-oxidative activity, contributing to the prevention of cardiac diseases [48].

Flavonols

These flavonoids are also known as flavon-3-ols or catechins. In these compounds, there is no double bond between C2 and C3. There is also an absence of C4 carbonyl in ring C of the structure. The presence of gallic acid residues has been observed in some Flavan-3-ols, which is attached to the C ring hydroxyl by an ester bond [39]. They exist as two diastereomers—catechin and epicatechin [32]. Flavanols present in cacao are found to have significant antihypertensive, anti-inflammatory and antioxidant activities. Cocoa consumption links with reduced blood pressure, as revealed by several studies on these compounds [49]. Tea, grapes, berries, apples, red wine, etc. are some major sources of flavonols.

Dietary Flavonoids that Reduce Obesity and Obesity-Induced CVDs

Luteolin

A trial was performed to assess the beneficial effects of luteolin on vascular dysfunction in a diet-induced obese mouse model (Table 16.2). Obesity was induced in 6 weeks old male mice using High fat-diet (60% kcal from fat) for 8 weeks and then treated with luteolin (10 mg/kg/day). On the evaluation of blood cholesterol, glucose, triglyceride levels and body weight it was determined that the animals displayed elevated levels of metabolic indexes in HFD mice and luteolin helped to reduce the increased weight and other metabolic changes. These findings can be attributed to the restoration of endothelial nitric oxide levels, decrease in reactive oxygen species and tumor necrosis factor levels and normalization of endothelial nitric oxide synthase (eNOS) and SOD1 gene expression demonstrating both anti-inflammatory and anti-oxidant properties [50].

Naringenin

Naringenin belongs to the class of citrus flavonoids which shows promising antioxidant properties. A clinical trial performed in rats sought to analyze the effect of naringenin on metabolic parameters and cardiovascular structure and function (Table 16.2). The rats were divided into 4 groups and each was fed with their respective diets for a period of 16 weeks and two of the groups were supplemented with naringenin 100 mg/kg/day. Analysis of both cardiovascular and metabolic criteria in control and test groups concluded in providing better insight into the multiple mechanisms of its cardio-protective activity. Naringenin showed antihypertensive, lipid-lowering and insulin-sensitizing properties in high-fat diet-fed rats.

Table 16.2 Pre-clinical and clinical studies done with flavonoids for the treatment of obesity induced CVD

Flavonoid	Test organism	Study duration	Study design	Results	Reference
Luteolin	Male C57BL/6 mice 6 weeks old	9 weeks	High-Fat diet induced	Prevents weight gain and metabolic changes Restores vascular endothelium nitric oxide levels owing to anti-inflammatory and antioxidant properties	Gentile et al. [15]
Naringenin	Male Wistar rats 9–10 weeks old	16 weeks	High-fat diet induced	Reduces hypertension, plasma lipid levels, mitochondrial function and oxidative stress	Alam et al. [51]
Cyanidin	Male Wistar rats 8–9 weeks old	16 weeks	High-fat diet induced	Improves cardiac and hepatic structure and function Reduction in metabolic parameters of body weight, visceral adiposity index and total body fat mass	Bhaswant et al. [54]
Quercetin	Obese male Zucker rats and lean rats 13 weeks old	Over 10 weeks	Lean and obese rats were selected previously	Reduces chances of hyperlipidemia, insulin resistance and hypertension The effects can be attributed to anti-inflammatory and antioxidant properties	Rivera et al. [55]
Rutin	Male Wistar rats 8–9 weeks old	16 weeks	High-fat diet induced	Shows significant anti-inflammatory and antioxidant properties Reduced pro-inflammatory cytokines, LDL and total cholesterol levels Improved endothelial function and increased nitric oxide levels	Panchal et al. [56]

(continued)

Table 16.2 (continued)

Flavonoid	Test organism	Study duration	Study design	Results	Reference
Apigenin	C57BL/6 J male mice Four weeks old	16 weeks	High-fat diet induced	Decreased total-cholesterol levels, apoB levels, apoB/apoA1 ratio, pro-inflammatory cytokines and chemokines in plasma Decreased expression of hepatic lipogenic and lipolytic genes Increased expression of hepatic genes involved in fatty acid oxidation	Jung et al. [57]
Baicalin	Male Sprague–Dawley rats 6–8 weeks old	16 weeks	High-fat diet induced	Decreased serum cholesterol, free fatty acid and insulin concentration Suppressed systemic inflammation by reducing serum levels of tumor necrosis factor-α Activates hepatic AMPK leading to protective effect	Guo et al. [58]
Diosmin	Male albino Wistar rats (180-200 g)	2 weeks	Isoproterenol induced	Decreased levels of serum cardiac marker enzymes, plasma lipid peroxidation and minimized alterations in lipid metabolism Inhibited the activity of liver HMG CoA reductase in the liver Free radical scavenging activity reduces inflammation and associated disorders	Queenthy et al. [59]

(continued)

Table 16.2 (continued)

Flavonoid	Test organism	Study duration	Study design	Results	Reference
Hesperidin	28 humans between 21–65 years suffering from metabolic syndrome	6 weeks and 3 days	Randomized, placebo-controlled, double-blind, crossover trial	Reduced total cholesterol, reduced insulin resistance Increased HDL levels Stimulates NO production in endothelial cells	Rizza et al. [52]
Genistein	120 post-menopausal women between 49–67 years	12 months	Randomized, double-blind and placebo-controlled study	Reduced total cholesterol, LDL, triglycerides and blood pressure Increased HDL levels Improved insulin resistance	Squadrito et al. (2013)

There are various possible mechanisms contributing to these properties like the reduction of oxidative stress and reduced inflammatory cell infiltration. It improved the functioning of liver mitochondria by increased peroxisome proliferator-activated receptor-γ expression and reduction of lipid peroxidation. The anti-hyperlipidemic effect is achieved by lowering of plasma cholesterol levels and biosynthesis of cholesterol. Further reduction in hepatic glycolysis and gluconeogenesis improves insulin resistance which improves metabolic syndrome conditions [51].

Hesperidin

Citrus flavonoids have been proved to show significant effects in comorbidities associated with metabolic disorders and hesperidin belongs to this category.

A trial was performed to understand the effect of hesperidin and its metabolite hesperetin on cardiovascular functions in patients with metabolic syndrome (Table 16.2). Adults with metabolic syndrome between 21 and 65 years of age were called in for the trial. The participants were divided into two groups and were given either hesperidin or the placebo for a 3-week period then after a 3-day washout period they were shifted to the other treatment arm. Various metabolic parameters and markers of cardiac and endothelial function were assessed after 3 weeks of periods.

The analysis showed improvement in endothelial cell function and favorable changes in both lipid profile and inflammation biomarkers that provide information on the beneficial effects of hesperidin on metabolic disorders [52].

Genistein

Genistein is one of the major components of soy flavonoids and has been shown to have various health benefits in cancer, cardiovascular and metabolic disorders. A trial that had 120 postmenopausal women between 49 and 67 years old was observed for a period of 12 months where they were randomly divided into two groups who were fed with genistein (54 mg/day) and placebo besides a Mediterranean diet (Table 16.2). Various parameters were observed namely inflammatory markers, insulin resistance, blood pressure, BMI index and other metabolic markers. Genistein administration in the given population showed improvement of both glucose and lipid metabolism hence a direct effect on insulin resistance and cardiovascular function [53].

Cyanidin

Cyanidin is the most common anthocyanin present in fruits and vegetables and is said to have beneficial effects in cases of metabolic syndrome due to their antioxidant properties. A trial consisting of 72 male rats where there were 6 experimental diets for a period of 16 weeks (Table 16.2). Initially, for a period of 8 weeks, they are fed with their respective diets and later for the next 8 weeks their diet was supplemented with cyaniding-3-glucoside and queen garnet plum juice. After the 16-week trial cardiovascular, histopathological, plasma and body measurements were taken for analysis. Both cyanidin-3-glucoside and queen garnet plum juice showed reduced body weight gain, fat mass, oxidative stress, plasma lipid levels and inflammatory cell infiltration. It increased insulin sensitivity and thus exhibited positive effects in cases of metabolic disorders [54].

Quercetin

Quercetin is a flavanol present in multiple dietary sources like tea, spices, herbs, fruits etc. and is said to have favorable anti-inflammatory properties. A study was attempted to analyze the effect of quercetin on abnormalities associated with metabolic disorders in male rats (Table 16.2). In this study, 3 groups of rats containing both obese and lean rats were randomly administered with quercetin either 2 mg/kg or 10 mg/kg of body weight. For analysis blood pressure and blood plasma levels were observed. This study concluded that chronic administration of quercetin helped in improving the inflammatory status and other metabolic parameters [55].

Rutin

Rutin is a flavanol present abundantly in plants, buckwheat and passionflower. It is a major component found in apples. It has been proved to show antioxidant, cytoprotective, vasoprotective, anti-carcinogenic, neuroprotective and cardio-protective activities. A study was done to characterize the hepatic, cardiovascular and metabolic response to rutin in a rat model (Table 16.2). Male Wistar rats were randomly divided into 6 experimental groups each containing 12 rats. Two groups were fed a high-carbohydrate diet and two groups were fed a high-fat diet for either 8 weeks or 16 weeks. Two further groups one with carbohydrate and the other fat diet were administered rutin (1.6 g/kg food) for the last 8 weeks of the 16-week period. The 8-week groups were used to study the change in the pathophysiology of the rats due to the diet.

After the 16-week trial systolic blood pressure measurements, echocardiography, vascular reactivity, erythrocyte reactive oxygen species production and plasma markers for oxidative stress were tested to analyze the effect of rutin. The study concluded that rutin had a significant role to play in improving inflammation, oxidative stress-related cardiovascular impairment and prevented metabolic changes due to fat diet [56].

Apigenin

Apigenin is a flavone present in chamomile, parsley, onions, grapefruit and oranges. It has shown to have many beneficial properties like anticancer, anti-diabetic, anti-obesity antioxidant and anti-inflammatory. A given study investigated the protective effects of apigenin against obesity and related metabolic disorders (Table 16.2). In the study, the animals were randomly divided into two groups and were fed an either high-fat diet or high-fat diet along with 0.005% w/w apigenin for 16 weeks. At the end of the study, blood glucose levels, plasma adipocytokines, lipids and amino-transferases levels were measured for analysis besides other histopathological analyses. It was concluded that apigenin lowered plasma levels of free fatty acid, total cholesterol, apolipoprotein B and hepatic dysfunction markers. It increased the expression of genes related to fatty acid oxidation, TCA cycle and cholesterol homeostasis. It down-regulated expression of lipolytic genes, lipogenic genes and decreased activity of TGE and cholesterol synthesis. It also showed a prominent effect on the reduction of pro-inflammatory mediators [57].

Baicalin

Baicalin is a flavonoid with significant anti-inflammatory and antioxidant properties. In a study, animals were divided into standard diet-fed animals and high-fat diet-fed animals (Table 16.2). The high-fat diet-fed animals were treated with baicalin (80 mg/kg) once per day. On analysis of both histopathological and serum parameters, it was found that baicalin decreased serum cholesterol and free fatty acid levels, suppressed systemic inflammation and activated hepatic AMPK leading to protective effect in case of metabolic disorders [58].

Diosmin

Diosmin is a flavonoid present mainly in citrus fruits. It has shown the presence of antioxidant, anti-hyperglycemic, anti-inflammatory, anti-mutagenic, and antiulcer properties. The aim of the study was to evaluate the protective effects of diosmin on an experimentally induced myocardial infarcted rat model (Table 16.2). The rats were pretreated with diosmin (5 mg/kg) before inducing isoproterenol (100 mg/day) at an interval of 24 h for 2 days. At the end of the study period cardiac markers, cardiac indices, lipid profile, lipid peroxidation products, electrocardiogram were estimated along with its radical scavenging abilities. It was concluded that treatment with diosmin provided cardioprotection due to antihyperlipidemic effects by preventing the accumulation of lipids by its free radical scavenging effects and also improves the status of lipid profile by HMG-CoA reductase inhibiting capacity [59]. Marketed products of various flavonoids available to target different disorders are summarized in Table 16.3.

Discussion and Future Perspectives

Obesity and cardiovascular dysfunction are major health concerns around the world. Obesity and CVD can lead to lifelong co-morbidities such as Type-II diabetes, hypertension, bacterial or microbial infections and mortality. The current lifestyle practiced and the amount of stress around the world, especially in the developing and the developed countries is enormous. As per the prevalence, the induction rate of obesity and cardiovascular dysfunction is shifted from the elderly population to the population in their late 30 s and 40 s. The allopathic medicines prescribed for obesity like Orlistat, Lorcaserin, Phentermine-topiramate, Liraglutide etc. have many side-effects like dizziness, anorexia, stomach pain, increased blood pressure and heart rate, raised pulse etc. As obesity leads to CVD manifestations, anti-obesity medications contribute more to these complications, thus leading to fatality. Also, medications for CVD are lifelong, and have too many complications when taken for an extended amount of time. Therefore, there is a need to explore phytoconstituents that can treat obesity, CVD; and will not cause untoward side-effects. There are several research studies undertaken for the treatment of obesity induced CVD. The preclinical data obtained from animal models support the beneficial role of flavonoids in the prevention of CVDs and distinctive cancer types. The current data also supports that the flavonoids are promising molecules to improve cardiometabolic function as well as health and wellbeing. Despite interacting with multiple targets, intake of flavonoids is considered to be safe. Healthful foods containing flavonoids and other bioactive polyphenols may be the best cost-effective strategy for the prevention of non-communicable and cardiometabolic diseases. Further, the bioavailability of flavonoids can be improved via the use of newer nanotechnology methods and drug delivery systems such as nano-delivery systems, microencapsulation, and micro-emulsions. The future drug delivery techniques would help in improving bioavailability by increasing the intestinal absorption, enhancing metabolic stability in the gut, and targeting a different absorption site [60]. Many investigators have reported clinically important adverse interactions between herbal remedies and fruit juices with orally administered synthetic or allopathic drugs. Studies are needed to understand any adverse interactions between flavonoid products with synthetic or prescription drugs. In addition, studies on the adverse interactions, if any, of flavonoids with gut microbiota are also warranted.

Table 16.3 Marketed Flavonoids and their indications

Flavonoid	Product	Dose	Indication
Luteolin	LutiMax (Luteolin 100 mg +Rutin 100 mg)	1 to 4 tablets × 200 mg	Neuro-protective, immuno-modulatory and ameliorates neurodegeneration
	Luteolin Complex (Luteolin 50 mg + Rutin 50 mg)	1 capsule/day	To treat Age-related cognitive decline
	Luteolin 100	1 capsule/day	Neuroprotection And for treatment of age-related memory defects
Hesperidin	HD Complex (Hesperidin 50 mg + Diosmin 450 mg)	1 tablet, twice daily	Improved and healthy vasculature
	Nuvaprin HD (Hesperidin 50 mg + Diosmin 450 mg)	2 capsules × 500 mg	For healthy and normal vasculature And reduction in inflammation
Genistein	Genistein 125 mg	2 capsules × 125 mg	For cardiac health And menopausal symptoms
	GeniKinoko (Genistein 180 mg)	4 capsules, 2 times a day	Improves immune health, cellular health and prostate health
Quercetin	Quercetin Bromelain Complex (Quercetin 500 mg + Bromelain 156 mg)	1 capsule 2–3 times a day × 500 mg	For immuno-modulatory diseases and cardiac health
Rutin	Rutin 250 mg	1 capsule × 250 mg	For absorption and assimilation of vitamin C For healthy vasculature
Apigenin	Apigenin 50 mg	1 capsule × 50 mg	Prostate health and Promotes glucose metabolism
Diosmin	Venostor (Diosmin 450 mg + Hesperidin 50 mg)	Depends on the indication type	Healthy vasculature and acute hemorrhoidal disease crisis

Flavonoids have proven to have a beneficial effect in ameliorating obesity and cardiovascular morbidity. Supplementary studies should be conducted in knowing the exact mechanism of action and molecular mechanisms of flavonoids in the treatment of obesity induced CVD. This will help in increasing the success rates with less side-effects and decreasing the mortality rate in the obese patients.

Conflict of interests The authors declare no conflict of interest.

References

1. Hertz RP, Unger AN, Cornell JA, Saunders E (2005) Racial disparities in hypertension prevalence, awareness, and management. Arch Intern Med 165(18):2098–2104
2. Kearney PM, Whelton M, Reynolds K, Whelton PK, He J (2004) Worldwide prevalence of hypertension: a systematic review. J Hypertens 22(1):11–19
3. Anchala R, Kannuri NK, Pant H, Khan H (2014) Hypertension in India: a systematic review and meta-analysis of prevalence, awareness, and control of hypertension. J Hypertens 32(6):1170–1177
4. Jeemon P, Reddy KS (2010) Social determinants of cardiovascular disease outcomes in Indians. Indian J Med Res 132(5):617–622
5. Cardiovascular diseases (CVDs)-Key Facts [Internet] (2017) Cited 2019 Mar 3. Available from: https://www.who.int/en/news-room/fact-sheets/detail/cardiovascular-diseases-(cvds)
6. Huffman MD, Bhatnagar D (2012) Novel treatments for cardiovascular disease prevention. Cardiovasc Ther 30(5):257–263
7. Vernarelli JA, Lambert JD (2017) Flavonoid intake is inversely associated with obesity and C-reactive protein, a marker for inflammation, US adults. Nutr Diab 7(5):e276
8. Kopustinskiene DM, Savickas A, Vetchý D, Masteikova R, Kasauskas A, Bernatoniene J (2015) Direct effects of (-)-Epicatechin and procyanidin B2 on the respiration of rat heart mitochondria. Biomed Res Int
9. Vita JA (2005) Polyphenols and cardiovascular disease: effects on endothelial and platelet function. Am J Clin Nutr 81(1 Suppl):292S-297S
10. Donato KA (1998) Executive summary of the clinical guidelines on the identification, evaluation, and treatment of overweight and obesity in adults. Arch Intern Med 158(17):1855–1867
11. Williams CL, Hayman LL, Daniels SR, Robinson TN, Steinberger J, Paridon S, Bazzarre T (2002) Cardiovascular health in childhood: a statement for health professionals from the Committee on Atherosclerosis, Hypertension, and Obesity in the Young (AHOY) of the Council on Cardiovascular Disease in the Young. American Heart Association. Circulation 106(1):143–160
12. Ortega FB, Lavie CJ, Blair SN (2016) Obesity and cardiovascular disease. Circ Res 118(11):1752–1770
13. Kenchaiah S, Evans JC, Levy D, Wilson PWF, Benjamin EJ, Larson MG, Kannel WB, Vasan RS (2002) Obesity and the risk of heart failure. N Engl J Med 347(5):305–313
14. Van Gaal LF, Mertens IL, De Block CE (2006) Mechanisms linking obesity with cardiovascular disease. Nature 444:875–880
15. Lai SW, Ng KC, Lin HF, Chen HL (2001) Association between obesity and hyperlipidemia among children. Yale J Biol Med 74(4):205–210
16. Greenberg AS, Obin MS (2006) Obesity and the role of adipose tissue in inflammation and metabolism. Am J Clin Nutr 83(2):461S-465S
17. Tilg H, Moschen AR (2006) Adipocytokines: Mediators linking adipose tissue, inflammation and immunity. Nat Rev Immunol 6(10):772–783
18. Cavalera M, Wang J, Frangogiannis NG (2014) Obesity, metabolic dysfunction, and cardiac fibrosis: pathophysiological pathways, molecular mechanisms, and therapeutic opportunities. Transl Res 164(4):323–335
19. Matsuzawa Y, Funahashi T, Kihara S, Shimomura I (2004) Adiponectin and metabolic syndrome. Arterioscler Thromb Vasc Biol 24(1):29–33
20. Hansen T, Ahlström H, Söderberg S, Hulthe J, Wikström J, Lind L, Johansson L (2009) Visceral adipose tissue, adiponectin levels and insulin resistance are related to atherosclerosis as assessed by whole-body magnetic resonance angiography in an elderly population. Atherosclerosis 205(1):163–167
21. Tilg H, Hotamisligil GS (2006) Nonalcoholic fatty liver disease: cytokine-adipokine interplay and regulation of insulin resistance. Gastroenterology 131(3):934–945
22. Ross R (1993) The pathogenesis of atherosclerosis: a perspective for the 1990s. Nature 362(6423):801–809

23. Ouchi N, Kihara S, Arita Y, Maeda K, Kuriyama H, Okamoto Y, Hotta K, Nishida M, Takahashi M, Nakamura T, Yamashita S, Funahashi T (1999) Novel modulator for endothelial adhesion molecules: adipocyte-derived plasma protein. Circulation 100:2473–2476
24. Ouchi N, Walsh K (2007) Adiponectin as an anti-inflammatory factor. Clin Chim Acta 380(1–2):24–30
25. Stofkova A (2009) Leptin and adiponectin: from energy and metabolic dysbalance to inflammation and autoimmunity. Endocr Regul 43(4):157–168
26. Flier JS (1995) The adipocyte: Storage depot or node on the energy information superhighway? Cell 80(1):15–18
27. Frederich RC, Hamann A, Anderson S, Löllmann B, Lowell BB, Flier JS (1995) Leptin levels reflect body lipid content in mice: evidence for diet-induced resistance to leptin action. Nat Med 1(12):1311–1314
28. Pelleymounter MA, Cullen MJ, Baker MB, Hecht R, Winters D, Boone T, Collins F (1995) Effects of the obese gene product on body weight regulation in ob/ob mice. Science 269(5223):540–543.
29. Wang TY, Li Q, Bi KS (2017) Bioactive flavonoids in medicinal plants: Structure, activity and biological fate. Asian J Pharm Sci 13(1):12–23. Available from: https://doi.org/10.1016/j.ajps.2017.08.004
30. Baldim JL, Alcantara BGV De, Domingos ODS, Soares MG, Caldas IS, Novaes RD, Oliveira TB, Lago JHG, Chagas-Paula DA (2017) The correlation between chemical structures and antioxidant, prooxidant, and antitrypanosomatid properties of flavonoids. Oxid Med Cell Longev
31. Kumar S, Pandey AK (2013) Chemistry and biological activities of flavonoids : an overview chemistry and biological activities of flavonoids : an overview. Sci World J
32. Tsao R (2010) Chemistry and biochemistry of dietary polyphenols. Nutrients 2(12):1231–1246
33. Nijveldt RJ, Van Nood E, Van Hoorn DEC, Boelens PG, Van Norren K, Van Leeuwen PAM (2001) Flavonoids: a review of probable mechanisms of action and potential applications. Am J Clin Nutr 74(4):418–425
34. Fraga CG, Croft KD, Kennedy DO, Tomás-Barberán FA (2019) The effects of polyphenols and other bioactives on human health. Food Funct 10(2):514–528. Available from: https://xlink.rsc.org/?DOI=C8FO01997E
35. Higdon JVJD (2016) BDAC. Flavonoids [Internet]. Cited 2019 Jan 28. Available from: https://lpi.oregonstate.edu/mic/dietary-factors/phytochemicals/flavonoids
36. Citrus fruits could help prevent obesity-related heart disease, liver disease, diabetes [Internet] (2016). Cited 2019 Mar 3. Available from: https://www.acs.org/content/acs/en/pressroom/newsreleases/2016/august/citrus-fruits-could-help-prevent-obesity-related-heart-disease-liver-disease-diabetes.html
37. Felgines C, Texier O, Morand C, Manach C, Scalbert A, Régerat F, Rémésy C (2000) Bioavailability of the flavanone naringenin and its glycosides in rats. Am J Physiol Gastrointest Liver Physiol 279(6):G1148–G1154
38. Sacks FM, Lichtenstein A, Van Horn L, Harris W, Kris-Etherton P, Winston M (2006) Soy protein, isoflavones, and cardiovascular health: an American Heart Association Science Advisory for professionals from the Nutrition Committee. Circulation 113(7):1034–1044
39. Iglesias J, Medina I, Pazos M (2013) Galloylation and polymerization: role of structure to antioxidant activity of polyphenols in lipid systems. Polyphenols Hum Heal Dis 1:323–338
40. Hostetler GL, Ralston RA, Schwartz SJ (2017) Flavones : food sources. Bioavailability 423–435
41. Gong WYL, Wu JF, Liu BJ, Zhang HY, Cao YUX, Sun J, Lv YUB, Wu X, Dong JC (2014) Flavonoid components in *Scutellaria baicalensis* inhibit nicotine-induced proliferation, metastasis and lung cancer-associated inflammation in vitro. Int J Oncol 44(5):1561–1570
42. Tomás-Navarro M, Vallejo F, Tomás-Barberán FA (2013) Bioavailability and metabolism of citrus fruit beverage flavanones in humans. Polyphenols Hum Heal Dis 1:537–551
43. Hernández-Aquino E, Muriel P (2017) Naringenin and the liver. Liver Pathophysiol Therapies Antioxidants

44. Di Majo D, Giammanco M, La Guardia M, Tripoli E, Giammanco S, Finotti E (2005) Flavanones in citrus fruit: structure-antioxidant activity relationships. Food Res Int 38(10):1161–1166
45. Anthony MS, Clarkson TB, Williams JK (1998) Effects of soy isoflavones on atherosclerosis: potential mechanisms. Am J Clin Nutr 68(6 Suppl):1390S-1393S
46. Teede HJ, Dalais FS, Kotsopoulos D, Liang YL, Davis S, Mcgrath BP (2001) Dietary soy has both beneficial and potentially adverse cardiovascular effects: a placebo-controlled study in men and postmenopausal women. J Clin Endocrinol Metab 86(7):3053–3060
47. Swanson BG (2003) Tannins and polyphenols. Encyclopedia Food Sci Nutr
48. Khoo HE, Azlan A, Tang ST, Lim SM (2017) Anthocyanidins and anthocyanins: colored pigments as food, pharmaceutical ingredients, and the potential health benefits. Food Nutr Res 61(1):1361779
49. Patel K, Watson RR (2018) Chocolate and its component's effect on cardiovascular disease. Lifestyle Heart Health Dis 255–266
50. Gentile D, Fornai M, Pellegrini C, Colucci R, Benvenuti L, Duranti E, Masi S, Carpi S, Nieri P, Nericcio A, Garelli F, Virdis A, Pistelli L, Blandizzi C, Antonioli L (2018) Luteolin prevents cardiometabolic alterations and vascular dysfunction in mice with HFD-induced obesity. Front Pharmacol
51. Alam MA, Kauter K, Brown L (2013) Naringin improves diet-induced cardiovascular dysfunction and obesity in high carbohydrate, high fat diet-fed rats. Nutrients 5(3):637–650
52. Rizza S, Muniyappa R, Iantorno M, Kim J, Chen H, Pullikotil P, Senese N, Tesauro M, Lauro D, Cardillo C, Quon MJ (2011) Citrus polyphenol hesperidin stimulates production of nitric oxide in endothelial cells while improving endothelial function and reducing inflammatory markers in patients with metabolic syndrome. J Clin Endocrinol Metab 96(5):E782–E792
53. Irace C, Marini H, Bitto A, Altavilla D, Polito F, Adamo EB, Arcoraci V, Minutoli L, Di Benedetto A, Di Vieste G, de Gregorio C, Gnasso A, Corrao S, Licata G, Squadrito F (2013) Genistein and endothelial function in postmenopausal women with metabolic syndrome. Eur J Clin Invest 43(10):1025–1031
54. Bhaswant M, Fanning K, Netzel M, Mathai ML, Panchal SK, Brown L (2015) Cyanidin 3-glucoside improves diet-induced metabolic syndrome in rats. Pharmacol Res 102:208–217
55. Rivera L, Morón R, Sánchez M, Zarzuelo A, Galisteo M (2008) Quercetin ameliorates metabolic syndrome and improves the inflammatory status in obese Zucker rats. Obesity 16(9):2081–2087
56. Panchal SK, Poudyal H, Arumugam TV, Brown L (2011) Rutin attenuates metabolic changes, nonalcoholic steatohepatitis, and cardiovascular remodeling in high-carbohydrate, high-fat diet-fed rats. J Nutr 141(6):1026–1029
57. Jung U, Cho Y-Y, Choi M-S (2016) Apigenin ameliorates dyslipidemia, hepatic steatosis and insulin resistance by modulating metabolic and transcriptional profiles in the liver of high-fat diet-induced obese mice. Nutrients 8(5):305
58. Guo HX, Liu DH, Ma Y, Liu JF, Wang Y, Du ZY, Wang X, Shen JK, Peng HL (2009) Long-term baicalin administration ameliorates metabolic disorders and hepatic steatosis in rats given a high-fat diet. Acta Pharmacol Sin 30(11):1505–1512
59. Queenthy SS, John B (2013) Diosmin exhibits anti-hyperlipidemic effects in isoproterenol induced myocardial infarcted rats. Eur J Pharmacol 718(1–3):213–218
60. Silverberg DS, Wexler D, Blum M, Iaina A, Sheps D, Keren G, Schwartz D (2005) Erythropoietin in heart failure. Semin Nephrol 25(6):397–403

Chapter 17
Homocysteine and Related B Vitamins in Pre-diabetes and Diabetes Mellitus

Slavica S. Mutavdzin and Dragan M. Djuric

Abstract Diabetes mellitus (DM) is the most common endocrine disorder and a global health problem with increasing prevalence. DM is associated with different organ dysfunction due to micro- and macrovascular complications. It should be taken into account that organ damages usually start earlier, in subjects with impaired glucose tolerance i.e. in the pre-diabetic state. There are different biomarkers that give information about organ damages. They are increased in pre-diabetic state, and indicate necessity to start preventing the development of DM and its complications on time. Homocysteine is also significant endothelial function biomarker. Its level is increased in both DM and pre-diabetes. DM is also characterized by a decrease in vitamin B-group levels, especially in vitamins B6, B12 and folic acid (folate). Reduced levels of these vitamins will lead to an additional increase in homocysteine levels, which will result in further impairment of organ functions. The results of numerous studies indicate the positive effects of administration of these vitamins in subjects with DM. Therefore, in order to prevent or delay the complications of DM, it is advised adequate dietary intake of vitamins. Since there is scientific evidence that metformin used in DM therapy reduces vitamin B12 and folic acid (folate) level, supplementation of these two vitamins is recommended in patients receiving metformin in therapy.

Keywords Diabetes mellitus · Pre-diabetes · Homocysteine · Vitamin B6 · Vitamin B12 · Folic acid (folate)

S. S. Mutavdzin · D. M. Djuric (✉)
Institute of Medical Physiology "Richard Burian", Faculty of Medicine, University of Belgrade, Visegradska 26/II, 11000 Belgrade, Serbia
e-mail: drdjuric@eunet.rs; dr_djuric@yahoo.com

P. S. Tappia et al. (eds.), *Biochemistry of Cardiovascular Dysfunction in Obesity*, Advances in Biochemistry in Health and Disease 20,
https://doi.org/10.1007/978-3-030-47336-5_17

Introduction

Diabetes mellitus (DM) is the most common endocrine disorder characterized by hyperglycemia and followed by numerous metabolic changes due to decreased insulin production or secretion, decreased insulin action, or both [1, 2]. It represents a global health problem. More than 6% of the world population is affected by DM. Current data demonstrate that 415 million adults suffer from diabetes mellitus; while 318 million people have impaired glucose tolerance or pre-diabetes [3]. It is considered that DM prevalence will increase to about 642 million in 2040 [4]. This prevalence does not contain the people with pre-diabetes. The prevalence of pre-diabetes is increasing worldwide. It is expected to be more than 400 million in 2030 [5].

Diabetes mellitus is defined as fasting blood glucose level of 126 mg/dl (7.0 mmol/l) and greater, while pre-diabetes or impaired glucose tolerance is defined as fasting plasma glucose levels 100 mg/dl (5.6 mmol/l) to 125 mg/dl (6.9 mmol/l) [6].

DM is associated with different organ dysfunction, such as retinopathy, renal and cardiovascular disease (CVD), gastrointestinal disturbance, sexual dysfunction and neuropathy [7, 8]. It is followed by macrovascular complications which are the most common cause of death among persons with diabetes mellitus [9]. CVD occur more frequently in patients with diabetes mellitus and represents one of macrovascular complications of DM. To prevent the onset of CVD, primary prevention should be implemented. However, it should be taken into account that cardiovascular risk is already increased in people in the pre-diabetic state or with impaired glucose tolerance [5]. It is shown that impaired fasting glucose is not just a precursor of diabetes; it is also an individual risk factor for CVD and all-cause mortality [10]. If left untreated, pre-diabetes would progress to diabetes and the associated complications [5]. In order to prevent diabetes development and its complications early detection and treatment of pre-diabetes is necessary.

It is important to discover pathological processes associated with asymptomatic or preclinical stages of the disease to prevent its progression. The onset of type 2 diabetes can be prevented or delayed by change in diet, introducing moderate physical activity and weight loss [11]. With the increasing technical development that enables more accurate investigation of biochemical processes occurring during diseases, the role of biomarkers is increasing. Biomarkers, such as blood glucose levels, hemoglobin A1c and cholesterol level have been measured as part of clinical assessment of DM and cardiovascular disease for many years [12]. Diabetes progression is, not only connected with changes in these biomarkers. It is reported that also inflammatory, endothelial dysfunction, oxidative stress and autonomic nervous system dysfunction markers are elevated in diabetes [12, 13]. Inflammatory markers, C reactive protein and interleukins (ILs), have been shown to predict development of DM type 2 [14] and are used routinely in medical practice to screen for cardiovascular disease [13, 15]. Other inflamatory biomarkers in relation with DM are tumor necrosis factor α (TNFα), antioxidants such as reduced glutathione and vitamin E (α-tocopherol), lipid peroxidation products such as malondialdehyde and isoprostanes. Biomarkers of endothelial dysfunction include 8-hydroxy-2-deoxy-guanosine (8-OHdG) and homocysteine (Hcy) [12, 13].

It is shown that increased oxidative stress and changes in antioxidant defense represent an important mechanism in the pathogenesis of DM and in the development of its complications [16]. Increased glucose level is followed by the synthesis of reducing sugars that can react with lipids and proteins and promote the production of reactive oxygen species (ROS) [17, 18]. Hyperglycaemia increases the production of peroxynitrite and other reactive nitrogen species, as well. It induces DNA and protein damage, activates polyADP ribose polymerase, and stimulates endothelial stress and cell death [19]. The oxidative stress alters total metabolism [20], and may cause damage to endothelial cells and promote the development of atherosclerosis. The association between oxidative stress and CVD in diabetic conditions via endothelial dysfunction and the resulting pathological changes in inflammation, coagulation status and cell-proliferation has been widely investigated [21–24]. In the case of pre-diabetes, the situation is not completely known, but studies suggest that there are changes in redox-status, and oxidative stress related DNA damage before DM diagnosis [25, 26]. It is revealed that central mechanism responsible for the increased cardiovascular risk in pre-diabetes is endothelial dysfunction. Endothelial dysfunction occurs due to the elevated formation of ROS and advanced glycation end products (AGEs) as well as increased lipid peroxidation under hyperglycaemic conditions [27, 28]. Endothelial dysfunction results in imbalance between coagulation and fibrinolysis, platelet activation, proliferation of vascular smooth muscle cell and stimulation of inflammatory processes. Consequently, pro-thrombotic environment is created, so if hyperglycaemia remains untreated it will cause CVD in the long-term [27, 28].

The association between biomarkers and disease progression from normal to impaired fasting blood glucose (pre-diabetes) has not been studied extensively. Recently, it has been shown that 8-OHdG is elevated already in the pre-diabetic state, suggesting an impact on pathological processes of even moderate increases in blood glucose level [26]. If reliable and specific biomarkers for prediabes can be identified, early detection and treatment of pre-diabetes could slow the diabetes epidemy.

Homocisteine—Biomarker of Endothelial Dysfunction

Homocysteine (Hcy) is a cytotoxic sulfur-containing amino acid. It is produced as an intermedier during the breakdown of methionine [29–31]. Butzand du Vigneud was the first to describe Hcy, while the association of Hcy with various diseases was first indicated by Carson and Neil in 1962 [32, 33]. Total homocysteine level between 12 and 30 mmol/L is considered mild hyperhomocysteinaemia, while Hcy level higher than 100 mmol/L severe hyperhomocysteinemia [34].

Hcy is removed by reactions of one carbon cycle. One carbon cycle is an important group of biochemical reactions that involves the utilization and production of methyl group (CH_3). The methylation cycle is a significant part of the one carbon cycle. During methylation, *S*-adenosylmethionine is formed from methionine and adenosine triphosphate (ATP). Methionine is formed from Hcy and 5-methyltetrahydrofolate in the presence of methionine synthase. 5-methylteterahydrofolate originates from dietary folate. Folate species that are not consumed in this reaction are used for the production of RNA/DNA bases and high energy molecules and cofactors such as ATP, nicotinamide adenine dinucleotide (NAD) and coenzyme A [35]. Vitamin B12 is cofactor in methionine synthase. Hcy is secreted into the plasma from endothelial and red blood cells as a product of incomplete conversion of methionine to cysteine [12].

Epidemiology of Hyperhomocysteineamia

The incidence of hyperhomocysteineamia is 5–10% in general population; it increases with aging and reaches 30–40% [34]. Hyperhomocysteineamia was found in 13–47% of patients with vascular diseases [36]. It has been reported that 58% patients with diabetes mellitus had hyperhomocysteinaemia, and higher incidence was found in males than in females [37]. Hcy level is increased under conditions of renal impairment, increased concentration of insulin, and administration of certain medications, such as metformin, glitazones, phenytoin, and methotrexate. Also, Hcy level increases in cases of the nutritional deficiency of vitamin cofactors: folate, vitamin B12, and vitamin B6 which are necessary for the homocysteine metabolism [38–41]. Deficiencies of these vitamins with resultant mild hyperhomocysteinemia can be risk factor of several diseases [42]. Estimated prevalence of vitamin B6 deficiency is 10.6% [43], folate <1% [44] and vitamin B12 about 6% [45] in the USA. Prevalence of these vitamins deficiency in Asia is higher, for instance vitamin B6 deficiency was reported in 52.8% and folate deficiency in 39.7% of healthy persons in Pakistan [46], while prevalence vitamin B12 deficiency was about 70% in India [45].

Homocisteine as a Risk Factor

Hcy is considered as a potential risk factor for many disorders such as cardiovascular diseases, atherosclerosis, venous thrombosis, vascular complications, and systemic diseases [47, 48]. Some authors demonstrated association between plasma homocysteine level and CVD risk [49], while others showed that reducing homocysteine level did not reduce cardiovascular risk and CVD incidence [49]. There is an agreement about the prothrombotic and prooxidant properties of homocysteine, which can cause the formation of ROS and contribute to endothelial dysfunction [50, 51]. Homocysteine can inhibit the expression of antioxidant enzyme glutathione peroxidase-1. The consequence of this inhibition is that the regeneration of reduced glutathione is limited [49].

It is evidenced that Hcy has an important role in the development of CVD and nervous system disorders [52]. An experimental research showed that D,L-homocysteine thiolactone strongly inhibited Na^{+}/K^{+}-ATPase activity in cortex, hippocampus and brain stem, while D,L-homocysteine induced just a moderate inhibition of hippocampal Na^{+}/K^{+}-ATPase of rats. Decreased Na^{+}/K^{+}-ATPase activity could be a factor that results in epileptic seizures in hyperhomocysteinemic conditions [53]. Hyperhomocysteinemia is considered to be a risk factor for atherosclerosis and cardiovascular diseases [52, 54]. However, the mechanism by which Hcy leads to atherosclerotic changes is not completely clear. It is demonstrated that the single administration of Hcy increased contractility of rat femoral artery smooth muscles. Also, 24-h-long incubation of rat femoral artery with Hcy induced an impairment of vascular endothelium, expressed as interruption of endothelial cells [55]. Experimental data suggested that Hcy produces an endothelial dysfunction through ROS production, decreases the production of endothelial nitric oxide, stimulates proliferation of vascular smooth-muscle cells, increases the formation of highly atherogenic oxycholesterols, stimulates lipid peroxidation, has a thrombogenic effect [29], and stimulates the expression of vascular endothelial growth factor (VEGF) [39], intracellular adhesion molecule-1, proinflammatory cytokines such as IL-1β, IL-6, TNF-α, and monocyte chemoattractant protein (MCP-1) [52]. Hcy influences cardiac function, its administration decreased cardiac contractility and reduced coronary flow of isolated rat hearts [56], inhibited the cardiac oxygen consumption and may lead to cardiotoxicity [57]. Hyperhomocysteinemia is also linked with oxidative stress and inflammation in DM type 2. It was demonstrated that Hcy contributed in atherosclerotic process of diabetes. Hyperhomocysteinemia was found in diabetic patients and represents strong risk factor [58, 59]. Other studies reported elevated or lower Hcy level in diabetic patients comparing to non-diabetics, demonstrating potential role of Hcy in development of vascular disorders in diabetic subjects [38, 60]. It was demonstrated that 58% of diabetic patients had increased homocysteine level [38]. One meta-analysis established that Hcy levels were higher in DM type 1 patients with complications, such as retinopathy or nephropathy, and it was not elevated in DM type 1 patients without any complications [61]. Also, it was confirmed that increased Hcy level is typically associated with increased risk of DM type 2 [62]. It is considered that increased Hcy has been involved in DM type 2 development and diabetic adverse outcomes [47, 58]. DM type 2 is characterized by increased oxidative stress. Increased plasma level of homocysteine reflects increased oxidative stress [50]. In addition to elevated Hcy levels, vitamin B12 and folate deficiency are linked to oxidative stress in diabetics [63].

Maschirow et al. demonstrated significantly increased homocysteine level in pre-diabetic patients comparing to the controls (14). The elevated homocysteine showed that increased oxidative stress was accompanied with changes in intracellular metabolism. Also, the elevated plasma homocysteine level indicated of endothelial dysfunction [12]. It is likely that homocysteine contributes itself to the oxidative stress by promotion the formation of ROS. The association between homocysteine level and oxidative stress is obvious, although the mechanism of this association is not completely clear. It was demonstrated that a relatively small increase in blood glucose levels certainly affects endothelial function, leading to increased DNA damage and release of homocysteine from endothelial cells [12].

Higher homocysteine levels are found in patients with diabetic nephropathy compared to patients with diabetes who have not developed nephropathy. It was demonstrated significantly higher level of plasma homocysteine in patients with diabetic nephropathy with macroalbuminuria than in patients with microalbuminuria and diabetic patients without albuminuria [64]. There are several explanations of alteration of Hcy level. The main way to remove Hcy from the body is through renal excretion [65, 66]. The deterioration of renal function may decrease the renal clearance of homocysteine leading to increase in plasma homocysteine concentration. Next explanation is that Hcy could promote oxidative stress, which could induce renal injury and impairment. It was shown that the auto-oxidation of homocysteine formed reactive oxygen species, and impaired the production of glutathione peroxidase [67]. Hcy can also have direct toxic effects on kidney tissue [68]. Experimentally induced chronic hyperhomocysteinemia demonstrated arterial and arteriolar thickening, and tubulointerstitial and podocyte injury in the kidney [69]. In addition, Hcy has ability to activate MAP kinases resulting in endoplasmic reticulum stress in mesangial cells [70]. Homocysteine activates nuclear factor-kappa B and increases the expression of monocyte chemoattractant protein-1 in the kidney [71], then, induces inflammatory transcriptional signaling in monocytes [72], enhances inflammatory status, supports proinflammatory cytokine production and macrophage infiltration [73]. Thus, in hyperhomocysteinemia inflammation is expected. Inflammation plays a significant role in the development and progression of diabetic nephropathy [74]. In one study plasma concentrations of methionine cycle intermediates S-Adenosyl-homocysteine (SAH), *S*-adenosyl-methionine (SAM), and homocysteine were measured in patients with renal failure and DM type 2. Increased plasma concentrations of thesse paramethers were related to the degree of renal insufficiency in patients with DM type 2 [75]. Patients with DM type 2 had significantly higher erythrocyte SAH concentrations than non-diabetic patients [76]. However, no study has proved that patients with DM and with increased plasma SAH are at a higher cardiovascular risk than patients with normal SAH levels. In addition, persons with DM type 2 are more susceptible to the damaging effects of Hcy than healthy persons [77].

Patients with insulin resistance and increased insulin level have elevated Hcy, as well. It was demonstrated proportional elevation in plasma homocysteine and plasma insulin levels [78]. Hyperhomocysteinemia is considered to exacerbate insulin resistance by leading to endoplasmic reticulum stress, increasing glucose output and upregulating phosphoenolpyruvate carboxykinase (PEPCK) [73, 79]. If insulin resistance worsens, it would result in more poor glucose control which is a risk factor for diabetic complications [64].

Obesity is a risk factor for many diseases as CVD, DM and fatty liver disease. Studies demonstrated that aberrant metabolism of methionine and its metabolites are associated with increased body fat and adiposity. One research showed that high plasma cysteine levels were associated with increased body mass index (BMI) and body mass both in children and adults [80]. Some studies found correlation between increased homocysteine levels and abdominal obesity, BMI, insulin level and HOMA-IR (Homeostatic Model Assessment for Insulin Resistance) [81]. Another research showed positive association between high intake of saturated fatty acids and higher homocysteine levels [82]. Hyperhomocysteinemia inhibits lipolysis and results in adipose tissue dysfunction. Also, the dietary administration of homocysteine during two weeks was shown to reduce the levels of glycerol and free fatty acids, as well as adiponectin, and increase plasma leptin levels [83]. Hyperhomocysteinemia and fatty liver are usually found simultaneously, however the mechanism is not completely known. A diet rich in fats resulted in increased total cholesterol and doubled homocysteine levels in animals [84]. Cysteine diet in mice induced activity of lipogenic enzymes, lowered metabolic rate, and increased visceral adiposity [85]. Also, total homocysteine level is related to obesity [86]. Studies had reported correlation of plasma SAH and SAM with BMI in female [87, 88]. Another study revealed that plasma SAM level is associated with fat mass and trunk adiposity in older adults. However, SAH values did not show this association [89]. An in vitro study, found that SAH impaired both basal and induced glucose uptake and lipolysis in 3T3-L1 preadipocytes [90]. This research demonstrated that increased intracellular SAH did not alter preadipocyte factor 1 and peroxisome proliferator activated receptor-γ 2. Their results indicated that SAH did not affect adipogenesis per se but altered adipocyte functionality that consequently leads to altered glucose disposal and lipolysis. Hyperhomocysteinemia inhibits lipolysis and results in adipose tissue dysfunction. Previous data indicated that increased homocysteine values due to cholesterol and fat intake may lead to a progression of atherosclerosis. It is suggested that homocysteine induces endoplasmic reticulum stress. This would lead to increased hepatic biosynthesis and uptake of cholesterol and triglycerides [91]. Another mechanism that may explain atherosclerosis and fatty liver development is that hypomethylation associated with hyperhomocysteinemia results in lipid accumulation, decreased synthesis of phosphatidylcholine required for very low-density lipoprotein assembly and homeostasis. Hcy is thought to block enzymes involved in high density lipoprotein cholesterol (HDL-C) synthesis, which in turn leads to decreased HDL-C levels [92].

Control of Hyperhomocysteinemia

Plasma Hcy level is negatively proportional to the folate (vitamin B9) and cobalamin (vitamin B12) [93], so Hcy can serve as an indicator of these vitamins level [94]. The association of low folate level with elevated plasma *S*-adenosylhomocysteine and lymphocyte DNA hypomethylation was firstly reported by Yi and Melnyk in 2000 [95]. It is considered that deficiency of vitamin B6, B9 and B12 and hyperhomocysteinemia may be a risk factor for coronary artery disease [96]. Hyperhomocysteinemia can be genetic or acquired. Mutations in gene for methylenetetrahydrofolate reductase are the main genetic defect that leads to hyperhomocysteinaemia. Deficiency of vitamins B6, folate, chronic kidney disorders, older age and use of antifolate drugs are the main acquired causes for hyperhomocysteinemia [97]. Deficiency of folate and vitamin B12 in patients with DM type 2 was linked to hyperhomocysteinemia [63]. It is demonstrated that folate, vitamin B6 and B12 intake control and reduce the Hcy level. Folate is considered that reduces Hcy level in plasma for 30%. Therefore, an economical and reasonable strategy to control plasma Hcy is possible by intake of these vitamins [48].

Homocysteine Related B Vitamins

Folate (Folic Acid, Vitamin B9)

Vitamin B9 is a carbon donor in the synthesis of amino acids, purines, and pyrimidine bases and a methyl donor in the synthesis of methylcobalamin and methionine [98]. Folate is a natural vitamin B9 found in foods, while folic acid is a synthetic derivative that is added to foods or supplements. The bioavailability of folic acid is greater than the bioavailability of folate [99].

Folate Functions

Folate is necessary for growth [100], DNA synthesis and erythropoiesis [101], and has an important role in methylation process. So, deficiency of folate can lead to damage both in DNA synthesis and in the methylation cycle [102]. This vitamin is essential for many important processes such as cell growth and proliferation; it has the role of co-factor in numerous metabolic reactions [103, 104]. The importance of folate is reflected in one of its major functions that is nucleotide synthesis, DNA production and reparation, and methionine production by homocysteine methylation. The obtained methionine is then converted to *S*-adenosylmethionine or it could be used in synthesis of proteins [105].

Folic acid is considered to have many important positive effects, such as reducing oxidative stress, improving endothelial function, and preventing apoptosis by reducing plasma homocysteine level [106, 107]. It was demonstrated that folic acid increased coronary flow, increased nitrite outflow and decreased superoxide anion production, however it increased lipid peroxidation index in isolated rat hearts. The effects of folic acid were reversed or blocked by N(ω)-nitro-L-arginine methyl ester (L-NAME), indicating involvement of NO in the mechanism of the folic acid effects [108].

Epidemiology of Folate Deficiency

In the USA only 0.1% people have a folate deficiency, thanks to a folic acid fortification program started in 1998 [109]. Similarly, folate deficiency is rare in countries with folic acid fortification program that is established in more than 75 countries [99]. Most developed countries have mandatory folic acid flour fortifications, and most European countries recommend folic acid supplementation before conception and during the first three months of pregnancy [110]. Deficiency of folate is related to low socioeconomic status and inadequate intake of green leafy vegetables, malnutrition, and mental status changes, medication intake, alcoholism and genetic defects [94, 111]. It was indicated that deficiency of vitamin B6 and folate are the major factors of hyperhomocysteinemia [96].

Folate Deficiency

Women of childbearing age and non-black Hispanic women, people with high BMI, celiac diseases and vitamin B12 deficiency may exhibit folate deficiency [103, 111]. In addition, it is known that folate deficiency also disturbs DNA and methylation cycles [112], and it can lead to many disturbances as anemia, neurological abnormalities and birth defects [94]. Also, it is demonstrated that folate deficiency is related to hyperhomocysteinemia [63]. Folic acid involvement in the pathogenesis of DM type 2 is associated with vitamin B12 deficiency and subsequent hyperhomocysteinemia. Although folic acid deficiency is not widespread, folic acid supplementation has been tested in people with DM. A case–control study found that low intakes of folate and vitamin B12 in patients with DM type 2 were followed by hyperhomocysteinemia [113]. Individuals with DM type 2 who received folic acid treatment demonstrated reverted DNA damage expressed as a significant decrease in micronuclei, and reverted oxidative stress comparing to the individuals with DM type 2 without folic acid treatment [114]. Furthermore, folate supplementation improves glycemic control. It reduces glycosylated hemoglobin, fasting blood glucose, serum insulin, insulin resistance and Hcy in DM type 2 patients [115]. An experimental research found that the application of folic acid in a diabetic rats lead to a cellular response characterized by cardiac antioxidative enzymes activity decrease and by a significantly reduced blood glucose level [116]. Another research in diabetic rats demonstrated that folic acid administration leaded to reduction in a liver damage parameters and cardiac matrix metalloproteinase 2 activity, marker of tissue remodeling, so it has potential hepato- and cardioprotective effects [117]. Folic acid, pyridoxine and vitamin B12 have positive effects on signs and symptoms of diabetic retinopathy [118]. Metformin administration can cause folate deficiency. Administration of folic acid for 8 weeks in patients with type 2 DM who received metformin therapy led to an improvement in homocysteine levels, total antioxidant capacity and malondialdehyde concentration [119].

Cobalamin (Vitamin B12)

Vitamin B12 belongs to the cobalamin group. The synthetic form of vitamin B12 is cyanocobalamin (CNCbl). When extracting vitamin B12 from bacterial cultures, a CN group is added. Naturally, there are two forms of vitamin B12, 50-deoxyadenosylcobalamin (coenzyme B12) and methylcobalamin (MeCbl) [120]. Conversion of vitamin B12 into coenzyme B12 occurs in mitochondria; while into MeCbl occurs in cytoplasm. Coenzyme B12 and MeCbl are essential for the metabolism of methylmalonic acid and homocysteine, respectively [45]. Vitamin B6, B9, B12 supplementation is recommended and used to control homocysteine levels and to reduce the risk of cardiovascular disease [121].

Vitamin B12 Function

Vitamin B12 is cofactor of methionine synthase. It is a folate-dependent enzyme that is essential for methionine production from homocysteine. Vitamin B12 is a major coenzyme for the methylation reaction in humans, including DNA and RNA methylation. It participates in the regeneration of methionine from homocysteine in the cytoplasm, as well as in the conversion of methylmalonic acid-coenzyme A (CoA) to succinyl-CoA in mitochondria. These reactions are important for the removal of toxic compounds, such as homocysteine and methylmalonic acid and for lipid metabolism [122]. Vitamin B12 has an important role in methylation and DNA cycle and cell metabolism, thus deficiency in this vitamin may cause an interruption in DNA production and damage of cell metabolism [45]. Its deficiency will lead to cell division and differentiation damage, and an increase in homocysteine levels. The resulting effects are similar to those caused by folate deficiency [100].

Epidemiology of Vitamin B12 Deficiency

Vitamin B12 deficiency is rare in the general population as it is present in most foods of animal origin. Deficiency of this vitamin can often occur in vegans [63]. The prevalence of vitamin B12 deficits in the UK and the United States is 6% for those under 60 years of age. With aging, the prevalence of this deficit increases to 20% in the population over 60 years of age. Vitamin B12 deficiency is far more common in Africa and Asia. Data show that 80% of pre-school children and 70% of adults in India have vitamin B12 deficiency [45]. Also, 65% of newborns have vitamin B12 deficiency and 27% of newborns have folate deficiency in India [123]. Vitamin B6, B12 and folate deficiency and their association with hyperhomocysteinemia has been shown in Pakistan [42].

Vitamin B12 Deficiency

Vitamin B12 deficiency is related to cell metabolism and it is very important for the conversion of the homocysteine to methionine. Vitamin B12 deficiency leads to elevated plasma Hcy. Thus, elevated plasma homocysteine is a sensitive marker of vitamin B12 deficiency [124–126]. Increased Hcy represents a risk factor for hypertension [127], insulin resistance [128], DM type 2 [129], diabetes-related complications [130] and coronary artery disease [131]. It was shown that vitamin B12 deficiency had high prevalence in adults with DM type 2 [63, 132–137]. Nevertheless, there are limited data about vitamin B12 levels and the prevalence of vitamin B12 deficiency in people with pre-diabetes. One recent study demonstrated that the levels of vitamin B12 are negatively correlated with the severity of glucose tolerance. That study confirmed that prevalence of vitamin B12 deficiency is higher in pre-diabetics than in the healthy population [122]. A negative correlation between vitamin B12 and homocysteine levels was shown not only in individuals with DM, but also in those with pre-diabetes and normal glucose tolerance [122]. Maternal vitamin B12 deficiency and high levels of folate before the child delivery may contribute to the onset and progression of DM type 2 and obesity [138, 139]. Studies demonstrated vitamin B12 deficiency in DM type 2 patients treated with metformin [140, 141]. Another research reported the association between deficit of vitamin B12, adiposity and gestational diabetes [142].

Subjects with a family history of diabetes or with risk factors such as obesity, hypertension, and impaired glucose tolerance should be screened for MTHFR C677T mutation. If mutation is confirmed, there is suggestion for vitamin B12, B6 and folic acid supplementation that might help reduce the risk in these individuals [143]. In contrast, systematic analysis of cohort studies has shown that there is limited evidence that cobalamin deficiency can be considered a risk factor for cardiovascular disease or diabetes mortality, therefore supplementation with this vitamin is not necessary [144]. So the use of vitamin B12 to reduce the risk of developing diabetes is still controversial. Alternatively, oxidative stress is one of the mechanisms in the pathogenesis of DM. Vitamin B12 and folic acid deficiency in people with DM is associated with oxidative stress and hyperhomocysteinemia [145]. Therefore, it is possible that vitamin B12 deficiency is a risk factor for developing DM complications. One of the most common complications of DM type 2 is peripheral neuropathy and its development is associated with hyperhomocysteinemia, which is more commonly found in patients with diabetes [146, 147]. Atherosclerosis is another frequent complication associated with DM. It was shown that *in milieu* of increased homocysteine level, arterial stiffness occured more frequently [148].

Prolonged use of metformin was shown to cause malabsorption of cobalamin, thereby increasing the risk of vitamin B12 deficiency [138, 149–155]. In elderly patients, short-term metformin therapy may reduce vitamin B12 levels, too [156]. On the other hand, vitamin B12 deficiency has also been reported in diabetic patients who did not take metformin [157]. Subjects with DM who used metformin showed worse cognitive performance than those without metformin in therapy or those without DM. To improve cognitive performance, the use of vitamin B12 supplements is suggested [158]. A meta-analysis demonstrated that supplementation with lipoic acid and methylcobalamin improved nerve conduction velocity and reduced diabetic neuropathy [159].

Thrombosis may be associated with hyperhomocysteinemia due to vitamin B12 deficiency [160]. Mild deficiency of vitamin B12 can cause fatigue and anemia without neurological features. Moderate deficiency can induce macrocytic anemia with some minor or clear neurological disturbances, such as loss of sensation in the distal parts of the body [47]. Chronic vitamin B12 deficiency leads to bone marrow suppression, neurological damage and increases the risk of cardiomyopathy [45].

Therefore, besides physical exercise, a vitamin B12-rich diet can be useful in lowering plasma homocysteine levels, decrease in insulin resistance, and reducing the risk of developing DM type 2 or cardiovascular disease [122].

Pyridoxine (Vitamin B6)

Vitamin B6 represents a group of three related compounds: pyridoxal, pyridoxine and pyridoxamine, and their corresponding phosphorylated derivatives. The active form of vitamin B6 is pyridoxal-5-phosphate (PLP). PLP is an aminotransferase, which acts as a coenzyme for more than 140 metabolic reactions, such as the interconversion of amino acids, neurotransmitters synthesis, regulation of energy homeostasis, and formation of heme [161]. It is coenzyme for the enzyme glucose-phosphorylase that is essential for the utilization of glycogen in the liver and muscles. In this way, PLP actively participates in glucose metabolism [162].

Recently, it has been demonstrated that pyridoxine (vitamin B6) has antioxidant effects, although it is not classified as an antioxidant [163, 164]. PLP plays important role in H_2S biogenesis [165], as well in the metabolism of proteins, fats, and carbohydrates [163, 166]. It influences polyunsaturated fatty acids synthesis, and in the conditions of its deficiency lipid peroxidation increases and antioxidant defense decreases, so there are suggestions, that pyridoxine deficiency is associated with atherogenesis [167, 168].

It was demonstrated that newly diagnosed DM patients had decreased concentrations of PLP compared with healthy subjects [169]. Other research reported that long-term co-administration of folic acid, pyridoxine and vitamin B12 led to a decrease in Hcy levels; however, it did not reduce the risk of developing DM type 2 in women at high risk of CVD [170].

Vitamin B6 levels are not fully and clearly associated with the development of DM type 2, however, there is evidence that a deficiency of this vitamin stimulates the progression of DM complications. In an experimental study it was shown that vitamin B6 supplementation reduced insulin concentration and insulin resistance without affecting blood glucose level [171]. Also it was demonstrated that the simultaneous administration of vitamins B6 and B1 suppressed DNA glycation in diabetic leukocytes, however, these effects were not obtained when vitamin B6 was administered alone [172]. After 6 months of vitamin B6 supplementation testing, retinal edema was decreased and light sensitivity increased in diabetics with non-proliferative retinopathy [118].

Other B-group Vitamins

Thiamine (Vitamin B1)

Thiamine or vitamin B1 is a coenzyme and plays an important role in the active transfer of aldehyde groups and glycation, and in neuro-transmission. It has been shown that vitamin B1 has the potential to influence the development of diabetic complications [173]. One recent study demonstrated that coadministration of thiamine, α-lipoic acid and carnosine effectively reduced glucose concentration in obese patients with DM type 2 [174]. In another study an increased DNA-glycation in leukocytes from diabetic patients with nephropathy was decreased after a 5-month thiamine and pyridoxine supplement trail [172]. Both DM type 1 and 2 are associated with decreased vitamin B1 levels and increased renal clearance [175]. In a comparative cross-sectional study, healthy subjects were compared with those with DM and microalbuminuria or macroalbuminuria. Thiamine was lower in diabetics, and especially in those with microalbuminuria. A negative correlation between thiamine and lipid profile was demonstrated in patients with DM and microalbuminuria [176]. Various studies show the positive effects of thiamine administration. Thiamine administration for one month reduced glucose and leptin in diabetics compared to controls [177]. Another study reports that there was a significant decrease in urinary albumin excretion in patients with DM and microalbuminuria after three months of vitamin B1 administration [178].

Niacin (Nicotinic Acid, Nicotinamide, Vitamin B3)

Nicotinic acid is a compound of NAD and reduced nicotinamide adenine dinucleotide (NADH), which are necessary for the production of ATP and to provide energy needs at the cellular level [179]. It is reported that niacin increased HDL-C, decreases low density lipoprotein cholesterol (LDL-C) and tryacilglycerides [180]. Due to its beneficial effects on lipid status, vitamin B3 is administered alone or in combination with other lipid-lowering drugs, nevertheless its effect on reducing cardiovascular risk is not completely clear [181]. Niacin supplementation reduced monocyte adhesion to endothelial cells from diabetic patients, suggesting that niacin has many effects apart from lipid parameter influence. These effects could be important in lowering of cardiovascular risk [182]. In one experimental study, the effect of niacin in DM rats was tested [183]. A significant decrease in oxidative stress was observed with a decrease in blood glucose level. Also, there was a recovery of the liver and kidney tissue damage, as well as a decrease in DNA damage. Therefore, it is considered that the use of niacin in DM may reduce the negative effects of this disease. On contrary, there are some negative effects of niacin supplementation. In patients with previous myocardial infarction and normoglycemia or impaired fasting glucose niacin administration increased risk for DM type 2 development [184]. It is demonstrated that both nicotinic acid and nicotinamide induce insulin resistance in human and in rats [185, 186]. However, despite a modest increase in risk of new onset DM type 2, there is a potential cardiovascular benefit of niacin administration [184].

Biotin

Biotin is a cofactor for various carboxylases, such as acetyl CoA, pyruvate, methylcrotonyl CoA, and propionyl CoA carboxylase. Even if mammals lack the ability to produce biotin, its deficiency is rare due to its presence in many animal and plant foods [187]. Biotin stimulates the synthesis of insulin, so it may have positive effects and therapeutic potential in the treatment of DM [188]. There is not much research regarding biotin and DM type 2. Experimental studies demonstrated that biotin and chromium piccolinate supplementation of rats with DM type 2 resulted in antidiabetic effects, preventing insulin resistance in skeletal muscle by increasing glucose transporter protein (GLUT4) expression [113, 189].

Conclusion

Diabetes mellitus is followed by numerous metabolic changes, as well as micro- and macrovascular complications. However, these changes often start earlier, in the pre-diabetic state. Therefore, it is important to monitor biomarkers that indicate the presence of pre-diabetes and to start preventing the development of DM and its complications in time. Homocysteine is a significant endothelial function biomarker which plasma level is increased in both DM and pre-diabetes. In addition to the increase in Hcy, diabetes is also characterized by a decrease in vitamin B-group levels, especially in vitamins B6, B12 and folic acid. Positive effects of administration of almost all B-group vitamins have been observed in people with DM. Administration of vitamins B12, B6 and folic acid will increase the conversion of Hcy to methionine, and thus reduce Hcy induced DM complications. Therefore, in order to prevent or delay the complications of DM, it is advised adequate dietary intake of vitamins. Since, there is scientific evidence that metformin used in DM therapy reduces vitamin B12 and folic acid level, these two vitamins supplementation is recommended in patients receiving metformin in therapy.

Acknowledgements This work was supported by the Ministry of Education, Science and Technological Development of the Republic of Serbia, Grant number 175043.

References

1. Chaiyasut C, Kusirisin W, Lailerd N et al (2011) Effects of phenolic compounds of fermented Thai indigenous plants on oxidative stress in streptozotocin-induced diabetic rats. Evid Based Complement Alternat Med 2011:10
2. Holt RI, Cockram C, Flyvbjerg A, Goldstein BJ (eds) (2011) Textbook of diabetes, vol 3, 4th edn. Wiley, New York
3. Lau DCW (2016) Diabetes in the elderly: a silent global tsunami. Can J Diabetes 40(1):2–3
4. Cavan D, da Rocha Fernandes J, Makaroff L et al (eds) (2015) IDF diabetes atlas, 7th edn. www.diabetesatlas.org.
5. Tabák AG, Herder C, Rathmann W et al (2012) Prediabetes: a high-risk state for diabetes development. The Lancet 379:2279–2290
6. American Diabetes Association (2010) Diagnosis and classification of diabetes mellitus. Diabetes Care 33(Suppl 1):S62–S69
7. Romeo JH, Seftel AD, Madhun ZT, Aron DC (2000) Sexual function in men with diabetes type 2: association with glycemic control. J Urol 163:788–791
8. Cade WT (2008) Diabetes-related microvascular and macrovascular diseases in the physical therapy setting. Phys Ther 88:1322–1335
9. Rahman S, Rahman T, Ismail AA-S, Rashid ARA (2007) Diabetes-associated macrovasculopathy: pathophysiology and pathogenesis. Diabetes Obes Metab 9:767–780
10. Barr ELM, Zimmet PZ, Welborn TA et al (2007) Risk of cardiovascular and all-cause mortality in individuals with diabetes mellitus, impaired fasting glucose, and impaired glucose tolerance. Circulation 116:151–157
11. Knowler WC, Barrett-Conner E, Fowler SE, et al. Diabetes Prevention Program Research Group (2002) Reduction in the incidence of type 2 diabetes with lifestyle intervention or metformin. N Engl J Med 346:393–403
12. Maschirow L, Khalaf K, Al-Aubaidy HA, Jelinek HF (2015) Inflammation, coagulation, endothelial dysfunction and oxidative stress in prediabetes-Biomarkers as a possible tool for early disease detection for rural screening. Clin Biochem 48(9):581–585
13. Bloomgarden ZT (2004) Consequences of diabetes. Diabetes Care 27:1825–1831
14. Pradhan A, Manson JE, Rifai N et al (2001) C-reactive protein, interleukin 6, and risk of developing type 2 diabetes mellitus. JAMA 286:327–334
15. Patel VB, Robbins MA, Topol EJ (2001) C-reactive protein: a "golden marker" for inflammation and coronary artery disease. Cleve Clin J Med 68:521–524
16. Samarghandian S, Azimi-Nezhad M, Farkhondeh T (2017) Catechin treatment ameliorates diabetes and its complications in streptozotocin-induced diabetic rats. Dose-Response 15(1):1559325817691158
17. Palmeira CM, Santos DL, Seiça R et al (2001) Enhanced mitochondrial testicular antioxidant capacity in Goto-Kakizaki diabetic rats: role of coenzyme Q. Am J Physiol Cell Physiol 281(3):C1023–C1028
18. Daniel OO, Adeoye AO, Ojowu J, Olorunsogo OO (2018) Inhibition of liver mitochondrial membrane permeability transition pore opening by quercetin and vitamin E in streptozotocin-induced diabetic rats. Biochem Biophys Res Commun 504(2):460–469
19. Hoffman R (2014) Vascular endothelial dysfunction and nutritional compounds in early type 1 diabetes. Curr Diabetes Rev 10(3):201–207
20. Shanmugam KR, Ramakrishana C, Mallikarjuna K, Reddy KS (2009) The impact of ginger on kidney carbohydrate metabolic profiles in STZ induced diabetic rats. Asian J Exp Sci 23(1):127–134
21. Dincer Y, Akcay T, Alademir Z, Ilkova H (2002) Effect of oxidative stress on glutathione pathway in red blood cells from patients with insulin-dependent diabetes mellitus. Metabolism 51:1360–1362
22. Serdar Z, Aslan K, Dirican M, Sarandol E, Yesilbursa D, Serdar A (2006) Lipid and protein oxidation and antioxidant status in patients with angiographically proven coronary artery disease. Clin Biochem 39:794–803

23. Tracy RP (2003) Thrombin, inflammation, and cardiovascular disease: an epidemiologic perspective. Chest 124:49S-57S
24. Al-Aubaidy HA, Jelinek HF (2011) Oxidative DNA damage: antioxidant response in postprandial hyperglycaemia in type 2 diabetes mellitus. Br J Diabetes Vasc Dis 11:87–91
25. Nwose EU, Jelinek HF, Richards RS, Kerr PG (2007) Changes in the erythrocyte glutathione concentration in the course of diabetes mellitus. Redox Rep 11(3):99–104
26. Al-Aubaidy HA, Jelinek HF (2010) 8-hydroxy-2-deoxy-guanosine identifies oxidative DNA damage in a rural prediabetes cohort. Redox Rep 5(4):155–160
27. Penckofer S, Schwertz D, Florczak K (2002) Oxidative stress and cardiovascular disease in type 2 diabetes: the role of antioxidants and pro-oxidants. J Cardiovasc Nurs 16:68–85
28. Beckman JA (2004) Endothelial dysfunction. In: Stern SPMDM, (ed). Diabetes and cardiovascular disease integrating science and clinical medicine. Lippincott Williams & Wilkins, Philadelphia, pp 239–250
29. Welch GN, Loscalzo J (1998) Homocysteine and atherothrombosis. N Engl J Med 338(15):1042–1050
30. Cho HC (2011) The relationship among homocysteine, bilirubin, and diabetic retinopathy. Diabetes Metab J 35(6):595–601
31. Kumar J, Ingelsson E, Lind L, Fall T (2015) No evidence of a causal relationship between plasma homocysteine and type 2 diabetes: a Mendelian randomization study. Front Cardiovasc Med 2:11
32. Seemungal T, Rios M, Wedzicha JA (2012) Homocysteine is elevated in COPD. In: Ong KC (ed) Chronic obstructive pulmonary disease: current concepts and practice. InTech, Rijeka, pp 21–32
33. Carson NAJ, Neill DW (1962) Metabolic abnormalities detected in a survey of mentally backward individuals in Northern Ireland. Arch Dis Child 37(195):505–513
34. Martí-Carvajal AJ, Sola I, Lathyris D (2017) Homocysteine-lowering interventions for preventing cardiovascular events. Cochrane Database Syst Rev 8:CD006612
35. Finer S, Saravanan P, Hitman G, Yajnik C (2014) The role of the one-carbon cycle in the developmental origins of Type 2 diabetes and obesity. Diabet Med 31(3):263–272
36. Ramachandran L, Negi N, Gupta B (2012) Prevalence of hyperhomocysteinaemia in type 2 diabetes mellitus and its correlation with its complications. J Indian Acad Clin Med 13:277–281
37. Moraba M, Sengwayo D, Motaung S (2013) Association of homocysteinaemia with hyperglycaemia, dyslipidaemia, hypertension and obesity. Cardiovasc J Afr 24(7):265–269
38. Shaikh MK, Devrajani BR, Shaikh A, et al (2012) Plasma homocysteine level in patients with diabetes mellitus. World App Sci J 16(9):1269–1273
39. Lim CP, Loo AV, Khaw KW et al (2012) Plasma, aqueous and vitreous homocysteine levels in proliferative diabetic retinopathy. Br J Ophthalmol 96(5):704–707
40. Brazionis L, Rowley K Sr, Itsiopoulos C et al (2008) Homocysteine and diabetic retinopathy. Diabetes Care 31(1):50–56
41. Simó-Servat O, Simó R, Hernández C (2016) Circulating biomarkers of diabetic retinopathy: an overview based on physiopathology. J Diabetes Res 2016:5263798
42. Iqbal M, Ishaq M, Kazmi K (2005) Role of vitamins B6, B12 and folic acid on hyperhomocysteinemia in a Pakistani population of patients with acute myocardial infarction. Nutr Metab Cardiovas Dis 15(2):100–108
43. Otten JJ, Pitzi Hellwig JJ, Meyers LD (eds) (2006) Dietary reference intakes: the essential guide to nutrient requirements. The National Academies Press, Washington
44. Pfeiffer CM, Sternberg MR, Hamner HC et al (2016) Applying inappropriate cutpoints leads to misinterpretation of folate status in the US population. Am J Clin Nutr 104:1607–1615
45. Hunt A, Harrington D, Robinson S (2014) Vitamin B12 deficiency. BMJ 349:g5226
46. Iqbal MP, Lindblad BS, Mehboobali N et al (2009) Folic acid and vitamin B6 deficiencies related hyperhomocysteinemia in apparently healthy Pakistani adults; is mass micronutrient supplementation indicated in this population? J Coll Physicians Surg Pak 19(5):308–312

47. Mursleen MT, Riaz S (2017) Implication of homocysteine in diabetes and impact of folate and vitamin B12 in diabetic population. Diabetes Metab Syndr 11(Suppl 1):S141–S146
48. Robinson K (ed) (2013) Homocysteine and vascular disease, 1st edn. Springer Science & Business Media
49. Schnabel R, Lackner KJ, Rupprecht HJ et al (2005) Glutathione peroxidase-1 and homocysteine for cardiovascular risk prediction: results from the AtheroGene study. J Am Coll Cardiol 45:1631–1637
50. Faeh D, Chiolero A, Paccaud F (2006) Homocysteine as a risk factor for cardiovascular disease: should we (still) worry about? Swiss Med Wkly 136(47–48):745–756
51. Dumaswala UJ, Zhuo L, Mahajan S et al (2001) Glutathione protects chemokine-scavenging and antioxidative defense functions in human RBCs. Am J Physiol Cell Physiol 280:C867-873
52. Djuric D, Jakovljevic V, Zivkovic VI, Srejovic IM (2018) Homocysteine and homocysteine-related compounds: an overview of the roles in the pathology of the cardiovascular and nervous systems. Can J Physiol Pharmacol 96(10):991–1003
53. Rasić-Marković A, Stanojlović O, Hrncić D et al (2009) The activity of erythrocyte and brain Na^{+}/K^{+} and Mg^{2+}-ATPases in rats subjected to acute homocysteine and homocysteine thiolactone administration. Mol Cell Biochem 327(1–2):39–45
54. McCully KS (1969) Vascular pathology of homocysteinemia: implications for the pathogenesis of arteriosclerosis. Am J Pathol 56(1):111–128
55. Radenković M, Djurić D, Janković R, Prostran M (2014) The analysis of transduction mechanisms associated with an acute action of homocysteine on isolated rat femoral artery. Acta Physiol Hung 101(4):448–460
56. Zivkovic V, Jakovljevic V, Djordjevic D et al (2012) The effects of homocysteine-related compounds on cardiac contractility, coronary flow, and oxidative stress markers in isolated rat heart. Mol Cell Biochem 370(1–2):59–67
57. Uzelac JJ, Stanić M, Krstić D et al (2018) Effects of homocysteine and its related compounds on oxygen consumption of the rat heart tissue homogenate: the role of different gasotransmitters. Mol Cell Biochem 444(1–2):143–148
58. Hoogeveen EK, Kostense PJ, Jakobs C et al (2000) Hyperhomocysteinemia increases risk of death, especially in type 2 diabetes: 5-year follow-up of the Hoorn Study. Circulation 101(13):1506–1511
59. Hoogeveen EK, Kostense PJ, Beks PJ et al (1998) Hyperhomocysteinemia is associated with an increased risk of cardiovascular disease, especially in non-insulin-dependent diabetes mellitus: a population based study. Arterioscler Thromb Vasc Biol 18(1):133–138
60. Schalinske KL (2003) Interrelationship between diabetes and homocysteine metabolism: Hormonal regulation of cystathionine beta-synthase. Nutr Rev 61(4):136–138
61. Feng Y, Shan M-Q, Bo L et al (2015) Association of homocysteine with type 1 diabetes mellitus: a meta-analysis. Int J Clin Exp Med 08(8):12529–12538
62. Huang T, Ren J, Huang J, Li D (2013) Association of homocysteine with type 2 diabetes: a meta-analysis implementing Mendelian randomization approach. BMC Genomics 14:867
63. Valdés-Ramos R, Guadarrama-López AL, Martínez-Carrillo BE, Benítez-Arciniega AD (2015) Vitamins and type 2 diabetes mellitus. Endocr Metab Immune Disord Drug Targets 15(1):54–63
64. Mao S, Xiang W, Huang S, Zhang A (2014) Association between homocysteine status and the risk of nephropathy in type 2 diabetes mellitus. Clin Chim Acta 431:206–210
65. House JD, Brosnan ME, Brosnan JT (1998) Renal uptake and excretion of homocysteine in rats with acute hyperhomocysteinemia. Kidney Int 54:1601–1607
66. T he Bogalusa heart study (2009) Ruan L, ChenW, Srinivasan SR, et al. Plasma homocysteine is adversely associated with glomerular filtration rate in asymptomatic black and white young adults. Eur J Epidemiol 24:315–319
67. Hayden MR, Tyagi SC (2004) Homocysteine and reactive oxygen species in metabolic syndrome, type 2 diabetes mellitus, and atherosclerotopathy: the pleiotropic effects of folate supplementation. Nutr J 3:4–10

68. Bridges CC, Zalups RK (2004) Homocysteine, system b0, + and the renal epithelial transport and toxicity of inorganic mercury. Am J Pathol 165:1385–1394
69. Kumagai H, Katoh S, Hirosawa K et al (2002) Renal tubulointerstitial injury in weanling rats with hyperhomocysteinemia. Kidney Int 62:1219–1228
70. Ingram AJ, Krepinsky JC, James L et al (2004) Activation of mesangial cell MAPK in response to homocysteine. Kidney Int 66:733–745
71. Hwang SY, Woo CW, Au-Yeung KK et al (2008) Homocysteine stimulates monocyte chemoattractant protein-1 expression in the kidney via nuclear factorkappaB activation. Am J Physiol Renal Physiol 294:F236-244
72. Meng S, Ciment S, Jan M et al (2013) Homocysteine induces inflammatory transcriptional signaling in monocytes. Front Biosci (Landmark Ed) 18:685–695
73. Li Y, Zhang H, Jiang C et al (2013) Hyperhomocysteinemia promotes insulin resistance by inducing endoplasmic reticulum stress in adipose tissue. J Biol Chem 288:9583–9592
74. Wada J, Makino H (2013) Inflammation and the pathogenesis of diabetic nephropathy. Clin Sci (Lond) 124:139–152
75. Herrmann W, Schorr H, Obeid R et al (2005) Disturbed homocysteine and methionine cycle intermediates *S*-adenosylhomocysteine and *S*-adenosylmethionine are related to degree of renal insufficiency in type 2 diabetes. Clin Chem 51(5):891–897
76. Becker A, Henry RM, Kostense PJ et al (2003) Plasma homocysteine and S-adenosylmethionine in erythrocytes as determinants of carotid intima-media thickness: different effects in diabetic and non-diabetic individuals. The Hoorn Study. Atherosclerosis 169(2):323–330
77. Xiao Y, Su X, Huang W et al (2015) Role of *S*-adenosylhomocysteine in cardiovascular disease and its potential epigenetic mechanism. Int J Biochem Cell Biol 67:158–166
78. Stehouwer CD, Gall MA, Hougaard P et al (1999) Plasma homocysteine concentration predicts mortality in non-insulin-dependent diabetic patients with and without albuminuria. Kidney Int 55:308–314
79. Yu X, Huang Y, Hu Q, Ma L (2009) Hyperhomocysteinemia stimulates hepatic glucose output and PEPCK expression. Acta Biochim Biophys Sin (Shanghai) 41:1027–1032
80. Elshorbagy AK, Valdivia-Garcia M, Graham IM et al (2012) The association of fasting plasma sulfur-containing compounds with BMI, serum lipids and apolipoproteins. Nutr Metab Cardiovasc Dis 22(12):1031–1038
81. Golbahar J, Aminzadeh MA, Kassab SE, Omrani GR (2007) Hyperhomocysteinemia induces insulin resistance in male Sprague-Dawley rats. Diabetes Res Clin Pract 76(1):1–5
82. Berstad P, Konstantinova SV, Refsum H et al (2007) Dietary fat and plasma total homocysteine concentrations in 2 adult age groups: the Hordaland Homocysteine Study. Am J Clin Nutr 85(6):1598–1605
83. Wang Z, Pini M, Yao T et al (2011) Homocysteine suppresses lipolysis in adipocytes by activating the AMPK pathway. Am J Physiol Endocrinol Metab 301(4):E703-712
84. Wang H, Jiang X, Yang F et al (2003) Hyperhomocysteinemia accelerates atherosclerosis in cystathionine beta-synthase and apolipoprotein E double knock-out mice with and without dietary perturbation. Blood 101(10):3901–3907
85. Elshorbagy AK, Church C, Valdivia-Garcia M et al (2012) Dietary cystine level affects metabolic rate and glycaemic control in adult mice. J Nutr Biochem 23(4):332–340
86. Elshorbagy AK, Nurk E, Gjesdal CG et al (2008) Homocysteine, cysteine, and body composition in the Hordaland Homocysteine Study: does cysteine link amino acid and lipid metabolism? Am J Clin Nutr 88(3):738–746
87. Inoue-Choi M, Nelson HH, Robien K et al (2012) One-carbon metabolism nutrient status and plasma *S*-adenosylmethionine concentrations in middle-aged and older Chinese in Singapore. Int J Mol Epidemiol Gene 3(2):160–173
88. van Driel LM, Eijkemans MJ, de Jonge R et al (2009) Body mass index is an important determinant of methylation biomarkers in women of reproductive ages. J Nutr 139(12):2315–2321

89. Elshorbagy AK, Nijpels G, Valdivia-Garcia M et al (2013) *S*-adenosylmethionine is associated with fat mass and truncal adiposity in older adults. J Nutr 143(12):1982–1988
90. Ngo S, Li X, O'Neill R et al (2014) Elevated *S*-adenosylhomocysteine alters adipocyte functionality with corresponding changes in gene expression and associated epigenetic marks. Diabetes 63(7):2273–2283
91. Werstuck GH, Lentz SR, Dayal S et al (2001) Homocysteine-induced endoplasmic reticulum stress causes dysregulation of the cholesterol and triglyceride biosynthetic pathways. J Clin Invest 107(10):1263–1273
92. Obeid R, Herrmann W (2009) Homocysteine and lipids: *S*-adenosyl methionine as a key intermediate. FEBS Lett 583(8):1215–1225
93. Bonetti F, Brombo G, Magon S, Zuliani G (2015) Cognitive status according to homocysteine and B-group vitamins in elderly adults. J Am Geriatr Soc 63(6):1158–1163
94. Barnabé A, Aléssio ACM, Bittar LF (2015) Folate, vitamin B12 and homocysteine status in the post-folic acid fortification era in different subgroups of the Brazilian population attended to at a public health care center. Nutr J 14:19
95. Yi P, Melnyk S, Pogribna M et al (2000) Increase in plasma homocysteine associated with parallel increases in plasma *S*-adenosylhomocysteine and lymphocyte DNA hypomethylation. J Biol Chem 275(38):29318–29323
96. Iqbal MP (2006) Hyperhomocysteinemia and coronary artery disease in Pakistan. J Pak Med Assoc 56:282–285
97. Nakashima MO (2014) Hypercoagulable states: an algorithmic approach to laboratory testing and update on monitoring of direct oral anticoagulants. Blood Res 49(2):85–94
98. Bayes J, Agrawal N, Schloss J (2019) The bioavailability of various oral forms of folate supplementation in healthy populations and animal models: a systematic review. J Altern Complement Med 25(2):169–180
99. Bailey RL, West KP Jr, Black RE (2015) The epidemiology of global micronutrient deficiencies. Ann Nutr Metab 66(Suppl. 2):22–33
100. Strand TA, Taneja S, Kumar T (2015) Vitamin B-12, folic acid, and growth in 6- to 30-month-old children: a randomized controlled trial. Pediatrics 135(4):e918-926
101. Rittner D, Bailey RA (2005) Encyclopedia of chemistry (Science encyclopedia)
102. Nazki FH, Sameer AS, Ganaie BA (2014) Folate: metabolism, genes, polymorphisms and the associated diseases. Gene 533(1):11–20
103. Riaz S (2014) Evaluation and analysis of human folate levels in Pakistani diabetic population. Int J Sci Eng Res 5(12):1572–1576
104. Boyles AL, Yetley EA, Thayer KA, Coates PM (2016) Safe use of high intakes of folic acid: research challenges and paths forward. Nutr Rev 74(7):469–474
105. Chan YM, Bailey R, O'Connor DL (2013) Folate. Adv Nutr 4(1):123–125
106. Peluso MR (2006) Flavonoids attenuate cardiovascular disease, inhibit phosphodiesterase, and modulate lipid homeostasis in adipose tissue and liver. Exp Biol Med (Maywood) 231(8):1287–1299
107. Hou J, Yan G, Liu B et al (2018) The protective effects of enalapril maleate and folic acid tablets against contrast-induced nephropathy in diabetic rats. Biomed Res Int 2018:4609750
108. Djuric D, Vusanovic A, Jakovljevic V (2007) The effects of folic acid and nitric oxide synthase inhibition on coronary flow and oxidative stress markers in isolated rat heart. Mol Cell Biochem 300:177–183
109. Odewole OA, Williamson RS, Zakai NA (2013) Near-elimination of folate-deficiency anemia by mandatory folic acid fortification in older US adults: reasons for Geographic and Racial Differences in Stroke study 2003–2007. Am J Clin Nutr 98(4):1042–1047
110. Khan KM, Jialal I (2019) Folic acid (folate) deficiency. StatPearls Publishing, Treasure Island (FL)
111. Green R, Miller JW (1999) Folate deficiency beyond megaloblastic anemia: hyperhomocysteinemia and other manifestations of dysfunctional folate status. Semin Hematol 36(1):47–64

112. Laanpere M, Altmäe S, Stavreus-Evers A et al (1999) Folate-mediated one-carbon metabolism and its effect on female fertility and pregnancy viability. Nutr Rev 68(2):99–113
113. Sasaki Y, Sone H, Kamiyama S et al (2012) Administration of biotin prevents the development of insulin resistance in the skeletal muscles of Otsuka Long-Evans Tokushima Fatty rats. Food Funct 3(4):414–419
114. Lazalde-Ramos BP, Zamora-Perez AL, Sosa-MacA-as M et al (2012) DNA and oxidative damages decrease after ingestion of folic acid in patients with type 2 diabetes. Arch Med Res 43(6):476–481
115. Gargari BP, Aghamohammadi V, Aliasgharzadeh A (2011) Effect of folic acid supplementation on biochemical indices in overweight and obese men with type 2 diabetes. Diabetes Res Clin Pract 94(1):33–38
116. Mutavdzin S, Gopcevic K, Stankovic S et al (2019) The effects of folic acid administration on cardiac oxidative stress and cardiovascular biomarkers in diabetic rats. Oxid Med Cell Longev 2019:1342549
117. Mutavdzin S, Gopcevic K, Stankovic S, et al (2019) The effects of folic acid administration on cardiac tissue matrix metalloproteinases activities and hepatorenal biomarkers in diabetic rats. Can J Physiol Pharmacol 97(9):893–901
118. Smolek MK, Notaroberto NF, Jaramillo AG, Pradillo LR (2013) Intervention with vitamins in patients with nonproliferative diabetic retinopathy: a pilot study. Clin Ophthalmol 7:1451–1458
119. Aghamohammadi V, Gargari BP, Aliasgharzadeh A (2011) Effect of folic acid supplementation on homocysteine, serum total antioxidant capacity, and malondialdehyde in patients with type 2 diabetes mellitus. J Am Coll Nutr 30(3):210–215
120. Martens J-H, Barg H, Warren M, Jahn D (2002) Microbial production of vitamin B12. Appl Microbiol Biotechnol 58(3):275–285
121. Thornalley PJ, Rabbani N (2010) Therapy: vitamin B6, B9 and B12 in diabetic nephropathy[mdash]beware. Nat Rev Endocrinol 6(9):477–478
122. Jayashri R, Venkatesan U, Rohan M et al (2018) Prevalence of vitamin B12 deficiency in South Indians with different grades of glucose tolerance. Acta Diabetol 55(12):1283–1293
123. Sukla KK, Nagar R, Raman R (2014) Vitamin-B12 and folate deficiency, major contributing factors for anemia: a population based study. e-SPEN J 9(1): e45–e48
124. Savage DG, Lindenbaum J, Stabler SP, Allen RH (1994) Sensitivity of serum methylmalonic acid and total homocysteine determinations for diagnosing cobalamin and folate deficiencies. Am J Med 96:239–246
125. Klee GG (2000) Cobalamin and folate evaluation: measurement of methylmalonic acid and homocysteinevs vitamin B12 and folate. Clin Chem 46:1277–1283
126. Gilfix BM (2005) Vitamin B12 and homocysteine. CMAJ 173(11):1360
127. van Guldener C, Nanayakkara PW, Stehouwer CD (2003) Homocysteine and blood pressure. Curr Hypertens Rep 5:26–31
128. Platt DE, Hariri E, Salameh P et al (2017) Type II diabetes mellitus and hyperhomocysteinemia: a complex interaction. Diabetol Metab Syndr. 9:19
129. Sudchada P, Saokaew S, Sridetch S et al (2012) Effect of folic acid supplementation on plasma total homocysteine levels and glycemic control in patients with type 2 diabetes: a systematic review and meta-analysis. Diabetes Res Clin Pract 98:151–158
130. Agullo-Ortuno MT, Albaladejo MD, Parra S et al (2002) Plasmatic homocysteine concentration and its relationship with complications associated to diabetes mellitus. Clin Chim Acta 326:105–112
131. Chambers JC, Obeid OA, Refsum H et al (2000) Plasma homocysteine concentrations and risk of coronary heart disease in UK Indian Asian and European men. Lancet 355:523–527
132. Pflipsen M, Oh R, Saguil A et al (2009) The prevalence of vitamin B12 deficiency in patients with type 2 diabetes: a cross-sectional study. J Am Board Fam Med 22:528–534
133. Akabwai GP, Kibirige D, Mugenyi L et al (2016) Vitamin B12 deficiency among adult diabetic patients in Uganda: relation to glycaemic control and haemoglobin concentration. J Diabetes Metab Disord 15:26

134. Shailendra D, Rizwan K, Narayan KMV et al (2018) Prevalence of vitamin B12 deficiency among individuals with type 2 diabetes mellitus in a South Indian rural community. Int J Basics Clin Pharmacol 7(2):309–314
135. Ko SH, Ko SH, Ahn YB et al (2014) Association of vitamin B12 deficiency and metformin use in patients with type 2 diabetes. J Korean Med Sci 29:965–972
136. Hermann L, Nilsson B, Wettre S (2004) Vitamin B12 status of patients treated with metformin: a cross-sectional cohort study. Br J Diabetes Vasc Dis 4(6):401–406
137. Kumthekar A, Gidwani H, Kumthekar A (2012) Metformin associated B12 deficiency. J Assoc Physicians India 60:58–60
138. Krishnaveni G, Hill J, Veena S (2009) Low plasma vitamin B12 in pregnancy is associated with gestational 'diabesity' and later diabetes. Diabetologia 52(11):2350–2358
139. Yajnik C, Deshpande S, Jackson A (2008) Vitamin B12 and folate concentrations during pregnancy and insulin resistance in the offspring: the Pune maternal nutrition study. Diabetologia 51(1):29–38
140. Iftikhar R, Qadir A, Iqbal Z, Usman H (2013) Prevalence of vitamin B12 deficiency in patients of type 2 diabetes mellitus on metformin: a case control study from Pakistan. Pan Afr Med J 16:67
141. Aroda VR, Edelstein SL, Goldberg RB (2016) Long-term metformin use and vitamin B12 deficiency in the diabetes prevention program outcomes study. J Clin Endocrinol Metab 101(4):1754–1761
142. Rush E, Katre P, Yajnik C (2014) Vitamin B12: one carbon metabolism, fetal growth and programming for chronic disease. Eur J Clin Nutr 68(1):2–7
143. Movva S, Alluri RV, Venkatasubramanian S et al (2011) Association of methylene tetrahydrofolate reductase C677T genotype with type 2 diabetes mellitus patients with and without renal complications. Genet Test Mol Biomarkers 15(4):257–261
144. Rafnsson SB, Saravanan P, Bhopal RS, Yajnik CS (2011) Is a low blood level of vitamin B12 a cardiovascular and diabetes risk factor? A systematic review of cohort studies. Eur J Nutr 50(2):97–106
145. Al-Maskari MY, Waly MI, Ali A et al (2012) Folate and vitamin B12 deficiency and hyperhomocysteinemia promote oxidative stress in adult type 2 diabetes. Nutrition 28(7–8):e23–e26
146. Molina M, Gonzalez R, Folgado J et al (2013) Correlation between plasma concentrations of homocysteine and diabetic polyneuropathy evaluated with the Semmes-Weinstein monofilament test in patients with type 2 diabetes mellitus. Med Clin (Barc) 141(9):382–386
147. Jianbo L, Yuche C, Ming S et al (2011) Association of homocysteine with peripheral neuropathy in Chinese patients with type 2 diabetes. Diabetes Res Clin Pract 93(1):38–42
148. Shargorodsky M, Boaz M, Pasternak S et al (2009) Serum homocysteine, folate, vitamin B12 levels and arterial stiffness in diabetic patients: which of them is really important in atherogenesis? Diabetes Metab Res Rev 25(1):70–75
149. Wile DJ, Toth C (2010) Association of metformin, elevated homocysteine, and methylmalonic acid levels and clinically worsened diabetic peripheral neuropathy. Diabetes Care 33(1):156–161
150. de Groot-Kamphuis DM, van Dijk PR, Groenier KH et al (2013) Vitamin B12 deficiency and the lack of its consequences in type 2 diabetes patients using metformin. Neth J Med 71(7):386–390
151. Sato Y, Ouchi K, Funase Y et al (2013) Relationship between metformin use, vitamin B12 deficiency, hyperhomocysteinemia and vascular complications in patients with type 2 diabetes. Endocr J 60(12):1275–1280
152. Sumitani S, Morita S, Utsu Y et al (2012) Effectiveness of metformin and lifestyle interventions as an initial treatment in Japanese patients with newly diagnosed type 2 diabetes: a prospective observational study. J Med Invest 59(1–2):166–173
153. Nervo M, Lubini A, Raimundo FV et al (2011) Vitamin B12 in metformin-treated diabetic patients: a cross-sectional study in Brazil. Rev Assoc Med Bras 57(1):46–49
154. Calvo Romero JM, Ramiro Lozano JM (2012) Vitamin B(12) in type 2 diabetic patients treated with metformin. Endocrinol Nutr 59(8):487–490

155. Kos E, Liszek MJ, Emanuele MA et al (2012) Effect of metformin therapy on vitamin D and vitamin B_{12} levels in patients with type 2 diabetes mellitus. Endocr Pract 18(2):179–184
156. Leung S, Mattman A, Snyder F et al (2010) Metformin induces reductions in plasma cobalamin and haptocorrin bound cobalamin levels in elderly diabetic patients. Clin Biochem 43(9):759–760
157. Jawa AA, Akram J, Sultan M et al (2010) Nutrition-related vitamin B12 deficiency in patients in Pakistan with type 2 diabetes mellitus not taking metformin. Endocr Pract 16(2):205–208
158. Moore EM, Mander AG, Ames D et al (2013) Increased risk of cognitive impairment in patients with diabetes is associated with metformin. Diabetes Care 36(10):2981–2987
159. Xu Q, Pan J, Yu J et al (2013) Meta-analysis of methylcobalamin alone and in combination with lipoic acid in patients with diabetic peripheral neuropathy. Diabetes Res Clin Pract 101(2):99–105
160. Stabler SP (2013) Vitamin B12 deficiency. New Engl J Med 368(2):149–160
161. Stover PJ, Field MS (2015) Vitamin B-6. Adv Nutr 6(1):132–133
162. Hellmann H, Mooney S (2010) Vitamin B_6: a molecule for human health? Molecules 15(1):442–459
163. Matxain JM, Ristilä M, Strid A, Eriksson LA (2006) Theoretical study of the antioxidant properties of pyridoxine. J Phys Chem 110(48):13068–13072
164. Higashi-Okai K, Nagino H, Yamada K, Okai Y (2006) Antioxidant and prooxidant activities of B group vitamins in lipid peroxidation. J UOEH 28(4):359–368
165. Singh S, Banerjee R (2011) PLP-dependent H2S biogenesis. Biochim Biophys Acta 1814(11):1518–1527
166. Taş S, Sarandöl E, Dirican M (2014) Vitamin B6 supplementation improves oxidative stress and enhances serum paraoxonase/arylesterase activities in streptozotocin-induced diabetic rats. Sci World J 2014:351598
167. Matxain JM, Padro D, Ristilä M et al (2009) Evidence of high *OH radical quenching efficiency by vitamin B6. J Phys Chem B 113(29):9629–9632
168. Keles M, Al B, Gumustekin K et al (2010) Antioxidative status and lipid peroxidation in kidney tissue of rats fed with vitamin B(6)-deficient diet. Ren Fail 32(5):618–622
169. Ahn HJ, Min KW, Cho YO (2011) Assessment of vitamin B(6) status in Korean patients with newly diagnosed type 2 diabetes. Nutr Res Pract 5(1):34–39
170. Song Y, Cook NR, Albert CM et al (2009) Effect of homocysteine-lowering treatment with folic acid and B vitamins on risk of type 2 diabetes in women: a randomized, controlled trial. Diabetes 58(8):1921–1928
171. Unoki-Kubota H, Yamagishi S, Takeuchi M et al (2010) Pyridoxamine, an inhibitor of advanced glycation end product (AGE) formation ameliorates insulin resistance in obese, type 2 diabetic mice. Protein Pept Lett 17(9):1177–1181
172. Polizzi FC, Andican G, Çetin E et al (2012) Increased DNA-glycation in type 2 diabetic patients: the effect of thiamine and pyridoxine therapy. Exp Clin Endocrinol Diabetes 120(6):329–334
173. Manzetti S, Zhang J, van der Spoel D (2014) Thiamin function, metabolism, uptake, and transport. Biochemistry 53(5):821–835
174. Karkabounas S, Papadopoulos N, Anastasiadou C et al (2018) Effects of α-lipoic acid, carnosine, and thiamine supplementation in obese patients with type 2 diabetes mellitus: a randomized. Double-Blind Study J Med Food 21(12):1197–1203
175. Al-Attas OS, Al-Daghri NM, Alfadda AA et al (2012) Blood thiamine and its phosphate esters as measured by high-performance liquid chromatography: levels and associations in diabetes mellitus patients with varying degrees of microalbuminuria. J Endocrinol Invest 35(11):951–956
176. Waheed P, Naveed AK, Ahmed T (2013) Thiamine deficiency and its correlation with dyslipidaemia in diabetics with microalbuminuria. J Pak Med Assoc 63(3):340–345
177. Gonzalez-Ortiz M, MartA-nez-Abundis E, Robles-Cervantes JA et al (2011) Effect of thiamine administration on metabolic profile, cytokines and inflammatory markers in drug-naïve patients with type 2 diabetes. Eur J Nutr 50(2):145–149

178. Rabbani N, Alam SS, Riaz S et al (2009) High-dose thiamine therapy for patients with type 2 diabetes and microalbuminuria: a randomised, double-blind placebo-controlled pilot study. Diabetologia 52(2):208–212
179. Maiese K, Chong ZZ, Hou J, Shang YC (2009) The vitamin nicotinamide: translating nutrition into clinical care. Molecules 14(9):3446–3485
180. Julius U, Fischer S (2013) Nicotinic acid as a lipid-modifying drug-a review. Atheroscler 14(Suppl 1):7–13
181. Song WL, FitzGerald GA (2013) Niacin, an old drug with a new twist. J Lipid Res 54(10):2586–2594
182. Tavintharan S, Woon K, Pek LT et al (2011) Niacin results in reduced monocyte adhesion in patients with type 2 diabetes mellitus. Atherosclerosis 215(1):176–179
183. Abdullah KM, Alam MM, Iqbal Z, Naseem I (2018) Therapeutic effect of vitamin B3 on hyperglycemia, oxidative stress and DNA damage in alloxan induced diabetic rat model. Biomed Pharmacother 105:1223–1231
184. Sazonov V, Maccubbin D, Sisk CM, Canner PL (2013) Effects of niacin on the incidence of new onset diabetes and cardiovascular events in patients with normoglycaemia and impaired fasting glucose. Int J Clin Pract 67(4):297–302
185. Zhou SS, Li D, Sun WP et al (2009) Nicotinamide overload may play a role in the development of type 2 diabetes. World J Gastroenterol 15(45):5674–5684
186. Chang AM, Smith MJ, Galecki AT et al (2006) Impaired beta-cell function in human aging: response to nicotinic acid-induced insulin resistance. J Clin Endocrinol Metab 91:3303–3309
187. Tong L (2013) Structure and function of biotin-dependent carboxylases. Cell Mol Life Sci 70(5):863–891
188. Dakshinamurti K (2015) Vitamins and their derivatives in the prevention and treatment of metabolic syndrome diseases (diabetes). Can J Physiol Pharmacol 93(5):355–362
189. Sahin K, Tuzcu M, Orhan C et al (2013) Anti-diabetic activity of chromium picolinate and biotin in rats with type 2 diabetes induced by high-fat diet and streptozotocin. Br J Nutr 110(2):197–205

Chapter 18
Cardiac, Hemodynamic and Electrophysiological Changes in Obesity and the Effects of Bariatric and Metabolic Surgery

Sjaak Pouwels, Elijah E. Sanches, Besir Topal, and Alper Celik

Abstract Obesity is associated with various diseases such as type 2 diabetes, hypertension, obstructive sleep apnoea syndrome (OSAS), metabolic syndrome, and cardiovascular diseases. It affects the function of several organ systems, including the pulmonary and cardiac systems. Furthermore, it induces pulmonary and cardiac changes that can result in right and/or left heart failure. Secondly it is associated with electrophysiological changes that can induce cardiac arrhythmias and electrocardiogram (ECG) abnormalities. In this chapter we will discuss a few components of the complex pathophysiology of obesity on the cardiovascular system; (1) hemodynamic and cardiac structural changes (2) electrophysiological influences of obesity and finally (3) the effects of bariatric and metabolic surgery.

Keywords Obesity · Bariatric surgery · Metabolic surgery · Atrial fibrillation · cardiac function · hemodynamic · QT interval · P-wave dispersion

Introduction

Obesity is a public health endemic according to the World Health Organisation (WHO) and numbers are still increasing [1]. Obesity is defined as a Body Mass Index (BMI) greater than or equal to 30 kg/m^2. Obesity can be subdivided in the following 3 classes; class I obesity 30–35 kg/m^2, class II obesity 35–40 kg/m^2 and class III obesity >40 kg/m^2 [1, 2]. In 2014, 39% of adults aged 18 years and over

S. Pouwels (✉) · E. E. Sanches
Department of Surgery, Haaglanden Medical Center, Lijnbaan 32, P.O. Box 432, The Hague 2501 CK, The Netherlands
e-mail: Sjaakpwls@gmail.com

B. Topal
Department of Cardiothoracic Surgery, Onze Lieve Vrouwe Gasthuis, Amsterdam, The Netherlands

A. Celik
Metabolic Surgery Clinic, Sisli, Istanbul, Turkey

P. S. Tappia et al. (eds.), *Biochemistry of Cardiovascular Dysfunction in Obesity*, Advances in Biochemistry in Health and Disease 20,
https://doi.org/10.1007/978-3-030-47336-5_18

were overweight (BMI > 25) and 13% were obese. Although obesity is preventable, there are 650 million obese individuals worldwide [1].

Obesity is associated with a wide range of co-morbidities such as cardiovascular disease (CVD), type 2 diabetes (T2DM), hypertension and obstructive sleep apnoea syndrome (OSAS) [2, 3]. Obesity also has detrimental physiological effects on cardiac structure, electrophysiological properties and cardiovascular hemodynamics [4–6]. This can result in several cardiac anomalies, which eventually can lead to serious cardiac arrhythmias and even mortality [7, 8]. Patients with obesity, that are weight stable have an increased risk of developing cardiac arrhythmias and even sudden cardiac death [7, 8]. According to the Framing study, the mortality rate due to sudden cardiac death in men and women with obesity was estimated to be about 40 times higher than the mortality rate in a matched lean population [7, 9]. In men with severe obesity a 6 - and 12-fold excess mortality rate was demonstrated [10]. The mechanism of unexplained deaths in obese patients is still unclear and may be related to repolarization abnormalities [5, 7, 10].

Due to cardiac structural changes repolarization abnormalities and cardiac rhythm disorders (like atrial fibrillation (AF)) are often seen on an electrocardiogram (ECG). Several studies have reported QTc prolongation and increased QTc dispersion in patients with left ventricular (LV) hypertrophy, particularly in association with hypertension and obesity [11]. LV hypertrophy occurs commonly in severely obese persons, even in those who are normotensive [11]. Some studies state that there is an improvement in ventricular repolarization in a population with obesity after weight loss [5, 12]. Regarding rhythm disorders, AF is the most common arrhythmia worldwide, with a prevalence of 0.5%, representing nearly 33,5 million individuals on the globe [13]. The AF numbers are probably underestimated, because a large part of the AF population is asymptomatic and another part of the AF population consists of individuals with transient symptoms who remain undiagnosed [13]. Obesity is an independent risk factor for developing AF [5, 14]. AF can have very serious medical consequences, which can lead to strokes, heart failure and an overall increased mortality [5, 15].

In this chapter, we will discuss a few components of the complex pathophysiology of obesity on the cardiovascular system: (1) hemodynamic and cardiac structural changes (2) electrophysiological influences of obesity and finally (3) the effects of bariatric and metabolic surgery on cardiac rhythm disorders and ECG abnormalities.

Hemodynamic and Cardiac Structural Changes

Patients with obesity are prone for developing complications in particular cardiovascular complications. Epidemiological data suggest that obesity is associated with 30% increased risk of developing heart failure [16, 17]. To this extent an increase of 1 point in body mass index increases the risk of heart failure by 5% and 7% in respectively men and women [18]. Furthermore there is a linear relationship between overall body weight and the heart. In more detail, long-term obesity is associated with significant morbidity (e.g. left ventricular hypertrophy and dilatation), which may result in heart failure [18, 19]. In patients with obesity left ventricular dilatation can be present in 8–40% and most of them have an increased mass of the left ventricle [20, 21].

Pathophysiology

As stated earlier, obesity has detrimental physiological effects on several organ systems. Most influences were seen on pulmonary and cardiac function [5, 22]. These induced pulmonary and cardiac changes can result in right and/or left heart failure [18]. Basically, the pathophysiological mechanism is multifactorial. In patients with obesity, significant increments in blood volume are found (this reflects an increased size of the vascular bed) and this paralleled with an increase in cardiac output [18]. As a direct consequence of the earlier mentioned physiological change, the renal and cerebral blood flow remain roughly the same (compared to 'ideal body weight') [18]. According to the results of inert wash out studies, excess body weight incorporates extra blood volume and flow. This is roughly an extra blood flow of 2−3 ml/min/100 g. This means that 100 kg of excess body fat would require as much as 3 L/min blood flows, that implicates an increase in cardiac output [18]. Actually, the blood volume and cardiac output of an individual of 170 kg are roughly twice those of a subject of 70 kg [18]. Clinically this can induce a series of symptoms like dyspnoea, fatigue and chest pains, but physiologically because most of these patients do not a have (significant) tachycardia, there must be some sort of cardiac remodelling [4, 5, 22]. This indicates that patients with obesity will have an augmented left ventricular preload (volume) and often increased afterload (hypertension), with maintenance of a high output state at the expense of elevated right and left ventricular filling pressures [18, 22, 23].

In obesity, the volume overload will lead to complex compensatory mechanisms leading to left ventricular (LV) enlargement and hypertrophy. This is mainly due to an increase in preload and afterload. The degree of change can be assessed with echocardiography measuring the basal interventricular septum (IVST) and posterior wall (PWT). With these variables a mathematical formula can be used to calculate the left ventricular mass (LVM) [24]. Regularly, we can state that an increase in wall thickness (defined by an increase in IVST, PWT) with a normal LVM leads to concentric remodelling. Concentric hypertrophy is present, when there is an increase in wall thickness and an increase in LVM [24]. Left ventricular remodelling is induced by volume overload (either through an absolute increase of blood volume or by the presence of hypertension) [23, 25–27]. Left ventricular dilatation is often present in patients with obesity that will lead to increased diastolic and systolic volumes [4, 22, 28]. Figure 18.1 gives an overview of cardiac structural changes.

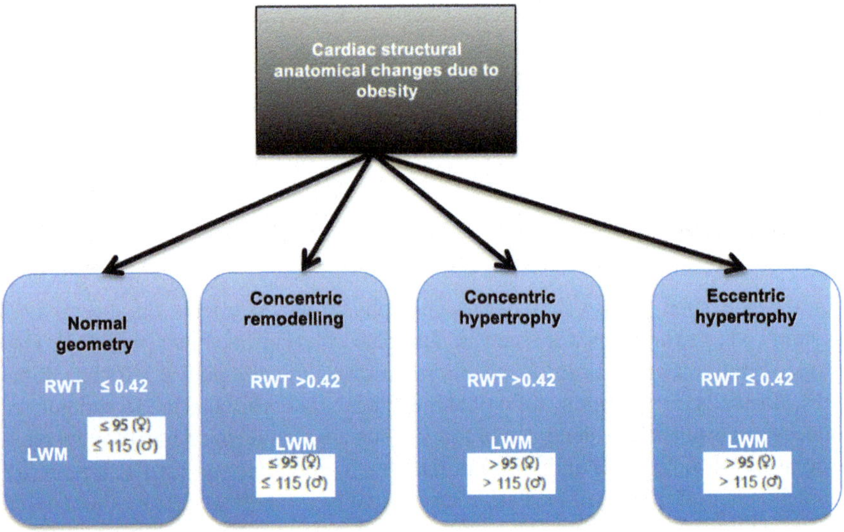

Fig. 18.1 Cardiac structural changes determined by left ventricular mass (LVM) index and relative wall thickness in gm/m^2 (RWT)

With regards to the effects of obesity on right ventricular dysfunction, there is less attention and interest for compared to the function of the left ventricle. Therefore the literature specifically investigating the right ventricular dysfunction is sparse. In the study by Kardassis et al. [29] an increase in right myocardial performance index was found. Also there was a significant difference in isovolumetric relaxation and contraction time in patients with obesity compared to lean subjects. Furthermore, patients with obesity had a significantly reduced RV ejection time compared to lean subjects [29]. Rider et al. [30] found similar results in patients with obesity compared to lean. They measured several parameters with MRI and found a significant greater RV mass in patients with obesity together with RV end-diastolic and end-systolic volumes. Interestingly there was no significant difference in RV ejection fraction between obese and lean patients.

Obstructive sleep apnoea (OSA) and related sleep disordered breathing patterns (like Obesity Hypoventilation syndrome (OHS)) can have a detrimental impact on cardiovascular hemodynamic function. These conditions induce vascular changes, in particular in the pulmonary vascular bed, mainly due to chronic hypoxia and hypercapnia [18]. This results in pulmonary vasoconstriction, which will eventually lead to higher pulmonary blood flow and finally pulmonary hypertension [18, 23]. These pathophysiological changes will lead to a significant trans pulmonary pressure gradient with elevated left ventricular filling pressures [18, 23]. Finally hypertrophy of the right heart will be induced and might lead to right heart failure [26]. Figure 18.2 gives an overview of the cardiovascular and hemodynamic changes caused by obesity.

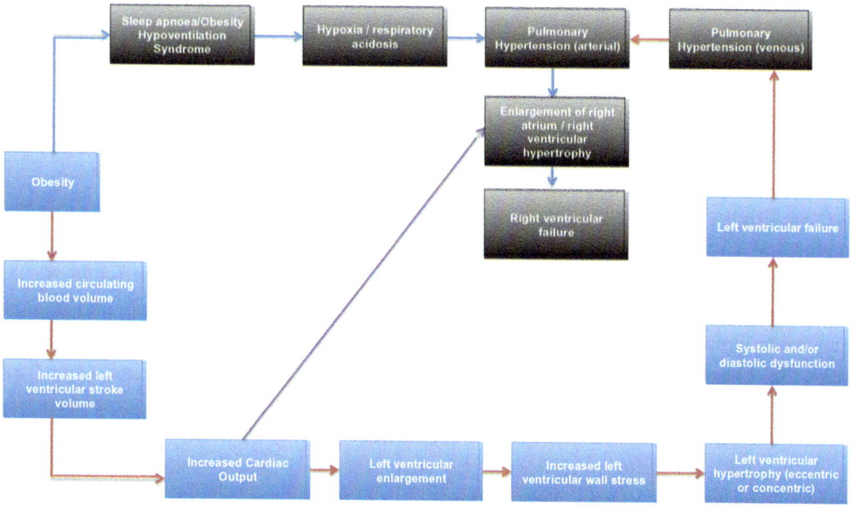

Fig. 18.2 Hemodynamic and structural changes due to obesity

Electrophysiological Influences of Obesity

Patients with obesity have an increased risk of developing ECG abnormalities and heart rhythm disorders of which atrial fibrillation (AF) is the most common [31, 32]. As described earlier in this chapter, patients with AF have an increased risk for developing cardiovascular morbidity and mortality [31, 32]. In the development of AF, obesity plays a significant role and it is estimated nearly half of the incidence of AF is due to obesity [33]. Pathophysiologically this is a complex multifactorial mechanism that includes epicardial adipose tissue biology, ventricular adaptation and well-known obesity related diseases like hypertension, obstructive sleep apnoea syndrome (OSAS), type 2 diabetes mellitus (T2DM) [32, 33].

Pathophysiology

In the pathophysiology of obesity and AF we have to consider two separate pathophysiological entities, namely the pathophysiology of AF in general and the pathophysiology of AF in patients with obesity [5]. Unfortunately, after more than 100 years of research we still not fully understand the pathophysiological mechanisms. One of the landmark studies in AF research is the work of Moe and Abildskov [34]. They formed the hypothesis that AF is basically the result of multiple coexisting electrical wavelets that move through the atria of the heart [34].

Secondly, AF is to be considered as a self-sustainable condition. This means that a minimum number of electrical wavelets must excite a tissue volume. This hypothesis was confirmed by Cox et al. [35] using the surgical MAZE procedure. However with these hypotheses many questions still remain about several aspects of which focal activity, wavebreak formation and re-entry are the most important. The above-mentioned questions and hypotheses have led to the formulation of two main theories that try to explain the complex mechanisms of wave propagation in AF. Several studies indicate that high frequency activation of AF depends on electrical dissociation. This leads to a complex multidirectional conduction between two or more layers of the atrial walls [36]. Secondly there is the 'rotor' theory and 'rotors' are considered as localized re-entrant sources that generate electrical spiral waves [37, 38]. These waves are induced with a high frequency and are moving away from the rotor to interact with tissue irregularities. This results in one or more complex patterns of 'fibrillatory' conduction waves [38].

When looking at the initiation of AF, there seems to be a different kind of mechanism. The study done by Haïssaguerre et al. [39] showed that atrial sleeves in pulmonary veins are responsible for the majority of the ectopic triggers that initiate AF. These findings were confirmed by many follow-up studies and even additional ectopic trigger locations were found (e.g., the superior vena cava) [40]. These pathophysiological studies increased knowledge and led to new treatment methods of which electrical isolation of the pulmonary veins (PV) is the most important one. This treatment method revolutionized the field and became the gold standard.

In pathophysiology of AF in obesity, there seems to be a clearer link between structural changes and electrophysiological changes. This was found in post-mortem studies of patients with obesity that showed enlargement of the left atrium was present in nearly all of them [41–43]. However there are no clear prevalence numbers of left atrial enlargement in patients with obesity. We can hypothesize that left atrial enlargement/remodelling is a multifactorial mechanism that can be induced by factors like duration of obesity, presence of comorbidities and a variety of metabolic effects [44–47]. Left atrial remodelling and the size of the left atrium are considered to be an important factor in AF in patients with obesity. There are multiple studies that show significant differences in left atrial size between patients with obesity and lean subjects [48–50]. Stritzke et al. [51] performed a 10-year longitudinal study that indicated that hypertension and obesity were independent predictors for left atrial enlargement. After adjustment for age and gender, obesity was shown to be an even more potent predictor [51]. These findings were substantiated in a study done by Tang and colleagues [52]. They investigated and followed a cohort of 3,248 patients with paroxysmal AF for 21 years and found that BMI and left atrial volume significantly predicted the development of permanent AF [52]. In studies investigating the correlation between obesity and left atrial enlargement and its relation to AF there might be discrepancies in the results. This is mainly due to the methods used to assess the size of the left atrium. Also the multifactorial mechanisms in obesity and AF development need to be taken into account [53, 54].

Treatment

The treatment of obesity and AF is a process benefitted from a multidisciplinary approach. Capehorn et al. [55] introduced a 4-tier framework to treat overweight and obesity and this framework can also be used as blueprint for the multidisciplinary treatment of patients with AF (and obesity). The first part is often a consultation of the general physician who can give a basic set of lifestyle adjustment and can start treatment of possible obesity related diseases. The second tier is often a more intensified lifestyle management program, very often in a group setting. Tier 3 is and intensified program. These programs incorporate physical exercise, cessation of smoking and alcohol, possible medication adjustments and a structured nutritional program. Tier 4 is basically a referral to a bariatric surgical team due to a complex and severe form of obesity. These patients have a BMI > 40 kg/m^2 or BMI > 35 kg/m^2 with obesity related comorbidities, which mandate surgical treatment [55].

In the treatment of AF, multidisciplinary treatment strategies also play a pivotal role, especially in patients with cardiac arrhythmias and obesity [50, 56]. Several studies indicated that AF might be reversible after multidisciplinary treatment [50]. In the study by Abed and colleagues weight reduction showed to be beneficial in regressing LV hypertrophy, reducing LA size and eventually reversal of AF [50]. This program showed to be effective in patients with obesity. Also in patients with overweight (BMI between 25 and 30 kg/m^2) aggressive cardiovascular risk reduction and weight management leads to a reduction in AF associated parameters (like left atrial volume, LV hypertrophy) and to greater arrhythmia-free survival after catheter ablation [57].

Lifestyle Management

Lifestyle management has been shown to be beneficial in the improvement of metabolic profiles of patients with obesity and other associated diseases [5, 58]. These programs should be focussed on optimizing blood pressure levels, glycaemic control and lipid profile and finally overall physical fitness [59, 60]. Studies by Pouwels indicated that there might be an added benefit of exercise programs in perioperative bariatric practice [58, 61]. Exercise programs can give a significant reduction in cardiovascular risk factors, but will also increase physical fitness. These programs can be easily used in the multidisciplinary treatment of patients with overweight or obesity and AF, not (yet) ready for bariatric surgery [58, 61]. Secondly, several comorbidities like cardiomyopathy and OSAS will improve if adequate weight reduction is achieved [62, 63]. We can postulate that improvement of these conditions might lead to reverse cardiac remodelling, however further studies are need to substantiate this hypothesis.

In multiple studies weight reduction is correlated with an improvement in cardiac function in particular improvements in oxidative stress [64], adipokine profile [65], inflammatory milieu [66, 67] and microvascular fibrosis [68]. Next to these effects, weight reduction can also be beneficial in inducing electrophysiological changes [44, 69]. Gaborit et al. [70] performed a magnetic resonance imaging study and showed a reduction in epicardial fat volume in patients with obesity that lost weight. It can be stated that moderate physical activity and possibly avoiding weight gain might be a preventative strategy for AF. This strategy can also be beneficial in reducing recurrences after the first episode AF [71]. The most important study in this matter is the CARDIO-FIT study by Pathak et al. [71]. In this study 1-MET higher cardiorespiratory fitness at baseline was correlated with a reduction of AF recurrence risk of 13% [71].

Bariatric and Metabolic Surgery

In the era of increasing numbers of patients with morbid obesity and rhythm disorders, there is a place for bariatric and metabolic surgery in its treatment. Several studies showed beneficial effects of bariatric and metabolic surgery on cardiovascular diseases, but the underlying mechanism is not entirely understood [4, 72]. It is postulated that the mechanism might have two components (1) a weight-dependent component (e.g. weight loss, circulating volume) and (2) a weight-independent component (inotropic hormones, like GLP-1) that might induce cardiac remodelling [4, 5].

In the last few years there has been an increasing number of studies investigating the incidence of ECG abnormalities in patients with obesity and its resolution after bariatric and metabolic surgery [11, 12, 44, 69, 73–77]. The majority of these studies show a significant reduction of ECG abnormalities after surgery, either with the QT interval, QT interval dispersion or P-Wave or P-wave dispersion, irrespective of the type of surgical procedure and duration of follow-up [11, 12, 44, 69, 73–77]. Peiser et al. [78] investigated the occurrence of cardiac arrhythmias in patients with OSA and obesity. They showed a drastic decrease in incidence of sinus bradycardia, premature atrial and ventricular complexes, sinus arrhythmias and heart blocks [78]. Despite impressive numbers, no statistical analysis was done on these numbers and therefore it is unknown whether these results were statistically significant.

Regarding AF conflicting results exist after bariatric and metabolic surgery. In several studies AF related morbidity and hospital admissions were investigated [14, 15, 79, 80]. Donnellan et al. [14] AF recurrence was investigated in two groups: (1) group of patients that had bariatric surgery prior to ablation and (2) group patients that only had ablation. They showed that the bariatric surgery group had an AF recurrence rate of 20% compared to 61% in the control group ($p < 0.0001$) [14]. The frequency of repeat ablation was also significantly different in favour of the bariatric surgery group (6 patients (12%) versus 77 patients (41%) in the control group ($p < 0.0001$)) [14]. Jamaly et al. [15] investigated the frequency of new onset AF in the patients that were originally included in the Swedish Obese Subjects (SOS) study. In 19 years of follow-up, 247 patients (12.4%) in the surgery group versus 340 (16.8%) in the non-surgery group developed new onset AF (HR 0.71; 95% CI 0.60–0.83; *p*

< 0.001) [15]. In the study by Shoemaker et al. [79] investigated the prevalence and predictors of AF in a cohort of 1341 patients who underwent bariatric surgery from 1/2008 to 10/2012. In this cohort the prevalence of AF was calculated and 1.9% of the study population developed AF. The patients with AF were significantly older (median 56 versus 46 years $p < 0.001$). More males developed AF ($p = 0.004$) and the patients with AF had a higher rate of comorbidities (DMII $p = 0.02$, HT $p = 0.002$, CAD $p = 0.002$, CHF $p < 0.0001$ and OSA $p < 0.001$) [79]. However, Shimada et al. [80] showed contradicting results. They investigated the AF related Emergency Department (ED) visits and rate of hospitalization due to AF in a population that had bariatric surgery. During the first 12 months after bariatric surgery there was an increased risk of ED visit of hospitalization for AF (22.8%; 95% CI: 19.1−26.4%) with an OR of 1.53 (95% CI: 1.13−2.07; $p = 0.006$). Between 13 and 24 months after bariatric surgery there are still elevated risks (21.2%; 95% CI: 17.7−24.7%), corresponding to an OR of 1.41 (95% CI: 1.03−1.91; $p = 0.03$) [80].

References

1. Ng M, Fleming T, Robinson M, Thomson B, Graetz N, Margono C et al (2014) Global, regional, and national prevalence of overweight and obesity in children and adults during 1980–2013: a systematic analysis for the Global Burden of Disease Study 2013. Lancet (London, England) 384(9945):766–781
2. Buchwald H, Avidor Y, Braunwald E, Jensen MD, Pories W, Fahrbach K, et al (2004) Bariatric surgery: a systematic review and meta-analysis. JAMA 292(14):1724–1737
3. Buchwald H (2014) The evolution of metabolic/bariatric surgery. Obes Surg 24(8):1126–1135
4. Lascaris B, Pouwels S, Houthuizen P, Dekker LR, Nienhuijs SW, Bouwman RA, et al (2018) Cardiac structure and function before and after bariatric surgery: a clinical overview. Clin Obes
5. Pouwels S, Topal B, Knook MT, Celik A, Sundbom M, Ribeiro R, et al (2019) Interaction of obesity and atrial fibrillation: an overview of pathophysiology and clinical management. Exp Rev Cardiovasc Therapy 17(3):209–223
6. Lavie CJ, De Schutter A, Parto P, Jahangir E, Kokkinos P, Ortega FB, et al (2016) Obesity and prevalence of cardiovascular diseases and prognosis-the obesity paradox updated. Prog Cardiovasc Dis 58(5):537–547.
7. Kannel WB, Plehn JF, Cupples LA (1988) Cardiac failure and sudden death in the Framingham Study. Am Heart J 115(4):869–875
8. Messerli FH, Nunez BD, Ventura HO, Snyder DW (1987) Overweight and sudden death. Increased ventricular ectopy in cardiopathy of obesity. Arch Internal Med 147(10):1725–1728.
9. Rabkin SW, Mathewson FA, Hsu PH (1977) Relation of body weight to development of ischemic heart disease in a cohort of young North American men after a 26 year observation period: the Manitoba Study. Am J Cardiol 39(3):452–458
10. Drenick EJ, Bale GS, Seltzer F, Johnson DG (1980) Excessive mortality and causes of death in morbidly obese men. JAMA 243(5):443–445
11. Alpert MA, Nusair MB, Mukerji R, Omran J, Mehra A, Ardhanari S et al (2015) Effect of weight loss on ventricular repolarization in normotensive severely obese patients with and without heart failure. Am J Med Sci 349(1):17–23
12. Mukerji R, Petruc M, Fresen JL, Terry BE, Govindarajan G, Alpert MA (2012) Effect of weight loss after bariatric surgery on left ventricular mass and ventricular repolarization in normotensive morbidly obese patients. Am J Cardiol 110(3):415–419
13. Patel NJ, Atti V, Mitrani RD, Viles-Gonzalez JF, Goldberger JJ (2018) Global rising trends of atrial fibrillation: a major public health concern. Heart 104(24):1989–1990

14. Donnellan E, Wazni OM, Kanj M, Baranowski B, Cremer P, Harb S, et al (2019) Association between pre-ablation bariatric surgery and atrial fibrillation recurrence in morbidly obese patients undergoing atrial fibrillation ablation. Eurospace
15. Jamaly S, Carlsson L, Peltonen M, Jacobson P, Sjostrom L, Karason K (2016) Bariatric surgery and the risk of new-onset atrial fibrillation in swedish obese subjects. J Am Coll Cardiol 68(23):2497–2504
16. He J, Ogden LG, Bazzano LA, Vupputuri S, Loria C, Whelton PK (2001) Risk factors for congestive heart failure in US men and women: NHANES I epidemiologic follow-up study. Arch Intern Med 161(7):996–1002
17. He J, Ogden LG, Bazzano LA, Vupputuri S, Loria C, Whelton PK (2002) Dietary sodium intake and incidence of congestive heart failure in overweight US men and women: first National Health and Nutrition Examination Survey Epidemiologic Follow-up Study. Arch Intern Med 162(14):1619–1624
18. Alexander JK (1985) The cardiomyopathy of obesity. Prog Cardiovasc Dis 27(5):325–334
19. Alpert MA, Lambert CR, Terry BE, Cohen MV, Mulekar M, Massey CV et al (1995) Effect of weight loss on left ventricular diastolic filling in morbid obesity. Am J Cardiol 76(16):1198–1201
20. Alpert MA, Lambert CR, Panayiotou H, Terry BE, Cohen MV, Massey CV et al (1995) Relation of duration of morbid obesity to left ventricular mass, systolic function, and diastolic filling, and effect of weight loss. Am J Cardiol 76(16):1194–1197
21. Alpert MA, Terry BE, Lambert CR, Kelly DL, Panayiotou H, Mukerji V et al (1993) Factors influencing left ventricular systolic function in nonhypertensive morbidly obese patients, and effect of weight loss induced by gastroplasty. Am J Cardiol 71(8):733–737
22. Pouwels S, Lascaris B, Nienhuijs SW, Bouwman AR, Buise MP (2017) Short-term changes in cardiovascular hemodynamics in response to bariatric surgery and weight loss using the Nexfin(R) non-invasive continuous monitoring device: a pilot study. Obes Surg 27(7):1835–1841
23. Alpert MA, Terry BE, Kelly DL (1985) Effect of weight loss on cardiac chamber size, wall thickness and left ventricular function in morbid obesity. Am J Cardiol 55(6):783–786
24. Lang RM (2015) Recommendations for cardiac chamber quantification by echocardiography in adults: an update from the American Society of Echocardiography and the European Association of Cardiovascular Imaging. Eur Heart J Cardiovasc Imaging. 16(2):233–270
25. Karason K, Wallentin I, Larsson B, Sjostrom L (1998) Effects of obesity and weight loss on cardiac function and valvular performance. Obes Res 6(6):422–429
26. Rider OJ, Francis JM, Ali MK, Petersen SE, Robinson M, Robson MD, et al (2009) Beneficial cardiovascular effects of bariatric surgical and dietary weight loss in obesity. J Am Coll Cardiol 54(8):718–726
27. Ikonomidis I, Mazarakis A, Papadopoulos C, Patsouras N, Kalfarentzos F, Lekakis J et al (2007) Weight loss after bariatric surgery improves aortic elastic properties and left ventricular function in individuals with morbid obesity: a 3-year follow-up study. J Hypertens 25(2):439–447
28. Aggarwal R, Harling L, Efthimiou E, Darzi A, Athanasiou T, Ashrafian H (2015) The effects of bariatric surgery on cardiac structure and function: a systematic review of cardiac imaging outcomes. Obes Surg
29. Kardassis D, Bech-Hanssen O, Schonander M, Sjostrom L, Petzold M, Karason K (2012) Impact of body composition, fat distribution and sustained weight loss on cardiac function in obesity. Int J Cardiol 159(2):128–133
30. Rider OJ, Petersen SE, Francis JM, Ali MK, Hudsmith LE, Robinson MR et al (2011) Ventricular hypertrophy and cavity dilatation in relation to body mass index in women with uncomplicated obesity. Heart (British Cardiac Society) 97(3):203–208
31. Rahman F, Kwan GF, Benjamin EJ (2016) Global epidemiology of atrial fibrillation. Nat Rev Cardiol 13(8):501
32. Ball J, Carrington MJ, McMurray JJ, Stewart S (2013) Atrial fibrillation: profile and burden of an evolving epidemic in the 21st century. Int J Cardiol 167(5):1807–1824

33. Wang TJ, Parise H, Levy D, D'Agostino RB Sr, Wolf PA, Vasan RS et al (2004) Obesity and the risk of new-onset atrial fibrillation. JAMA 292(20):2471–2477
34. Moe GK, Abildskov JA (1959) Atrial fibrillation as a self-sustaining arrhythmia independent of focal discharge. Am Heart J 58(1):59–70
35. Cox JL, Boineau JP, Schuessler RB, Ferguson TB Jr, Cain ME, Lindsay BD et al (1991) Successful surgical treatment of atrial fibrillation. Rev Clin Update JAMA 266(14):1976–1980
36. de Groot NM, Houben RP, Smeets JL, Boersma E, Schotten U, Schalij MJ et al (2010) Electropathological substrate of longstanding persistent atrial fibrillation in patients with structural heart disease: epicardial breakthrough. Circulation 122(17):1674–1682
37. Narayan SM, Krummen DE, Enyeart MW, Rappel WJ (2012) Computational mapping identifies localized mechanisms for ablation of atrial fibrillation. PLoS ONE 7(9):e46034
38. Jalife J, Berenfeld O, Mansour M (2002) Mother rotors and fibrillatory conduction: a mechanism of atrial fibrillation. Cardiovasc Res 54(2):204–216
39. Haissaguerre M, Jais P, Shah DC, Takahashi A, Hocini M, Quiniou G et al (1998) Spontaneous initiation of atrial fibrillation by ectopic beats originating in the pulmonary veins. N Engl J Med 339(10):659–666
40. Enriquez A, Liang JJ, Santangeli P, Marchlinski FE, Riley MP. Focal Atrial Fibrillation from the Superior Vena Cava. J Atrial Fibrillation 9(6):1593.
41. Alpert MA, Omran J, Bostick BP (2016) Effects of obesity on cardiovascular hemodynamics, cardiac morphology, and ventricular function. Curr Obes Rep 5(4):424–434
42. Alpert MA, Omran J, Mehra A, Ardhanari S (2014) Impact of obesity and weight loss on cardiac performance and morphology in adults. Prog Cardiovasc Dis 56(4):391–400
43. Alpert MA, Lavie CJ, Agrawal H, Aggarwal KB, Kumar SA (2014) Obesity and heart failure: epidemiology, pathophysiology, clinical manifestations, and management. Transl Res 164(4):345–356
44. Russo V, Ammendola E, De Crescenzo I, Docimo L, Santangelo L, Calabro R (2008) Severe obesity and P-wave dispersion: the effect of surgically induced weight loss. Obes Surg 18(1):90–96
45. Tomita T, Takei M, Saikawa Y, Hanaoka T, Uchikawa S, Tsutsui H et al (2003) Role of autonomic tone in the initiation and termination of paroxysmal atrial fibrillation in patients without structural heart disease. J Cardiovasc Electrophysiol 14(6):559–564
46. Garza CA, Pellikka PA, Somers VK, Sarr MG, Collazo-Clavell ML, Korenfeld Y et al (2010) Structural and functional changes in left and right ventricles after major weight loss following bariatric surgery for morbid obesity. Am J Cardiol 105(4):550–556
47. Garza CA, Pellikka PA, Somers VK, Sarr MG, Seward JB, Collazo-Clavell ML et al (2008) Major weight loss prevents long-term left atrial enlargement in patients with morbid and extreme obesity. Eur J Echocardiogr 9(5):587–593
48. Kalman JM, Nalliah CJ, Sanders P (2016) Surgical weight loss and atrial fibrillation: a convenient paradigm to evaluate a complex problem. J Am Coll Cardiol 68(23):2505–2507
49. Abed HS, Samuel CS, Lau DH, Kelly DJ, Royce SG, Alasady M et al (2013) Obesity results in progressive atrial structural and electrical remodeling: implications for atrial fibrillation. Heart Rhythm 10(1):90–100
50. Abed HS, Wittert GA, Leong DP, Shirazi MG, Bahrami B, Middeldorp ME et al (2013) Effect of weight reduction and cardiometabolic risk factor management on symptom burden and severity in patients with atrial fibrillation: a randomized clinical trial. JAMA 310(19):2050–2060
51. Stritzke J, Markus MR, Duderstadt S, Lieb W, Luchner A, Doring A et al (2009) The aging process of the heart: obesity is the main risk factor for left atrial enlargement during aging the MONICA/KORA (monitoring of trends and determinations in cardiovascular disease/cooperative research in the region of Augsburg) study. J Am Coll Cardiol 54(21):1982–1989
52. Tsang TS, Barnes ME, Miyasaka Y, Cha SS, Bailey KR, Verzosa GC et al (2008) Obesity as a risk factor for the progression of paroxysmal to permanent atrial fibrillation: a longitudinal cohort study of 21 years. Eur Heart J 29(18):2227–2233

53. Lavie CJ, McAuley PA, Church TS, Milani RV, Blair SN (2014) Obesity and cardiovascular diseases: implications regarding fitness, fatness, and severity in the obesity paradox. J Am Coll Cardiol 63(14):1345–1354
54. Lavie CJ, Milani RV, Ventura HO (2009) Obesity and cardiovascular disease: risk factor, paradox, and impact of weight loss. J Am Coll Cardiol 53(21):1925–1932
55. Capehorn MS, Haslam DW, Welbourn R (2016) Obesity treatment in the UK health system. Curr Obes Rep 5(3):320–326
56. Ausma J, van der Velden HM, Lenders MH, van Ankeren EP, Jongsma HJ, Ramaekers FC et al (2003) Reverse structural and gap-junctional remodeling after prolonged atrial fibrillation in the goat. Circulation 107(15):2051–2058
57. Pathak RK, Middeldorp ME, Lau DH, Mehta AB, Mahajan R, Twomey D et al (2014) Aggressive risk factor reduction study for atrial fibrillation and implications for the outcome of ablation: the ARREST-AF cohort study. J Am Coll Cardiol 64(21):2222–2231
58. Pouwels S, Wit M, Teijink JA, Nienhuijs SW (2015) Aspects of exercise before or after bariatric surgery: a systematic review. Obes Facts 8(2):132–146
59. Smith PL, Gold AR, Meyers DA, Haponik EF, Bleecker ER (1985) Weight loss in mildly to moderately obese patients with obstructive sleep apnea. Ann Internal Med 103(6(Pt 1)):850–855.
60. Stevens VJ, Obarzanek E, Cook NR, Lee IM, Appel LJ, Smith West D et al (2001) Long-term weight loss and changes in blood pressure: results of the Trials of Hypertension Prevention, phase II. Ann Intern Med 134(1):1–11
61. Pouwels S, Smelt HJM, Smulders JF (2017) The underestimated effect of perioperative exercise interventions in bariatric surgery: increasing need for large impact studies. Obes Surg 27(10):2690–2691
62. de las Fuentes L, Waggoner AD, Mohammed BS, Stein RI, Miller BV, 3rd, Foster GD, et al (2009) Effect of moderate diet-induced weight loss and weight regain on cardiovascular structure and function. J Am Coll Cardiol 54(25):2376–2381.
63. Tuomilehto H, Seppa J, Uusitupa M, Peltonen M, Martikainen T, Sahlman J et al (2014) The impact of weight reduction in the prevention of the progression of obstructive sleep apnea: an explanatory analysis of a 5-year observational follow-up trial. Sleep Med 15(3):329–335
64. Roberts CK, Vaziri ND, Barnard RJ (2002) Effect of diet and exercise intervention on blood pressure, insulin, oxidative stress, and nitric oxide availability. Circulation 106(20):2530–2532
65. Garcia de la Torre N, Rubio MA, Bordiu E, Cabrerizo L, Aparicio E, Hernandez C, et al (2008) Effects of weight loss after bariatric surgery for morbid obesity on vascular endothelial growth factor-A, adipocytokines, and insulin. J Clin Endocrinol Metab 93(11):4276–4281.
66. Laimer M, Kaser S, Kranebitter M, Sandhofer A, Muhlmann G, Schwelberger H, et al (2005) Effect of pronounced weight loss on the nontraditional cardiovascular risk marker matrix metalloproteinase-9 in middle-aged morbidly obese women. Int J Obes 29(5):498–501
67. Esposito K, Pontillo A, Di Palo C, Giugliano G, Masella M, Marfella R et al (2003) Effect of weight loss and lifestyle changes on vascular inflammatory markers in obese women: a randomized trial. JAMA 289(14):1799–1804
68. De Ciuceis C, Rossini C, Porteri E, La Boria E, Corbellini C, Mittempergher F et al (2013) Circulating endothelial progenitor cells, microvascular density and fibrosis in obesity before and after bariatric surgery. Blood Press 22(3):165–172
69. Russo V, Ammendola E, De Crescenzo I, Ricciardi D, Capuano P, Topatino A et al (2007) Effect of weight loss following bariatric surgery on myocardial dispersion of repolarization in morbidly obese patients. Obes Surg 17(7):857–865
70. Gaborit B, Jacquier A, Kober F, Abdesselam I, Cuisset T, Boullu-Ciocca S et al (2012) Effects of bariatric surgery on cardiac ectopic fat: lesser decrease in epicardial fat compared to visceral fat loss and no change in myocardial triglyceride content. J Am Coll Cardiol 60(15):1381–1389
71. Pathak RK, Elliott A, Middeldorp ME, Meredith M, Mehta AB, Mahajan R et al (2015) Impact of CARDIO respiratory FITness on arrhythmia recurrence in obese individuals with atrial fibrillation: the CARDIO-FIT Study. J Am Coll Cardiol 66(9):985–996

72. Omran J, Firwana B, Al-Dadah A, Alpert M (2015) Effect of weight loss on ventricular repolarization: a meta-analysis of clinical studies. J Am Coll Cardiol 1:A283
73. Al-Salameh A, Allain J, Jacques A, Verhaeghe P, Desailloud R (2014) Shortening of the QT interval is observed soon after sleeve gastrectomy in morbidly obese patients. Obes Surg 24(1):167–170
74. Bezante GP, Scopinaro A, Papadia F, Campostano A, Camerini G, Marinari G et al (2007) Biliopancreatic diversion reduces QT interval and dispersion in severely obese patients. Obesity (Silver Spring, Md) 15(6):1448–1454
75. Papaioannou A, Michaloudis D, Fraidakis O, Petrou A, Chaniotaki F, Kanoupakis E et al (2003) Effects of weight loss on QT interval in morbidly obese patients. Obes Surg 13(6):869–873
76. Pontiroli AE, Pizzocri P, Saibene A, Girola A, Koprivec D, Fragasso G (2004) Left ventricular hypertrophy and QT interval in obesity and in hypertension: effects of weight loss and of normalisation of blood pressure. Int J Obes Related Metabolic Disord 28(9):1118–1123
77. Yilmaz M, Altin C, Tekin A, Erol T, Arer I, Nursal TZ et al (2018) Assessment of atrial fibrillation and ventricular arrhythmia risk after bariatric surgery by P wave/QT interval dispersion. Obes Surg 28(4):932–938
78. Peiser J, Ovnat A, Uwyyed K, Lavie P, Charuzi I. Cardiac arrhythmias during sleep in morbidly obese sleep-apneic patients before and after gastric bypass surgery. Clin Cardiol 8(10):519–521.
79. Shoemaker MB, Gidfar S, Pipilas DC, Tamboli RA, Savio Galimberti E, Williams DB et al (2014) Prevalence and predictors of atrial fibrillation among patients undergoing bariatric surgery. Obes Surg 24(4):611–616
80. Shimada YJ, Tsugawa Y, Camargo CA Jr, Brown DFM, Hasegawa K (2017) Effect of bariatric surgery on emergency department visits and hospitalizations for atrial fibrillation. Am J Cardiol 120(6):947–952

Chapter 19
Estradiol Benzoate Ameliorates Obesity-Induced Renal Dysfunction in Male Rats: Biochemical and Morphological Observations

Amrit Pal Singh, Manjinder Singh, Tajpreet Kaur, Harpal Singh Buttar, Sarvpreet Singh Ghuman, and Devendra Pathak

Abstract Obesity-induced renal dysfunction is a potential risk factor for causing cardiometabolic diseases, but the underlying mechanism remains unclear. Present study was designed to evaluate the protective role of estradiol benzoate in high fat diet (HFD)-induced renal dysfunction in male rats. Six groups of rats (7 animals/group) were randomly assigned to different treatment groups. Adult male and female rats were fed high fat diet (HFD) containing 30% fat for 12 consecutive weeks. One group of male rats simultaneously received daily injections of estradiol benzoate (50 and 100 μg/kg/day, i.p.) over 12 weeks. HFD-induced obesity was assessed by calculating obesity index, adiposity index, and serum lipid profile. Renal function was determined by measuring creatinine clearance, serum urea, uric acid, electrolytes, and microproteinuria. Serum estradiol level and systolic blood pressure (SBP) were measured using standardized techniques. Hydroxyproline content was quantified in the kidneys to estimate collagen deposition. Renal oxidative stress was measured through quantification of thiobarbituric acid reactive substances, superoxide anion generation and reduced glutathione levels. Hematoxylin and Eosin and special Picrosirius red staining of the isolated kidney tissues was done to observe

A. P. Singh (✉) · M. Singh
Department of Pharmaceutical Sciences, Guru Nanak Dev University, Amritsar 143005, India
e-mail: apsingh82@gmail.com

T. Kaur
Department of Pharmacology, Khalsa College of Pharmacy, Amritsar, India

H. S. Buttar
Faculty of Medicine, Department of Pathology & Laboratory Medicine, University of Ottawa, Ottawa, Canada

S. S. Ghuman
Department of Gynaecology and Obstetrics, Guru Angad Dev Veterinary and Animal Sciences University, Ludhiana, India

D. Pathak
Department of Veterinary Anatomy, Guru Angad Dev Veterinary and Animal Sciences University, Ludhiana, India

P. S. Tappia et al. (eds.), *Biochemistry of Cardiovascular Dysfunction in Obesity*,
Advances in Biochemistry in Health and Disease 20,
https://doi.org/10.1007/978-3-030-47336-5_19

changes in gross morphology, glomerular volume and collagen content, respectively. As opposed to the control group, HFD-fed rats demonstrated significant increase in obesity and adiposity indices, lipid profile, SBP and renal dysfunction along with increased hydroxyproline content and oxidative stress in the renal tissues. HFD also caused marked increase in SBP in both sexes. Biochemical and histological studies revealed that the males exposed to HFD were more susceptible to renal dysfunction than females. However, estradiol benzoate administration to male rats showed protection against HFD-induced renal dysfunction accompanied by significant reductions of SBP, renal oxidative stress and fibrosis. It is concluded that estradiol protects against HFD-induced hypertension and renal dysfunction in male rats. To the best of our knowledge, we are the first ones to report the renal protective action of estradiol in male rats exposed to chronically fed HFD. While the findings of this study cannot be directly extrapolated to humans, nevertheless, the renoprotective effects of estradiol warrant verification in obese men suffering from acute renal malfunction.

Keywords Estradiol benzoate · High fat diet · Renal dysfunction · Hypertension · Oxidative stress · Obesity-induced nephropathy · Renoprotection by estradiol

Introduction

According to the WHO estimates, global prevalence of obesity has doubled in the last 25 years [1]. Obesity is a rapidly growing public health problem both in developed and developing countries affecting around 600 million people that constitutes nearly 13% of adult population worldwide [1]. One-third of adults and one-sixth of young obese children live in America alone [2]. There is an escalating trend of obesity among adult men and women in south Asian countries [3]. In India, obesity has become an important public health issue affecting around 25–44% of adult urban population [4]. Obesity-related cardio-metabolic disorders, including nephropathy and poor quality of life, and employee absenteeism put high economic burden on the healthcare costs worldwide.

While the total amount of white adipose tissue (WAT) in lean adult men or women consists of about 20%, the quantity of WAT can increase >40% in obese humans. WAT secretes a wide range of adipokines, inflmmatory cytokines and interleukins (IL6, IL8). Adipokines regulate appetite, insulin sensitivity, angiogenesis, blood pressure, and immune response. Obesity-induced up-regulation of inflammatory cytokines is linked with pathological conditions such as atherosclerosis, hypertension, hyperlipidemia, heart attack, stroke, and various types of cancers.

The unhealthy dietary habits, increased intake of calorie-rich foods, high consumption of sugar and sugar loaded drinks, and sedentary lifestyles are considered the main reasons for causing obesity. The unmanaged obesity leads to various chronic disorders such as type II diabetes, hyperlipidemia, cardiometabolic diseases, stroke, fatty liver, breathing problems, neurodegenerative disorder and some cancers

[5]. Obesity is also one of the well-recognized risk factor for causing renal complications [6, 7]. Obese men and women not only suffer from an enhanced incidence of oxidative stress, but also the large amount of adipose tissues in the body produce high levels of pro-inflammatory cytokines like tumor necrosis factor-α (TNF-α), interleukin-6 (IL-6), adipokines, leptin, resistin, and a reduction in adiponectin that provokes renal damage [8–10]. It has been reported that obesity-induced hyperlipidemia, hyperglycemia, and hypertension cause pathological changes in the renal vascular endothelial cells, and produce glomerular interstitial cell proliferation and nephropathy [10]. As mentioned earlier, the white adipocytes secrete a wide array of bioactive substances that cause activation of renin-angiotensin-aldosterone system, enhance sympathetic activity, and produce hemodynamic alterations which invoke obesity-related cardiovascular and renal complications [6]. Some biochemical factors considered to be involved in obesity-induced nephropathy and renoprotection by estradiol are displayed in Fig. 19.5.

In women, ovarian hormones are essential for reproduction and bone development. Estrogens strongly influence adipocyte differentiation and body fat distribution. Also, estrogen hormones play an important role in maintaining body homeostasis and kidney function and renal hemodynamics. Postmenopausal women are more susceptible to body fat accumulation, development of metabolic syndrome, and onset of cardiovascular diseases. Obesity-induced lesser nitric oxide production can influence the local renin-angiotensin system, release of growth factors, cytokines as well as proliferation of mesangial cells [11]. Two distinct types of estrogen receptors ER-α and ER-β have been identified in various parts of the nephron, suggesting the role of estrogen in regulation of renal function. ER-α is present on the podocytes, whereas ER-β is expressed on the ascending loop and distal convolute tubules of the nephron. Clinically speaking, men are more prone than women to renal complications arising from multifactorial etiologies. It appears that in women, estrogens play an important function in combating renal damage or kidney dysfunction [12–14]. Estrogen deficiency in menopausal women and in experimental animals (e.g. mice) has been noted to induce obesity and skeletal abnormalities [15, 16]. However, information regarding the renoprotective effects of estrogen in obesity-induced renal dysfunction in obese men and women is lacking. Animal studies have indicated that estrogen treatment protects against surgical- and drug-induced renal injuries, which are attributed mainly to the anti-inflammatory and anti-oxidant properties of estrogens [14, 17, 18]. The renoprotective actions of estrogen in high fat diet (HFD)-induced renal pathology has never been explored. Therefore, the present study was designed to investigate the renoprotective role of estradiol benzoate against HFD-induced renal dysfunction in male and female rats.

Materials and Methods

Adult Wistar albino rats of both sexes weighing 225–250 g were employed in the present study. Animals were kept under standard animal husbandry conditions and were maintained on group specific diet and water ad libitum. Normal diet contained fat (5%), carbohydrates (85%) and proteins (10%) as macronutrients, whereas the HFD consisted of fat (30%), carbohydrates (60%) and proteins (10%) as macronutrients.

Drugs and Chemicals

Estradiol benzoate was procured from Macmillon Pharmaceuticals Ltd., India. Picrosirius red was purchased from Sigma Aldrich, India. All other reagents used were of analytical grade.

Experimental Protocol

Six groups of rats, each containing 7 animals, were randomly assigned to various study groups. Group 1 and 2 (Control males and females) were fed on normal diet. Group 3 and 4 (HFD males and females) were given HFD for 12 consecutive weeks. Group 5 and 6 (HFD males + estradiol) were treated with estradiol benzoate (50 and 100 μg/kg/day, i.p.) and fed on HFD for 12 weeks.

At the end of 12th week, the rat urine was collected by placing them in metabolic cages. Then, the rats were removed from cages, weighed and their naso-anal length was recorded. Animals were anaesthetized with ketamine, the blood samples were collected using retro-orbital puncture and were sacrificed by cervical dislocation. The perirenal fat, retroperitoneal fat and epididymal fat was isolated and weighed. Serum separated from coagulated blood and urine samples were used for various biochemical estimations. Immediately after the animal's sacrifice, the kidneys were removed, weighed and washed with 1.17% potassium chloride (KCl) solution. Small portions from each kidney were used for estimation of superoxide anion generation (SAG) and hydroxyproline content, and the rest of tissue was minced and homogenized in 1.17% KCl solution (10% w/v) using teflon homogenizer. The contents were centrifuged at 800 $\times$ *g* for 20 min to remove cellular debris and then re-centrifuged at 11,000 $\times$ *g* for 20 min. The clear supernatant was used to estimate lipid peroxides and reduced glutathione (GSH) levels. A small part of kidneys was preserved in 10% neutral buffered formalin (NBF) for histological studies.

Morphological Assessment of Obesity

Obesity index and adiposity index was calculated. Obesity index was calculated as cube root of body weight (g)/nasoanal length (mm) $\times$ 10^4. Adiposity index = (perirenal fat + retroperitoneal fat + epididymal fat/body weight) $\times$ 100.

Estimation of Lipid Profile and Glucose Level

Serum cholesterol, triglyceride, high density lipoprotein (HDL) and glucose levels were estimated in serum samples by using commercially available kit by Delta Lab and Transasia Biomedicals Ltd, India. Results were expressed as milligram per decilitre of serum.

Estimation of Renal Parameters

Estimation of creatinine in serum and urine samples was done by using commercially available kit by Avecon Healthcare Pvt. Ltd., India. Creatinine clearance (CrCl) was calculated as: CrCl = [Urine creatinine $\times$ urine volume/serum creatinine $\times$ 24 $\times$ 60 $\times$ animal wt] and the results were expressed as millilitre per minute per kilogram of rat body weight. Urea and uric acid levels were estimated in serum by using commercially available kit by Span Diagnostics Ltd. and Precision Biomed Pvt Ltd. India, respectively and the values were expressed as milligram per decilitre of serum. Potassium and sodium level in serum and urinary microproteins and sodium was assayed using commercially available kits by Crest Biosystems, India. Potassium level was expressed as millimoles per litre of serum.

The Fe_{Na} was calculated by using formula [Fe_{Na} = urine sodium $\times$ serum creatinine $\times$ 100/serum sodium $\times$ urine creatinine]. Results of Fe_{Na} were expressed as percentage change in the values. The microproteinuria was expressed as milligram per day excretion.

Estimation of Serum Estradiol

Serum estradiol was estimated by using commercially available kit by DIA-source Immunoassays, Belgium. The results were expressed as nanogram estradiol/ml of serum.

Assessment of Systolic Blood Pressure (SBP)

The SBP in rats was measured at the start and end of experimental period using non-invasive blood pressure measuring apparatus (Kent Scientific, USA).

Estimation of Renal Hydroxyproline Content and Oxidative Stress in Renal Tissues

Renal hydroxyproline content was estimated using well established method [19]. The hydroxyproline concentration was expressed as milligram per gram of tissue. Quantification of lipid peroxides measured in terms of thiobarbituric acid reactive substances (TBARS), SAG and GSH was done by methods described elsewhere [19].

Histological Studies

Kidneys sections preserved in 10% NBF were dehydrated in graded concentrations of ethanol, immersed in xylene and then embedded in paraffin. The 5 μm sections were cut and stained with haematoxylin–eosin to observe gross histological changes and to measure glomerular volume as an index of glomerular hypertrophy as per standardized procedure [20]. The picrosirius red staining was done to witness fibrosis in renal tissues.

Statistical Analysis

Data obtained from various groups were statistically analyzed using one way analysis of variance followed by Tukey–Kramer post hoc test. The $p < 0.05$ was considered to be statistically significant. Results were expressed as mean ± standard error of mean.

Results

No significant differences were observed in the serum and tissue parameters between control female and male rats. Therefore, the data of control male rats were used for statistical comparison with the treated group.

Effect of Estradiol Treatment on Morphological Parameters

The HFD exposure produced significant increases in obesity and adiposity indices in male and female rats. However, the HFD-induced increase in both obesity parameters was markedly lesser in females than that of male rats. Estradiol treatment profoundly attenuated HFD-induced escalation in obesity and adiposity index in a dose dependent manner in male rats (Table 19.1).

Effect of Estradiol Treatment on Lipid Profile and Glucose Level

Both males and female rats fed on HFD demonstrated significant changes in lipid profile showing higher levels of cholesterol and triglycerides accompanied by reduction in HDL level. Results displayed in Table 19.1 show that HFD exposure caused more profound changes in lipid profile of males than females, and this effect was markedly attenuated with estradiol treatment in male rats. Similarly, gender related trend was witnessed in the glucose concentration as observed in lipid profile (Table 19.1).

Effect of Estradiol Treatment on Renal Parameters

As compared to control group, rats fed on HFD for 12 weeks depicted significant reduction in CrCl along with increase in serum urea, uric acid, potassium, Fe_{Na} and microproteinuria (Fig. 19.1). The HFD-induced changes in renal parameters were more pronounced in males than females. Interestingly, the estradiol treatment abolished HFD-induced renal damage in a dose dependent manner in male rats.

Serum Concentration of Estrogen

As was expected, the control females showed significantly higher levels of estrogen in comparison to control males. HFD exposure did not alter the serum estrogen levels in neither males nor females. However, estradiol injection in male rats caused a significant increase in serum estradiol as compared to their control counterparts (Fig. 19.2).

Effect of Estradiol Treatment on Systolic Blood Pressure (SBP)

Figure 19.2 shows that SBP of both male and female rats was significantly increased after 12 weeks of HFD exposure. No statistically significant difference was observed in the SBP of HFD fed males and females. Estradiol treatment significantly attenuated SBP in HFD fed males in a dose dependent manner (Fig. 19.2).

Table 19.1 Effects of estradiol benzoate treatment on the morphological parameters and lipid profile in rats

Groups → Parameters ↓	Control male	Control female	HFD male	HFD female	HFD + Estradio benzoate (50 μg/kg/day)	HFD + Estradiol benzoate (100 μg/kg/day)
Obesity index	190.41 ± 6.27	204.12 ± 4.21	305.21 ± 5.34[a]	271.23 ± 4.73[a,b]	261.47 ± 2.58[a,b]	229.52 ± 3.49[a,b,c,d]
Adiposity index	0.97 ± 0.06	0.91 ± 0.15	3.69 ± 0.19[a]	3.04 ± 0.17[a,b]	2.70 ± 0.16[a,b]	1.75 ± 0.12[a,b,c,d]
Total cholesterol (mg/dL)	33.46 ± 0.95	35.12 ± 0.84	56.74 ± 0.39[a]	52.01 ± 0.94[a,b]	48.97 ± 1.21[a,b]	41.01 ± 0.43[a,b,c,d]
Triglycerides (mg/dL)	41.70 ± 3.21	37.10 ± 2.89	190.12 ± 4.22[a]	162.32 ± 8.23[a,b]	145.21 ± 7.14[a,b]	91.14 ± 5.12[a,b,c,d]
HDL (mg/dL)	22.40 ± 0.45	20.54 ± 0.59	7.12 ± 0.24[a]	8.71 ± 0.69[a,b]	9.12 ± 0.42[a,b]	14.52 ± 0.41[a,b,c,d]
Glucose (mg/dL)	89.65 ± 3.56	83.56 ± 4.35	165.32 ± 8.36[a]	137 ± 6.52[a,b]	128 ± 5.63[a,b]	112 ± 4.65[a,b,c,d]

Values represent mean ± S.E.M from seven rats in each group. a = $p < 0.05$ versus control male; b $p < 0.05$ versus HFD male; c = $p < 0.05$ versus HFD female; d = $p < 0.05$ versus HFD + estradiol benzoate (50 μg/kg/day)

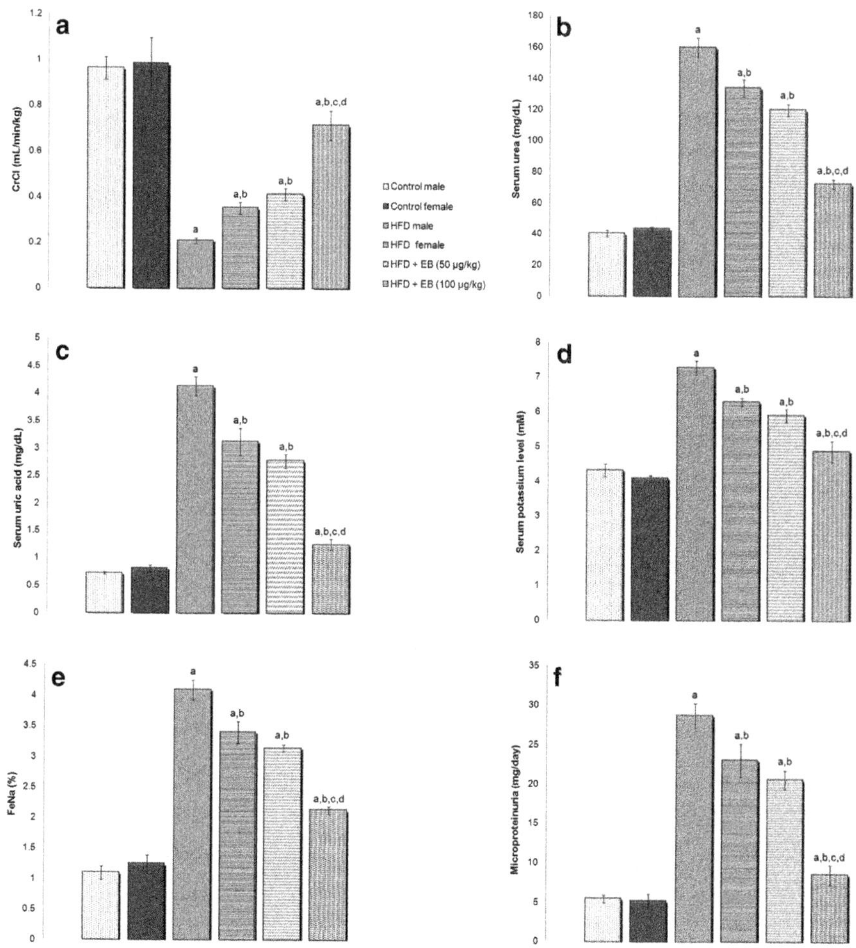

Fig. 19.1 Effects of estradiol benzoate treatment on renal parameters in male and female rats. Values represent mean ± S.E.M. a = $p < 0.05$ versus control male; b = $p < 0.05$ versus HFD male; c = $p < 0.05$ versus HFD female; d = $p < 0.05$ versus HFD + estradiol benzoate male (50 μg/kg/day)

Effect of Estradiol on Renal Hydroxyproline and Oxidative Stress Parameters

Renal hydroxyproline level was markedly increased in HFD fed rats as compared to control group. Obese males had more hydroxyproline content in their kidneys than obese females. Marked reduction in renal tissue hydroxyproline was noted after the concomitant administration of estradiol in HFD fed animals (Fig. 19.2).

When compared with the control group, the obese rat kidneys revealed significant oxidative stress as depicted by increase in TBARS and SAG accompanied by reduction in GSH content. As observed in other parameters, the renal oxidative stress in

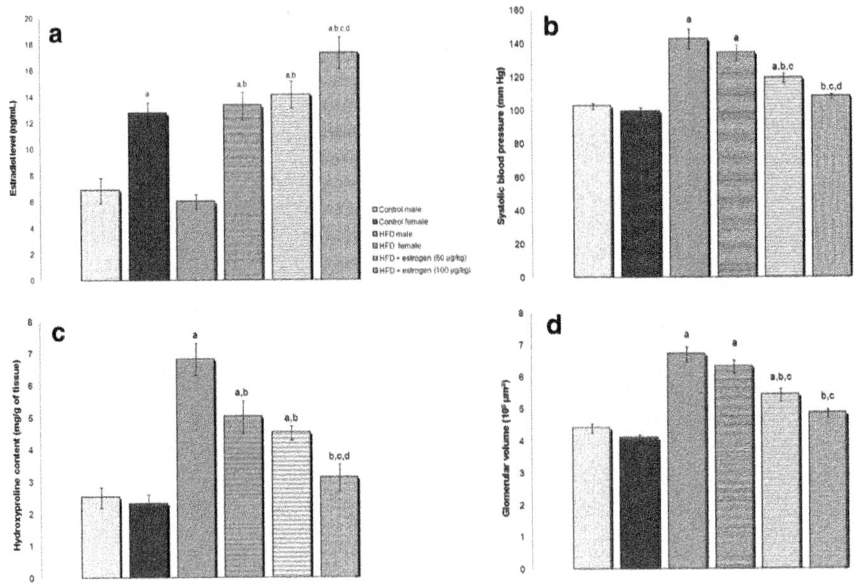

Fig. 19.2 Effects of estradiol benzoate treatment on serum estradiol level, SBP, renal hydroxyproline content and glomerular volume in male and female rats. Values represent mean ± S.E.M. a = $p < 0.05$ versus control male; b = $p < 0.05$ versus HFD male; c = $p < 0.05$ versus HFD female; d = $p < 0.05$ versus HFD + estradiol benzoate male (50 μg/kg/day)

obese females was markedly lesser than obese males. Simultaneous treatment with estradiol benzoate significantly attenuated HFD-induced oxidative stress in male rats (Table 19.2).

Histological Examination of Renal Tissues Collected from Various Groups

Hematoxylin and eosin stained renal tissues from control rats showed normal integrity of glomerulus surrounded by Bowman's capsule and convoluted tubules. On the other hand, renal tissues collected from the HFD fed rats showed histological changes such as detachment of basement membrane from glomeruli, neutrophil accumulation and increase in tubulointerstitial space along with tubular dilation. Profound glomerular hypertrophy measured in terms of glomerular volume was observed in HFD fed males and females as compared to their control counterparts. Estrogen treatment attenuated HFD-induced increase in glomerular volume and other histological changes in rat kidneys (Figs. 19.2 and 19.3). Renal tissues stained with picrosirius red demonstrated marked deposition of collagen at glomeruli and convoluted tubules that was ameliorated with estradiol treatment in male rats (Fig. 19.4).

Table 19.2 Effects of estradiol benzoate treatment on renal oxidative stress parameters in rats

Groups → Parameters ↓	Control male	Control female	HFD male	HFD female	HFD + Estradiol benzoate (50 μg/kg/day)	HFD + Estradiol benzoate (100 μg/kg/day)
TBARS (nM/mg of protein)	0.29 ± 0.02	0.31 ± 0.01	1.04 ± 0.06^{a}	$0.87 \pm 0.04^{a,b}$	$0.78 \pm 0.03^{a,b}$	$0.44 \pm 0.02^{a,b,c,d}$
SAG (pm/min/mg of tissue)	6.23 ± 0.51	5.80 ± 0.56	23.94 ± 1.65^{a}	$19.35 \pm 1.89^{a,b}$	$17.64 \pm 0.92^{a,b}$	$9.32 \pm 0.47^{a,b,c,d}$
GSH (μM/mg of protein)	7.35 ± 0.14	6.87 ± 0.19	2.86 ± 0.13^{a}	$4.12 \pm 0.17^{a,b}$	$4.47 \pm 0.25^{a,b}$	$6.12 \pm 0.34^{a,b,c,d}$

Values represent mean ± S.E.M. from seven animals in each group. a = $p < 0.05$ versus control male; b = $p < 0.05$ versus HFD male; c = $p < 0.05$ versus HFD female; d = $p < 0.05$ versus HFD + estradiol benzoate (50 μg/kg/day)

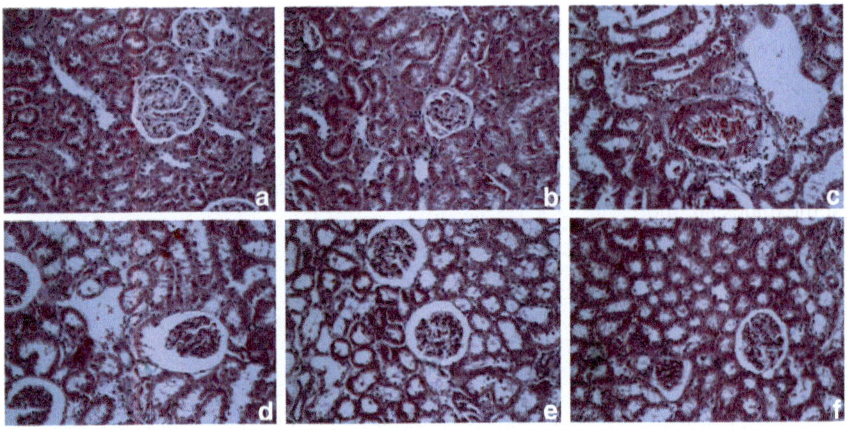

Fig. 19.3 Histopathology of H & E stained renal sections at 200× magnification. **a** control male; **b** control female; **c** HFD male; **d** HFD female; **e** HFD + estradiol benzoate male (50 μg/kg/day); **f** HFD + estradiol benzoate male (100 μg/kg/day)

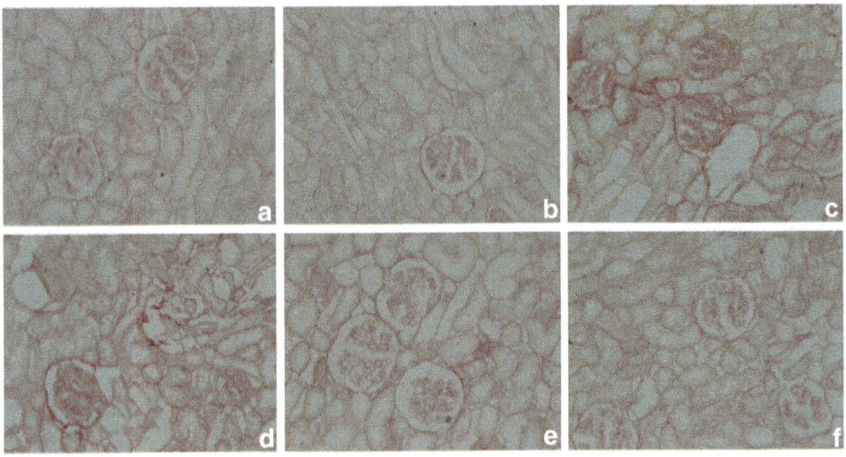

Fig. 19.4 Histopathology of picrosirius red stained renal sections at 200× magnification. **a** control male; **b** control female; **c** HFD male; **d** HFD female; **e** HFD + estradiol benzoate (50 μg/kg/day); **f** HFD + estradiol benzoate male (100 μg/kg/day)

Discussion

Obesity is considered a major risk factor for cardiovascular, renal, and cardiometabolic disorders. Renal injury has been reported in rats given fat rich diet for 12 weeks [21]. Our results corroborate the findings of previous investigators, since rats given 30% HFD for 12 weeks showed marked renal dysfunction and fibrosis with significant increase in lipid parameters, SBP and kidney oxidative stress both in male and female rats. However, the renal toxicity was more pronounced in males than that of females. Exogenous administration of estradiol to males demonstrated protection against HFD-induced renal injury. To the best of our knowledge, we are the first ones to report the renal protective action of estradiol in male rats exposed to chronically fed high fat diet.

The gender related results of various preclinical and clinical studies have shown that the males are more vulnerable than females towards renal complications arising from various etiologies [12, 13]. The estrogen is considered to have anti-oxidant and anti-inflammatory effect. Estradiol's protective effect on mesangial cells and in glomerulosclerosis was noted by decreased growth factors in rat remnant kidney [17]. Estradiol also protected ischemia reperfusion induced kidney injury in rats by curtailing overproduction of endothelin-1 [22].

The present study showed greater obesity index, adiposity index and lipid profile in obese males than obese female rats (Table 19.2). Body fat is primarily stored in two type of adipose tissues, namely white adipose tissues (WAT) and brown adipose tissue (BAT). The WAT serves as a depot for release of free fatty acids and adipokines consisting of leptin and resistin that cause suppression of appetite and reduce insulin sensitivity, respectively. The BAT has large number of mitochondria containing uncoupling protein-1 (UCP-1), and is involved in body heat production [15]. The BAT also activates peroxisome proliferator activated receptor-γ, thereby improving metabolism of body fat [23]. An imbalance between WAT and BAT leads to obesity by accumulating fat in visceral and retroperitoneal tissues. Generally, obese females demonstrate higher expression of BAT than males [24]. Such phenomenon may have accounted for low obesity index and adiposity index observed in obese females as compared to obese males in the present study. Estrogen hormone increases mitochondrial BAT and over expression of UCP-1 that affords suitable justification for reduction of obesity and adiposity index in estradiol treated male rats as compared to control group in our study [25]. However, further studies focusing on quantification of WAT and BAT content and expression of UCP-1 with estrogen treatment in obese animals are required to prove this hypothesis. The DNA microarray analysis has revealed that females have greater fat clearing capacity in their skeletal muscles through activation of genes encoding for enzymes involved in fatty acid beta-oxidation [26].

Our results demonstrated higher serum glucose levels in obese rats than controls (Table 19.1). The increase in resistin and decrease in adiponectin level contributes towards obesity-induced insulin resistance and down regulation of the expression of glucose transporters, thereby preventing glucose entry into cells that leads to hyperglycemia [9, 27]. Obese females showed lesser hyperglycemia than obese males that may be attributed to the fact that significantly decreased levels of adiponectin are reported in obese females as compared to males [28]. Estradiol treatment resulted in lowering of glucose levels in male rats (Table 19.1). The estrogen is documented to reduce the resistin level, thereby increasing insulin sensitivity of the cells [29].

Our results highlighted marked increase in systolic blood pressure (SBP) in obese rats as compared to normal rats (Fig. 19.2B). Various factors such as activation of renal sympathetic system, renin angiotensin system, angiotensin converting enzyme

(ACE) as well as of angiotensin receptors and increased oxidative stress are documented to account for obesity-induced increase in blood pressure in various studies [30, 31]. The increase in endothelin-1 is noted to contribute towards endothelial damage and consequent increase in blood pressure [32]. Treatment with estradiol markedly attenuated SBP in obese rats. The estrogen treatment is noted to reduce activity of ACE activity and AT-1 receptors along with increase in synthesis of vasodilator peptides [33, 34]. Estradiol up-regulates the expression of endothelial nitric oxide synthase (eNOS), which leads to vasodilatation through enhanced production of nitric oxide (NO) and reducing sympathetic tone [34]. The estrogen increases the level of prostaglandin E2 and cGMP level that mediated vasodilatation in kidneys [35]. Increased leptin levels in obesity leads to activation of sympathetic nervous system that contributes to obesity-induced hypertension [36]. Estrogen is noted to reduce leptin level, improve leptin sensitivity and to modulate leptin receptors [37]. Such action might have accounted for estrogen mediated anti-hypertensive effect in obese rats. However, a separate study is needed to evaluate this hypothesis.

A marked decrease in CrCl along with increase in serum urea, uric acid, electrolytes, and microproteinuria demonstrated obesity-induced renal dysfunction in male and female rats, and these parameters were significantly attenuated by estradiol treatment of male rats (Fig. 19.1). The hyperlipidemia leads to renal damage by executing changes in glomeruli and renal vasculature with increase of intimal thickening and narrowing of the lumen of renal blood vessels. The activation of sympathetic nervous system and the renin-angiotensin-alodeosterone system contributes towards hypertension in renal disease that further damages the kidneys. The hyperlipidemia and hypertension are noted to produce glomerular injury and podocyte damage [38]. The estrogen receptors are present on podocytes and their activation leads to improvement of renal function, marked by decreased proteinuria [38, 39].

The serum levels of estrogen did not change significantly with HFD in both male and female rats (Fig. 19.2A). Interestingly, we observed that obese rats treated with low dose of estradiol demonstrated almost similar levels of serum estrogen levels as compared to obese females with no significant change in various parameters between the two groups. The results in both groups, *viz,* HFD fed females and HFD fed males treated with low dose of estradiol benzoate were markedly less than HFD fed obese males, which highlight the role of estrogen as renoprotective agent in both sexes.

The elevation in TBARS and SAG, and reduction of GSH in kidneys of obese rats indicated marked oxidative stress as compared to the controls (Table 19.2). The generation of highly reactive oxygen species (ROS) in kidneys contributes causing damage in glomerular, podocyte and tubulo-interstitial cells [40, 41]. The activated RAS elevates the activity of NAD(P)H oxidase enzyme that in turn increases the production of ROS [42]. It has been reported that obesity associated hyperglycemia and raised angiotensin-II levels alter the level of cytokines through activation of protein kinase C, mitogen activated protein kinase and nuclear factor kappa-B that further lead to oxidative stress [7]. The adipocytes interact with macrophages and are able to secrete inflammatory cytokines such as TNF-α, IL-1 and IL-6, that are well documented to produce oxidative stress through multiple pathways [7, 9]. The obesity is considered an inflammatory disorder that provokes the infiltration of neutrophils.

Hematoxylin and Eosin staining demonstrated marked neutrophil infiltration in renal tissues of obese rats (Fig. 19.3). The mast cell stabilizers are known to ameliorate obesity-induced renal dysfunction in rats, thereby adding to the inflammatory aspect of obesity associated renal complications [43]. The HFD fed females seem to suffer from lesser oxidative stress than males as seen in our study. The estradiol administration attenuated HFD mediated renal oxidative stress in male rats (Table 19.2). The elevated oxidative stress observed in post-menopausal women clearly indicates that estrogens play an important role in controlling oxidative stress [44]. The estrogens are also reported to reduce renal NAD(P)H oxidase activity [45].

Nearly 12.5% of collagen is constituted by hydroxyproline and its increased level in tissues is an indicator of tissue fibrosis. Our results demonstrated marked increase in renal fibrosis in HFD fed rats as compared to the controls. This effect was ameliorated in estradiol treated male rats (Figs. 19.2 and 19.4). Transforming growth factor-β (TGF-β) is a cytokine that is involved in causing fibrosis, and obesity is documented to increase TGF-β level in kidneys [46]. on the other hand, estradiol is reported to curtail TGF-β level in other models of renal injury [47].

Obesity and hypertension play synergistic role in the development of renal complications. The bioactive molecules released from white adipocytes appear to cause hypertension through renin-angiotensin induced constriction of renal blood vessels, activation of RAS, and enhanced sympathetic tone [30]. The obesity-induced hypertension in turn causes damage in renal tissues. Estrogen treatment is reported to reduce blood pressure in animal models of hypertension through reduction of endothelins and increased production of eNOS [48, 49]. The estrogens ameliorate renal injury through their anti-inflammatory and anti-oxidant potential [17]. The results of this study also showed that estradiol treatment not only caused reduction in systolic blood pressure but also lessened obesity-induced renal injury in male rats. Further studies are warranted to enhance our understanding about the biochemical and morphological mechanisms involved in the estradiol-induced amelioration of kidney failure observed in obese patients as well as reduction of obesity-related cardiovascular complication.

In summary, continuous exposure to HFD for 12-weeks produced marked hypertension, oxidative stress, and significant morphological injury especially in the kidneys of male rats than their female counterparts. Simultaneous daily administration of estradiol benzoate (50 and 100 μg/kg/day, i.p.) to male rats for 12-weeks protected against HFD-induced hypertension and renal dysfunction. Figure 19.5 summarizes the hypothetical mechanism of renoprotective action of estradiol benzoate in male rats. Further studies are needed to understand the underlying cellular and molecular mechanisms of estradiol's renoprotective actions.

While the findings of this study provide important clues about the renoprotective effects of estradiol, these findings cannot be directly extrapolated to humans. Nevertheless, the renoprotective effects of estradiol warrant verification in obese men suffering from acute renal dysfunction malady.

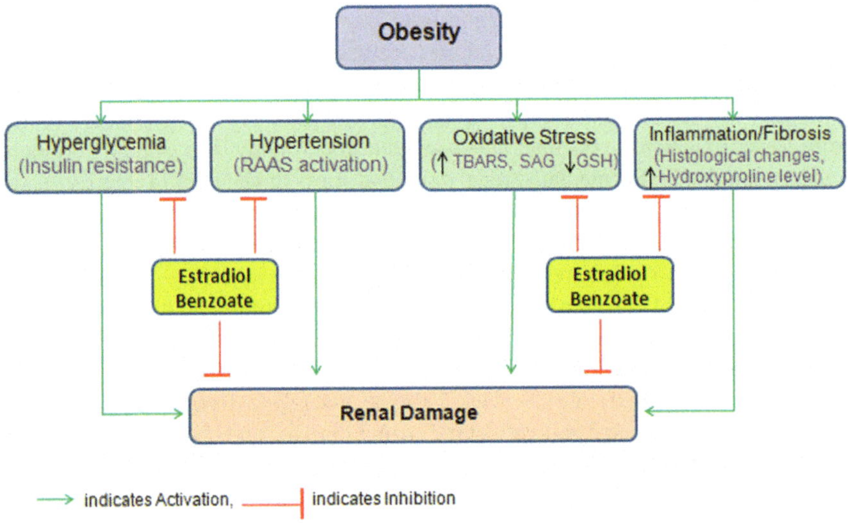

Fig. 19.5 Diagrammatic illustration of pathophysiological mechanisms involved in obesity-induced renal damage and reversal by estradiol benzoate treatment in male rats

Author Contributions APS and HSB conceived the idea. APS designed and supervised the project. MS performed pharmacological treatments. MS, TK did biochemical studies. SSG did estradiol assay. DP did histological studies. MS, TK and APS analyzed the data. HSB and APS wrote manuscript. HSB and APS critically screened the manuscript prior to the submission to editors. All authors approved the manuscript before submission to editors.

Compliance with Ethical Standards The authors did not receive any funding for this research work. The authors declare no conflict of interest.

Animal experiments were performed in accordance with the guidelines framed by Committee for the Purpose of Control and Supervision of Experiments on Animals (CPCSEA), Ministry of Environment, Forests and Climate Change, Government of India. The protocol was approved by Institutional Animal Ethics Committee of Guru Nanak Dev University, Amritsar (226/CPCSEA/2014/14).

References

1. World Health Organization. Obesity and overweight. Fact sheet number 311, January 2015
2. Ogden CL, Carroll MD, Kit KB et al (2014) Prevalence of childhood and adult obesity in the United States. JAMA 311:806–814
3. Misra A, Shrivastava U (2013) Obesity and dyslipidemia in South Asians. Nutrients 5:2708–2733
4. Gupta R, Deedwania PC, Gupta A et al (2004) Prevalence of metabolic syndrome in an Indian urban population. Int J Cardiol 97:257–261
5. Segula D (2014) Complications of obesity in adults: a short review of the literature. Malawi Med J 26:20–24

6. Kovesdy CP, Furth SL, Zoccali C; World Kidney Day Steering Committee (2017) Obesity and kidney disease: hidden consequences of the epidemic. Kidney Dis (Basel) 3:33–41
7. Felizardo RJ, da Silva MB, Aguiar CF et al (2014) Obesity in kidney disease: a heavyweight opponent. World J Nephrol 3:50–63
8. Bravo PE, Morse S, Borne DM et al (2006) Leptin and hypertension in obesity. Vasc Health Risk Manag 2:163–169
9. Fernández-Sánchez A, Madrigal-Santillán E, Bautista M et al (2011) Inflammation, oxidative stress, and obesity. Int J Mol Sci 12:3117–3132
10. Kim SY, Lim AY, Jeon SK et al (2011) Effects of dietary protein and fat contents on renal function and inflammatory cytokines in rats with adriamycin-induced nephrotic syndrome. Mediators Inflamm 945123
11. Gluhovschi G, Gluhovschi A, Anastasiu D et al (2012) Chronic kidney disease and the involvement of estrogen hormones in its pathogenesis and progression. Rom J Intern Med 50:135–144
12. Neugarten J, Acharya A, Silbiger SR (2000) Effect of gender on the progression of nondiabetic renal disease: a meta-analysis. J Am Soc Nephrol 11:319–329
13. Nematbakhsh M, Ebrahimian S, Tooyserkani M et al (2013) Gender difference in cisplatin-induced nephrotoxicity in a rat model: greater intensity of damage in male than female. Nephrourol Mon 5:818–821
14. Tanaka R, Tsutsui H, Kobuchi S et al (2012) Protective effect of 17β-estradiol on ischemic acute kidney injury through the renal sympathetic nervous system. Eur J Pharmacol 683:270–275
15. Lizcano F, Guzmán G (2014) Estrogen deficiency and the origin of obesity during menopause. Biomed Res Int 757461
16. Danilovich N, Babu PS, Xing W et al (2000) Estrogen deficiency, obesity, and skeletal abnormalities in follicle-stimulating hormone receptor knockout (FORKO) female mice. Endocrinology 141:4295–4308
17. Antus B, Hamar P, Kokeny G et al (2003) Estradiol is nephroprotective in the rat remnant kidney. Nephrol Dial Transplant 18:54–61
18. Pezeshki Z, Nematbakhsh M, Nasri H et al (2013) Evidence against protective role of sex hormone estrogen in Cisplatin-induced nephrotoxicity in ovarectomized rat model. Toxicol Int 20:43–47
19. Singh R, Singh AP, Singh M et al (2011) Impact of obesity on hypertension-induced cardiac remodeling: role of oxidative stress and its modulation by gemfibrozil treatment in rats. Free Radic Biol Med 50:363–370
20. Wang F, Li M, Cheng L et al (2008) Intervention with cilostazol attenuates renal inflammation in streptozotocin-induced diabetic rats. Life Sci 83:828–835
21. Altunkaynak ME, Ozbek E, Altunkaynak BZ et al (2008) The effects of high-fat diet on the renal structure and morphometric parametric of kidneys in rats. J Anat 212:845–852
22. Iran-Nejad A, Nematbakhsh M, Eshraghi-Jazi F et al (2015) Preventive role of estradiol on kidney injury induced by renal ischemia-reperfusion in male and female rats. Int J Prev Med 6:22
23. Nedergaard J, Petrovic N, Lindgren EM et al (2005) PPAR gamma in the control of brown adipocyte differentiation. Biochim Biophys Acta 1740:293–304
24. Choi DK, Oh TS, Choi JW et al (2011) Gender difference in proteome of brown adipose tissues between male and female rats exposed to a high fat diet. Cell Physiol Biochem 28:933–948
25. Vega RB, Kelly DP (1997) A role for estrogen-related receptor alpha in the control of mitochondrial fatty acid beta-oxidation during brown adipocyte differentiation. J Biol Chem 272:31693–31699
26. Oh TS, Yun JW (2012) DNA microarray analysis reveals differential gene expression in the soleus muscle between male and female rats exposed to a high fat diet. Mol Biol Rep 39:6569–6580
27. Pedersen O, Kahn CR, Kahn BB (1992) Divergent regulation of the Glut 1 and Glut 4 glucose transporters in isolated adipocytes from Zucker rats. J Clin Invest 89:1964–1973

28. Song HJ, Oh S, Quan S et al (2014) Gender differences in adiponectin levels and body composition in older adults: Hallym aging study. BMC Geriatr 14:8
29. Huang SW, Seow KM, Ho LT et al (2005) Resistin mRNA levels are down regulated by estrogen in vivo and in vitro. FEBS Lett 579:449–454
30. Dobrian AD, Davies MJ, Prewitt RL et al (2000) Development of hypertension in a rat model of diet-induced obesity. Hypertension 35:1009–1015
31. Thethi T, Kamiyama M, Kobori H (2012) The link between the renin-angiotensin-aldosterone system and renal injury in obesity and the metabolic syndrome. Curr Hypertens Rep 14:160–169
32. Cardillo C, Campia U, Iantorno M et al (2004) Enhanced vascular activity of endogenous endothelin-1 in obese hypertensive patients. Hypertension 43:36–40
33. Gallagher PE, Li P, Lenhart JR et al (1999) Estrogen regulation of angiotensin-converting enzyme mRNA. Hypertension 33:323–328
34. Maranon R, Reckelhoff JF (2013) Sex and gender differences in control of blood pressure. Clin Sci (Lond) 125:311–318
35. Baiardi G, Macova M, Armando I et al (2005) Estrogen upregulates renal angiotensin II AT1 and AT2 receptors in the rat. Regul Pept 124:7–17
36. Simonds SE, Cowley MA (2013) Hypertension in obesity: is leptin the culprit? Trends Neurosci 36:121–132
37. Meli R, Pacilio M, Raso GM et al (2004) Estrogen and raloxifene modulate leptin and its receptor in hypothalamus and adipose tissue from ovariectomized rats. Endocrinology 145:3115–3121
38. Gross ML, Adamczak M, Rabe T et al (2004) Beneficial effects of estrogens on indices of renal damage in uninephrectomized SHRsp rats. J Am Soc Nephrol 15:348–358
39. Doublier S, Lupia E, Catanuto P et al (2011) Estrogens and progression of diabetic kidney damage. Curr Diabetes Rev 7:28–34
40. Hodgkins KS, Schnaper HW (2012) Tubulointerstitial injury and the progression of chronic kidney disease. Pediatr Nephrol 27:901–909
41. Chen S, Meng XF, Zhang C (2013) Role of NADPH oxidase-mediated reactive oxygen species in podocyte injury. Biomed Res Int 839761
42. Kobori H, Nangaku M, Navar LG et al (2007) The intrarenal renin-angiotensin system: from physiology to the pathobiology of hypertension and kidney disease. Pharmacol Rev 59:251–287
43. Reena, Kaur T, Kaur A et al (2016) Mast cell stabilizers obviate high fat diet-induced renal dysfunction in rats. Eur J Pharmacol 777:96–103
44. Doshi SB, Agarwal A (2013) The role of oxidative stress in menopause. J Midlife Health 4:140–146
45. Sullivan JC, Pardieck JL, Brinson K et al (2009) Effects of estradiol on renal cyclic guanosinemonophosphate and oxidative stress in spontaneously hypertensive rats. Gend Med 6:498–510
46. Aihara K, Ikeda Y, Yagi S et al (2011) Transforming growth factor-β1 as a common target molecule for development of cardiovascular diseases, renal insufficiency and metabolic syndrome. Cardiol Res Pract 175381
47. Dixon A, Maric C (2007) 17 beta-Estradiol attenuates diabetic kidney disease by regulating extracellular matrix and transforming growth factor-beta protein expression and signaling. Am J Physiol Renal Physiol 293:F1678–F1690
48. Montezano AC, Callera GE, Mota AL et al (2005) Endothelin-1 contributes to the sexual differences in renal damage in DOCA-salt rats. Peptides 26:1454–1462
49. Brosnihan KB, Li P, Figueroa JP et al (2008) Estrogen, nitric oxide, and hypertension differentially modulate agonist-induced contractile responses in female transgenic (mRen2)27 hypertensive rats. Am J Physiol Heart Circ Physiol 294:H1995–H2001

Chapter 20
Biochemical, Metabolic and Clinical Effects of Intermittent Fasting

Meghana Gore, Amudhan Jyothidasan, and Vijaiganesh Nagarajan

Abstract Periods of voluntary abstinence from food and drink has been practiced by different civilizations around the world for religious and other reasons for centuries. Obesity is a global pandemic and prevalence is on the rise. It has been well proven that the obesity increases the risk of cardiovascular disease, malignancy and premature death. Multiple strategies for weight loss have been investigated and energy restriction is one of the popular and effective method. Two common types of energy restriction are caloric restriction and intermittent fasting. Restriction of energy consumption by fasting would likely help in weight loss despite the regimen used and adhering to the most practical regimen may be the appropriate decision. Overall effects of energy restriction are decrease in inflammation, blood pressure, blood glucose levels and increase in insulin sensitivity and antioxidants. Increase in physical activity has enormous health benefits and must be combined with caloric restriction to achieve clinically significant weight loss and to sustain weight loss. Although the positive effects of energy restriction are remarkable, we should not forget undernutrition component of malnutrition continues to remain as a global cause of death and disease. Overall, appropriate intake of calories and nutrients necessary to meet our daily requirements would promote health and longevity and either excess or lower amount of calorie intake could result in adverse events.

Keywords Obesity · Intermittent fasting · Caloric restriction · Weight loss · Cardiovascular disease markers · Diabetes · Inflammation

---

M. Gore
Department of Internal Medicine, Memorial Hermann Katy Hospital, Katy, TX, USA

A. Jyothidasan
Department of Cardiology, Cape Fear Valley Hospital, Fayetteville, NC, USA

V. Nagarajan (✉)
Interventional Cardiology, The University of Texas Health Sciences Center at Houston, 6400 Fannin Street, Suite 2350, Houston, TX 77030, USA
e-mail: doctorvijai@yahoo.co.uk

P. S. Tappia et al. (eds.), *Biochemistry of Cardiovascular Dysfunction in Obesity*, Advances in Biochemistry in Health and Disease 20,
https://doi.org/10.1007/978-3-030-47336-5_20

Introduction

Periods of voluntary abstinence from food and drink has been practiced by different civilizations around the world for religious and other reasons for centuries. Obesity is a global pandemic and the prevalence is expected to rise in the next few decades. One of the major causes of obesity is passive overconsumption of energy due to affordable, readily available, processed calorie dense food along with significant drop in physical activity. It has been well proven that obesity increases the risk of cardiovascular disease, malignancy and premature death [1]. Multiple strategies for weight loss have been investigated which include fat restriction, increasing physical activity, behavior modification, high fiber intake and importantly energy restriction [2]. Two common types of energy restriction are caloric restriction and intermittent fasting. Caloric restriction is reduction in daily caloric intake by 20–40% over a long period of time without any prolonged periods of fasting. Intermittent fasting is periods of voluntary abstinence from food and drinks as a measure to decrease overall energy intake. Dietary energy restriction has been associated with excellent health benefits in multiple early animal studies performed few decades ago [3, 4]. Similarly, health benefits of intermittent fasting have been described in rats as early as 1980s [5]. Although we use the broad term of 'intermittent fasting', different studies have used different regimens to demonstrate the health benefits.

Regimens of Intermittent Fasting

Two common types of intermittent fasting are alternate day fasting and time restricted feeding, although there are multiple modified intermittent fasting regimens used in different studies. Alternate day fasting is abstinence from food for 20–28 h followed by a day of ad libitum feeding. Time restricted feeding is a concept where there is 12–16 h fasting from food every day to restrict the number of meals in a day. James Neel's 'thrifty gene hypothesis' claims ancient humans went through cycles of feast and famine and hence humans are genetically efficient at storing excess energy and utilize it during famine. Halberg et al. based his study on this hypothesis and tested whether the fluctuations in energy intake is required in humans for optimal metabolic function. 20 h of fasting in healthy individuals showed increase in insulin mediated glucose uptake [6]. This is an example of alternate day fasting. Stote et al. performed a randomized crossover design study where participants were assigned to 1 meal/day or 3 meals/day group. Health effects of time restricted feeding on this study was mixed. Some other studies have used modified regimens like Harvie et al. where one group used intermittent energy restriction (~2700 kJ/day for 2 days a week) and the other group used continuous energy restriction (25% restriction for 7 days a week). Authors concluded that intermittent energy restriction almost had the same effects as continuous energy restriction and could be used as an alternative [7]. Patterson et al. reviewed 16 different fasting interventions and noted 11 of them reported statistically significant weight loss [8]. Overall, restriction of energy consumption by fasting would likely help in weight loss despite the regimen used and adhering to the most practical regimen may be the appropriate decision.

Effects of Fasting in Animal Studies

Multiple animal studies have interrogated different health effects of fasting. Funes et al. noted rapid changes in gastrointestinal tract within 36 h of fasting in sparrows and atrophic changes noted on small intestine could be adaptive response expecting refeeding [9]. Intermittent fasting increased the mRNA expression and IgA levels in intestines promoting bacterial clearance in mice [10]. Immune response to viral mimic intraperitoneal injection was exaggerated when measured using IL-6, IL-10, IFN-□ and TNF-alpha [11]. Telemetric analysis of heart rate in rats were performed by Mager and team [12]. Decrease in body weight, heart rate, systolic and diastolic blood pressure along with increased parasympathetic activity and decreased sympathetic activity was noted. Incidence of diabetes was significantly lower in rats which were in the fasting group compared to normally fed rats. Rats which had alternate day fasting and two days a week fasting didn't have any difference in diabetes incidence [13]. Significant reduction in accumulation of diacylglycerols (DAG) was noted in intermittent fasted mice and was associated with decreased Protein Kinase C activation may explain improved insulin sensitivity [14]. Increase in adiponectin levels along with decrease in visceral fat and increase in triglyceride accumulation in subcutaneous fat was noted with both caloric restriction and intermittent fasting groups. This proves the beneficial modulation of lipid metabolism and insulin sensitivity was present in both types of fasting [15]. One of the very interesting studies showed myocardial infarction caused by ligation of coronary arteries in rats was smaller in size resulting in better left ventricular function in intermittent fed rats compared to the control group [16]. Animal studies cannot be directly extrapolated to human for various reasons. Metabolic response of humans to fasting could be very different from animals and the feasibility and long-term sustainability of the prolonged periods of fasting in humans is less.

Metabolic Effects of Fasting in Humans

Multiple human studies have looked at different aspects of intermittent fasting. One of the very early observations were by Storm et al. who looked at mortality rates before and after second world war and noted cardiovascular mortality in Norway was lower during the World War II, but increased back to pre-war levels after the war. Authors noted a coincidence with the caloric and fat restriction happened during World War II in Norway [17]. Another study has noted that 83% of weight loss in 10-week caloric restriction was fat mass and happened even with decrease in resting metabolic rate in energy restriction group [18]. Mattson has performed detailed study involving 10 overweight asthma patients who were subject to alternate day calorie restriction for 8 weeks and lost 8% of total body weight. Increased serum beta-hydroxybutyrate levels and decreased leptin levels confirmed compliance and switch to fatty acid metabolism on fasting days. Cholesterol, triglycerides, markers of oxidative stress like 8-isoprostane, nitrotyrosine, protein carbonyls, markers of inflammation like serum tumor necrosis factor-alpha and brain-derived neurotrophic factor were significantly decreased suggesting a beneficial effect from alternate day fasting apart from improvement in their asthma symptoms [19]. Although the sole focus of this article is to discussion the overnutrition part, undernutrition part of malnutrition needs to be mentioned briefly as well. Undernutrition is insufficient intake of energy and nutrients to maintain good health in an individual. Undernutrition continues to be a major health care problem in developing countries and affect mainly children, pregnant women and elderly. Common examples are protein energy malnutrition, iron deficiency anemia, vitamin A deficiency, etc. Malnutrition is well associated with childhood mortality and some authors consider up to 50% of childhood deaths could be related to malnutrition [20]. Malnutrition is also significantly associated with increased mortality in adults irrespective of cause of death [21]. One of the other studies noted the hazard ratio for all-cause mortality in patients over 65 years of age were 3.7 higher in patients who were malnourished and 1.5 in patients who were at risk of malnutrition [22]. Overall, appropriate intake of calories and nutrients necessary to meet our daily requirements would promote health and longevity and either excess or lower amount of calorie intake could result in adverse events.

Fasting and Diabetes

Fasting promotes certain biochemical changes in carbohydrate and lipid metabolism. Anabolic process help building macromolecules like glycogen and lipids while catabolic process involves breakdown of macromolecules to produce energy. Diabetes is a condition with high blood glucose level attributed because of insulin resistance or deficiency or absence of insulin. Type 2 diabetes mellitus is associated with obesity, adiposity and insulin resistance [23]. In fed state, as food starts to breakdown, blood sugar level increase leading to insulin secretion, which activates anabolic pathway. Glucose homeostasis is maintained by liver, muscle and adipose tissues with help of Insulin by means of glucose transport, glycogen synthesis and lipogenesis. When high insulin levels are needed to maintain glucose homeostasis, it is called as 'insulin resistance' and is a common occurrence with overfeeding and obesity. In fasting state, insulin activity is low and glucagon activity is high as blood sugar levels starts lowering. Low insulin state activates catabolic pathway which involves breakdown of macromolecules like glycogen into glucose which is used to produce energy [24]. Calorie restrictions by means of fasting is known to reverse insulin resistance and visceral obesity in animal models [25]. Furmli et al. demonstrated that intermittent fasting in insulin-dependent adults help them to loose weight, reverse their insulin resistance and decrease HbA1c. It eventually resulted in cessation on insulin therapy [26]. Caution is needed given intermittent fasting increases the risk of hypoglycemic episodes despite reducing the diabetes medications [27].

Fasting and Cancer

Obesity has been identified as one of the important risk factors for cancer. There are studies linking obesity to at least 13 different types of cancer including breast, colorectal, esophageal, pancreatic cancer, etc [28]. Obesity reportedly could account for approximately 20% of all cancer cases. Weight loss and exercise have been associated with decreased risk of breast cancer [29]. Bariatric surgery and subsequent weight loss were linked to lower incidence of cancer diagnosis in morbidly obese patients [30]. In humans, Intermittent fasting has shown to reduce abdominal obesity and inflammatory markers which could potentially have preventive effects for obesity, inflammation and cancer [31]. There is no human data available to show direct effect of Intermittent fasting on cancer prevention, however its effect on metabolic pathways leading to low IGF-level, low blood sugar levels and promoting ketogenesis may reduce DNA damage and carcinogenesis [32]. American Cancer Society recommends patients receiving chemotherapy to increase calories and protein intake. However recent studies and preclinical animal trials have shown advantage of intermittent fasting on cancer prevention and treatment [33–35]. Fasting makes cancer cells vulnerable and normal cells resistant to chemotherapy which is a desired outcome for the cancer treatment [36, 37]. Overall, intermittent fasting has shown significant anticancer role, even during chemotherapy as it increases susceptibility of cancer cells to chemotherapy, prevent resistance to chemotherapy, protects normal cells from toxic effects of chemotherapy [38–40].

Fasting and Cardiovascular Diseases

Cardiovascular disease is leading cause of mortality and morbidity in the world. Obesity, diabetes mellitus, hypertension, dyslipidemia, smoking and lack of physical activity are considered as major modifiable risk factors for development of cardiovascular illnesses. Control of these risk factors reduce incidence of cardiovascular diseases [41]. Intermittent fasting helps with weight loss and prevent/improve obesity, helps control blood glucose levels and lowers insulin resistance and decrease the incidence of diabetes mellitus. During first 12 h of fasting, glycogenolysis remains the primary mode of energy. In later stages, when glycogen stores are depleted, metabolism switches to lipolysis which result in conversion of fat to free fatty acids and ketones. This is called as 'metabolic switch'. This metabolic switch reduces the concentration of total cholesterol, triglycerides, and LDL cholesterol and thereby alters plasma lipid profile favorably [42]. Periodic fasting also has beneficial effects on circulatory system—it reduces systolic blood pressure, diastolic blood pressure and heart rate by activating parasympathetic system [12]. Brain-derived neurotrophic factor (BDNF) is produced by activation of glutamatergic receptors which is stimulated by intermittent fasting. BDNF stimulates synthesis and release of acetylcholine by cholinergic neurons and activates parasympathetic system [43]. Atherosclerosis is a chronic inflammatory condition which is a leading cause of vascular diseases. LDL accumulation in arterial wall under oxidative stress, triggers inflammatory response and cause plaque formation with adhesion of inflammatory blood cells and proliferation of vascular smooth muscle. Proinflammatory factors such as IL-6, homocysteine and CRP are known to be reduced by intermittent fasting [44].

Religious Fasting and Ramadan

Purpose of intermittent fasting could be for spiritual, cultural and health benefits. Fasting has been popular in different cultures and religions since ages. Hinduism calls fasting 'Vrata' which is observed to test penance or honor god. Islam has 'Ramadan' and Judaism has 'Yom Kippur'. Many churches follow 'Lent', which last for 40 days—starts on Ash Wednesday and ends on Holy Saturday prior to Easter, when religious fasting is observed. Ramadan is the holy month for Muslims when religious fasting is practiced. Muslims abstain from eating any food, drinks or any oral intake from dawn to sunset for whole month. Nature of the fasting during Ramadan allow up to 12–16 h of fasting period. This type of intermittent fasting is being practiced across the globe for centuries by Muslims. During early hours of fasting, glycogen acts as the energy source, but later during the day, metabolism goes into catabolic phase where fatty acid utilization occurs.

In general, all the benefits of intermittent fasting are seen in Ramadan fasting, including improvement in insulin sensitivity, decreased atherogenic risk, oxidative stress, and decrease in various inflammatory markers [44]. Ramadan fasting is associated with positive impact on lipid profile, leukocyte count and blood coagulation which may help prevent atherogenicity [45]. In a meta-analysis by Fernando et al., reduction in weight and fat percentage with the Ramadan fasting, especially in people with overweight or obesity is observed. They also noted return to pre-Ramadan weight and body composition was observed in 2–5 weeks [46]. Other studies showed reduction in BMI, lipid profile, blood pressure, blood glucose, and HbA1C level in diabetic population [47]. But given people restrict not only food, but also fluids can have deleterious effects like hypoglycemia, dizziness, hyperglycemia, dehydration, etc. Dehydration may cause irritability, headaches, sleep deprivation and lassitude [48]. Though Ramadan fasting is considered safe in healthy individuals, it might not be safe in individuals with medical illness. Islam allows certain excuses for not fasting Ramadan, includes children, menstruating women, pregnant or breastfeeding women, travelers. It is believed that fasting for long hours may also result in harm [49].

Major Studies of Fasting in Humans

Most of the research about intermittent fasting is done on animal models and most of the limited number of human trials are observational. When the published human trials are analyzed, many trials have small sample size (hundred or less participants) and there is significant variation in duration and protocols of intermittent fasting and the parameters analyzed. Many observational trials are based upon religious beliefs and not primarily motivated by health reasons [50]. Observational studies of Ramadan fasting showed various benefits from intermittent fasting like weight loss and metabolic health benefits [46, 51, 52]. Meta-analysis has investigated studies of intermittent fasting and concluded weight loss happens during Ramadan, but weight returns to baseline after fasting ceases [53]. Human studies have been reviewed by Barnosky et al. which suggested the possible benefit of diabetes risk reduction in overweight and obese population [54]. Another randomized control trial enrolled 107 overweight or obese individuals to compare intermittent fasting and continuous energy restriction and concluded both are comparable strategies. We could not find any large randomized clinical trials in humans to study intermittent fasting. Wegman et al. recruited 24 healthy individuals in a double-crossover, double-blinded, randomized clinical trial. This study suggests that the intermittent fasting is acceptable in healthy individuals, however additional research is needed to further assess the potential benefits and risks [55]. Mattson et al. reported that intermittent fasting could lead to a reduction in blood pressure, heart rate, cholesterol, and triglycerides in humans [43]. Group of researchers randomized 100 healthy participants into 2 groups, one exposed to unrestricted eating and others had fasting mimicking diet—low in calories, sugars and protein for 5 days a month. Later, they crossed over these groups. Findings were clear revealed that the group that fasted lost weight, lost some body fat, lowered their blood pressure, and decreased their IGF-1, a genetic marker for diseases such as cancer [31]. Though popularity of intermittent fasting to gain health benefits from weight loss is on rise, we still do not have strong evidence to support and recommend various fasting regimens as a health intervention to prevent and manage certain medical diseases [56]. More human subject research trials and stronger data is needed to recommend intermittent fasting regimens to general population.

Other Options for Weight Loss

Obesity is well associated with increase in morbidity and mortality. Weight loss in morbidly obese individuals reduces mortality [57]. To achieve weight loss, various prescriptions are available includes lifestyle intervention, pharmacotherapy and bariatric surgeries. In general, typical response with lifestyle modification yields 5% reduction in total body weight, 10% reduction in weight is considered successful weight loss [58]. Lifestyle intervention for weight loss includes various dietary regimens, exercise and behavioral therapy.

Pharmacotherapy can be considered for individuals with BMI $\geq$ 30 or BMI $\geq$ 27 with comorbidities. A specific diet plan should be attempted first with moderated physical activity as tolerated to maintain weight loss. Pharmacotherapy is recommended for those who are unable to achieve weight loss of goal of 5% with lifestyle interventions within six months. Choice of the drug depends of individual preference, side effects, cost and comorbid conditions [59]. Candidates for bariatric surgery include adults with a BMI $\geq$ 40 kg/m^2 with one or more obesity related morbid condition. When more than 30% reduction in body weight is goal—bariatric surgeries are recommended choice [60]. Device therapy is considered for those who are unable to tolerate pharmacotherapy and bariatric surgery. Gastric banding to decrease size of stomach, intragastric balloon systems to take up stomach space and give feeling of satiety. Decision of surgery should be based on multiple factors including patient motivation, compliance, operative risk, and comorbid conditions. First and predominant way to maintain healthy weight is lifestyle change—adaptation in daily life should include proper management of eating behavior, appropriate dietary modification and moderate and as tolerated physical activity.

Conclusions

Given obesity is a growing global pandemic, multiple strategies against weight gain has been investigated. Basic idea should be to limit the intake of calories and nutrients necessary to meet our daily requirements and strictly avoid 'caloric overdose or caloric toxicity'. We cannot overemphasize the importance of moderate physical activity in promoting health and preventing illness. It is well proven that obesity increases the morbidity and mortality and there is reasonable amount of evidence to suggest weight loss in obese individuals promote health. Intermittent fasting is comparable to continuous caloric restriction in weight loss and positive health benefits. Multiple biochemical and metabolic changes have been analyzed in different studies and most of them are positive as mentioned above. We strongly recommend lifestyle modifications and moderate physical activity as the first line. Although we have mentioned multiple animal and small human studies which showed the benefits of intermittent fasting, large size randomized trials are necessary before we prescribe intermittent fasting regimens to general population to promote health.

References

1. Van Gaal LF, Mertens IL, De Block CE (2006) Mechanisms linking obesity with cardiovascular disease. Nature 444(7121):875–880
2. Ramage S, Farmer A, Eccles KA, McCargar L (2014) Healthy strategies for successful weight loss and weight maintenance: a systematic review. Appl Physiol Nutr Metab 39(1):1–20
3. Sprott RL (1997) Diet and calorie restriction. Exp Gerontol 32(1–2):205–214

4. Houthoofd K, Braeckman BP, Lenaerts I et al (2002) Axenic growth up-regulates mass-specific metabolic rate, stress resistance, and extends life span in caenorhabditis elegans. Exp Gerontol 37(12):1371–1378
5. Goodrick CL, Ingram DK, Reynolds MA, Freeman JR, Cider NL (1982) Effects of intermittent feeding upon growth and life span in rats. Gerontology 28(4):233–241
6. Halberg N, Henriksen M, Soderhamn N et al (2005) Effect of intermittent fasting and refeeding on insulin action in healthy men. J Appl Physiol (1985) 99(6):2128–2136
7. Harvie MN, Pegington M, Mattson MP et al (2011) The effects of intermittent or continuous energy restriction on weight loss and metabolic disease risk markers: a randomized trial in young overweight women. Int J Obes (Lond) 35(5):714–727
8. Patterson RE, Sears DD (2017) Metabolic effects of intermittent fasting. Annu Rev Nutr 37:371–393
9. Funes SC, Filippa VP, Cid FD, Mohamed F, Caviedes-Vidal E, Chediack JG (2014) Effect of fasting in the digestive system: histological study of the small intestine in house sparrows. Tissue Cell 46(5):356–362
10. Godinez-Victoria M, Campos-Rodriguez R, Rivera-Aguilar V et al (2014) Intermittent fasting promotes bacterial clearance and intestinal IgA production in salmonella typhimurium-infected mice. Scand J Immunol 79(5):315–324
11. Zenz G, Jacan A, Reichmann F, Farzi A, Holzer P (2019) Intermittent fasting exacerbates the acute immune and behavioral sickness response to the viral mimic poly(I:C) in mice. Front Neurosci 13:359
12. Mager DE, Wan R, Brown M et al (2006) Caloric restriction and intermittent fasting alter spectral measures of heart rate and blood pressure variability in rats. FASEB J 20(6):631–637
13. Pedersen CR, Hagemann I, Bock T, Buschard K (1999) Intermittent feeding and fasting reduces diabetes incidence in BB rats. Autoimmunity 30(4):243–250
14. Baumeier C, Kaiser D, Heeren J et al (2015) Caloric restriction and intermittent fasting alter hepatic lipid droplet proteome and diacylglycerol species and prevent diabetes in NZO mice. Biochim Biophys Acta 1851(5):566–576
15. Varady KA, Allister CA, Roohk DJ, Hellerstein MK (2010) Improvements in body fat distribution and circulating adiponectin by alternate-day fasting versus calorie restriction. J Nutr Biochem 21(3):188–195
16. Ahmet I, Wan R, Mattson MP, Lakatta EG, Talan M (2005) Cardioprotection by intermittent fasting in rats. Circulation 112(20):3115–3121
17. Strom A, Jensen RA (1951) Mortality from circulatory diseases in norway 1940–1945. Lancet 1(6647):126–129
18. Velthuis-te Wierik EJ, Westerterp KR, van den Berg H (1995) Impact of a moderately energy-restricted diet on energy metabolism and body composition in non-obese men. Int J Obes Relat Metab Disord 19(5):318–324
19. Johnson JB, Summer W, Cutler RG et al (2007) Alternate day calorie restriction improves clinical findings and reduces markers of oxidative stress and inflammation in overweight adults with moderate asthma. Free Radic Biol Med 42(5):665–674
20. Pelletier DL, Frongillo EA Jr, Schroeder DG, Habicht JP (1994) A methodology for estimating the contribution of malnutrition to child mortality in developing countries. J Nutr 124(10 Suppl):2106S–2122S
21. Soderstrom L, Rosenblad A, Thors Adolfsson E, Bergkvist L (2019) Malnutrition is associated with increased mortality in older adults regardless of the cause of death. Br J Nutr 117(4):532–540
22. Soderstrom L, Rosenblad A, Adolfsson ET, Saletti A, Bergkvist L (2014) Nutritional status predicts preterm death in older people: a prospective cohort study. Clin Nutr 33(2):354–359
23. Wilcox G (2005) Insulin and insulin resistance. Clin Biochem Rev 26(2):19–39
24. Schwartsburd PM (2017) Catabolic and anabolic faces of insulin resistance and their disorders: a new insight into circadian control of metabolic disorders leading to diabetes. Future Sci OA 3(3):FSO201-2017-0015. eCollection

25. Barzilai N, Banerjee S, Hawkins M, Chen W, Rossetti L (1998) Caloric restriction reverses hepatic insulin resistance in aging rats by decreasing visceral fat. J Clin Invest. 101(7):1353–1361
26. Furmli S, Elmasry R, Ramos M, Fung J (2018) Therapeutic use of intermittent fasting for people with type 2 diabetes as an alternative to insulin. BMJ Case Rep. https://doi.org/10.1136/bcr-2017-221854.
27. Corley BT, Carroll RW, Hall RM, Weatherall M, Parry-Strong A, Krebs JD (2018) Intermittent fasting in type 2 diabetes mellitus and the risk of hypoglycaemia: a randomized controlled trial. Diabet Med 35(5):588–594
28. Avgerinos KI, Spyrou N, Mantzoros CS, Dalamaga M (2019) Obesity and cancer risk: emerging biological mechanisms and perspectives. Metabolism 92:121–135
29. Wolin KY, Carson K, Colditz GA (2010) Obesity and cancer. Oncologist 15(6):556–565
30. Christou NV, Lieberman M, Sampalis F, Sampalis JS (2008) Bariatric surgery reduces cancer risk in morbidly obese patients. Surg Obes Relat Dis 4(6):691–695
31. Wei M, Brandhorst S, Shelehchi M et al (2017) Fasting-mimicking diet and markers/risk factors for aging, diabetes, cancer, and cardiovascular disease. Sci Transl Med 9(377). https://doi.org/10.1126/scitranslmed.aai8700
32. Shevah O, Laron Z (2007) Patients with congenital deficiency of IGF-I seem protected from the development of malignancies: a preliminary report. Growth Horm IGF Res 17(1):54–57
33. De Lorenzo MS, Baljinnyam E, Vatner DE, Abarzua P, Vatner SF, Rabson AB (2011) Caloric restriction reduces growth of mammary tumors and metastases. Carcinogenesis 32(9):1381–1387
34. Dirx MJ, Zeegers MP, Dagnelie PC, van den Bogaard T, van den Brandt PA (2003) Energy restriction and the risk of spontaneous mammary tumors in mice: a meta-analysis. Int J Cancer 106(5):766–770
35. Descamps O, Riondel J, Ducros V, Roussel AM (2005) Mitochondrial production of reactive oxygen species and incidence of age-associated lymphoma in OF1 mice: effect of alternate-day fasting. Mech Ageing Dev 126(11):1185–1191
36. Nencioni A, Caffa I, Cortellino S, Longo VD (2018) Fasting and cancer: molecular mechanisms and clinical application. Nat Rev Cancer 18(11):707–719
37. Lee C, Raffaghello L, Brandhorst S et al (2014) Fasting cycles retard growth of tumors and sensitize a range of cancer cell types to chemotherapy. Sci Transl Med 4(124):124ra27
38. de Groot S, Vreeswijk MP, Welters MJ et al (2015) The effects of short-term fasting on tolerance to (neo) adjuvant chemotherapy in HER2-negative breast cancer patients: a randomized pilot study. BMC Cancer 15:652-015-1663-5
39. Bauersfeld SP, Kessler CS, Wischnewsky M et al (2018) The effects of short-term fasting on quality of life and tolerance to chemotherapy in patients with breast and ovarian cancer: a randomized cross-over pilot study. BMC Cancer 18(1):476-018-4353-2
40. Dorff TB, Groshen S, Garcia A et al (2016) Safety and feasibility of fasting in combination with platinum-based chemotherapy. BMC Cancer 16:360-016-2370-6
41. Yusuf S, Hawken S, Ounpuu S et al (2004) Effect of potentially modifiable risk factors associated with myocardial infarction in 52 countries (the INTERHEART study): case-control study. Lancet 364(9438):937–952
42. Bhutani S, Klempel MC, Kroeger CM, Trepanowski JF, Varady KA (2013) Alternate day fasting and endurance exercise combine to reduce body weight and favorably alter plasma lipids in obese humans. Obesity (Silver Spring) 21(7):1370–1379
43. Mattson MP, Longo VD, Harvie M (2017) Impact of intermittent fasting on health and disease processes. Ageing Res Rev 39:46–58
44. Aksungar FB, Topkaya AE, Akyildiz M (2007) Interleukin-6, C-reactive protein and biochemical parameters during prolonged intermittent fasting. Ann Nutr Metab 51(1):88–95
45. Saleh SA, El-Kemery TA, Farrag KA et al (2004) Ramadan fasting: relation to atherogenic risk among obese muslims. J Egypt Public Health Assoc 79(5–6):461–483
46. Fernando HA, Zibellini J, Harris RA, Seimon RV, Sainsbury A (2019) Effect of Ramadan fasting on weight and body composition in healthy non-athlete adults: a systematic review and meta-analysis. Nutrients 11(2). https://doi.org/10.3390/nu11020478

47. Bener A, Yousafzai MT (2014) Effect of Ramadan fasting on diabetes mellitus: a population-based study in Qatar. J Egypt Public Health Assoc 89(2):47–52
48. Leiper JB, Molla AM, Molla AM (2003) Effects on health of fluid restriction during fasting in Ramadan. Eur J Clin Nutr 57(Suppl 2):S30–S38
49. Abolaban H, Al-Moujahed A (2017) Muslim patients in Ramadan: a review for primary care physicians. Avicenna J Med 7(3):81–87
50. Patterson RE, Laughlin GA, LaCroix AZ et al (2015) Intermittent fasting and human metabolic health. J Acad Nutr Diet 115(8):1203–1212
51. Lessan N, Ali T (2019) Energy metabolism and intermittent fasting: the Ramadan perspective. Nutrients 11(5). https://doi.org/10.3390/nu11051192
52. Nematy M, Alinezhad-Namaghi M, Rashed MM et al (2012) Effects of Ramadan fasting on cardiovascular risk factors: a prospective observational study. Nutr J 11:69-2891-11-69
53. Sadeghirad B, Motaghipisheh S, Kolahdooz F, Zahedi MJ, Haghdoost AA (2014) Islamic fasting and weight loss: A systematic review and meta-analysis. Public Health Nutr 17(2):396–406
54. Barnosky AR, Hoddy KK, Unterman TG, Varady KA (2014) Intermittent fasting vs daily calorie restriction for type 2 diabetes prevention: a review of human findings. Transl Res 164(4):302–311
55. Wegman MP, Guo MH, Bennion DM et al (2015) Practicality of intermittent fasting in humans and its effect on oxidative stress and genes related to aging and metabolism. Rejuvenation Res 18(2):162–172
56. Horne BD, Muhlestein JB, Anderson JL (2015) Health effects of intermittent fasting: hormesis or harm? A systematic review. Am J Clin Nutr 102(2):464–470
57. Poobalan AS, Aucott LS, Smith WC, Avenell A, Jung R, Broom J (2007) Long-term weight loss effects on all cause mortality in overweight/obese populations. Obes Rev 8(6):503–513
58. Look AHEAD Research Group, Gregg EW, Jakicic JM et al (2016) Association of the magnitude of weight loss and changes in physical fitness with long-term cardiovascular disease outcomes in overweight or obese people with type 2 diabetes: a post-hoc analysis of the look AHEAD randomised clinical trial. Lancet Diabetes Endocrinol 4(11):913–921
59. Jensen MD, Ryan DH, Apovian CM et al (2014) 2013 AHA/ACC/TOS guideline for the management of overweight and obesity in adults: a report of the american college of cardiology/american heart association task force on practice guidelines and the obesity society. Circulation 129(25 Suppl 2):S102–S138
60. Gloy VL, Briel M, Bhatt DL et al (2013) Bariatric surgery versus non-surgical treatment for obesity: a systematic review and meta-analysis of randomised controlled trials. BMJ 347:f5934

Index

P. S. Tappia et al. (eds.), *Biochemistry of Cardiovascular Dysfunction in Obesity*,
Advances in Biochemistry in Health and Disease 20,
https://doi.org/10.1007/978-3-030-47336-5